Handbook of Probabilistic Models

Handbook of Probabilistic Models

Edited by

Pijush Samui
Department of Civil Engineering, NIT Patna, Bihar, India

Dieu Tien Bui
Department of Business and IT, School of Business
University of South-Eastern Norway(USN), Telemark, Norway

Subrata Chakraborty
Department of Civil Engineering, Indian Institute of
Engineering Science and Technology, Howrah, India

Ravinesh C. Deo
University of Southern Queensland
Springfield, QLD, Australia

ELSEVIER

Butterworth-Heinemann
An imprint of Elsevier

Butterworth-Heinemann is an imprint of Elsevier
The Boulevard, Langford Lane, Kidlington, Oxford OX5 1GB, United Kingdom
50 Hampshire Street, 5th Floor, Cambridge, MA 02139, United States

Library of Congress Cataloging-in-Publication Data
A catalog record for this book is available from the Library of Congress

British Library Cataloguing-in-Publication Data
A catalogue record for this book is available from the British Library

ISBN: 978-0-12-816514-0

For information on all Butterworth-Heinemann publications
visit our website at https://www.elsevier.com/books-and-journals

Publisher: Matthew Deans
Acquisition Editor: Matthew Deans
Editorial Project Manager: Joshua Mearns
Production Project Manager: Kamesh Ramajogi
Cover Designer: Miles Hitchen

Typeset by TNQ Technologies

Working together
to grow libraries in
developing countries

www.elsevier.com • www.bookaid.org

Dedicated to my grandfather and grandmother

Contents

3. Monthly rainfall forecasting with Markov Chain Monte Carlo simulations integrated with statistical bivariate copulas

Mumtaz Ali, Ravinesh C. Deo, Nathan J. Downs and Tek Maraseni

4. A model for quantitative fire risk assessment integrating agent-based model with automatic event tree analysis

Farid Wajdi Akashah, Rachid Ouache, Jianping Zhang and Michael Delichatsios

5. Prediction capability of polynomial neural network for uncertain buckling behavior of sandwich plates

R.R. Kumar, Tanmoy Mukhopadhya, K.M. Pandey and S. Dey

6. Development of copula-statistical drought prediction model using the Standardized Precipitation-Evapotranspiration Index

Kavina S. Dayal, Ravinesh C. Deo and Armando A. Apan

7. An efficient approximation-based robust design optimization framework for large-scale structural systems

Tanmoy Chatterjee and Rajib Chowdhury

8. Probabilistic seasonal rainfall forecasts using semiparametric D-vine copula-based quantile regression

Thong Nguyen-Huy, Ravinesh C. Deo, Shahbaz Mushtaq and Shahjahan Khan

9. Geostatistics: principles and methods

Saman Maroufpoor, Omid Bozorg-Haddad and Xuefeng Chu

10. Adaptive H_∞ Kalman filtering for stochastic systems with nonlinear uncertainties

Yuankai Li

11. R for lifetime data modeling via probability distributions

Vikas Kumar Sharma

12. Probability-based approach for evaluating groundwater risk assessment in Sina basin, India

Thendiyath Roshni, Sourav Choudhary, Madan K. Jha and Nehar Mandal

13. Novel concepts for reliability analysis of dynamic structural systems

J. Ramon Gaxiola-Camacho, Hamoon Azizsoltani, Achintya Haldar, S. Mohsen Vazirizade and Francisco Javier Villegas-Mercado

14   Probabilistic neural networks: a brief overview of theory, implementation, and application

Behshad Mohebali, Amirhessam Tahmassebi, Anke Meyer-Baese and Amir H. Gandomi

17. Stochastic optimization: stochastic diffusion search algorithm

Saman Maroufpoor, Rahim Azadnia and Omid Bozorg-Haddad

18. Resampling methods combined with Rao-Blackwellized Monte Carlo Data Association algorithm

Soheil Sadat Hosseini and Mohsin M. Jamali

19. Back-propagation neural network modeling on the load—settlement response of single piles

*Zhang Wengang, Anthony Teck Chee Goh, Zhang Runhong,
Li Yongqin and Wei Ning*

20. A Monte Carlo approach applied to sensitivity analysis of criteria impacts on solar PV site selection

Hassan Z. Al Garni and Anjali Awasthi

Contributors

Farid Wajdi Akashah, Centre for Building, Construction and Tropical Architecture (BuCTA), Faculty of Built Environment, University of Malaya, Kuala Lumpur, Malaysia

Hassan Z. Al Garni, Department of Electrical and Electronic Engineering Technology, Jubail Industrial College, Jubail Industrial City, Saudi Arabia

Mumtaz Ali, School of Agricultural Computational and Environmental Sciences, University of Southern Queensland, Springfield, QLD, Australia; Deakin-SWU Joint Research Centre on Big Data, School of Information Technology, Deakin University, Burwood, VIC, Australia

Armando A. Apan, School of Civil Engineering and Surveying, University of Southern Queensland, Toowoomba, QLD, Australia

Anjali Awasthi, Concordia University, CIISE, Montreal, QC, Canada

Rahim Azadnia, Department of Agricultural Machinery Engineering, University of Tehran, Tehran, Iran

Hamoon Azizsoltani, Department of Computer Science, North Carolina State University, Raleigh, NC, United States

Mourad Belgasmia, Department of Civil engineering, Setif University, Setif, Algeria

Omid Bozorg-Haddad, Department of Irrigation and Reclamation Engineering, Faculty of Agricultural Engineering and Technology, College of Agriculture and Natural Resources, University of Tehran, Tehran, Iran

Tanmoy Chatterjee, College of Engineering, Swansea University, Bay Campus, Swansea, United Kingdom; Department of Civil Engineering, Indian Institute of Technology Roorkee, Roorkee, Uttarakhand, India

Sourav Choudhary, Department of Civil Engineering, National Institute of Technology Patna, Patna, Bihar, India

Rajib Chowdhury, Department of Civil Engineering, Indian Institute of Technology Roorkee, Roorkee, Uttarakhand, India

Xuefeng Chu, Department of Civil and Environmental Engineering, North Dakota State University, Fargo, ND, United States

Kavina S. Dayal, School of Agricultural Computational and Environmental Sciences, University of Southern Queensland, Springfield, QLD, Australia

Michael Delichatsios, Fire Safety Engineering Research and Technology Centre (FireSERT), University of Ulster, Newtownabbey, United Kingdom

Ravinesh C. Deo, School of Agricultural Computational and Environmental Sciences, University of Southern Queensland, Springfield, QLD, Australia

S. Dey, Department of Mechanical Engineering, National Institute of Technology Silchar, Silchar, Assam, India

Nathan J. Downs, School of Agricultural, Computational and Environmental Sciences, University of Southern Queensland, Springfield, QLD, Australia

Subhrajit Dutta, Assistant Professor, Department of Civil Engineering, National Institute of Technology Silchar, Silchar, Assam, India; National Institute of Technology Silchar, Department of Civil Engineering, Silchar, Assam, India

Amir H. Gandomi, Faculty of Engineering and Information Technology, University of Technology Sydney, Ultimo, NSW, Australia; School of Business, Stevens Institute of Technology, Hoboken, NJ, United States

J. Ramon Gaxiola-Camacho, Department of Civil Engineering, Autonomous University of Sinaloa, Culiacan, Sinaloa, Mexico

Aboubaker Gherbi, Department of Civil engineering, Constantine University, Constantine, Algeria

Achintya Haldar, Department of Civil and Architectural Engineering and Mechanics, University of Arizona, Tucson, AZ, United States

Soheil Sadat Hosseini, Department of Electrical Engineering, Capitol Technology University, Laurel, MD, United States; Department of Electrical Engineering and Computer Science, The University of Toledo, Toledo, OH, United States

Pravin Jagtap, Department of Civil Engineering, Indian Institute of Technology (IIT) Delhi, New Delhi, India

Mohsin M. Jamali, College of Engineering, The University of Texas of the Permian Basin, Odessa, TX, United States

Madan K. Jha, Agricultural and Food Engineering Department, Indian Institute of Technology Kharagpur, Kharagpur, West Bengal, India

Shahjahan Khan, School of Agricultural, Computational and Environmental Sciences, Centre for Applied Climate Sciences, University of Southern Queensland, Toowoomba, QLD, Australia

Anoop Kodakkal, Chair of Structural Analysis, Department of Civil, Geo and Environmental Engineering, Technische Universität München (TUM), Munich, Germany

R.R. Kumar, Department of Mechanical Engineering, National Institute of Technology Silchar, Silchar, Assam, India

Yuankai Li, University of Electronic Science and Technology of China, Chengdu, China

Nehar Mandal, Department of Civil Engineering, National Institute of Technology Patna, Patna, Bihar, India

Tek Maraseni, School of Agricultural, Computational and Environmental Sciences, University of Southern Queensland, Springfield, QLD, Australia

Saman Maroufpoor, Department of Irrigation and Reclamation Engineering, Faculty of Agricultural Engineering and Technology, College of Agriculture and Natural Resources, University of Tehran, Tehran, Iran

Vasant Matsagar, Department of Civil Engineering, Indian Institute of Technology (IIT) Delhi, New Delhi, India

Anke Meyer-Baese, Department of Scientific Computing, Florida State University, Tallahassee, FL, United States

Behshad Mohebali, Department of Scientific Computing, Florida State University, Tallahassee, FL, United States

Tanmoy Mukhopadhya, Department of Aerospace Engineering, Indian Institute of Technology Kanpur, Kanpur, India

Shahbaz Mushtaq, Centre for Applied Climate Sciences, University of Southern Queensland, Toowoomba, QLD, Australia

S. Naskar, School of Engineering, University of Aberdeen, Aberdeen, United Kingdom

Thong Nguyen-Huy, School of Agricultural, Computational and Environmental Sciences, Centre for Applied Climate Sciences, University of Southern Queensland, Toowoomba, QLD, Australia

Wei Ning, School of Civil Engineering, Chongqing University, Chongqing, China

Rachid Ouache, School of Engineering, Faculty of Applied Science, University of British Columbia, Okanagan campus, Kelowna, BC, Canada

K.M. Pandey, Department of Mechanical Engineering, National Institute of Technology Silchar, Silchar, Assam, India

Thendiyath Roshni, Department of Civil Engineering, National Institute of Technology Patna, Patna, Bihar, India

Zhang Runhong, School of Civil Engineering, Chongqing University, Chongqing, China

Vikas Kumar Sharma, Institute of Infrastructute Technology Research and Management (IITRAM), Department of Mathematics, Ahmedabad, Gujarat, India

S. Sriramula, School of Engineering, University of Aberdeen, Aberdeen, United Kingdom

Amirhessam Tahmassebi, Department of Scientific Computing, Florida State University, Tallahassee, FL, United States

Anthony Teck Chee Goh, School of Civil and Environmental Engineering, Nanyang Technological University, Singapore

S. Mohsen Vazirizade, Department of Civil and Architectural Engineering and Mechanics, University of Arizona, Tucson, AZ, United States

Francisco Javier Villegas-Mercado, Department of Civil and Architectural Engineering and Mechanics, University of Arizona, Tucson, AZ, United States

Zhang Wengang, Key Laboratory of New Technology for Construction of Cities in Mountain Area, Chongqing University, Chongqing, China; School of Civil Engineering, Chongqing University, Chongqing, China; National Joint Engineering Research Center of Geohazards Prevention in the Reservoir Areas, Chongqing University, Chongqing, China

Li Yongqin, School of Civil Engineering, Chongqing University, Chongqing, China

Jianping Zhang, Fire Safety Engineering Research and Technology Centre (FireSERT), University of Ulster, Newtownabbey, United Kingdom

Chapter 1

Fundamentals of reliability analysis

Achintya Haldar

Department of Civil and Architectural Engineering and Mechanics, University of Arizona, Tucson, AZ, United States

1. Introduction

The presence of uncertainty in every aspect of engineering analysis and design has been under consideration over a long period of time. In fact, a famous mathematician Pierre-Simon Laplace (1749–1827) wrote "… the principal means of ascertaining truth — induction, and analogy — are based on probabilities; so that the entire system of human knowledge is connected with the theory (of probability). …. It leaves no arbitrariness in the choice of opinions and sides to be taken; and by its use can always be determined the most advantageous choice. Thereby it supplements most happily the ignorance and weakness of the human mind." (Laplace, 1951).

The aforementioned statements by a well-known scholar clearly justify the need for this handbook. More recently, Freudenthal (1956), Ang and Tang (1975), Shinozuka (1983), and Haldar and Mahadevan (2000a) made similar comments justifying the needs for structural safety and reliability analyses. The related areas grew exponentially in the 1970s and 1980s. It appears that most of the design guidelines and codes either have been or are in the process of incorporating the risk-based design concept, at least in the US. This handbook is expected to be extremely valuable in moving in that direction.

However, before moving forward, it is important to figure out what is the concept of uncertainty, probability, reliability, stochasticity, etc. and how they implicate the engineering analysis and design. Stochos is a Greek word for stochasticity or uncertainty. In general, most observable phenomena of interest to engineers produce multiple outcomes and cannot be predicted with certainty. Multiple outcomes may not have any pattern, and some outcomes may occur more frequently than others covering different regions of interest. Testing of identical specimens may not produce identical outcomes. The design wind velocity or rain fall at a site during the lifetime of a structure

Handbook of Probabilistic Models. https://doi.org/10.1016/B978-0-12-816514-0.00001-1

cannot be predicted with certainty. We may know their upper and lower limits or bounds and the most likely value but not the design value for which a specific structure needs to be designed. This type of unpredictability is generally represented by uncertainty or randomness. Complexity of a problem may not have anything to do with uncertainty. There is no doubt that landing on the moon consists of numerous complicated and complex processes, but we never missed the moon when we attempted to land on it. Even the outcome of a very simple task of tossing a coin cannot be predicted with certainty. Uncertainty is associated with most of the analysis and design of interest to engineers. Considering the unpredictability of most of the design variables, the basic challenge is to assure satisfactory performance of engineering systems in the presence of uncertainty. The presence of uncertainty cannot be completely eliminated, but with reasonable efforts, its impact on the design can be appropriately managed. This observation clearly indicates that engineering systems cannot be designed "full-proof" or "risk-free." There will always be some risk or probability of failure. For acceptable design, the amount of underlying risk needs to be minimized to an acceptable level or mitigated appropriately. Risk and reliability are complementary terms and need to be mathematically estimated using probability theory. Probability and statistics are not synonymous terms. Statistics is the mathematical quantification of uncertainty. Probability theory uses statistical information to estimate the likelihood of specific events. With this introduction, fundamentals of reliability analysis are briefly discussed in this introductory chapter.

2. Important steps in reliability evaluation

Most engineering problems consist of multiple random variables (RVs). The first step is then the quantification of randomness in them, one variable at a time and jointly when possible. The uncertainty in an RV is generally described pictorially in terms of histogram and frequency diagrams or probability density function (PDF) and analytically with the help of mean, variance, standard deviation, coefficient of variation (COV), skewness, etc. They are sometimes collectively denoted as the statistics of an RV. For multiple RVs, it is described as joint PDF, correlation coefficients, etc. Mean is the central tendency, variance and standard deviation indicate the dispersion from the mean, COV is the ratio of standard deviation and mean and represents the amount of uncertainty in a nondimensional way, and skewness represents the symmetry in the data, generally expressed in terms of skewness coefficient. For a known PDF, mean is the centroidal distance from the origin; it is also known as the first moment. Variance is the moment of inertia of the PDF about the mean; it is also known as the second moment. Skewness is the third moment of the PDF about the mean. For symmetric data, the skewness coefficient will be zero; more spread in the data above the mean will have positive skewness coefficient, etc. In the context of quantifying randomness in one RV, three additional parameters are commonly used. They are mode, median, and percentile value of an RV. The mode or modal value of an RV is the value of the highest PDF. The median value of an RV is the value for which it is equally likely to be above or below or the 50th percentile value. The percentile value is the probability, expressed as a percentage, the value of the RV will be less than the specified value. For more complete discussions on the topics, readers are requested to refer to Haldar and Mahadevan (2000a).

When dealing with multiple RVs, the information on covariance and correlation is very important. The covariance of two RVs X and Y, denoted as Cov (X, Y), is the second moment about their respective means. Normalizing the covariance with respect to the standard deviations of the two RVs will result the correlation coefficient. The correlation coefficient represents the linear dependency between two RVs. It varies between ± 1.0 and zero when two are uncorrelated. The value of correlation coefficient exactly zero is rarely obtained from observed data. Haldar and Mahadevan (2000a) suggested that if it is less than ± 0.3, the two variables can be considered uncorrelated.

The second step in the reliability analysis will be the performance requirements that need to be satisfied. The performance of engineering systems will depend on the uncertainty in the RVs present in the formulation; however, they will not affect the performance equally. The propagation of uncertainty from the variable or parameter to system level depends on many factors including the individual nature of uncertainty in an RV, functional relationship between variables, types of analysis; linear versus nonlinear, load path to failure, etc. In general, a system needs to be designed satisfying performance requirements at the element and system levels, and they are expected to be functions of one or multiple RVs. This will require to propagate the uncertainties in RVs to the corresponding level of interest. Only then, the underlying risk or the probability of failure corresponding to a specific performance requirement can be estimated. The uncertainty in some of the variables may be amplified at the system level, some of them may be deamplified, and some other will have very minor effect. Reliability-based engineering analysis and design are expected to be more involved or complicated in comparison with the deterministic approaches commonly used by the profession in the past without changing the basic underlying mathematical principles. It is a major challenge to the reliability community to make sure that what they are doing is similar to practices followed by the deterministic community. The most recent trend is to extract the reliability information by conducting multiple deterministic analyses. However, to maintain the basic simplicity in the formulation, effort should be exercised to reduce the size of the problem as much as possible by considering less significant RVs as deterministic at their mean values. It needs to be pointed out that in a typical deterministic analysis and design, nominal values (several standard deviations above the mean for the load-related variables and several standard deviations below the mean value for the resistance-related variables) are used, indirectly introducing the safety factor concept. If formulated properly, it may also satisfy the required unknown underlying risk. In the risk-based design, the uncertainty in a variable is explicitly expressed in terms of the statistical information (mean, variance or standard deviation, PDF, etc.).

The performance level can be at the element or system level, as mentioned earlier. Element-level performance requirements are generally strength related. The system level requirements generally reflect serviceability-related performances causing excessive lateral (interstory or overall) deflection or drift caused by failure of structural elements or inadequate stiffness of elements, etc. The performance requirements are generally expressed in the form of performance or limit state functions/equations. A typical performance function consists of all load- and resistance-related RVs present in the formulation. Performance requirements are generally suggested in the design guidelines or specified by the owner. Sometimes, performance functions can be explicit, but in most cases of practical interest, they are implicit in nature.

The third step is the estimation of risk/reliability corresponding to a performance requirement. Some of the estimation procedures will be discussed in more detail later in this chapter. Essentially, the information informs the designer the underlying risk for the specific design. If the risk is not acceptable even when all design requirements were satisfied, the design needs to be altered to satisfy the owner or the stakeholder. This step makes the reliability-based analysis and design more attractive than the conventional deterministic evaluation. It helps to compare different design alternatives. Using information on risk of two design alternatives, the most desirable or appropriate option can be selected.

3. Elements of set theory

Essential steps in estimating risk have been discussed in the previous section. It is now necessary to quantify the risk associated with a design. For the reliability evaluation, the concept of set theory instead of conventional algebra needs to be used. The concept of set theory is very briefly discussed in this section.

A typical engineering problem must have a sample space, discrete or continuous, consisting of mutually exclusive sample points. An event of interest must contain at least one sample point. Events can be combined using union or intersection operation, and they can be mutually exclusive, statistically independent, or correlated. The information on risk or reliability can be extracted with the help of set theory and the three axioms of probability. Three axioms of probability are (1) the probability of an event E, generally written as $P(E)$, will always be nonnegative, (2) the probability of the sample space, S, or $P(S)$, will be 1.0; thus, the probability of an event will be between zero and one, and (3) if the two events E_1 and E_2, are mutually exclusive, the probability of their union will be the summation of their individual probability. It is generally expressed as

$$P(E_1 \cup E_2) = P(E_1) + P(E_2) \tag{1.1}$$

The complement of an event, denoted as E, can be shown to be

$$P(\overline{E}) = 1.0 - P(E) \tag{1.2}$$

Eq. (1.2) indicates that the probability of survival can be estimated as $1.0 - \textit{risk}$.

In general, events are not mutually exclusive. In that case, Eq. (1.1) can be shown to be

$$P(E_1 \cup E_2) = P(E_1) + P(E_2) - P(E_1 E_2) \tag{1.3}$$

The probability of intersection or joint occurrence of events, $P(E_1 E_2)$ as in Eq. (1.3), needs to be calculated using the multiplication rule. The general multiplication rule for two events can be shown to be

$$P(E_1 E_2) = P(E_1 | E_2) P(E_2) = P(E_2 | E_1) P(E_1) \tag{1.4}$$

$P(E_1 | E_2)$ or $P(E_2 | E_1)$ is generally known as the probability of the conditional events indicating the occurrence of one event given the occurrence of the other event. It reduces the size of the original sample space. To estimate the probability with respect to the original sample space, the conditional probability needs to be multiplied by the probability of the event conditioned on with respect to the original sample space.

If the events are statistically independent, indicating the occurrence of one does not depend on the other, Eq. (1.4) becomes

$$P(E_1 E_2) = P(E_1) P(E_2) \tag{1.5}$$

If the events are mutually exclusive, i.e., if the occurrence of one precludes the occurrence of the other, the probability of joint occurrences of them will be zero. If the unions of all the events constitute the sample space, they are called collectively exhaustive events. More discussions on set theory, axioms of probability, multiplication rules, and other operations can be found in the study by Haldar and Mahadevan (2000a).

4. Quantification of uncertainties in random variables

The next essential task in the reliability estimation is the quantification of uncertainties in all the RVs present in the formulation. The quantification of uncertainty in an RV requires the collection of data on it from many sources as possible. The sample size is very important in the uncertainty quantification. Larger sample size is always preferable. It may be impractical to collect data from available sources from all over the world. The collected data can be graphically represented with the help of the histogram and PDF. PDF is a histogram with the area of 1.0. In most cases, a PDF can be generated by fitting a polynomial whose parameters can be estimated from the samples used to generate it. In most engineering applications, two-parameter PDFs are routinely used. It can be shown that these parameters can be estimated from the first two moments, the mean and variance, of the samples. The most encouraging part of reliability-based analysis and design is that most of the design variables used in routine engineering applications are already researched and widely available in the literature. The information on the PDF and its parameters of an RV of interest can be easily obtained with a casual literature search.

For the completeness of this discussion, commonly used continuous RVs are represented by normal, lognormal, Beta, Rayleigh, and exponential distributions. A normal distribution will be valid when an RV is valid from minus infinity to plus infinity. A lognormal distribution is used when an RV is valid from zero to plus infinity. A Beta distribution can be used when an RV is valid between a lower and upper limits. Some of the commonly used discrete RVs are binomial, geometric, Poisson, etc. Parameters required to define these RVs are tabulated in many books (Ang and Tang, 1975; Haldar and Mahadevan, 2000a).

5. Transformation of uncertainty from parameter to the system level

As will be discussed later, if a performance or limit state equation (LSE) or function is a linear function of one RV, it will be relatively simple to estimate the risk using the available statistical information on the RV. However, in most cases of practical interest, the performance of a structural element may depend on nonlinear function of one RV or multiple RVs. For a structural system consisting of numerous structural elements, the LSE will be a function of numerous RVs. The underlying risk will depend on the formulations and the statistical characteristics of all the RVs involved in the formulation. For the sake of discussion, the axial load carrying capacity of a steel column will be denoted as the response variable and will depend on the load acting on it, grade of steel used, cross-sectional area, radius of gyration, length of the column, support conditions, etc. Collectively, they will be denoted hereafter as the basic RVs. Uncertainties in these basic variables will dictate the overall uncertainty in the axial load carrying capacity. Functional relationship between these design variables to the load carrying capacity is generally suggested in design codes or guidelines or can be mathematically derived. When the functional relationship of all RVs and the appropriate performance requirement is available, it will be considered as an explicit LSE. A steel frame may consist of numerous such columns. The lateral deflection at the top of the frame (response variable) will depend on the properties of these columns and other structural elements present in the frame, essentially all the variables involved in the estimation of the lateral deflection and the loads acting on it. However, an explicit expression for the lateral deflection in terms of all the RVs may not be practical to develop; it may be available in an algorithmic form, for an example, in the form of finite elements representation. In this situation, the LSE can be considered as implicit in nature.

Transformation of uncertainty from the basic RVs to the response variable will depend on many factors including the types of the functional relationship, linear or nonlinear, the total number of RVs present in the formulation, their statistical and correlation characteristics, etc. Obviously, the underlying risk estimation procedure will be different for each situation. Each scenario will produce different types of challenges and will require different levels of sophistication or expertise to extract the reliability information. Only in few cases, the propagation of uncertainties from the variable to system level can be accomplished exactly. In some cases, the propagation can be achieved partially; in some other cases, the propagation can be achieved approximately; in some other cases, the information can be generated numerically; and in some cases, the unknown relationship needs to be generated from the available data. They are discussed briefly in the following sections.

For the ease of presentation and considering the space limitation, the discussions are subdivided into two parts: when functional relationship is available and when it is not. Haldar and Mahadevan (2000a) discussed many related issues in more detail and should be referenced if desired.

5.1 Explicit functional relationship

Even when explicit functional relationships are available, they may be of different forms. Some of them are discussed in the following sections.

5.1.1 Exact solution

Suppose the functional relationship is linear, the response variable Y is a function of one basic random variable X. The relationship can be expressed as

$$Y = a + bX \tag{1.6}$$

where a and b are known constants. It can be shown that the distribution of Y will not change, but its mean $E(Y)$ and variance $Var(Y)$ need to be recalculated as

$$E(Y) = a + b\,E(X) \tag{1.7}$$

and

$$Var(Y) = b^2 Var(X) \tag{1.8}$$

Suppose the known functional relationship is nonlinear, response variable Y is a function of one basic random variable X. The relationship can be expressed as

$$Y = g(X) \tag{1.9}$$

The PDF of Y can be shown to be

$$f_Y(Y) = \sum_{i=1}^{n} f_X\left[g_i^{-1}(y)\right] \left| \frac{dg_i^{-1}(y)}{dy} \right| \tag{1.10}$$

where $x = g^{-1}(y)$ and n is the number of terms after the inversion; n will be two for a quadratic relationship.

Suppose Y is a known function of multiple RVs, the exact solution can be obtained only for few special cases. Some of them are discussed in the following.

Case 1 — Multiple RVs with known joint PDF

This case is expected to be rare, but the procedure discussed for the single basic RV using Eq. (1.10) can be followed, as discussed in Haldar and Mahadevan (2000a).

Case 2 — Sum of Independent normal variables

For this case, the response variable Y can be written as

$$Y = c_1 X_1 + c_2 X_2 + \cdots + c_n X_n \tag{1.11}$$

where $c_i's$ are known constants and $X_i's$ are statistically independent normal RVs with mean and standard deviation of μ_{X_i} and σ_{X_i}, respectively.

The mean and variance of Y can be shown to be

$$\mu_Y = \sum_{i=1}^{n} c_i \mu_{X_i} \tag{1.12}$$

and

$$\sigma_Y^2 = \sum_{i=1}^{n} c_i^2 \sigma_{X_i}^2 \tag{1.13}$$

Case 3 — Product of independent lognormal variables

In this case,

$$Y = X_1 X_2 \cdots X_n = \prod_{1}^{n} X_i \tag{1.14}$$

Where $X_i's$ are statistically independent lognormal RVs with parameters λ_{X_i} and ξ_{X_i}.

It can be shown that

$$\lambda_Y = \sum_{i=1}^{n} \lambda_{X_i} \tag{1.15}$$

and

$$\xi_Y^2 = \sum_{i=1}^{n} \xi_{X_i}^2 \tag{1.16}$$

Case 4 − Sum of independent Poisson RVs

Suppose $X_i's$ are statistically independent Poisson RVs with parameters v_{X_i}, they are related to Y as

$$Y = X_1 + X_2 + \cdots + X_n \tag{1.17}$$

It can be shown that Y is also a Poisson RV with parameter

$$\lambda_Y = \sum_{i=1}^{n} v_{X_i} \tag{1.18}$$

Case 5 − Central limit theorem

The aforementioned discussions will not be complete without briefly discussing the central limit theorem. It states that the sum of a large number of RVs, if none of them dominate the sum, regardless of their initial distributions, tends to the normal distribution as the number increases. Similarly, the product of a large number of RVs, if none of them dominate the product, regardless of their initial distributions, tends to the lognormal distribution as the number increases.

5.1.2 Partial Solutions

The exact solutions discussed in the previous section can be very limited from the application point of view. For some of the cases not discussed previously, it will be desirable if partial solutions can be obtained to propagate uncertainties from the variable to the system level.

Partial Solutions − linear function of multiple RVs

Consider Eq. (1.11) again. Suppose the information on distribution of $X_i's$ is unknown or they have different distributions, it will not be possible to determine the exact distribution on Y. However, its mean and variance can be obtained as

$$\mu_Y = \sum_{i=1}^{n} c_i \mu_{X_i} \tag{1.19}$$

$$Var(Y) = \sum_{i=1}^{n} \sum_{j=1}^{n} c_i c_j Cov(X_i, X_j) \tag{1.20}$$

since

$$Cov(X_i, X_i) = Var(X_i) = \sigma_{X_i}^2 \tag{1.21}$$

5.1.3 Approximate solutions—general function of multiple RVs

If a partial solution is not possible, it will be reasonable to obtain an approximate solution. Some of them are discussed in the following sections.

Suppose the relationship between Y and X_i's is known, the means and variance of X_i's are known without their distribution information. The relationship can be represented as

$$Y = g(X_1, X_2, ..., X_n) \tag{1.22}$$

It will not be possible to obtain the distribution information on Y; however, the approximate mean and variance can be obtained by expanding the function in Taylor series about the mean values of X_i's.

The approximate first-order and second-order means, denoted as $E(Y')$ and $E(Y'')$, can be shown to be

$$E(Y') \approx g(\mu_{X_1}, \mu_{X_2}, ..., \mu_{X_n}) \tag{1.23}$$

and

$$E(Y'') \approx g(\mu_{X_1}, \mu_{X_2}, ..., \mu_{X_n}) + \frac{1}{2} \sum_{i=1}^{n} \left(\frac{\partial^2 g}{\partial X_i^2} \right) Var(X_i) \tag{1.24}$$

In most cases, only the first-order variance, denoted as $Var(Y')$, can be estimated as

$$Var(Y') \approx \sum_{i=1}^{n} \sum_{j=1}^{n} E_i E_j Cov(X_i, X_j) \tag{1.25}$$

where E_i and E_j are partial derivatives $\partial g / \partial X_i$ and $\partial g / \partial X_j$, respectively, evaluated at the mean values of X_i's. They are constants. The coefficients E_i's can be considered as amplification or deamplification factors based on their numerical values. They will show the role of each RV in propagating uncertainty from the variable to the system level. If they are very much smaller than 1.0, they can be considered as deterministic constants evaluated at their mean values. If they are much larger than 1.0, resources should be invested to reduce their uncertainties with the help of quality control and more inspections.

5.2 Multiple random variables with unknown relationship

In many practical cases, the exact functional relationship of Y and X_i's may not be known; it may only be known in an algorithmic form, for example, with the help of the finite element algorithm. In this situation, the first-order mean and variance of Y, without calculating the partial derivatives as in Eqs. (1.24) and (1.25), can be obtained using the Taylor series finite difference procedure. In this case, Y values need to be calculated two more times for each RV as

$$Y_i^+ = g\left[\mu_{X_1}, \mu_{X_2}, \ldots, \left(\mu_{X_i} + \sigma_{X_i}\right), \ldots, \right] \tag{1.26}$$

and

$$Y_i^- = g\left[\mu_{X_1}, \mu_{X_2}, \ldots, \left(\mu_{X_i} - \sigma_{X_i}\right), \ldots, \right] \tag{1.27}$$

Eq. (1.26) suggests that Y_i^+ to be estimated at the mean values of all the RVs except the ith one; it has to be at the mean plus one standard deviation value. Similarly, Eq. (1.27) suggests that Y_i^- to be estimated at the mean values of all the RVs except the ith one; it has to be mean minus one standard deviation value. Using the central difference approximation, it can be shown that E_i in Eq. (1.25) can be estimated as

$$E_i = \frac{\partial g}{\partial X_i} \approx \frac{Y_i^+ - Y_i^-}{2\sigma_{X_i}} \tag{1.28}$$

The corresponding $Var(Y')$ can be calculated as

$$Var(Y') \approx \sum_{i=1}^{n} \left(\frac{Y_i^+ - Y_i^-}{2}\right)^2 \tag{1.29}$$

5.3 Regression analysis

Regression analysis is generally conducted to develop probabilistic relationship between the response variable and one or multiple basic RVs. Developing a linear relationship between the response variable and one basic RV is called the simple linear regression analysis. When linear relationship is required for multiple RVs, it is called the multiple regression analysis. When nonlinear relationship is required, it is called nonlinear regression analysis. The concept behind regression analysis is relatively simple; however, there is a potential for misuse. The basic concept is briefly discussed in the following section for simple linear regression. The interested readers are suggested to refer to Haldar and Mahadevan (2000a) and Montgomery et al. (2012) for more advanced discussions on regression analysis.

5.3.1 Simple linear regression

Denoting Y as the response variable and X as the basic RV, n pairs of data (x_1, y_1), (x_2, y_2), ..., (x_n, y_n) are available to develop a linear regression-based probabilistic relationship. A typical scatter diagram of the data is shown in Fig. 1.1. It can be expressed as

$$E(Y|X = x) = E(Y|x) = b_0 + b_1 x \tag{1.30}$$

where b_0 and b_1 are the intercept and the slope of the line, respectively, and they are known as the regression coefficients. Eq. (1.30) represents the mean value of the relationship. Some of the basic assumptions of the regression analysis represented by Eq. (1.30) are that the scatter diagram indicates the relationship is linear, the spread of the data about the equation is uniform represented by an error term ε with a zero mean and unknown but constant variance, and the errors are uncorrelated and normally distributed. The basic assumptions in regression analysis are shown in Fig. 1.1.

Generally, the regression coefficients are estimated using the method of least squares. They are estimated by minimizing ε^2, also known as the error sum of squares (SS_E) representing the differences between the observed and predicted values using the regression equation. Eq. (1.30) represents the mean value of the relationship. It has also a variance, denoted as $Var(Y|X = x) = S^2_{Y/x}$, known as the error mean square or the residual mean square (MS_E). It can be estimated as

$$MSE = S^2_{Y/x} = \frac{\varepsilon^2}{n-2} = \frac{SS_E}{n-2} \tag{1.31}$$

where SS_E can be estimated as

$$SS_E = \sum_{i=1}^{n} (y_i - b_0 - b_i - b_i x_i)^2 \tag{1.32}$$

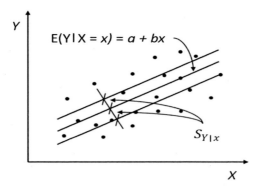

FIGURE 1.1   Scatter diagram of the data.

The appropriateness of the regression equation is evaluated by estimating the coefficient of determination, R^2. Denoting the total variability in Y as S_{yy} and the amount of uncertainty accounted for by the regression equation, SS_R, the following relationship can be obtained:

$$S_{yy} = SS_R + SS_E \tag{1.33}$$

Then, R^2 can be shown to be

$$R^2 = \frac{SS_R}{S_{yy}} = 1 - \frac{SS_E}{S_{yy}} \tag{1.34}$$

R^2 will have a value between 0 and 1. When it is close to 1.0, it represents that the regression equation is appropriate or reasonable. It is important to mention that the extrapolation of the equation beyond the range of the observed data is not recommended.

For multiple or nonlinear regression analyses, the interested readers are referred to the references cited earlier.

6. Fundamentals of reliability analysis

It is expected that the readers now developed the basic understanding of how to quantify uncertainty in an RV and how to propagate it from the variable level to the system level. As mentioned earlier, the basic intention is to estimate the underlying risk or reliability of engineering designs. The concept has been developed over time, and it is considered to be matured at present. Several risk estimation procedures with various levels of sophistication and complexity are available. However, before discussing them, the fundamentals of reliability analysis are briefly discussed in the following paragraph.

The basic intent of any engineering design is that the resistance R is greater than the most critical load effect S during the lifetime of a structure. Because both R and S cannot be predicted with certainty, they are random in nature. Representing uncertainty in them with the help of their corresponding PDFs in terms of mean μ and standard deviation σ, the basic concept is graphically shown in Fig. 1.2. The shaded area is related to the possibility of R less than S, indicating the underlying risk of the design. Obviously, it should be as small as possible satisfying the intent of the design guidelines. The figure also indicates that the overlapped tail ends of R and S cannot totally be eliminated, indicating it will be difficult if not impossible to design "fail-proof" systems.

The figure also indicates several ways to reduce the size of the overlapped area. The reduction can be achieved by increasing the distance between the two PDFs or separating the mean values of the two RVs by reducing the uncertainty in them in terms of σ value (may be with the help of additional inspections or quality control), or by changing the shape of the two PDFs representing the nature of uncertainty in them. The nominal values of R and S are indicated in the figure as R_N and S_N, respectively. The nominal values were used in the past by the deterministic community. The nominal value of the resistance R_N is several standard deviations below the mean, indicating the amount of underestimation of it. On the other hand, the nominal load effect S_N is several standard deviations above the mean, indicating the amount of overestimation of them. According to deterministic practices, if the ratio of R_N/S_N is greater than 1.0, it is an acceptable design and if it is less than 1.0, say even 0.999, it is not acceptable. By underestimating the resistance and overestimating the load effect, the deterministic community introduced the safety factor concept without estimating the underlying risk. The figure clearly indicates the basic weakness in the old practice.

The same design codes or guidelines for steel, concrete, or other materials are used to design nuclear power plants, hospitals, fire stations, buildings, etc. in the US. However, the engineering profession claims that a nuclear power plant is much safer than an ordinary building. It is true because for nuclear power plants, R_N and S_N are selected much more conservatively by considering them several standard deviations below and above the mean corresponding to different return periods, respectively. Denoting p as the annual probability of occurrence of an event, the return period, T, can be defined as $T = 1/p$ (Haldar and Mahadevan, 2000a). For 50 years of return period, on an average, the annual probability of occurrence will be $p = 1/50 = 0.02$. In routine applications, ordinary structures are designed for 50-year return period events, but nuclear power plants are designed for thousands of years return period events.

Incorporating the presence of uncertainty in the design, and referring to Fig. 1.2, the probability of failure, p_f, can be defined as $P(R < S)$ and can be estimated as

$$p_f = P(R < S) = \int_0^\infty [f_R(r)dr]f_S(s)ds \tag{1.35}$$

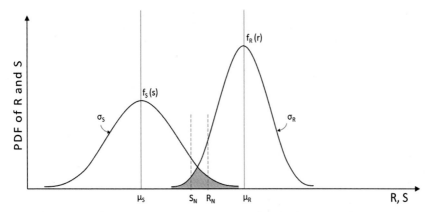

FIGURE 1.2   Fundamental concept of risk evaluation. PDF, probability density function. *Adopted from Haldar, A., Mahadevan, S., 2000. Probability, Reliability, and Statistical Methods in Engineering Design New York, NY. With permission from John Wiley and Sons, Inc.*

Numerical evaluation of Eq. (1.35) can be very challenging in most cases. In general, both R and S are expected to be functions of multiple RVs, and their joint PDFs may not be available. When R and S are assumed to have specific known PDFs and the information on the corresponding parameters to define them is available, closed-form numerical solution of Eq. (1.35) may require considerable effort. However, for few cases, p_f can be discussed in the following.

R and S are assumed to be normal RVs with known parameters, i.e., $N(\mu_R, \sigma_R)$ and $N(\mu_S, \sigma_S)$, respectively. It is reasonable to assume that they are statistically independent. Then, p_f can be estimated as

$$p_f = P(R < S) = P(R - S < 0) = 1 - \Phi\left[\frac{\mu_R - \mu_R}{\sqrt{\sigma_R^2 + \sigma_S^2}}\right] = 1 - \Phi(\beta) \quad (1.36)$$

In the aforementioned equation, Φ is the cumulative distribution function (CDF) of the standard normal distribution and β is generally known as the reliability index. The aforementioned equation indicates that when β is large, the underlying risk will be smaller. In most designs, the value of β between 3 and 5 is expected.

If R and S are statistically independent lognormal RVs with the same means and variances, p_f can be approximately estimated as

$$p_f = P\left(\frac{R}{S} < 1\right) \approx 1 - \Phi\left[\frac{ln\left(\frac{\mu_R}{\mu_S}\right)}{\sqrt{\delta_R^2 + \delta_S^2}}\right] \quad (1.37)$$

where δ is the COV of RVs. In the aforementioned equation, COV of an RV is considered to be not large, say it is less than 0.3 (Haldar and Mahadevan, 2000a). For the lognormal distribution, the reliability index β is the term in the square bracket.

In most cases, R and S are not expected to be normal or lognormal variables. Even one of them could be normal, and the other could be lognormal. In general, R and S are expected to be functions of multiple RVs, and the functional relationship could be of any form. Then, the evaluation of risk using Eq. (1.35) can be very challenging. For these cases, several risk evaluation techniques with various degrees of sophistication were proposed. However, before discussing them, the concept of limit state needs additional discussions.

6.1 Limit state equations or functions

Risk is always estimated for a specific limit state function or equation, it will be denoted hereafter as LSE. As mentioned earlier, they are functions of

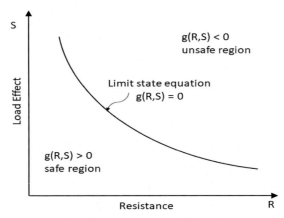

FIGURE 1.3 Limit state concept.

load- and resistance-related parameters and the required performance limit or acceptable behavior. The concept is graphically shown in Fig. 1.3.

A typical LSE can be an explicit or implicit function of the design variables and the design requirements explicitly suggested in design guidelines or codes. They are generally related to the strength and/or serviceability requirements. In most cases, multiple LSEs need to be satisfied, indicating multiple requirements need to be satisfied under the same desired risk. This is very important to eliminate the weakest-link failure mode. The requirements may be at the structural element level or based on the overall whole structural behavior. For framed structures, some of the commonly used LSEs are pure bending of members, combined axial and compression, and bending of frame members, sides way at the top of the frame, interstory drift, maximum defection along the length of a member, etc. It is extremely important to formulate the appropriate LSE properly.

6.1.1 Serviceability limit state equation or function

The serviceability LSE for the interstory, overall, or the maximum deflection along the length of a member can be expressed as

$$g(\mathbf{X}) = \delta_{allow} - \delta_{max}(\mathbf{X}) = \delta_{allow} - \widehat{g}(\mathbf{X}) \tag{1.38}$$

where δ_{allow} is the allowable value of the specific deflection of interest most of the time; suggested in design guidelines or commonly practiced and $\delta_{max}(\mathbf{X})$ is the maximum value obtained by the analysis using the unfactored loads. It is interesting to note that some design guidelines permit using higher δ_{allow} values depending on the procedures used to estimate them. For example, ASCE/SEI 7−10 permits increasing it by 125% if it is estimated for the seismic loading applying the excitation in time domain.

6.1.2 Strength limit state equation or function

Suppose a strength LSE of a steel member of a frame needs to be defined, the following interaction equations are used to design them according to the load and resistance factor design (LRFD) guidelines for two-dimensional elements of the American Institute of Steel Construction (AISC)'s (AISC, 2011):

$$\frac{P_u}{\phi P_n} + \frac{8}{9}\frac{M_u}{\phi_b M_n} \leq 1.0, \quad if \;\; \frac{P_u}{\phi P_n} \geq 0.2 \tag{1.39}$$

$$\frac{P_u}{2\phi P_n} + \frac{M_u}{\phi_b M_n} \leq 1.0, \quad if \;\; \frac{P_u}{\phi P_n} < 0.2 \tag{1.40}$$

where ϕ is the resistance factor, P_u is the required tensile/compressive strength, P_n is the nominal tensile/compressive strength, M_u is the required flexural strength, and M_n is the nominal flexural strength. P_u and M_u are the critical factored load effects. The corresponding LSEs can be expressed as

$$g(\mathbf{X}) = 1.0 - \left(\frac{\widehat{P}_u}{P_n} + \frac{8}{9}\frac{\widehat{M}_{ux}}{M_{nx}} \right); \quad if \;\; \frac{P_u}{\phi P_n} \geq 0.2 \tag{1.41}$$

$$g(\mathbf{X}) = 1.0 - \left(\frac{\widehat{P}_u}{2P_n} + \frac{\widehat{M}_{ux}}{M_{nx}} \right); \quad if \;\; \frac{P_u}{\phi P_n} < 0.2 \tag{1.42}$$

where $\widehat{P}_u$ and $\widehat{M}_{ux}$ are unfactored load effects. It is important to note that in LSEs, the load and resistance factors are not used.

6.2 Reliability evaluation methods

The stage is now set to estimate risk or reliability. To evaluate Eq. (1.35), several reliability evaluation methods with different levels of sophistication were proposed. Initially, the developments were limited to linear or tangent to the nonlinear limit state or performance function, leading to the development of the first-order reliability methods (FORMs). In Eqs. (1.36) and (1.37), the limit state equations (LSEs) are defined as $Z = R - S$ and $Z = R/S$, respectively. In general, R and S are functions of multiple random variables. To generalize the discussions, a typical LSE can be represented as

$$Z = g(\mathbf{X}) = g(X_1, X_2, \ldots, X_n) \tag{1.43}$$

Because $X_i's$ have different statistical characteristics, it will be difficult to estimate exactly the statistical information on Z. However, as discussed earlier, the approximate mean and variance of Z can be estimated by expanding Z in Taylor series about the mean of RVs. The safety index β can be calculated by taking the ratio of the mean and the standard deviation. However, because the Taylor series expansion is made at the mean values of all the RVs, it is known as the mean value first-order second moment (FOSM) or simply MVFOSM. Most of the first-generation reliability-based design guidelines are based on this concept.

The aforementioned concept has several major weaknesses as discussed by Haldar and Mahadevan (2000a). The procedure fails to incorporate the information on distribution of RVs even when it is available. Because the LSE represented by Eq. (1.43) is linearized at the mean values of the RVs, it can introduce significant error by not considering the higher order terms. It was documented in the study by Haldar and Mahadevan (2000a) that for the same mechanically equivalent LSEs defined as $(R-S < 0)$ or $\left(R/_S < 1\right)$, the procedure does not give the same reliability index. It needs improvements.

6.2.1 Advanced FOSM for statistically independent normal variables (Hasofer–Lind method)

During this phase of development, Hasofer and Lind (1974) gave a geometric definition of the reliability index. They observed that in the reduced variable space, as defined in the following, the reliability index, denoted as β_{HL}, is the shortest distance from the origin to the LSE. This observation is valid when all RVs are normal. The reduced variable of X_i, denoted as X_i', can be defined as

$$X_i' = \frac{X_i - \mu_{X_i}}{\sigma_{X_i}} \quad (i = 1, 2, ..., n) \tag{1.44}$$

The closet point on an LSE in the reduced variable space from the origin is generally denoted as the design or checking point. If its coordinates are expressed in a vectorial form as $\mathbf{x}'^*$, it can be shown that

$$\beta_{HL} = \sqrt{(x'^*)^t (x'^*)} \tag{1.45}$$

The concept is shown in Fig. 1.4.

The significance of Hasofer and Lind (1974) observation can be easily demonstrated with the help a linear LSE consisting of two independent RVs, as shown in Fig. 1.5. Consider Fig. 1.5A first. The LSE is plotted in the original coordinate system. It will be a straight line at a 45 degrees angle. The safe and unsafe regions are also shown in the figure. In MVFOSM, the function is evaluated at the mean values of RVs, indicating the coordinates of the most probable failure point (MPFP) or checking point or design point. It is not on LSE and is a major source of error in the risk estimation. The coordinates of the MPFP, denoted as (r^*, s^*), are also shown in the figure. It is on LSE. The author observed that error in the risk estimation increases with the separation distance between the MPFP and the mean values.

Consider Fig. 1.5B now. The same LSE is now plotted in the reduced coordinates in the figure by line AB. The coordinates of Point A are

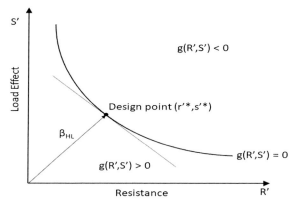

FIGURE 1.4 Hasofer–Lind reliability index for nonlinear LSE. *Adopted from Haldar, A., Mahadevan, S., 2000. Probability, Reliability, and Statistical Methods in Engineering Design Sons New York, NY. With permission from John Wiley and Sons, Inc.*

$\left(-\frac{\mu_R - \mu_S}{\sigma_R}, 0\right)$ and Point B is $\left(0, \frac{\mu_R - \mu_S}{\sigma_S}\right)$ It will be relatively simple and without any iteration, and using basic trigonometry, β_{HL} can be calculated as

$$\beta_{HL} = \frac{\mu_R - \mu_S}{\sqrt{\sigma_R^2 + \sigma_S^2}} \tag{1.46}$$

The reliability index according to MVFOSM and Hasofer and Lind will be identical. Haldar and Mahadevan (2000a) have shown that the coordinates of the checking point in the original coordinate system will be as

$$r^* = s^* = \frac{\mu_R \sigma_S^2 + \mu_S \sigma_R^2}{\sigma_R^2 + \sigma_S^2} \tag{1.47}$$

Eq. (1.47) indicates that the MPFP will be on the LSE, as shown in Fig. 1.5A. Fig. 1.5 clearly indicates the benefit of the Hasofer-Lind method. Haldar and Mahadevan (2000a) called it advanced FOSM or AFOSM.

The reliability index estimated by the Hasofer-Lind method is valid when the LSE is linear and all the variables are normal. For most cases of practical interest, these requirements cannot be satisfied. These issues were addressed by Rackwith and Fiessler (1976, 1978) and others. The related works led to the development of the full distributional approach.; specifically first-order reliability method (FORM) and second-order reliability Method (SORM). FORM is used when an LSE is a linear function of uncorrelated normal RVs or when it is nonlinear function represented by a first-order (linear; tangent at the checking point) approximation using equivalent normal variables as discussed in the next section. SORM is used when an LSE is nonlinear function of basic RVs including linear LSE with correlated non-normal variables by a second-order representation (tangent and curvature). The related issues are discussed in more detailed by Haldar and Mahadevan (2000a). Most design

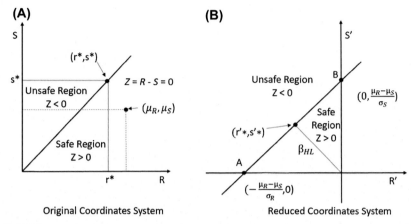

FIGURE 1.5 Hasofer–Lind reliability index for linear LSE. *Adopted from Haldar, A., Mahadevan, S., 2000a. Probability, Reliability, and Statistical Methods in Engineering Design John Wiley and Sons, New York, NY. With permission from John Wiley and Sons, Inc.*

variables for routine engineering applications can be represented by two parameters. FORM in the context of two-parameter distribution is briefly discussed in the following section.

6.2.2 First-order reliability method with two-parameter distributions

A FORM-based reliability evaluation procedure is implemented in the normal variable space, and the variables are considered to be statistically independent. In general, any nonnormal variable cannot be transformed to a normal variable over the range of the variable. For an example, a lognormal variable, valid from zero to plus infinity, cannot be transformed to a normal variable valid between minus to plus infinity. However, without changing the basic characteristic of a two-parameter RV, two conditions can be imposed. Rackwitz and Fiessler (1976) suggested that the PDF and CDF of a nonnormal variable and an equivalent normal variable should be equal at the checking point. The two conditions can be expressed as

$$\mu_{X_i}^N = x_i^* - \Phi^{-1}\left[F_{X_i}\left(x_i^*\right)\right]\sigma_{X_i}^N \qquad (1.48)$$

and

$$\sigma_{X_i}^N = \frac{\phi\left\{\Phi^{-1}\left[F_{X_i}\left(x_i^*\right)\right]\right\}}{f_{X_i}\left(x_i^*\right)} \qquad (1.49)$$

where $\Phi()$ and $\phi()$ are the CDF and PDF of the standard normal variable, respectively. The functions $F_{X_i}\left(x_i^*\right)$ and $f_{X_i}\left(x_i^*\right)$ represent the CDF and PDF of the original nonnormal variables at the checking point x_i^*, respectively. For further discussion on the topic, please refer to the study by Haldar and Mahadevan (2000a). It is also important to note that all the variables in the formulation also need to be statistically independent to each other. If they are not, they should be changed as discussed briefly later.

Reliability evaluation using FORM is an iterative procedure. The procedure originally proposed by Rackwitz and Fiessler (1978), improved by Ayyub and Haldar (1984), can be implemented with the help of the following steps.

Step 1 — Appropriate LSEs need to be defined at the initiation of any risk analysis.

Step 2 — Assume an initial value of β of 3.0. Any other reasonable value is also acceptable.

Step 3 — Initiate the iterative process by assuming the coordinates of the checking point at the mean values of all RVs.

Step 4 — Compute the equivalent mean and standard deviation of all nonnormal RVs at the checking point.

Step 5 — Compute the partial derivatives of the LSE with respect to the ith RV at the checking point.

Step 6 — Compute the direction cosine α_{X_i} for the ith RV at the checking point using the following equation:

$$\alpha_{X_i} = \frac{\left(\dfrac{\partial g}{\partial X_i}\right)^* \sigma_{X_i}^N}{\sqrt{\sum_{i=1}^{n}\left(\dfrac{\partial g}{\partial X_i}\sigma_{X_i}^N\right)^{2*}}} \tag{1.50}$$

Step 7 — Compute the new checking point value for the ith RV as

$$x_i^* = \mu_i^* = \mu_{X_i}^N - \alpha_i \beta\, \sigma_{X_i}^N \tag{1.51}$$

Repeat Steps 4 through 7 until α_{X_i} converges with a predetermined tolerance. A tolerance level of 0.005 is common.

Step 8 — Update β by satisfying the LSE at the checking point.

Step 9 — Repeat Steps 3 through 8 until β converges to a predetermined tolerance level.

The aforementioned nine steps FORM procedure converges very rapidly, most of the time within five cycles. For a linear LSE, no iteration is necessary. For nonlinear LSEs, a computer program can be written to implement the procedure.

6.2.3 Examples

The aforementioned FORM procedure is explained further with several examples.

Example 1 — linear LSE with two independent normal RVs

Suppose R and S are assumed to be normal RVs with known parameters, i.e., $N(\mu_R, \sigma_R)$ and $N(\mu_S, \sigma_S)$, respectively. Assuming LSE is of the form $Z = R - S$, p_f can be estimated using the nine steps discussed above. The information is summarized in Table 1.1.

TABLE 1.1 Risk estimation for linear LSE with two independent normal RVs.

Step 1	$Z = R - S$; $R \sim N(150, 22.5)$, $S \sim N(62.5, 15)$			
Step 2	β	Iteration is not required for this example.		
Step 3	f_R^*	150.0		
Step 4	f_S^*	62.5		
	μ_R^N	150.0		
	σ_R^N	22.5		
	μ_S^N	62.5		
	σ_S^N	15.0		
Step 5	$\left(\frac{\partial Z}{\partial R}\right)^*$	1		
	$\left(\frac{\partial Z}{\partial S}\right)^*$	-1		
Step 6	α_R	0.832		
	α_S	-0.555		
Step 7	After α converges, compute coordinates for the new checking point.			
Step 8	β	3.236	$P_f = 6.066 \times 10^{-4}$	
Step 9	Repeat steps 3 to 8 until β converges			

Coordinates of the failure point; $r^* = 89.423$, $s^* = 89.423$. RV, random variable.

Example 2 — Linear LSE with independent non-normal RVs

Consider the same linear LSE as in Example 1 but R is lognormally and S is normally distributed. They are independent with the same mean and standard deviation. In this case, an iterative procedure is necessary to estimate the risk. All steps are summarized in Tables 1.1 and 1.2.

TABLE 1.2 Risk estimation for linear LSE with independent nonnormal RVs.

Step											
Step 1	$Z = R - S$; $R \sim LN(150, 22.5)$, $S \sim N(62.5, 15)$										
Step 2	β	3.000				3.63			3.60		
Step 4	f_R^*	150.00	92.58	108.16	106.88	99.79	101.31	101.03	102.70	101.03	101.36
	f_S^*	62.50	87.56	95.61	93.14	99.79	101.13	100.84	102.70	100.30	100.62
	μ_R^N	148.33	136.22	142.33	141.92	139.35	139.94	139.83	140.46	139.84	139.96
	σ_R^N	22.37	13.81	16.13	15.94	14.89	15.11	15.07	15.32	15.07	15.12
	μ_S^N	62.50	62.50	62.50	62.50	62.50	62.50	62.50	62.50	62.50	62.50
	σ_S^N	15.00	15.00	15.00	15.00	15.00	15.00	15.00	15.00	15.00	15.00
Step 5		1.00	1.00	1.00	1.00	1.00	1.00	1.00	1.00	1.00	1.00
		-1.00	-1.00	-1.00	-1.00	-1.00	-1.00	-1.00	-1.00	-1.00	-1.00
Step 6	α_R	-0.83	-0.68	-0.73	-0.73	-0.70	-0.71	-0.71	-0.71	-0.71	-0.71
	α_S	0.56	0.74	0.68	0.69	0.71	0.70	0.71	0.70	0.71	0.70
Step 7	After α converges, compute coordinates for the new checking point										
Step 8	β	3.63				3.60			3.64		
Step 9	Repeat steps 3 to 8 β until converges										

Coordinates of the failure point; $r^* = 101.30$, $s^* = 100.56$. RV, random variable.

Example 3 – Nonlinear LSE with independent normal RVs

This example is very similar to the previous two examples with linear LSE. The linear LSE is modified to nonlinear LSE by replacing resistance R by the product of the yield stress F_y and the plastic section modulus Z. To avoid any confusion, LSE is now defined as $G = R - S$. LSE consists of three independent normal RVs. It will also require an iterative procedure to extract the risk information. All the steps are summarized in Table 1.3.

The reliability indexes and the corresponding risks are different for the three examples. The aforementioned exercise clearly indicates the importance of properly defining LSE and the statistical characteristics of all RVs in the formulation. It is important to note that in all the aforementioned examples, the RVs are assumed to be independent.

In the iterative process of implementing FORM, it is desirable to have an explicit expression for an LSE. The information can be difficult to obtain for complicated nonlinear LSE. Also, in many real applications, it may not even be practical to develop functional relationships of them. This will make it difficult to implement FORM. An alternative procedure, using Newton–Raphson type recursive algorithm, denoted as FORM Method 2, was presented by Haldar and Mahadevan (2000a). It is not possible to present FORM Method 2 here. If implemented properly, the original FORM and FORM Method 2 will give identical risk information. In fact, the author and his team used Method 2 in developing all computer programs representing structures by finite elements (Haldar and Mahadevan, 2000b).

6.3 Reliability analysis with correlated variables

The algorithm presented previously is based on the assumption that all the RVs are uncorrelated or statistically independent. If they are not, they need to be made uncorrelated and the LSEs need to be redefined in terms of the uncorrelated variables.

Suppose $X_1, X_2, ..., X_n$ are correlated RVs, their covariance matrix can be expressed as

$$[C] = \begin{bmatrix} \sigma_{X_1}^2 & Cov(X_1, X_2) & ... & Cov(X_1, X_n) \\ Cov(X_2, X_1) & \sigma_{X_2}^2 & ... & Cov(X_2, X_n) \\ \vdots & \vdots & \vdots & \vdots \\ Cov(X_n, X_1) & Cov(X_n, X_2) & ... & \sigma_{X_n}^2 \end{bmatrix} \tag{1.52}$$

TABLE 1.3 Risk estimation with nonlinear LSE with three independent normal RVs.

Step 1	$G = R-S = F_y Z - S$; $F_y \sim N(52, 5.2)$, $Z \sim N(2.88, 0.14)$, $S \sim N(62.5, 15)$						
Step 2	β	3.000			3.994		3.993	3.993
Step 4	$f^*_{F_y}$	52.000	41.575	41.650	38.158	38.181	38.154	38.153
	f^*_z	2.880	2.744	2.766	2.729	2.740	2.740	2.740
	f^*_s	62.500	92.619	93.890	104.135	104.635	104.549	104.550
	$\mu^N_{F_y}$	52.000	52.000	52.000	52.000	52.000	52.000	52.000
	$\sigma^N_{F_y}$	5.200	5.200	5.200	5.200	5.200	5.200	5.200
	μ^N_z	2.880	2.880	2.880	2.880	2.880	2.880	2.880
	σ^N_z	0.140	0.140	0.140	0.140	0.140	0.140	0.140
	μ^N_s	62.500	62.500	62.500	62.500	62.500	62.500	62.500
	σ^N_s	15.000	15.000	15.000	15.000	15.000	15.000	15.000
Step 5	$\left(\dfrac{\partial G}{\partial F_y}\right)^*$	52.000	41.575	41.650	38.158	38.181	38.153	38.152
	$\left(\dfrac{\partial G}{\partial Z}\right)^*$	2.880	2.744	2.766	2.729	2.740	2.740	2.740
	$\left(\dfrac{\partial G}{\partial S}\right)^*$	−1.000	−1.000	−1.000	−1.000	−1.000	−1.000	−1.000

Continued

TABLE 1.3 Risk estimation with nonlinear LSE with three independent normal RVs.—cont'd

Step 6	α_{f_y}	0.668	0.663	0.666	0.665	0.667	0.667	0.667
	α_z	0.325	0.271	0.270	0.250	0.250	0.250	0.250
	α_s	-0.669	-0.698	-0.695	-0.703	-0.702	-0.702	-0.702
Step 7	After α converges, compute coordinates for the new checking point							
Step 8	β			3.994		3.993		3.993
Step 9	Repeat steps 3 to 8 β until converges							

Coordinates of the failure point; $F_y^* = 38.152$, $Z^* = 2.740$, $S^* = 104.549$. RV, random variable.

In the reduced coordinates, the aforementioned covariance matrix can be written as

$$[C'] = \begin{bmatrix} 1 & \rho_{X_1,X_2} & \cdots & \rho_{X_1,X_n} \\ \rho_{X_2,X_1} & 1 & \cdots & \rho_{X_2,X_n} \\ \vdots & \vdots & \vdots & \vdots \\ \rho_{X_n,X_1} & \rho_{X_n,X_2} & \cdots & 1 \end{bmatrix} \qquad (1.53)$$

The correlated RVs can be transformed into uncorrelated reduced $\mathbf{Y}$ as

$$\{\mathbf{X}\} = [\sigma_x^N][\mathbf{T}]\{\mathbf{Y}\} + \{\mu_x^N\} \qquad (1.54)$$

where $[\mathbf{T}]$ is an orthogonal transformation matrix consisting of the eigenvectors of the correlation matrix $[C']$ and can be shown to be

$$[\mathbf{T}] = \begin{bmatrix} \theta_1^{(1)} & \theta_1^{(2)} & \cdots & \theta_1^{(n)} \\ \theta_2^{(1)} & \theta_2^{(2)} & \cdots & \theta_2^{(n)} \\ \vdots & \vdots & \vdots & \vdots \\ \theta_n^{(1)} & \theta_n^{(2)} & \cdots & \theta_n^{(n)} \end{bmatrix} \qquad (1.55)$$

where $\{\theta^i\}$ is the eigenvector of the ith mode. Using Eq. (1.54), one can write Eq. (1.43) in terms of reduced uncorrelated normal $\mathbf{Y}$ variables.

Example 4 — Linear LSE with correlated normal RVs

Consider Example1 again. However, R and S are considered to be correlated normal variables with correlation coefficient $\rho_{R,S} = 0.7$. For this example, Eq. (1.53) becomes

$$[c'] = \begin{bmatrix} 1 & 0.7 \\ 0.7 & 1 \end{bmatrix}$$

The two eigenvalues are $\lambda_1 = 0.3$ and $\lambda_2 = 1.7$. The transformation matrix $\mathbf{T}$ can be shown to be

$$[\mathbf{T}] = \begin{bmatrix} 0.707 & 0.707 \\ -0.707 & 0.707 \end{bmatrix}$$

Using Eq. (1.54), one can write

$$\begin{Bmatrix} R \\ S \end{Bmatrix} = \begin{bmatrix} 22.5 & 0 \\ 0 & 15 \end{bmatrix} \begin{bmatrix} 0.707 & 0.707 \\ -0.707 & 0.707 \end{bmatrix} \begin{Bmatrix} Y_1 \\ Y_2 \end{Bmatrix} + \begin{Bmatrix} 150 \\ 62.5 \end{Bmatrix}$$

Or,

$$R = 15.9075(Y_1 + Y_2) + 150$$

$$S = 10.605(Y_2 - Y_1) + 62.5$$

where $Y_1 \sim N(0, \sqrt{0.3})$ and $Y_2 \sim N(0, \sqrt{1.7})$.

The modified LSE in terms of uncorrelated normal variables can be shown to be

$$Z = g() = 26.625\, Y_1 + 5.3025\, Y_2 + 87.5$$

The mean value of Z, $\mu_Z = 87.5$. The corresponding standard deviation can be estimated as

$$\sigma_Z = \sqrt{26.625^2 \times 0.3 + 5.3025^2 \times 1.7} = 16.1389$$

The reliability index $\beta = \frac{87.5}{16.1389} = 5.42$

The reliability index has increased significantly in this case over when R and S were considered statistically independent.

Reliability evaluation using LSEs with correlated nonnormal variables can be complicated. To implement the FORM procedure, the equivalent normal mean and standard deviation need to be evaluated at the checking point at each iteration and the corresponding LSE needs to be redefined. It can be very time-consuming; a computer program needs to be used for the reliability evaluation.

6.4 Reliability analysis for implicit limit state analysis

When LSEs are explicit in nature, it is relatively straightforward to estimate the risk or reliability using FORM because the derivatives of them with respect to all design variables will be readily available. However, if they are implicit in nature, the risk evaluation process gets very complicated as discussed in the following paragraph.

For the reliability analysis of realistic structures, LSEs are expected to be implicit. To study their realistic linear or nonlinear, static or dynamic behavior, they are generally represented by finite elements. With this representation, it is relatively simple to consider complicated geometric arrangements of structural elements, realistic support and connection conditions, various sources of nonlinearity, applying the loads as realistically as possible, and following the load path of interest to failure. One of the major implications of this representation is that the LSEs will be implicit. To clarify this statement, the overall lateral drift at the top of a frame is of interest. It can be estimated using the finite element approach, but it cannot be explicitly expressed in terms of cross-sectional properties of all the elements and the applied loadings. For nonlinear and dynamic problems, this may not have any physical meaning. However, because the finite element−based formulation fails to incorporate the sources of uncertainty, it cannot be used for the reliability analysis. There are several ways information on uncertainty can be incorporated in the formulation. One approach, developed by the research team of the author, is commonly referred to in the literature as the stochastic finite element method (SFEM) (Haldar and Mahadevan, 2000b). The other two approaches, based on their essential philosophy, are Monte Carlo simulation (MCS) and the response surface−based methods. The basic concept behind MCS will be briefly discussed in this chapter.

The basic concept behind the response surface method (RSM) is to approximately generate a required LSE in the failure region in terms of the basic RVs in the formulation. It approximately generates an explicit expression for an implicit LSE. However, the basic RSM-based concept fails to incorporate the distribution information of RVs even when it is available. Furthermore, the success of an RSM-based procedure depends on the accuracy in identifying the failure region where the response surface should be generated. Essentially, an approximate expression of an LSE is expressed by a polynomial. The form of the polynomial, linear, quadratic, or any other types, and the total and optimal number of data needed to fit a surface or equation through them will affect the information on risk extracted. These are few complicated but challenging issues that need to be addressed before RSM-based procedure can be used for the reliability evaluation. It is a kind of meta-modeling concept and will be discussed in more detail in Chapter 13.

7. Performance-based seismic design

Before completing the discussions on reliability evaluation methods, the following additional discussions are necessary. The basic reliability-based concept generally used in codes or design guidelines is commonly denoted as the LRFD replacing the older allowable stress design (ASD) concept. Both of them were developed to protect life. In mid-1990s, several earthquakes caused significant amount of damage to infrastructures and properties. The observation prompted the profession to propose the performance-based seismic design (PBSD) to avoid adverse economic consequences replacing the life safety concept. It is also a very sophisticated risk-based design concept. It gives the option to the engineer/owner to accept certain consequences under predefined risk they are willing to take. The ASCE 41-13 standard (2014) defines four performance levels ranging from operational to collapse prevention. They are defined as the operational (OP) level; it represents the damage is minor or no damage occurred in structural and nonstructural components, the immediate occupancy (IO); it indicates no damage in structural elements and only minor damage to its nonstructural components, the life safety (LS); it represents extensive damage to structures and the risk to life is low, but repairs may be required before reoccupancy, and the collapse prevention (CP); it represents the structural damage is extensive; total collapse is avoided, but structural repairs are uneconomical. For further discussions on the topic, please refer to the study by Azizsoltani and Haldar (2018, 2017), Azizsoltani et al. (2018), and Gaxiola-Camacho et al. (2018, 2017a, 2017b).

8. Monte Carlo simulation

Simulation is a simple technique to extract reliability information without using extensive mathematical or statistical expertise. It is also routinely used to verify a new reliability estimation procedure. The concept is very simple. Each RV in a formulation is sampled several times satisfying the underlying statistical characteristics. Each realization of all the RVs produces a set of realization for the problem. It is generally termed as a simulation run, cycle, or trial, N. When the problem is defined properly, by conducting a large number of simulation cycles, the underlying risk can be extracted, particularly when N tends to infinity. Simulation is an inexpensive analytical technique compared with laboratory testing to study the implications of the presence of uncertainty in the problem. With the advancement of computational power, simulation is becoming an integral part of the reliability analysis.

The basic technique commonly used by the engineering profession is known as MCS. It can be implemented with the help of six essential steps.

Step 1 − Define the problem to be simulated. If the information on risk is of interest, the corresponding LSE needs to be considered. The LSE can be explicit or implicit in nature. However, for implicit LSE, algorithmic representation of the problem, for example, the finite element formulation, is necessary.

Step 2 − Information on the probabilistic characteristics of all RVs in the formulations needs to be available.

Step 3 − Generate N random numbers for each RV according to its probabilistic characteristics defined in Step 2, producing N sets of random numbers. Almost all computers currently available in the market can generate such random numbers. N should not be an extremely large number. Otherwise, based on the computer being used, the number can repeat itself defeating the purpose of simulation. In most cases, N should be limited to few millions, as will be discussed further later.

Step 4 − Deterministically analyze problem N times, most likely using a computer. Depending on the capabilities of the computer program, any form or type of computational algorithm can be used. It can be static or dynamic, linear or nonlinear. The only limitation could be the required time of a deterministic run. Suppose N is one million, and each run takes about 1 hour, it will require about one million hours or about 114 years of continuous running of a computer to extract the reliability information. Obviously, the basic MCS will be impractical for this class of problems, as will be discussed later.

Step 5 − Extract reliability information from N numbers generated using simulation.

Step 6 − Evaluate the accuracy and the efficiency of the information generated for practical implementation.

The MCS procedure, with the help of six steps, is discussed in more detail and elaborated further with the help of informative examples by Haldar and Mahadevan (2000a). It is important to point out that both explicit and implicit LSEs addressing any form of behavior can be simulated. If the estimation of risk is the primary objective, then an appropriate LSE is necessary. If an LSE is formulated properly, it will produce a negative value for a specific set of random numbers indicating failure. If out of a total of N simulations, N_f times the LSE produce negative values, then the probability of failure p_f can be estimated as N_f/N. However, one major drawback of MCS is to figure out the appropriate value of N to estimate the unknown risk. If N is too small, it may not be adequate to predict the underlying risk. If it is too large, it becomes an undesirable option considering the inefficiency of the simulation process. Haldar and Mahadevan (2000a) made the following observation and suggestion. For most engineering problems, the probability of failure p_f is expected to be smaller than 10^{-5}, indicating that on an average, only 1 out of 100,000 trials will show a failure. It indicates that at least 100,000 trials are necessary to predict this behavior. For reliable estimate of p_f, they suggested at least 10 times this minimum, or about one million trials. Obviously, the total number of simulations N could be different for different values of p_f; however, this information will be unknown at the initiation of simulation.

One of the major challenges to the profession is to use as few simulations as practicable to extract reliability information without compromising the accuracy. To increase the efficiency, several variance-reduction techniques (VRTs) were proposed. The success of a VRT depends on the problem under consideration, and it will be unknown before initiating the simulation process. The basic concept behind developing a VRT relies on by altering the input scheme, by altering the model, or by special analysis of the output. The essential building blocks for these schemes are sampling schemes, correlation methods, and special methods. Some of the major sampling schemes include systematic sampling, importance sampling, stratified sampling, Latin hypercube sampling, adaptive sampling, randomization sampling, conditional expectation sampling, etc. Some of the commonly used correlation methods are common random numbers, antithetic variates, and control variates. Concepts used in special methods include the partition of the region, random quadratic method, biased estimator, indirect estimator, etc. Haldar and Mahadevan (2000a) discussed these concepts in more detail.

The basic concept behind simulation is relatively simple, but simplicity also causes unfortunate misuses. The use of sophisticated VRTs requires a lot of expertise, not expected from a practicing engineer. In addition, the attractive feature of simulation, its basic simplicity, is lost. The author and his team recently developed an alternative to simulation by combining multiple mathematical schemes, producing compounding beneficial effects. This advanced concept is not appropriate for this introductory chapter. However, it is briefly presented in Chapter 13.

9. Alternative to simulations

As mentioned earlier, it may not be economically possible to design a risk-free structure. For practicing engineers without extensive expertise in the risk-based design, a more manageable alternative to estimate the underlying risk will be the basic MCS method discussed in Section 8. For dynamic problems, especially when it needs to be applied in time domain, for example, for the earthquake loading, one nonlinear FEM-based deterministic analysis may require about 1 h of computer time. For low probability events such as strong earthquakes, one may need to carry out at least 10,000 simulation cycles requiring about 10,000 h or 1.14 years of continuous running of a computer. To address the problem, parallel processing or other techniques can be followed, but they may require other forms of expertise. To address the situation, the author and his research team proposed to extract the reliability information using few dozens of runs at very intelligently selected points as an alternative to MCS. They developed the concept for structural engineering applications. However, they documented its application to multidisciplinary areas including the offshore mechanics and arctic engineering (Farag et al., 2016; Farag and Haldar, 2016), geotechnical engineering (Villegas-Mercado et al., 2017), solders in electronic packaging subjected to thermomechanical loading (Azizsoltani and Haldar, 2018), etc. The author believes that the team proposed an alternative to simulation. The concept is very advanced in nature. However, it will be briefly discussed in Chapter 13 of this handbook.

10. Computer programs

Readers will be interested in computer programs readily available for the risk and reliability evaluation. The author has been working on the related areas for over 4 decades. He, with the help with his students, developed numerous computer programs. In developing the stochastic finite element concept, they used stress-based finite element method not commonly used in the profession. They published numerous technical articles on the topic. However, these computer programs are not available for routine use outside the author's research team members.

The author would like to make few suggestions. A casual internet search will identify numerous computer programs currently available for the reliability evaluation. Because the author does not use them, it is not possible for him to comment on their accuracy and efficiency. There are also few open-access computer programs. They can also be used. Based on his personal experience, the author observed that it is not difficult to write a simple FORM-based reliability evaluation program using available computational tool such as MATLAB. Obviously, these programs need to be verified using example problems given in many text books including in Haldar and Mahadevan (2000a, b). They also need to be verified using simulation.

At this stage of his career, it may not be desirable for the author to try to distribute all the programs developed with the help of his students. In addition, these programs may not work with different computer platforms. To address this shortcoming, the author proposed an alternative to simulation and it can be used for many different purposes or applications. An interested person can use any computer program to deterministically study a problem. Of course, the information on uncertainties should be available. Then, by solving the problem deterministically, may be few dozens of times, the reliability information can be extracted, even for low probability events. The concept is briefly discussed in Chapter 13.

11. Education

Risk-based design concept is relatively new. However, most of the design guidelines have been modified to reflect the implication of the uncertainty related issues, at least in the US. However, the education-related activities, i.e., teaching the concept to the students, can be at best described as nonuniform. The Accreditation Board of Engineering and Technology (ABET) in the US now requires that all civil engineering undergraduate students take a course on the subject, indicating its importance in the civil engineering education. However, most countries around the world do not have such requirement. At the graduate level, we routinely offer several courses related to the risk-based design. For seismic designs, the use of multiple earthquake time histories is now suggested in design codes to address uncertainty-related issues. The author recently completed a study on the topic supported by the US National Science Foundation. The author strongly believes that integrating the risk-based design concept in the undergraduate and graduate levels will be very beneficial and will permanently change the currently used engineering design paradigm, making the final products more damage or defect tolerant.

Acknowledgments

The author would like to thank all the team members for their help in developing the overall research concept over a long period of time. They include Prof. Bilal M. Ayyub, Prof. Sankaran Mahadevan, Dr. Hari B. Kanegaonkar, Dr. Duan Wang, Dr. Yiguang Zhou, Dr. Liwei Gao, Dr. Zhengwei Zhao, Prof. Alfredo Reyes Salazar, Dr. Peter H. Vo, Dr. Xiaolin Ling, Prof. Hasan N. Katkhuda, Dr. Rene Martinez-Flores, Dr. Ajoy K. Das, Prof. Abdullah Al-Hussein, Prof. Ali Mehrabian, Dr. Seung Yeol Lee, Prof. J. Ramon Gaxiola-Camacho, Dr. Hamoon Azizsoltani, Ph.D. candidates Francisco Javier Villegas-Mercado and Sayyed Mohsen Vazirizade. It is not possible to list numerous other students who helped the author in various stages of the development process presented here. The team received financial supports from numerous sources including the National Science Foundation, the American Institute of Steel Construction, the US Army Corps of Engineers, the Illinois Institute of Technology, the Georgia Institute of Technology, the University of Arizona, and many other industrial sources that provided matching funds a Presidential award the author received.

Most recently, the author's study is partially supported by the National Science Foundation under Grant No. CMMI-1403844. Any opinions, findings, or recommendations expressed in this chapter are those of the writer and do not necessarily reflect the views of the sponsors.

References

American Institute of Steel Construction (AISC), 2011. Steel Construction Manual, fourteenth ed.

Ang, A.H.S., Tang, W.H., 1975. Probability Concepts in Engineering Planning and Design, vol. 1. John Wiley and Sons, New York, N.Y.

ASCE/SEI 41-13, 2014. Seismic Evaluation and Retrofit of Existing Buildings. American Society of Civil Engineers., Reston, Virginia 20191.

Ayyub, B.M., Haldar, A., 1984. Practical structural reliability techniques. Journal of Structural Engineering, ASCE 110 (8), 1707−1724.

Azizsoltani, H., Gaxiola-Camacho, J.R., Haldar, A., September 2018. Site-specific seismic design of damage tolerant structural systems using a novel concept. Bulletin of Earthquake Engineering 16 (9), 3819−3843. https://doi.org/10.1007/s10518-018-0329-5.

Azizsoltani, H., Haldar, A., 2018. Reliability analysis of lead-free solders in electronic packaging using a novel surrogate model and kriging concept. Journal of Electronic Packaging, ASME 140 (4), 041003-1−11. https://doi.org/10.1115/1.4040924.

Azizsoltani, H., Haldar, A., 2017. Intelligent computational schemes for designing more seismic damage-tolerant structures. Journal of Earthquake Engineering 1−28. https://doi.org/10.1080/13632469.2017.1401566. published online on 11/17/2017.

Farag, R., Haldar, A., 2016. A novel concept for reliability evaluation using multiple deterministic analyses. INAE Letters 1 (Issue), 85−97. https://doi.org/10.1007/s41403-016-0014-4.

Farag, R., Haldar, A., El-Meligy, M., May 11 2016. Reliability analysis of piles in multi-layer soil in mooring dolphin structures, offshore mechanics and arctic engineering. ASME. https://doi.org/10.1115/1.4033578.

Freudenthal, A.M., 1956. Safety and the probability of structural failure. American Society of Civil Engineers Transsactions 121, 1337−1397.

Gaxiola-Camacho, J.R., Haldar, A., Reyes-Salazar, A., Valenzuela-Beltran, F., Vazquez-Becerra, G.E., Vazquez-Hernandez, A.O., 2018. Alternative reliability-based methodology for evaluation of structures excited by earthquakes. Earthquakes and Structures 14 (4), 361−377.

Gaxiola-Camacho, J.R., Azizsoltani, H., Villegas-Mercado, F.J., Haldar, A., 2017a. A novel reliability technique for implementation of performance-based seismic design of structures. Engineering Structures 142, 137−147. http://doi.org/10.1016/j.engstruct.2017.03.076.

Gaxiola-Camacho, J.R., Haldar, A., Azizsoltani, H., Valenzuela-Beltran, F., Reyes-Salazar, A., 2017b. Performance-based seismic design of steel buildings using rigidities of connections. ASCE-ASME Journal of Risk and Uncertainty in Engineering Systems, Part A: Civil Engineering 4 (1), 04017036.

Haldar, A., Mahadevan, S., 2000a. Probability, Reliability, and Statistical Methods in Engineering Design. John Wiley & Sons, New York, NY.

Haldar, A., Mahadevan, S., 2000b. Reliability Assessment Using Stochastic Finite Element Analysis. John Wiley & Sons, New York, NY.

Hasofer, A.M., Lind, N.C., 1974. Exact and invariant second-moment code format. Journal of the Engineering Mechanics Division 100 (1), 111−121.

Laplace, P.S.M., 1951. A Philosophical Essay on Probabilities (Translated from the Sixth French Edition by F.W. Truscott and F.L. Emory). Dover Publications, N.Y.

Montgomery, D.C., Peck, E.A., Vining, G.G., 2012. Introduction to Linear Regression Analysis, fifth ed. Wiley, New York, N.Y.

Rackwitz, R., Flessler, B., 1978. Structural reliability under combined random load sequences. Computers and Structures 9 (5), 489—494.

Rackwitz, R., Flessler, B., 1976. Note on Discrete Safety Checking when Using Non-normal Stochastic Model for Basic Variables, Load Project Working Session. MIT, Cambridge, MA.

Shinozuka, M., 1983. Basic analysis of structural safety. Journal of Structural Engineering 109 (3), 721—740.

Villegas-Mercado, F.J., Azizsoltani, H., Gaxiola-Camacho, J.R., Haldar, A., 2017. Seismic reliability evaluation of structural systems for different soil conditions. International Journal of Geotechnical Earthquake Engineering 8 (2), 23—38. https://doi.org/10.4018/IJGEE.2017070102.

Chapter 2

Modeling wheat yield with data-intelligent algorithms: artificial neural network versus genetic programming and minimax probability machine regression

Mumtaz Ali[1,2], Ravinesh C. Deo[1]

[1]*School of Agricultural Computational and Environmental Sciences, University of Southern Queensland, Springfield, QLD, Australia;* [2]*Deakin-SWU Joint Research Centre on Big Data, School of Information Technology, Deakin University, Burwood, VIC, Australia*

1. Introduction

Culminating knowledge on the dynamics of cropping, agricultural management, and modeling of the amount and potential region of cropping is considered as useful tenets used by agricultural scientists in formulating accurate and timely information on the future yield that could benefit agriculture-reliant nations (Gonzalez-Sanchez et al., 2014; Raorane and Kulkarni, 2012). The present study is focused in Pakistan, a nation that is expecting a significant spurt in population over the coming decades. Droughts and floods exacerbated by climate change are the likely threats that could moderate the availability of staple food. Pakistan is considered to be one of the top 10 nations facing water and food security issues (Briscoe, 2005). Agriculture, which includes crops such as wheat, sugarcane, cotton, and rice (supplying 30% of the world's paddy output), contributes to 21% of the gross domestic product (GDP) (the total value of goods produced and services provided in a country during 1 year) (Sarwar, 2014). Out of these, wheat is a very important crop. According to the Food and Agriculture Organization, Pakistan was placed in the eighth position as a global wheat producer from 2007 to 2009 (FAO, 2013). Hence, the modeling of wheat production with intelligent learning systems that incorporate historical

Handbook of Probabilistic Models. https://doi.org/10.1016/B978-0-12-816514-0.00002-3
37

knowledge to predict the future yield can provide new strategic frameworks for forewarning systems in respect to future food security issues.

In Pakistan, wheat is cultivated in Rabi (i.e., winter) season in the agricultural lands of the province of Punjab, the largest wheat providing jurisdiction (Khan et al., 2012). Wheat and rice crops are the largest food commodities generated in Punjab, while the other important crops include rice, sugarcane, millet, corn, oilseeds, pulses, vegetables, and fruits. Owing to rich agricultural area, Punjab shares 76% of its annual food production with the nation (Punjab, 2015). Wheat accounts for 2.6% of Pakistan's GDP, and it also contributes 12.5% to the GDP in the agriculture sector (Survey, 2012). Considering the vitality of wheat, accurately predicting annual yield can assist government and agricultural-climate policy experts in making decisions on national imports and exports as well as maintaining sufficient reservations of wheat for national food security and setting the prices for agriculture markets. In 2005, the actual yield was very low compared with the predicted yield, and as such, poor estimation moderated the market price and prompted the government to export grains to the international market (Dorosh and Salam, 2008; Niaz, 2014). In 1980, the government adopted a system for collecting agricultural statistics to improve food security (Akhtar, 2014) which are limited in their ability to make inferences about future productions and implications.

In the past, Pakistan has faced significant crises of wheat supply particularly in the period of 2012—13 which occurred because of the failure of province of Punjab, the present study region, to meet its target production value. The harvested wheat was 17.5 million tonnes, undermining an initial target of 19.2 million tonnes, which was due to poor planning and inaccurate estimations (Bokhari, 2013). A report published in the Express Tribune (Sajjad, 2017) indicates that Pakistan is likely to face wheat shortages in the future. Owing to such uncertainties, governments and policymakers require improved models to allow them to estimate the potential reductions and associated risks of wheat yield. To surmount this issue, agricultural policymakers require sufficient knowledge about future yield and its consequence on prices, imports, and exports. This justifies the pivotal role of data-intelligent models for the prediction of yield not only in Pakistan but also in other agricultural nations.

One viable approach for prediction of crop is to use dynamic (physical) models that must be fed with initial conditions from site-specific experiment, calibrations, and pilot-tested yield although such information may only be available for some crops and the process can be costly too (Drummond et al., 2003). Overcoming such limitations, data-intelligent algorithms that represent high-level information are built empirically without a deep knowledge of the physical mechanisms and are used to generate crop yield. Indeed, such models have great adaptability for crop planning compared with physically based crop models because of their friendly implementation, competitive performance, and the evolution data analytic techniques (Irmak et al., 2006). Data-intelligent algorithms are an attractive tool for policymakers to enable them use a systematic way for estimating the future yield (Bauer, 1975). However, such models in agriculture, in general, are limited, and most studies, if any, have been restricted to statistical models with the assumptions of linearity between the predictor and target variable. Dempewolf et al. (2014) predicted wheat yield in Punjab using the vegetation index and crop statistics. Hamid et al. (1987) investigated the wheat economy and its likely future prospects, and Muhammad (1989) studied historical background of wheat improvement in Balochistan region. Specifically, Iqbal et al. (2005) applied the statistical model autoregressive moving average (ARIMA) to project wheat areas and productions until 2022 in Pakistan. A study performed by Saeed et al. (2000) predicted wheat using an ARIMA, Qureshi et al. (1992) presented an analysis of the relative contribution of an area yield to the total production of wheat and maize, and Amin et al. (2014) applied time series to predict wheat data. Memon et al. (2015) investigated energy consumption for wheat production in Sher and Ahmad (2008) developed a study on the prediction of wheat through a Cobb–Douglas function and an ARIMA model for each input and province. Muhammed et al. (1992) used ARIMA to predict rice yield, Sabur and Haque (1993) predicted rice prices in the market town of Mymensingh, and Sohail et al. (1994) predicted food grains (including wheat) for Pakistan with an ARIMA model. However, these studies have applied regressions (e.g., ARIMA) that assume linear relationship between wheat yield and its predictor variable.

Other than statistical models constructed through an ARIMA approach, there have been some studies based on rainfall, temperature, fertilizers, and related variables. Azhar et al. (1973, 1974) developed a model for the prediction of wheat production with rainfall for November to January in Punjab, and Qureshi (1963) used the idea of three seasonal periods to capture rainfall from July–September, October–December, and January–March. Rainfall in January–March period was found to have a maximum effect on the production. Chaudhry and Kemal (1974) conducted a study using alternative production functions, finding that the rainfall deviation for the July–January era was most suitable for wheat prediction in the irrigated region of Pakistan. Sabir and Tahir (2012) developed a model based on exponential smoothing for wheat yield, production area, and population growth for the year 2011–12. Ali et al. (2017) applied feasible generalized least square (FGLS) and heteroscedasticity and autocorrelation (HAC) methods to study the impact of climate change on the crop yield in Pakistan. In spite of being used quite profusely, these types of models embraced general ideas, without incorporating expert information to capitalize on the attributes in historical data, and were largely restricted to linear methods without optimally extracted features. Notwithstanding this, the advent of data-intelligent models at an astonishing rate in the current era can be useful for decision-makers to develop automatic expert systems containing optimal rules based on historical knowledge in farming strategies and crop management (McQueen et al., 1995, Dimitriadis and Goumopoulos, 2008, and Coopersmith et al., 2014).

Although the literature on application of data-intelligent algorithms for wheat (and other crop) prediction is relatively sparse, some studies show that such a contemporary approach can be an effective way to model future yield. The study of Balakrishnan and Muthukumarasamy, (2016) designed a crop-ensemble model for India, while Kumar et al. (2015b) used crop selection to maximize crop yield rate. Rahman et al. (2014) facilitated a machine learning model for rice prediction in Bangladesh, while a neural network—integrated model was used by Chen and McNairn (2006) for rice crop monitoring. Co and Boosarawongse (2007) forecasted rice export in Thailand using statistical techniques and artificial neural network (ANN), and Kaul et al. (2005) used ANN for corn and soybean prediction. A few other studies (Khamis et al., 2006; Paswan and Begum, 2013; Stathakis et al., 2006) used machine learning for crop prediction, although they were based primarily on rainfall and temperature. Crop prediction could be a difficult task because many variables are interrelated, so the yield can be affected by human decisions or activities (e.g., irrigated water, land, and crop rotation) and incontrollable, natural factors (e.g., weather) (Gonzalez-Sanchez et al., 2014). Therefore, crop planners could potentially use previous yield as an estimation of the future yield. Despite this, no study has used wheat yield of one region to predict the yield in a neighboring site that shares a common boundary, although this sort of strategic modeling can assist in decision-making and management zones about the efficiency of production and developing precise agricultural practices.

Techniques utilizing yield data from the nearest neighboring study sites to predict the objective station data is practically useful because it can enable the modelers to extract similar features and patterns prevalent at predictor site to be analyzed to estimate the yield at an objective site. This approach can enable agricultural experts to develop farming protocols by comparing site-specific yield and make appropriate deductions in respect to the presence of favorable (or unfavorable) environmental or soil fertility conditions, and also implement better management decisions necessary to generate optimal yield. In China, for example, multiple cropping systems and the location of weather stations buffer zones across neighboring provinces required model simulations based on rice cropping systems in each respective province (Van Wart et al., 2013). Ittersum et al. (2013) emphasized the need for accurate agronomic and yield data including calibrated and validated crop models and upscaling methods where bottom-up application of such a protocol could allow the verification of estimated yield gaps with on-farm data and experiments. Considering such needs, the modeling of crop yield in terms of their nearest neighbor yield data can provide a comparative framework for different farmers in identifying more cost-effective and productive agricultural management practices, such as ways to optimize fertilizer usage and reduce operational costs, improve irrigation systems remodeled from those in neighboring suites, and to use such information to make better farm management decisions.

In this chapter, we apply an ANN model for the prediction of wheat at a set of target stations trained with the nearest neighborhood station data in the agricultural district of Punjab. Using a total of eight station-based yield data (i.e., six sets for model development and two sets for testing) for the period 1981−2013, this study uses yield from neighboring stations where ANN is trained with a number of learning algorithms and hidden and output functions in accordance with previous studies (Deo and Sahin, 2015a; Deo et al., 2017). The performances are evaluated with respect to alternative methods: the minimax probability machine regression (MPMR) and genetic programming (GP) algorithms, judged in accordance with statistical metrics (Deo and Sahin, 2016; Legates and McCabe, 1999; Nash and Sutcliffe, 1970; Willmott, 1981, 1984). The rest of this chapter is structured as follows. In the next section, the theory pertaining to data-intelligent models are described, while in Section 3, the experimental data, study area, model strategy, and performance assessment metrics are stated. In Section 4, the results are presented, followed by Section 5 where discussions are made and Section 6 where conclusions and remarks for closing this chapter are made.

2. Theoretical framework

In this section, an overview of the ANN-based predictive model with its counterparts, the MPMR and the GP applied in a problem of wheat yield prediction, are presented. ANN is a straightforward model to implement, works well in most circumstances, and is relatively easy to develop. The MPMR model maximizes the minimum probability that a future-predicted output of the regression model will be within some bound of the true regression function. A major disadvantage of the MPMR model is the loss of control over the future prediction (Strohmann and Grudic, 2003), whereas the GP model is user-friendly to implement, but coding such an algorithm from the scratch can be a painstaking task because its development depends on the nature of the input data. All three data-intelligent models applied in this study were evaluated as the potential possibilities for modeling the wheat yield data because their application in the area of crop prediction have not been fully explored. In this section, the basis theory of each data-intelligent model is provided.

2.1 Artificial neural network

In this chapter, an ANN model is developed with wheat yield inputs from the neighboring stations to predict the yield at a set of two reference (or target) stations. This approach has the advantage of estimating the yield in a neighboring remote locality with limited statistical data. The structure of an ANN model is based on nonlinear modeling with a network architecture that resembles the biological structure of the nervous system (McCulloch and Pitts, 1943). The interconnected inputs are related to the output (Y) which have the ability to transfer information by weighted connections (i.e., functional neurons) that map the nonlinear predictive data features to a high-dimensional hyperplane. Fig. 2.1A shows a schematic structure of an ANN model.

Following earlier studies (Alvarez, 2009; Faramarzi et al., 2010; Ghodsi et al., 2012; Guo et al., 2010; Safa and Samarasinghe, 2011; Safa et al., 2015), feed-forward back-propagation (FFBP) generating a multilayer perceptron neural network is used in this problem. To validate the ANN model's parameters (i.e., weights and biases), the FFBP is considered to be better algorithm than the other categories of ANNs (Abbot and Marohasy, 2012; Adamowski et al., 2012; Deo and Sahin, 2015a; Keskin and Terzi, 2006; Mekanik et al., 2013) where FFBP neuronal architecture is constructed to fine-tune the errors of the forward passing of the updated parameters and backward propagation. ANN is formulated as (Deo and Sahin, 2015a; Deo and Sahin, 2016; Kim and Valdés, 2003)

$$Y(X) = F \left(\sum_{i=1}^{L} w_i\ (t) \cdot X_i(t) + b \right) \tag{2.1}$$

In Eq. (2.1), $X_i(t)$ represents the input variable(s) in discrete time space t, $Y(X)$ is the predicted wheat yield in the (test) set, L is the number of hidden neurons determined iteratively, and $w_i(t)$ is the weight which connects the ith neuron in the input layer. Note that b is used as a neuronal bias and $F(.)$ is the function of hidden transformation.

It should be noted that the training algorithm does not identify the best ANN without an iterative model identification process where several models with different algorithms should be trialed. The common algorithms, implemented in MATLAB (R2015b, The Mathworks), can be categorized into five classes. Quasi-Newton (Kim and Valdés, 2003) is based on *trainlm* and *trainbfg* (Dennis and Schnabel, 1996; Marquardt, 1963) functions that minimize the mean-squared error to locate the minimum inputs in the form of the sum of squares of nonlinear real-valued functions (HariKumar et al., 2009). The issue of memory overhead with very large networks (Anusree and Binu, 2011) is due to the computation of gradients and the Hessian matrix (Pham and Sagiroglu, 2001). The function *trainbfg* uses Newton's method

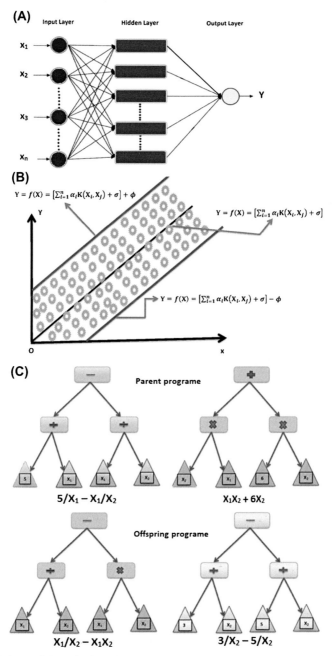

FIGURE 2.1 (A) Schematic view of ANN model architecture adopted for the prediction of wheat yield with the predictors (inputs, X) comprising of yield of neighboring sites together with objective (output) site yield Y. (B) MPMR models with X, bias (σ) and bound of error fluctuation (ϕ). (C) Graphical representation of a GP model. *ANN*, artificial neural network; *MPMR*, minimax probability machine regression.

following a hill-climbing optimization to determine a twice continuously differentiable stationary point for the problem of nonsmooth optimization (Avriel, 2003). In the gradient descent (Ali and Smith, 2006; Hestenes and Stiefel, 1952) category, (*traingda*) uses new weights and biases for training that are in the negative gradient direction. The (*traingdm*) network responds not only to the local gradient but also to recent trends in the error surface due to momentum and ignores small features, acting like a low-pass filter. The (*traingdx*) function is a combination of adaptive learning and momentum which includes the coefficients as training parameters (Anusree and Binu, 2011). The conjugate gradient (Fletcher and Reeves, 1964; Møller, 1993; Powell, 1977) consists of (*trainscg*), (*traincgf*), (*traincgb*), and (*traincgp*) algorithms that allows a change in weights and biases to avoid large computations (Møller, 1993), while (*traincgf*) (Fletcher and Reeves, 1964) and (*traincgp*) (Powell, 1977) have the ability to train with weights and the derivatives of inputs and transfer functions. The one step secant (Battiti, 1992) category uses (*trainoss*) algorithm for network training, and the category of Bayesian regularization (Foresee and Hagan, 1997) utilizes (*trainbr*) to train the network for the minimization of squared errors and weights that occur. For further reading on different training algorithms, readers can refer to the study of Deo and Sahin 2017.

In an ANN model, the first task is to determine a suitable transfer function that is not known in advance. A MATLAB toolbox provides some equations of $F(.)$ which can be trialed (Vogl et al., 1988):

$$\text{Tangent Sigmoid} \Rightarrow F(X) = \frac{2}{1 + \exp(-2X)} - 1 \tag{2.2}$$

$$\text{Log Sigmoid} \Rightarrow F(X) = \frac{1}{1 + \exp(-X)} \tag{2.3}$$

$$\text{Soft Max} \Rightarrow F(X) = \frac{\exp(X)}{\text{sum}(\exp(X))} \tag{2.4}$$

$$\text{Hard -Limit} \Rightarrow F(X) = 1, \quad \text{if } X > 0 \tag{2.5}$$

$$\text{Positive Linear} \Rightarrow F(X) = X \quad \text{if } X \geq 0, \text{ or } 0 \text{ otherwise} \tag{2.6}$$

$$\text{Triangular Basis} \Rightarrow F(X) = 1 - \text{abs}(X), \quad \text{if } -1 \leq X \leq 1, \text{ or } 0 \text{ otherwise} \tag{2.7}$$

$$\text{Radial Basis} \Rightarrow F(X) = \exp(-X^2) \tag{2.8}$$

$$\text{Symmetric Hardlimit} \Rightarrow F(X) = 1, \quad \text{if } X \geq 0 \tag{2.9}$$

$$\text{Saturating Linear} \Rightarrow F(X) = 0, \quad \text{if } X \leq 0$$
$$\Rightarrow F(X) = X, \quad \text{if } 0 < X \leq 1 \tag{2.10}$$
$$\Rightarrow F(X) = 1, \quad \text{if } X \geq 1$$

$$\text{Symmetric Saturating Linear} \Rightarrow F(X) = -1, \quad \text{if } X \leq -1,$$
$$\Rightarrow F(X) = X, \quad \text{if } -1 < X \leq 1, \tag{2.11}$$
$$\Rightarrow F(X) = 1, \quad \text{if } X > 1$$

where X represents the input (predictor) data set analyzed in accordance with the function $F(X)$ that has the ability to map the predictive features by constructing a hidden layer weight.

2.2 Minimax probability machine regression

The MPMR is a nonlinear probabilistic model with the capability of maximizing the least probabilities within the interval of true regression of the objective function (Deo and Samui, 2017; Kumar et al., 2013b; Strohmann and Grudic, 2003). The MPMR algorithm is implemented in the MATLAB programming environment (R2015b, The Mathworks). Fig. 2.1B shows a sketch of the MPMR model.

To attain an optimal level of prediction, MPMR relies on linear discriminant and convex optimization (Strohmann and Grudic, 2003) that makes it an advanced and improved version of support vector machines (Strohmann and Grudic, 2003). The features in the predictor variable are analyzed by shifting all the regression-based inputs between $+\phi$ and $-\phi$ along the axis of the dependent variable. The area in between is termed as a regression surface where the upper and lower bounds of probability identifies the region for misclassifying a point without making distributional assumptions by the model (Bertsimas and Sethuraman, 2000). The learning (D-dimensional inputs) generated from an unknown regression is a function of the form (Strohmann and Grudic, 2003)

$$f : \Re^D \rightarrow \Re$$

$$Y = f(\mathbf{X}) + \sigma \tag{2.12}$$

where $X \in R^D$ is the input vector according to a bounded distribution Ω, $\mathbf{Y} \in \Re$ is an output vector and variance $(\sigma) = \rho^2(\rho) = \sigma^2 \in \Re$. The function $\widehat{f}$ is an approximated function in MPMR where for X_i generated from Ω, $\widehat{Y}$ is given by

$$\widehat{Y} = \widehat{f}(\mathbf{X}) \tag{2.13}$$

The estimated bounds calculated by the model is based on the minimum probability (Ψ) that $\widehat{f}(\mathbf{X})$ is within ϕ of Y (Lanckriet et al., 2002; Strohmann and Grudic, 2003).

$$\Psi = \inf \ \Pr\left\{ \left| \widehat{Y} - Y \right| \leq \phi \right\} \tag{2.14}$$

The predictive ability of a true regression is evaluated from Eq. (2.14) by a bound based on this minimax probability to deduce Ψ directly between ϕ of the following true function:

$$\widehat{Y} = \widehat{f}(\mathbf{X}) = \sum_{i=1}^{N} \alpha_i K(X_i, X_j) + \sigma \tag{2.15}$$

In Eq. (2.15), $K_{i,j} = \theta(X_i, X_j)$ is a kernel function that must satisfy the mercer condition, vector X_i is from the learned data and α_i and σ are the output parameters.

2.3 Genetic programming

GP is a heuristic modeling tool which has the ability to attain accurate solutions without the user specifying a lot of information about the problem that is at hand (Koza, 1992) where the GP algorithm is implemented in MATLAB (R2015b, The Mathworks). In GP, evolutionary principles are used to acquire persistent patterns in the data structure without requiring prior knowledge. An organized domain-independent method is used to breed a population to get computers to solve the problem that starts from a high-level statement of what needs to be done (McPhee et al., 2008). Fig. 2.1C shows a structure of a GP model.

In a GP model, a population is transformed iteratively to produce new generations of programs by using similar operations that occur naturally. The genetic operations are divided into five components: crossover (sexual recombination), mutation; reproduction; gene duplication; and gene deletion. A GP model has the skill of self-parameterizing to extract features bypassing the user, tuning the model, and due to this capability resembles to some extent the Extreme Learning Machine model (Huang et al., 2006).

In a GP model, the input data transit through a number of routes where (1) analyzation of attributes occurs; (2) selection of the best fitness functions is made for the purpose of minimizing the mean-squared error; (3) functional and terminal sets are generated; and (4) parameterization of genetic operations occurs (Sreekanth and Datta, 2011). The GP model is optimized by the emulation of an evolutionary process to an adequate agreement between the response and input variable. A functional node performs the arithmetic operations $(+, -, \times, \div)$, Boolean logic functions (AND, OR, NOT), conditionals (IF, THEN, ELSE), or any other functions (SIN, EXP) that may be used. A random deduction is performed by these functions/terminals to develop a tree structure which consist of a root node and branch (Mehr et al., 2013).

In this chapter, the GP model is developed by (1) randomly creating the initial population (i.e., computer program); (2) performing the execution of the program with the best fitness values; (3) based on reproduction, mutation, crossover, and generation of a new population of computer programs; (4) comparison and evaluation of fitness; and (5) finally the selection of the best program by an evolutionary process (Mehr et al., 2013). For this purpose, a randomly equated population is created, and the fitness is determined where "parents" are chosen individually and the "offspring" are developed from the parents through the processes of reproduction, mutation, and crossover.

3. Materials and method

In this section, the description of the acquired wheat yield data, the present study regions, design of the predictive models, and the performance criteria are presented.

3.1 Wheat yield data set

In this chapter, the model development data has been sourced from the Federal Bureau of Statistics (i.e., Economic Wing) in Pakistan and the Agriculture Marketing Information Services, Directorate of Agriculture (Economics and Marketing) (Districts, 2008; Service, 2012, 2014) which can be verified in Fig. 2.2B. To construct this data set, the areal (district level) productions of wheat were acquired through the provincial Crop Reporting Services which had been compiled by the Economic Wing of the devolved Ministry of Food and Agriculture and later by the Pakistan Federal Bureau of Statistics. The acquired data had some missing wheat yield values for 2009. To overcome this issue, average of all other data for the period 1981−2013 is used to recover the missing data of the predictor and the corresponding target stations.

3.2 Study region

The present study sites include the eight agriculture-intensive districts, situated in Punjab, Pakistan (31.17°N, 72.70°E), which is the most populous province with an estimated head count of 101.4 million as of 2015 (Punjab, 2015). The economy is the largest compared with the other provinces, contributing a significant amount to the GDP of Pakistan. Since 1972, the Punjab province's economy has quadrupled (Statistics, 2015). The share of this province's GDP was 54.7% in 2000, and but it reached approximately 59% as of 2010. The agricultural sectors plays a vital role in the economy with a contribution ranging from 56.1% to 61.5% (Punjab, 2015). Extensive irrigation systems make the region a rich agricultural locality, despite its subtropical wet and dry climates. Canal-irrigation system, developed by the

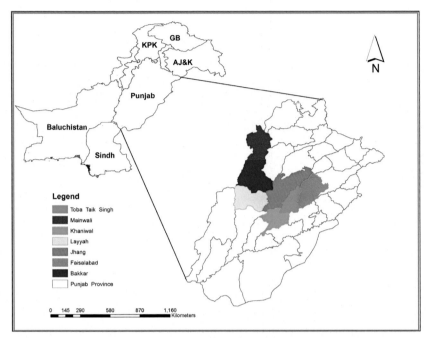

FIGURE 2.2 The two groups of training and cross-validation (i.e., objective) study sites (Group 1, centered on objective station 1 and Group 2, centered on objective station 2). Each of the two groups have three nearby site-specific yield incorporated in models to forecast wheat yield in the neighboring fourth objective station.

British Government in 1930s, is one of the largest systems in the world. Considering the region as a major agricultural belt, the development of data-intelligent models for the prediction of wheat yield, an important national crop, is an interesting research endeavor.

In Fig. 2.2, we illustrate the present study sites (i.e., the major districts) for the two groups of objective stations for which the predictive models for wheat yield simulation were constructed. Each group has four neighboring study sites where the yield data for the three adjacent (or nearest neighboring) agricultural sites have been used for the training of the model to predict the wheat yield for the fourth site that shares a common agricultural boundary. The study sites in Group 1 comprises the districts of Faisalabad, Jhang, and Khanewal (utilized in the training process) and the district of Toba Taik Sing (utilized in the evaluation process), while Group 2 includes the sites for the districts of Mianwali, Layyah, Jhang (utilized in the training), and Bakkar (utilized in the evaluation). For simplicity, the yield data for the first group of three training (predictor) station are denoted as (Y_F, Y_J, Y_K) versus its objective station 1 (Y_T), while those for the second training (predictor) group are denoted as (Y_M, Y_L, Y_J) versus the objective station 2 (Y_B).

In Table 2.1, the basic statistical properties of wheat yield (including the sowing time [October−December] and harvesting time [March−April]) for the predictor (i.e., input) and target (objective) stations (for Group 1 and Group 2) have been enumerated. To develop robust predictive models, in this chapter, a total of 33 years of data from 1981 to 2013 are acquired from the Islamabad Federal Bureau of Statistics (Districts, 2008; Statistics, 2012). To develop the ANN (and the comparative MPMR and GP based models), all available data were first partitioned into a subset of 22 years to be used for training purposes, 5 years for validation purposes, and 6 years for testing purposes.

Before the construction of the predictive models for wheat yield simulation, a scatterplot diagram was constructed using the training data to verify the causal relationships, if any, between the predictor (or input) variables, $X = (Y_F, Y_J,$ and $Y_K)$ versus the objective station 1 (Y_T) in Group 1. Similarly, a scatterplot was constructed for the predictor variables, $X = (Y_M,$ Y_J and $Y_L)$ versus the objective station 2 (Y_B) in Group 2. Fig. 2.3 shows the result. Note that the data for the three stations on the x-axis have been averaged for each of the training stations, presented separately for each of the two groups. It is clearly evident that the relationship between the training variable plotted against the target, as modeled by a coefficient of determination (r^2), is visible through a positively correlated line. The statistical relationship of Group 1 stations, i.e., (Y_F, Y_J, Y_K), with the objective station 1, i.e., Y_T, appears to be slightly stronger than Group 2 station (Y_M, Y_J, Y_L) versus objective station 2 (Y_B). The coefficient of determination for mean data for (Y_F, Y_J, Y_K) versus Y_T is found to be approximately 0.908, indicating that the covariance between the predictor stations and the objective station is relatively small, and so, the data patterns of the predictor stations agree well with those of the objective station. By contrast, the mean data for (Y_M, Y_J, Y_L) versus Y_B yield an r^2 value of 0.861, which is slightly lower than the value obtained for (Y_F, Y_J, Y_K) versus Y_T. This difference concurs with the maximum deviation for the stations of each group, with a value of approximately 5.04 for Group 1 and a value of 8.84 for Group 2. Overall, the relationship between the predictor data versus the objective data appears to be relatively strong for Group 1, as seen in Fig. 2.3A,B.

To verify the additive role of each wheat yield input station, X utilized in the prediction of the objective station wheat yield, the cross-correlation function was applied to determine the statistical similarity between each input and a shifted (lagged) copy of the objective data. In this regard, the covariance was computed for a pair of input variables X with the objective variable (Adamowski et al., 2012).

$$\Theta_k = \sum_{j=\max(0,k)}^{\min(M-1+k,N-1)} X_{j-k}, \quad k = -(M+1),...,0,...,(N-1)), \qquad (2.16)$$

TABLE 2.1 Description of the seasonal wheat yield data (kg per hectare) of neighboring (i.e., input) and target sites for the period 1981–2013 utilized in the design of an ANN and the comparative (MPMR and GP) model.

Station	Variable andDesignation	Average	Minimum	Maximum	St. Dev	Skewness	Kurtosis
Group 1							
Input site 1: Faisalabad	Wheat yield, Y_F	2490.03	1565	3257	511.18	−0.0969	−1.3433
Input site 2: Jhang	Wheat yield, Y_J	2357.88	1637	3089	431.23	−0.0018	−1.0650
Input site 3: Khanewal	Wheat yield, Y_K	2451.46	1000	3600	571.04	−0.8344	1.3752
Target site: Toba Taik Singh	Wheat yield, Y_S	2491.58	999	3440	538.88	−1.0551	2.0054
Group 1							
Input site 1: Mianwali	Wheat yield, Y_M	1720.94	1037	2510	423.30	0.4242	−0.8108
Input site 2: Jhang	Wheat yield, Y_J	2357.88	1637	3089	431.23	−0.0018	−1.0650
Input site 3: Layyah	Wheat yield, Y_L	1081.58	0	2038	404.68	0.4836	1.9688
Target site: Bakkar	Wheat yield, Y_B	1880.42	0	2517	535.07	−1.3461	3.1751

ANN, artificial neural network; GP, genetic programming; MPMR, minimax probability machine regression.

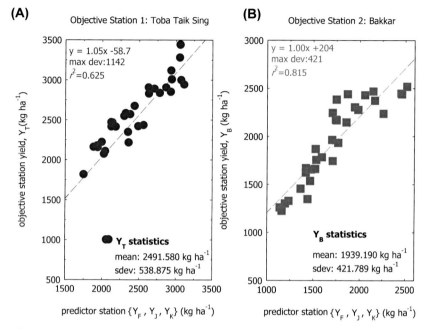

FIGURE 2.3 Scatterplot of the training stations versus target station with a linearly fitted equation of the form $y = mx + c$ with the least square fitting line and its respective correlation coefficient. (A) Group 1. (B) Group 2. The data for three stations on x-axis have been averaged for all three training stations.

By means of Eq. (2.16), the cross-correlation coefficient r was defined as

$$r(t) = \frac{\Theta_{XY}(t)}{\sqrt{\Theta_{XX}(t)\Theta_{YY}(0)}} \tag{2.17}$$

In Eq. (2.17), $r(t)$ lies in the interval $[-1, 1]$, and when $r = 1$, the correlation between X and Y is an ideal positive, while for $r = -1$, the correlation becomes negative. Note that a larger positive correlation indicates a greater potential for extracting the predictive features for an accurate model. Fig. 2.4A displays the values of r versus the lags presented at 95% confidence interval. At a lag of zero (i.e., no shifting of the input data), a relative value of r, about 90%, is accomplished for the correlation between Khanewal and Toba Taik Singh, followed by about 40% for Faisalabad and Toba Taik Singh and Jhang and Toba Taik Singh at about 30%.

For the stations in Group 2, Fig. 2.4B shows that the correlation is also relatively high for the yield of both Jhang and Layyah versus that of the objective station Bakkar, which stands at about 90%, followed by Mianwali versus objective station Bakkar at about 50%. This confirms the pivotal role of the predictor variables in contributing to the yield at the objective station in

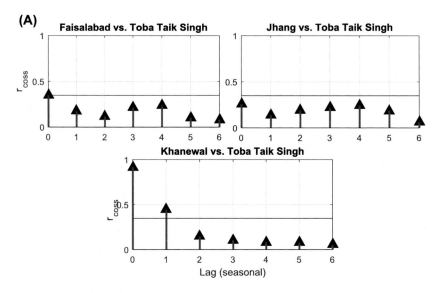

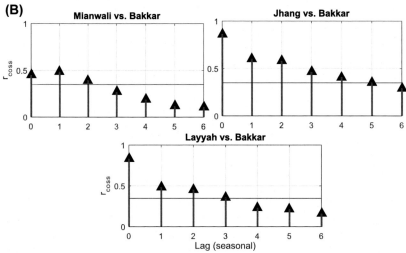

FIGURE 2.4 Cross-correlation coefficients (r_{cross}) between (A) predictor stations (Faisalabad, Jhang, and Khanewal) in respect to the objective station in Group 1 (Toba Taik Singh) and (B) predictor stations (Mianwali, Layyah, and Jhang) in respect to the objective station Group 2 (Bakkar). Ninety-five percent confidence interval is indicated by the blue lines.

each target group. Moreover, the cross-correlation analysis can assist us in verifying that the selected variables used in the model development are appropriate and thus comprise the inherent features that explain the seasonal evaluation of wheat yield for the present study region.

3.3 Development of data-intelligent models

For all models, appropriate normalizations of the data were adopted where the scaling of input-target matrix was performed to address any numerical issues presented by the data attributes, patterns, or fluctuations, following earlier works (Hsu et al., 2003).

$$X_{\text{normalized}} = \frac{(X - X_{\text{min}})}{(X_{\text{max}} - X_{\text{min}})} \tag{2.18}$$

In Eq. (2.18), X represents any datum (i.e., input or output) value, X_{min} is the minimum value of the entire data set, X_{max} is the maximum value of the entire data set, and $X_{\text{normalized}}$ is the normalized value of the datum point.

Because ANN model does not identify the network architecture and the most appropriate training algorithm and hidden transfer functions, a total of 2700 trial ANN networks were constructed. Following earlier work, (Deo and Sahin, 2015a,b, Prasad et al., 2017a), this was achieved by taking different combinations of the hidden transfer and output equations (i.e., 2.2−2.11), but for each combination, different training algorithms and a range of hidden neurons (from 1 to 30) were applied in the trial models. These models were sorted on the basis of the LM Index (Legates and McCabe, 1999, 2013) attained in the testing set to screen the optimal ANN model. Note that the optimization of these models was attained by testing training algorithm and hidden transfer functions in different combinations, as outlined in Table 2.2(A−B).

In each phase of the modeling process, a three-layer neuronal network was adopted to feed the predictor data for the learning process, where transfer functions were incorporated to extract the features for a model formulation with the minimum mean-squared error between input space and target wheat yield. For each case, a total of three input neurons, denoted as X, were used in both groups, with wheat yield prediction denoted as a single output neuron. As there were no set rules to attain the best transfer function (because the predictive features were not available before the development of the model), we followed an iterative modeling process, utilizing several trial transfer functions (Eqs. 2.2−2.11) and neuronal learning algorithms to optimize the model's accuracy (Table 2.3A−B) (Deo and Sahin, 2015a,b; Deo and Sahin, 2016). The number of neurons in the hidden layer was incremented in an interval of one for each trial including an adjustment adopted subsequently in the model. The nearly optimal neurons with the corresponding transfer function were intrinsically disparate for each model combination, determined by mean-squared error (Deo and Sahin, 2015a, 2016). The correlation coefficient (r) was used to verify the performance in training and validation phases.

It is noticeable that for Group 1, the ANN model (denoted as M_1), constructed with Levenberg−Marquardt (*trainlm*) algorithm and logarithmic

TABLE 2.2 Neuronal structure and design of ANN model for study sites (Groups 1 and 2) including the correlation coefficient (r) and root mean square error (RMSE) attained in the validation phase. The optimum model is boldfaced (red)

(A) Objective station: Toba Taik Singh.

Trial model number	Training algorithm	Hidden transfer function	Output transfer function	Neuronal architecture (input–hidden–output)	Validation	
					r	RMSE (kg/ha)
M_1	trainlm	logsig	tansig	3-14-1	0.842	105.97
M_2	trainlm	logsig	logsig	3-23-1	0.706	91.23
M_3	traincgp	purelin	logsig	3-15-1	0.723	119.79
M_4	trainlm	logsig	purelin	3-18-1	0.784	198.50
M_5	trainbfg	tansig	purelin	3-23-1	0.002	436.86
M_6	traincgf	purelin	logsig	3-25-1	0.753	113.54
M_7	traingda	tansig	tansig	3-23-1	0.664	196.05
M_8	traingda	purelin	tansig	3-1-1	0.719	83.84
M_9	trainbfg	logsig	purelin	3-5-1	0.485	237.56
M_{10}	traincgf	purelin	purelin	3-8-1	0.749	151.97
M_{11}	trainlm	tansig	logsig	3-3-1	0.774	60.94
M_{12}	traincgb	purelin	purelin	3-8-1	0.751	150.77
M_{13}	trainoss	purelin	purelin	3-15-1	0.725	143.15
M_{14}	traincgp	tansig	tansig	3-10-1	0.403	134.62

M_{15}	trainlm	tansig	purelin	3-10-1	0.653	45.57
M_{16}	traingda	tansig	purelin	3-7-1	0.229	75.35
M_{17}	trainscg	logsig	logsig	3-5-1	-0.049	105.53
M_{18}	trainoss	logsig	tansig	3-18-1	0.712	175.57
M_{19}	traincgf	logsig	purelin	3-21-1	0.720	474.24
M_{20}	traincgb	logsig	purelin	3-21-1	0.768	501.72
M_{21}	trainlm	tansig	tansig	3-14-1	0.575	180.97
M_{22}	traincgp	purelin	purelin	3-15-1	0.751	159.69
M_{23}	trainscg	purelin	purelin	3-8-1	0.759	144.99
M_{24}	traincgp	logsig	purelin	3-18-1	0.768	268.51
M_{25}	trainscg	tansig	tansig	3-23-1	0.684	187.48
M_{26}	traincgb	tansig	purelin	3-3-1	0.462	150.20
M_{27}	trainoss	purelin	tansig	3-16-1	0.437	132.66
M_{28}	trainoss	tansig	tansig	3-29-1	0.215	138.53
M_{29}	traincgb	logsig	logsig	3-20-1	-0.108	89.29
M_{30}	traincgf	logsig	logsig	3-30-1	0.844	141.54
M_{31}	traincgp	tansig	logsig	3-20-1	0.036	75.65
M_{32}	traingda	logsig	purelin	3-7-1	0.252	307.93
M_{33}	trainoss	logsig	purelin	3-21-1	0.690	459.55
M_{34}	traincgb	tansig	logsig	3-5-1	-0.017	214.87
M_{35}	traincgp	logsig	logsig	3-5-1	0.818	244.19

Continued

TABLE 2.2 Neuronal structure and design of ANN model for study sites (Groups 1 and 2) including the correlation coefficient (r) and root mean square error (RMSE) attained in the validation phase. The optimum model is boldfaced (red)—cont'd

(A) Objective station: Toba Taik Singh.

Trial model number	Training algorithm	Hidden transfer function	Output transfer function	Neuronal architecture (input-hidden-output)	Validation	
					r	RMSE (kg/ha)
M_{36}	trainbfg	logsig	logsig	3-13-1	0.787	150.83
M_{37}	traincgb	logsig	tansig	3-13-1	−0.674	125.97
M_{38}	traincgp	tansig	purelin	3-23-1	−0.029	551.01
M_{39}	trainscg	logsig	purelin	3-2-1	0.469	139.26
M_{40}	trainbfg	purelin	purelin	3-15-1	0.759	152.56
M_{41}	trainbfg	tansig	tansig	3-17-1	−0.417	174.26
M_{42}	trainoss	tansig	purelin	3-10-1	0.715	161.39
M_{43}	trainlm	purelin	logsig	3-25-1	0.880	53.81
M_{44}	trainbfg	logsig	tansig	3-13-1	−0.487	133.37
M_{45}	traincgf	tansig	logsig	3-17-1	0.781	33.46
M_{46}	trainscg	tansig	purelin	3-16-1	0.491	57.89
M_{47}	traincgf	tansig	tansig	3-30-1	0.606	132.81
M_{48}	traingdm	tansig	purelin	3-8-1	0.856	571.30
M_{49}	traingdx	tansig	tansig	3-30-1	0.685	106.25

M50	logsig	trainoss	logsig	3-5-1		0.504	126.99
M51	tansig	traincgf	purelin	3-7-1		0.326	59.03
M52	tansig	trainbfg	logsig	3-17-1		-0.142	69.17
M53	tansig	traingdx	purelin	3-3-1		0.233	0.06
M54	tansig	traincgb	tansig	3-30-1		0.580	151.83
M55	logsig	traingdm	purelin	3-7-1		0.207	313.05
M56	purelin	trainscg	tansig	3-1-1		0.721	75.51
M57	purelin	traingda	purelin	3-20-1		0.614	150.07
M58	tansig	traingdx	logsig	3-29-1		-0.025	151.05
M59	purelin	traingdx	logsig	3-8-1		0.729	75.02
M60	purelin	traingdx	logsig	3-28-1		0.729	75.02
M61	purelin	trainscg	logsig	3-28-1		0.729	76.63
M62	purelin	traingdm	logsig	3-28-1		0.729	74.55
M63	purelin	traingda	logsig	3-28-1		0.729	74.75
M64	purelin	traingdx	tansig	3-28-1		0.725	113.87
M65	purelin	traingdx	tansig	3-28-1		0.725	113.87
M66	tansig	traingdm	logsig	3-29-1		-0.026	155.33
M67	logsig	traingda	logsig	3-19-1		-0.022	152.96
M68	logsig	traingda	tansig	3-30-1		0.891	101.49
M69	tansig	traingda	logsig	3-29-1		-0.012	153.44
M70	logsig	trainbfg	purelin	3-28-1		0.702	100.01

Continued

TABLE 2.2 Neuronal structure and design of ANN model for study sites (Groups 1 and 2) including the correlation coefficient (r) and root mean square error (RMSE) attained in the validation phase. The optimum model is boldfaced (red)—cont'd

(A) Objective station: Toba Taik Singh.

Trial model number	Training algorithm	Hidden transfer function	Output transfer function	Neuronal architecture (input-hidden-output)	Validation r	RMSE (kg/ha)
M_{71}	traincgp	logsig	tansig	5-13-1	−0.629	232.56
M_{72}	traingdm	purelin	tansig	3-28-1	0.725	102.56
M_{73}	trainoss	tansig	logsig	3-17-1	−0.660	170.14
M_{74}	traingdm	logsig	tansig	3-17-1	−0.730	269.56
M_{75}	trainoss	purelin	logsig	3-1-1	0.679	64.07
M_{76}	trainscg	logsig	tansig	3-30-1	0.902	98.52
M_{77}	trainscg	tansig	logsig	3-17-1	−0.696	244.20
M_{78}	trainbfg	purelin	tansig	3-1-1	0.697	76.59
M_{79}	traincgf	logsig	tansig	3-13-1	−0.625	256.11
M_{80}	traingdm	tansig	tansig	3-30-1	0.788	109.67
M_{81}	traincgp	purelin	tansig	3-1-1	0.684	79.18
M_{82}	traincgb	purelin	logsig	3-1-1	0.675	95.33
M_{83}	trainlm	purelin	tansig	3-23-1	0.724	49.27
M_{84}	traingdx	purelin	purelin	3-26-1	0.496	145.17
M_{85}	traingdx	purelin	purelin	3-26-1	0.496	145.17

					r	RMSE
M86	traincgf	purelin	tansig	3-20-1	−0.541	134.67
M87	traingdm	purelin	purelin	3-8-1	0.733	61.21
M88	trainlm	purelin	purelin	3-6-1	0.677	127.08
M89	traincgb	purelin	tansig	3-23-1	0.658	79.83
M90	traingdm	logsig	logsig	3-29-1	−0.185	678.06

(B) Objective station: Bakkar.

Trial model number	Training algorithm	Hidden transfer function	Output transfer function	Neuronal architecture (input-hidden-output)	Validation	
					r	RMSE (kg/ha)
M1	trainlm	purelin	tansig	3-20-1	0.826	103.58
M2	traincgf	purelin	purelin	3-24-1	0.597	151.98
M3	trainscg	logsig	tansig	3-16-1	0.937	43.15
M4	trainlm	tansig	tansig	3-26-1	0.952	70.40
M5	traingdm	tansig	tansig	3-16-1	0.936	43.14
M6	traincgf	logsig	tansig	3-16-1	0.937	44.34
M7	traincgp	logsig	tansig	3-16-1	0.937	44.34
M8	traincgb	logsig	tansig	3-16-1	0.937	44.34
M9	traingda	purelin	purelin	3-4-1	0.595	149.49
M10	trainbfg	tansig	purelin	3-6-1	0.859	74.70
M11	traincgf	tansig	tansig	3-16-1	0.937	42.69
M12	traincgp	tansig	tansig	3-16-1	0.937	42.69

Continued

TABLE 2.2 Neuronal structure and design of ANN model for study sites (Groups 1 and 2) including the correlation coefficient (r) and root mean square error (RMSE) attained in the validation phase. The optimum model is boldfaced (red)—cont'd

(B) Objective station: Bakkar.

Trial model number	Training algorithm	Hidden transfer function	Output transfer function	Neuronal architecture (input-hidden-output)	Validation	
					r	RMSE (kg/ha)
M_{13}	traincgb	tansig	tansig	3-16-1	0.937	42.69
M_{14}	traingdx	tansig	tansig	3-16-1	0.936	43.15
M_{15}	traingdx	logsig	tansig	3-16-1	0.937	45.25
M_{16}	traincgp	purelin	tansig	3-21-1	0.865	79.68
M_{17}	traincgb	purelin	tansig	3-21-1	0.865	79.68
M_{18}	traincgf	purelin	logsig	3-19-1	0.935	67.87
M_{19}	traincgp	purelin	logsig	3-19-1	0.935	67.87
M_{20}	traincgb	purelin	logsig	3-19-1	0.935	67.87
M_{21}	traincgf	purelin	tansig	3-24-1	0.833	116.23
M_{22}	trainoss	purelin	logsig	3-2-1	0.911	92.24
M_{23}	trainbfg	purelin	logsig	3-2-1	0.910	91.05
M_{24}	trainbfg	purelin	tansig	3-10-1	0.840	127.03
M_{25}	trainscg	purelin	tansig	3-10-1	0.843	129.80
M_{26}	traingda	purelin	tansig	3-27-1	0.825	117.79

Continued

M$_{27}$	trainbfg	logsig	logsig	3-16-1	0.929	78.45
M$_{28}$	trainlm	purelin	logsig	3-18-1	0.881	58.29
M$_{29}$	trainoss	purelin	tansig	3-10-1	0.843	104.97
M$_{30}$	trainscg	purelin	logsig	3-30-1	0.841	96.48
M$_{31}$	trainoss	purelin	purelin	3-28-1	0.511	187.07
M$_{32}$	trainoss	tansig	purelin	3-1-1	0.419	305.34
M$_{33}$	traingdx	purelin	logsig	3-21-1	0.837	96.72
M$_{34}$	traingdm	purelin	logsig	3-21-1	0.828	108.44
M$_{35}$	traingda	purelin	logsig	3-21-1	0.824	109.27
M$_{36}$	traincgf	purelin	purelin	3-20-1	-0.339	229.10
M$_{37}$	trainoss	logsig	logsig	3-16-1	0.929	71.93
M$_{38}$	traingdm	logsig	tansig	3-21-1	0.867	66.15
M$_{39}$	traingdm	purelin	tansig	3-24-1	0.817	70.25
M$_{40}$	traincgb	logsig	logsig	3-4-1	0.968	44.68
M$_{41}$	traincgb	logsig	purelin	3-6-1	0.816	80.51
M$_{42}$	trainoss	tansig	tansig	3-16-1	0.931	55.70
M$_{43}$	traingdx	logsig	tansig	3-21-1	0.867	65.61
M$_{44}$	trainbfg	logsig	tansig	3-16-1	0.932	51.65
M$_{45}$	trainlm	logsig	tansig	3-13-1	0.953	56.11
M$_{46}$	traincgp	logsig	logsig	3-4-1	0.955	64.65
M$_{47}$	traincgb	purelin	purelin	3-4-1	0.578	160.92

TABLE 2.2 Neuronal structure and design of ANN model for study sites (Groups 1 and 2) including the correlation coefficient (r) and root mean square error (RMSE) attained in the validation phase. The optimum model is boldfaced (red)—cont'd

(B) Objective station: Bakkar.

Trial model number	Training algorithm	Hidden transfer function	Output transfer function	Neuronal architecture (input-hidden-output)	Validation	
					r	RMSE (kg/ha)
M_{48}	traingdx	tansig	logsig	3-1-1	0.932	51.30
M_{49}	traingda	tansig	tansig	3-16-1	0.933	49.79
M_{50}	trainlm	logsig	logsig	3-4-1	0.969	53.34
M_{51}	trainscg	logsig	logsig	3-4-1	0.962	53.12
M_{52}	traincgp	tansig	logsig	3-16-1	0.925	89.95
M_{53}	traingdx	logsig	logsig	3-18-1	0.833	66.17
M_{54}	trainscg	tansig	tansig	3-16-1	0.933	47.90
M_{55}	traingda	logsig	'ogsig	3-18-1	0.832	66.26
M_{56}	traincgf	logsig	logsig	3-16-1	0.929	54.18
M_{57}	traingdm	logsig	logsig	3-18-1	0.832	66.17
M_{58}	trainoss	tansig	logsig	3-16-1	0.926	79.69
M_{59}	traincgb	tansig	logsig	3-16-1	0.925	88.75
M_{60}	trainlm	tansig	logsig	3-1-1	0.850	96.62
M_{61}	traingda	tansig	logsig	3-16-1	0.931	52.99
M_{62}	trainbfg	tansig	logsig	3-16-1	0.927	69.32

Continued

M_{63}	traingdx	tansig	purelin	3-6-1	0.766	241.61
M_{64}	traingdx	logsig	purelin	3-20-1	−0.109	190.63
M_{65}	traingda	logsig	tansig	3-18-1	0.800	70.91
M_{66}	traincgf	tansig	logsig	3-9-1	0.905	83.99
M_{67}	traingdm	tansig	logsig	3-16-1	0.932	56.61
M_{68}	trainoss	logsig	purelin	3-19-1	0.439	125.78
M_{69}	trainbfg	purelin	purelin	3-16-1	0.543	219.76
M_{70}	traincgp	logsig	purelin	3-1-1	0.761	116.52
M_{71}	traingdm	Logsig	purelin	3-20-1	−0.149	194.22
M_{72}	traingda	tansig	purelin	3-6-1	0.806	222.25
M_{73}	traingda	logsig	purelin	3-20-1	−0.159	195.23
M_{74}	traingdx	purelin	purelin	3-4-1	0.551	132.16
M_{75}	traingdm	tansig	purelin	3-6-1	0.807	222.17
M_{76}	trainscg	tansig	logsig	3-16-1	0.941	73.86
M_{77}	trainlm	tansig	purelin	3-19-1	0.697	96.16
M_{78}	traincgf	tansig	purelin	3-6-1	0.832	63.17
M_{79}	trainscg	logsig	purelin	3-9-1	0.825	72.57
M_{80}	trainoss	tansig	tansig	3-28-1	0.663	106.70
M_{81}	trainbfg	tansig	tansig	3-3-1	0.934	88.51
M_{82}	trainscg	tansig	purelin	3-1-1	0.808	140.37

TABLE 2.2 Neuronal structure and design of ANN model for study sites (Groups 1 and 2) including the correlation coefficient (r) and root mean square error (RMSE) attained in the validation phase. The optimum model is boldfaced (red)—cont'd

(B) Objective station: Bakkar.

Trial model number	Training algorithm	Hidden transfer function	Output transfer function	Neuronal architecture (input-hidden-output)	Validation	
					r	RMSE (kg/ha)
M_{83}	traincgp	tansig	purelin	3-4-1	0.609	114.11
M_{84}	traincgb	logsig	purelin	3-20-1	−0.271	210.32
M_{85}	traingdm	purelin	purelin	3-4-1	0.540	157.09
M_{86}	trainbfg	logsig	purelin	3-20-1	−0.330	240.74
M_{87}	trainlm	logsig	purelin	3-4-1	0.925	99.28
M_{88}	traincgp	purelin	purelin	3-24-1	0.576	182.25
M_{89}	trainscg	purelin	purelin	3-4-1	0.533	202.47
M_{90}	trainlm	purelin	purelin	3-1-1	0.653	192.97

trainbfg, BFGS quasi-Newton; *trainrp*, resilient; *trainscg*, scaled conjugate gradient; *trainlm*, Levenberg—Marquardt; *traincgb*, conjugate gradient BP with Powell-Beale restarts; *traincgf*, conjugate gradient BP with Fletcher-Reeves update; *trainoss*, one-step secant; *traincgp*, conjugate gradient backpropagation with Polak-Ribiére updates; *traingda*, gradient descent with adaptive learning rate backpropagation; *traingdm*, gradient descent with momentum backpropagation; *traingdx*, gradient descent with momentum and adaptive learning are the machine learning algorithms with *tansig*, tangent sigmoid; *logsig*, logarithmic sigmoid and *purelin*, purely linear equation are the hidden transfer functions.

TABLE 2.3 (A–B): Development of the MPMR and GP models and the performance in the validation phase for the present study sites.

(A) MPMR model design

Group	Objective station	Model	ϕ	Scale	Kernel name	Ker.p$_1$	Ker.p$_2$	Validation	
								r	RMSE (kg/ha)
1	Toba Taik Singh	MPMR	1.0	None	linear	2	23	0.930	125.22
2	Bakkar	MPMR	1.0	None	linear	2	23	0.858	203.97

(B) GP model design

Group	Objective station	Model	Population size	Maximum tree depth	Parameters				Validation	
					a_1	a_2	a_3	a_4	r	RMSE (kg/ha)
1	Toba Taik Singh	GP	40	5	0.2	30	0.05	0	0.871	167.20
2	Bakkar	GP	40	5	0.2	30	0.05	0	0.962	143.89

ANN, artificial neural network; GP, genetic programming; MPMR, minimax probability machine regression.

sigmoid-based hidden transfer and tangent sigmoid output functions, generated the best architecture (3-14-1). Out of all trial models, the optimal model **M1** yielded *RMSE* value of 220,606 kg/ha and an *r* value of 0.961 (training period), whereas its performance in the validation period had an *r* value of 0.842 and *RMSE* value of 105.97 kg/ha. Similarly, for Group 2, the optimal model **M₁** constructed with Levenberg–Marquardt (*trainlm*) algorithm and pure linear as a hidden transfer function with tangent sigmoid as the output function yielded a network architecture of 3-20-1.

The comparative model, based on the MPMR algorithm, was dependent on the selection of the kernel function (i.e., Eq. 2.15) which was based on the inner product of the feature support vectors constructed from training and target data (Deo and Samui, 2017). In our study, the linear kernel was found to be most appropriate. Moreover, the value of σ which was the kernel width and the size error threshold ϕ can also affect the performance of the model. The optimal width of the error tube ϕ was found to be 1.0 for the best MPMR model, whereas the kernel parameters ker.p_1 and ker.p_2 were found to be approximately 2.0 and 23.0, respectively. In target Group 1, the *RMSE* of the optimal MPMR model was approximately 125.22 kg/ha with an *r* value of 0.930 attained in the training period, while for target Group 2, the model generated an *RMSE* value of 203.97 kg/ha with an *r* value of approximately 0.858 (Table 2.3A).

Construction of the GP model required a setting for a population size, maximum tree depth, and several parameters to generate the best results (Deo and Samui, 2017). In this chapter, GP was set at an optimal population size 40 with the maximum tree depth of five and the parameters $a_1 = 0.2, a_2 = 30.0$, $a_3 = 0.05$, and $a_4 = 0$ to develop the final model. In the optimal case for the training period, the *RMSE* value was approximately 167.20 kg/ha for Group 1 target station compared with 143.89 kg/ha for Group 2 target station, while an *r* value of 0.962 was obtained for Group 2 target station compared with a lower value of 0.871 for Group 1 target station. Table 2.3B shows the design parameters of the optimal GP model.

3.4 Model performance evaluation criteria

To evaluate the performance of ANN versus MPMR and the GP models applied for wheat yield prediction, the American Society for Civil Engineering (ASCE) guidelines (Yen, 1995) present two categories of evaluation processes, which comprise statistical and standardized metrics. The statistical metrics are used to investigate the differences in minimum, maximum, mean, variance, standard deviation skewness, and kurtosis between predicted and observed yield, while the standardized metrics are used for the validation of the predicted yield in respect to the observed data. The mathematical formulations of these assessment metrics are given as follows (Dawson et al., 2007; Deo et al., 2016; Legates and McCabe, 1999; Willmott, 1981, 1982, 1984).

I. Correlation coefficient (r) is expressed as

$$r = \left(\frac{\sum\limits_{i=1}^{N} \left(Y_{OBS,i} - \overline{Y}_{OBS,i}\right)\left(Y_{PRED,i} - \overline{Y}_{PRED,i}\right)}{\sqrt{\sum\limits_{i=1}^{N}\left(Y_{OBS,i} - \overline{Y}_{OBS,i}\right)^2}\sqrt{\sum\limits_{i=1}^{N}\left(Y_{PRED,i} - \overline{Y}_{PRED,i}\right)^2}} \right) \tag{2.19}$$

II. Willmott's Index (*WI*) is expressed as

$$WI = 1 - \left[\frac{\sum\limits_{i=1}^{N}\left(Y_{PRED,i} - Y_{OBS,i}\right)^2}{\sum\limits_{i=1}^{N}\left(\left|Y_{PRED,i} - \overline{Y}_{OBS,i}\right| + \left|Y_{OBS,i} - \overline{Y}_{OBS,i}\right|\right)^2} \right], \; 0 \le d \le 1 \tag{2.20}$$

III. Nash-Sutcliffe coefficient (*EV*) is expressed as

$$EV = 1 - \left[\frac{\sum\limits_{i=1}^{N}\left(Y_{OBS,i} - Y_{PRED,i}\right)^2}{\sum\limits_{i=1}^{N}\left(Y_{OBS,i} - \overline{Y}_{PRED,i}\right)^2} \right] \tag{2.21}$$

IV. root mean square error (*RMSE*) is expressed as

$$RMSE = \sqrt{\frac{1}{N}\sum\limits_{i=1}^{N}\left(Y_{PRED,i} - Y_{OBS,i}\right)^2} \tag{2.22}$$

V. mean absolute error (*MAE*) is expressed as

$$MAE = \frac{1}{N} \sum_{i=1}^{N} \left| \left(Y_{PRED,i} - Y_{OBS,i} \right) \right| \qquad (2.23)$$

VI. Legates-McCabe's (*LM*) is expressed as

$$LM = 1 - \left[\frac{\sum_{i=1}^{N} \left| Y_{PRED,i} - Y_{OBS,i} \right|}{\sum_{i=1}^{N} \left| Y_{OBS,i} - \overline{Y}_{OBS,i} \right|} \right] \qquad (2.24)$$

VII. Relative root mean square error (*RRMSE*, %) is expressed as

$$RRMSE = \frac{\sqrt{\frac{1}{N} \sum_{i=1}^{N} \left(Y_{PRED,i} - Y_{OBS,i} \right)^2}}{\frac{1}{N} \sum_{i=1}^{N} \left(Y_{OBS,i} \right)} \times 100 \qquad (2.25)$$

VIII. Relative mean absolute percentage error (*RMAE*, %) is expressed as

$$RMAE = \frac{1}{N} \sum_{i=1}^{N} \left| \frac{\left(Y_{PRED,i} - Y_{OBS,i} \right)}{Y_{OBS,i}} \right| \times 100 \qquad (2.26)$$

where Y_{OBS} and Y_{PRED} are the observed and predicted ith value of the wheat yield Y, $\overline{Y}_{OBS}$ and $\overline{Y}_{PRED}$ are the observed and predicted mean of Y in the (test) set, respectively, and N is the number of tested data points.

The physical interpretation of the metrics is as follows. The correlation coefficient (r) is bounded by [0, 1] and demonstrates the proportion of variance in the observed yield that can be explained by the data-intelligent model

(Dawson et al., 2007). Owing to standardization of the observed and predicted means and variance, the robustness of r is limited. The goodness of fit relevant to the high values is measured by the *RMSE*, while in contrast *MAE* evaluates all deviations from observed data both in the same manner regardless of sign. However, the performance can be reduced to partial peaks and higher magnitudes that can exhibit larger error and obtuse to small magnitudes (Dawson et al., 2007). Hence the WI was used to address this issue by considering the ratio of the mean-squared error instead of the differences (Mohammadi et al., 2015; Willmott, 1981, 1982; 1984; Willmott et al., 2012). Nash-Sutcliffe efficiency (*EV*) is another normalized metric that determines the relative magnitude of the residual variance of predicted data in comparison with the measured variance of the observed data (Nash and Sutcliffe, 1970). In this chapter, we also used the *LM*, a more advanced and powerful metric than both *WI* and *EV*, that aims to address the weaknesses of *WI* and *EV*. As such, the *LM* was found to be the best metric in evaluating the results by ignoring r and using *WI* and *EV* as baseline-adjusted indices together with an evaluation of *RMSE* and *MAE* (Legates and McCabe, 1999).

Owing to geographic differences among the region of this district-based study (Table 2.1; Fig. 2.2), the *RRMSE* and *RMAE* were also calculated (Dawson et al., 2007; Mohammadi et al., 2015) to evaluate and compare the model performances over geographically diverse sites. It should be noted that if the (*RRMSE, RMAE*) <10%, the performance of the model can be rated as outstanding, while the model is considered to be good if 10%< (*RRMSE, RMAE*) <20%, fair if 20%< (*RRMSE, RMAE*) <30%, and is poor if the (*RRMSE, RMAE*) >30% (Ertekin and Yaldiz, 2000).

4. Results

Results are presented for assessing the prescribed data-intelligent models using the wheat yield data from 1981 to 2013 for the wheat yield prediction for two groups (comprised of eight sites) in Punjab, Pakistan. ANN is appraised in comparison with MPMR and GP models, using statistical metrics, diagnostic plots, and error distributions (Eqs. (2.19)−(2.26)) between predicted and observed yield. To accomplish a desired accuracy level of a parsimonious ANN model with few possible predictors, a process of iteration was performed to tune the model's parameters using input combinations, training algorithms, and hidden transfer functions by sorting out the largest value of LM agreement for the selection of an optimum model (Section 3.2). Using statistical performance metrics, the prediction accuracy of seasonal wheat yield is investigated in this section.

Fig 2.5A−B plots a comparison of the predicted wheat yield, generated by the ANN model together with the MPMR and GP models in the testing phase (2008−13). Moreover, the predicted wheat yield is also plotted in comparison with the observed yield. There is compelling evidence that the

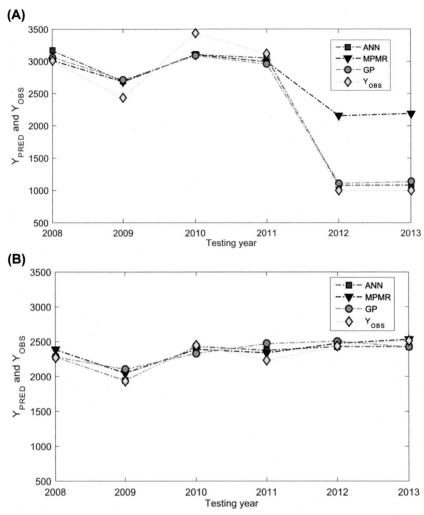

FIGURE 2.5 The predicted yield (Y_{PRED}) generated by ANN, MPMR, and GP models together with the observed yield (Y_{OBS}) (ha) for the objective stations: (A) Toba Taik Sing and (B) Bakkar. *ANN*, artificial neural network; *GP*, genetic programming; *MPMR*, minimax probability machine regression.

well-tuned ANN model performs very accurately for all tested years, in comparison with the MPMR and GP model. On a year-by-year basis, the ANN model is seen to exhibit a reasonable accuracy by predicting the yield that is very close to the observed wheat yield in the year 2011–13, but there is a slight increase from 2008 to 09 and then decrease in the prediction accuracy for the year 2010–11 in the objective station of Group 1. MPMR model, on the other hand, predicted a large amount of yield in the tested year

2008–09, which suddenly dropped in 2009–11 and slightly increased in 2011–13. The GP model predicted the testing period almost similar to ANN model. However, the prediction of ANN in Group 2 relative to its counterpart models was more accurate than that in Group 1. The predictions generated by ANN can be seen almost similar to the observed yield throughout the testing years 2008–13. MPMR started with slightly a high prediction in comparison with the predicted wheat yield of the ANN model and observed wheat yield which was almost similar in 2009 and was seen similar in 2010–13, while the prediction by the GP model was high in 2008–09 and then the improvement reduced in the following year as compared to the ANN model and then again rise in 2011–12. Although there is generally a good agreement between predicted and observed yield by all three tested models, this plot reveals that the predicted values generated by ANN are more accurate than those of the MPMR and the GP models.

In Table 2.4, the preciseness of ANN is evaluated in relation to the MPMR and the GP model where the results for each set of predictor stations corresponding to the objective station in each group are shown on the basis of the best performance in the testing period. It is noteworthy that the number of inputs are the same in each trial ANN model as well as MPMR and GP models and the performances are assessed on the basis of r, WI, EV, $RMSE$, and MAE. Accuracy of ANN for the study sites appears to be more improved by feeding the wheat yield into the algorithm than that of MPMR and GP, which can be confirmed by the remarkable increase in r calculated between the predicted and observed yield (in terms of observed decrease in the model generalization error).

ANN model M_1 applied at objective station in Group 1 site attained the highest correlation coefficient ($r \approx 0.983$) as compared with the MPMR ($r \approx 0.957$) and the GP ($r \approx 0.982$) model. Moreover, the largest WI was achieved by the ANN model ($WI \approx 0.984$), followed by the GP ($WI \approx 0.980$) and the MPMR ($WI \approx 0.544$). Similarly the Nash-Sutcliffe coefficient ($EV \approx 0.962$) acquired by ANN was the highest, followed by the GP ($EV \approx 0.955$) and the MPMR model ($EV \approx 0.527$). The optimum ANN model M_1 is also shown to have the smallest root mean square error ($RMSE \approx 192.02$ kg/ha) and mean absolute error ($MAE \approx 162.75$ kg/ha) as compared with the GP ($RMSE \approx 209.25$ kg/ha), ($MAE \approx 182.84$ kg/ha), and MPMR model ($RMSE \approx 614.46$ kg/ha), ($MAE \approx 431.29$ kg/ha) respectively. This is a clear indication that the ANN model can be considered to be a better data-intelligent tool for wheat yield estimation than the MPMR and the GP model for the Group 1 objective station.

In the case of Group 2 stations, overall, the ANN model $\mathbf{M_1}$ again generated better results than the MPMR and the GP model. In terms of the correlation coefficient attained for Group 2, ANN ($r \approx 0.945$), MPMR ($r \approx 0.845$), and GP ($r \approx 0.752$) ranked such that the ANN model had clearly a better correlation coefficient than the MPMR and the GP models.

TABLE 2.4 Evaluation of seasonal wheat production prediction models in the testing phase using correlation coefficient (r), Willmott index (WI), Nash-Sutcliffe coefficient (EV), root mean square error (RMSE; kg/ha), and mean absolute error (MAE; kg/ha) in the testing period. Note that the boldfaced magnitudes represents the performance of best mode.

Inputs	Model	r	WI	EV	RMSE (kg/ha)	MAE (kg/ha)
Group 1: Faisalabad and Jhung, Khanewal data used to predict wheat yield for Toba Taik Singh						
Faisalabad Jhung Khanewal	ANN	**0.983**	**0.984**	**0.962**	**192.02**	**162.75**
Faisalabad Jhung Khanewal	MPMR	0.957	0.544	0.527	614.46	431.29
Faisalabad Jhung Khanewal	GP	0.982	0.980	0.955	209.25	182.84
Group 2: Mianwali, Layyah, and Jhang data used to predict wheat yield for Bakkar						
Mianwali Layyah Jhang	ANN	**0.945**	**0.954**	**0.888**	**65.19**	**41.31**
Mianwali Layyah Jhang	MPMR	0.845	0.731	0.438	134.58	102.51
Mianwali Layyah Jhang	GP	0.752	0.694	0.503	137.56	116.43

Furthermore, the WI attained by ANN was higher than that obtained by both of the MPMR and the GP model for Group 2. In order of the performances, the best acquired models were ANN ($WI \approx 0.954$), followed by MPMR ($WI \approx 0.731$) and the GP model ($WI \approx 0.694$). Similarly, the values of Nash-Sutcliffe metric were high for the ANN case ($EV \approx 0.888$) than those for the MPMR ($EV \approx 0.438$) and the GP model ($EV \approx 0.503$). The mean-squared error in these models in order of their predictive performances were such that ANN ($RMSE \approx 65.19$ kg/ha), MPMR ($RMSE \approx 134.58$ kg/ha), and GP (RMSE ≈ 137.56 kg/ha), whereas the mean absolute error showed the same pattern with ANN ($MAE \approx 41.31$kg/ha), outperforming MPMR ($MAE \approx 102.51$ kg/ha) and the GP model ($MAE \approx 116.43$ kg/ha). From Table 2.4, it is clear that overall, the performance of the ANN model was quite better than the MPMR and the GP model. The ANN model had better results when applied to Group 1 study sites than to Group 2 study sites.

By inspecting Fig. 2.6A−B, a more concrete and conclusive argument about the predictive ability of the ANN model can be described, through a

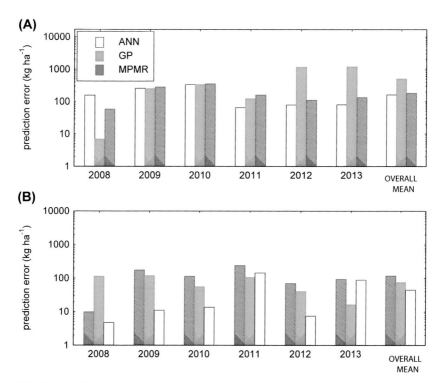

FIGURE 2.6 Prediction error (in ha) between observed and simulated wheat yield in the test period, 2008−13, with overall average generated by ANN, MPMR and GP models for objective stations: (A) Toba Taik Singh and (B) Bakkar. *ANN*, artificial neural network; *GP*, genetic programming; *MPMR*, minimax probability machine regression.

bar graph that depicts the forecasting error for both objective stations in Group 1 and Group 2. ANN model produced the smallest forecasting error in wheat yield. The errors for ANN were very low in the years of 2008−11, while they were slightly higher in the years of 2012−13 than those of the MPMR and the GP model in the same yearly bracket. Similarly, the forecasting error of the ANN model in relation to the MPMR and the GP model was much lower from 2008 to 12 except in 2013 in Group 2 station. Overall, the forecasting error of the ANN model was smaller than that of both the MPMR and the GP model.

Further assessment of model performances shows that the achieved accuracy of the ANN model was better than that of MPMR and GP model, as clearly evident from the marked difference in the assessment metrics. The ANN model obtained an optimal performance for the Group 1 objective station followed by the GP and MPMR models. The achieved LM Index for the ANN model (i.e., $LM \approx 0.817$), GP (i.e., $LM \approx 0.794$), and MPMR (i.e., $LM \approx 0.427$) reveals the dominancy of the ANN over the other counterpart models examined in this case study region. Furthermore, the relative percentage root mean square error of the ANN model was reasonably low (i.e., $RRMSE \approx 8.23\%$) compared with that of the GP model (i.e., $RRMSE \approx 8.97\%$) and the MPMR model (i.e., $RRMSE \approx 24.78\%$). Similarly, the relative percentage mean absolute error was also smaller in the case of the ANN model ($RMAE \approx 7.29\%$) in comparison with the GP ($RMAE \approx 8.90\%$) and the MPMR model ($RMAE \approx 34.03\%$).

In terms of site-specific accuracy, the results of Group 2 objective stations showed that the ANN model attained a high value of LM Index ($LM \approx 0.742$), which was followed by the MPMR ($LM \approx 0.310$) and then the GP model ($LM \approx 0.274$), while the relative percentage errors were ($RRMSE \approx 2.83\%$, $RMAE \approx 1.79\%$) for the ANN, ($RRMSE \approx 5.87\%$, $RMAE \approx 4.54\%$) for the MPMR, and ($RRMSE \approx 5.97\%$, $RMAE \approx 5.22\%$) for the GP model. In terms of the site-averaged performance, the ANN model was found to yield the highest value of the LM Index for Group 1 objective station while lowest relative percentage error in predicted wheat yield for Group 2 objective station. But overall, the predicted errors generated by ANN model were low in terms of their relative error values, but more importantly, they were within the recommended range of 10% threshold for an excellent model classification.

To illustrate the additive role of single-input stations to be used as the predictors, and also to assist in identifying the contribution of features to the model's yield for an objective station, we plot Fig. 2.7, where the relative sensitivity, via the predictive features (i.e., in terms of the error encountered) per year, is shown for the case of Bakker. Note that, through an assessment of the errors encountered, we can investigate which station contributes the most and which station contributes the least toward solving the prediction problem for the respective objective station. Hypothetically, if model error is low for a

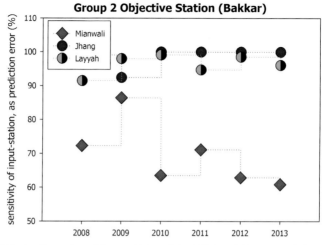

FIGURE 2.7   An illustration of the sensitivity of the ANN model in terms of single-input predictor station yield (Mianwali, Layyah, and Jhang) incorporated to predict the crop yield of objective station 2 (Bakkar). *ANN*, artificial neural network.

given input station, the contribution for wheat yield prediction for the objective stations by that predictor site is considered to be the greatest.

5. Discussion: limitations and opportunity for further research

To strategically address food scarcity issues, there is a growing desire to develop versatile data-intelligent models to enable agricultural policymakers in forecasting crop production, particularly in agriculture-based nations (e.g., Pakistan) where the population growth parallels a rising need for crop management. This chapter is dedicated to the modeling of district-level wheat production in remote regions in Punjab, Pakistan, with historical wheat data. This study aimed for the first time to validate an ANN model, a popular data-intelligent tool applied previously in agriculture (Alvarez, 2009; Kaul et al., 2005; Noh et al., 2006; Ye et al., 2006) in respect to its predictive ability against MPMR and GP models, for wheat yield prediction using neighboring station data. The approach is practically useful for crop management in respect to the predictor and the neighboring objective stations, in developing better agricultural practices with more efficient crop precision technologies. While the present study is only a case study in Pakistan, this research can be extended to any other location where wheat yield data are available to provide an accurate estimation of the wheat yield in neighboring sites. It is important to note that the current model coupled with wheat yield was successfully evaluated to yield less than 10% uncertainty based on

RRMSE and *RMAE,* respectively, with a reasonably large statistical correlation of LM (*LM* ≈ 0.817 (objective station 1) and 0.746 (objective station 2)) between predicted and observed yield (Table 2.5). The performance was remarkable, according to the achieved relative percentage errors which were less than 10%. Thus, we propose a suitably tuned ANN model can be used to predict wheat yield in the future where the prediction of such a commodity will likely become even more important due to an increase of population and for economic growth in terms of exporting to international markets.

In spite of the superior performance, we acknowledge that this study has some limitations that are likely to seed new research in future. For example, to increase its practicality in the field of agricultural precision, the ANN model should be trialed on larger data sets of wheat yield and regional farms which could be of interest to the government's national policymaking and agricultural engineers to help minimize crop estimation uncertainties (Akhtar, 2014; Niaz, 2014). Owing to the aforementioned qualities of the ANN, it is possible to apply such a model for the prediction of other crops not only in Pakistan but also elsewhere because no previous studies have been conducted with the present approach. As Pakistan is considered to be an agricultural country, other important national crops such as rice, maize, cotton, sugarcane, oilseeds, and the other coarse grains and pulses could also be utilized to generate similar predictions in a follow-up study. Another possible extension to the present study may use several different types of national crops (e.g., rice, maize, cotton, sugarcane, oilseeds, and the other coarse grains and pulses) trained with different data sets (including satellite data) and the prediction of crop yield in different seasons.

Despite the merits of ANN, there are other shortcomings that could also be addressed in a follow-up research work. In this chapter, the yield data of neighboring stations have been used to predict the yield in a centrally located, objective station that shares a common geographic boundary to the predictor station. However, meteorological data such as land surface and air temperatures, rainfall, soil moisture, wind, solar radiation, etc. could also be used to predict the crop yield as each of these variables are likely to impact the crop production amounts. Such predictor variables (whose data could be remotely sensed through satellites or atmospheric simulation models) (Bauer, 1975; Chen and McNairn, 2006; Dempewolf et al., 2014; Kumar et al., 2015a; Stathakis et al., 2006) are likely to be greatly valuable for modeling crop yield in remotely located agricultural areas. The incorporation of fertilizers (and the other relevant soil properties) could also be coupled with meteorological data to explore their use in the data-intelligent models. Drought is a major factor driving the agriculture sector, so using this parameter as a likely important variable together with meteorological parameters can improve the predictive accuracy. In the same respect, the use of climate scenarios downscaled from global climate models into potential crop yield predictive models with an ANN or another hybrid model downscaled

TABLE 2.5 Evaluation of the seasonal wheat yield prediction models in the testing phase using the normalized performance errors: Legates and McCabes (LM), relative root mean-squared error (RRMSE, %), and the relative mean absolute error (RMAE, %) computed within the test sites. The boldfaced magnitudes represents the performance of best mode.

Group	Predictor (input) station	Objective station	ANN			MPMR			GP		
			LM	RRMSE, %	RMAE, %	LM	RRMSE, %	RMAE, %	LM	RRMSE, %	RMAE, %
1	Faisalabad Jhung Khanewal	Toba Taik Singh	**0.817**	**8.23**	**7.29**	0.427	24.78	34.03	0.794	8.97	8.90
2	Mianwali Layyah Jhang	Bakkar	**0.742**	**2.83**	**1.79**	0.310	5.87	4.54	0.274	5.97	5.22

ANN, artificial neural network; GP, genetic programming; MPMR, minimax probability machine regression; RMAE, relative mean absolute percentage error; RRMSE, relative root mean square error.

for specific agricultural sites is yet another interesting research. Irrigation statistics (e.g., water supply) could also be used to improve crop yield and photosynthetically active radiation that governs crop production could be the focus of an independent research study.

It is expressly noted that in this chapter, ANN has been trialed through an iterative modeling process, but in terms of the model's optimization, the hybridization of different data-intelligent algorithms could yield more promising results than a standalone model. Therefore, the ANN model could be optimized by an ensemble modeling method (with confidence intervals of the yield simulations) to possibly achieve more accurate results, improving on our 10% level of uncertainty. An ensemble of model simulations could lead to a more strategic decision-making model, with the provision of how accurate a simulated yield could be, in terms of the uncertainties between several forecasted data. As an alternative tool, ANFIS model (integrating the neural network with a fussy set model) could be another optimization approach (Abdulshahed et al., 2015; Goyal et al., 2014; Karthika and Deka, 2015; Yaseen et al., 2018). Other advanced optimization methods applied could include particle swarm optimization (PSO), quantum-behaved PSO, genetic algorithm (GA), and firefly algorithm which have been tested to improve ANN models (Ghorbani et al., 2017; Hoang et al., 2014; Kayarvizhy et al., 2014; Kumar et al., 2013a; Pal et al., 2012; Raheli et al., 2017; Sedki and Ouazar, 2010; Taormina and Chau, 2015; Taormina et al., 2015), particularly in identifying accurately the weights and biases of neuronal arrangement. While the aforementioned approaches are well established, a new, more generalized framework for hybridizing the ANN model with intuitionistic fuzzy logic and neutrosophic logic can be achieved instead of using fuzzy logic integrator. The latter are known to handle uncertainty, indeterminacy, incompleteness, and inconsistency in the predictor-target data.

Furthermore, in a follow-up study, random selection of training and validation/test sets can be performed to develop an ensemble of model for crop yield prediction. Because the standard statistical approaches tend to avoid the hurdle of model uncertainty that potentially leads to overconfident inferences and risky agricultural decisions, Bayesian Model Averaging (BMA) (Raftery et al., 1997) is another data-intelligent tool to model uncertainties and ranking the models. Multiresolution analysis (e.g., empirical wavelet transform (Gilles, 2013), empirical mode composition (Huang et al., 1998)), maximum overlap wavelets (Prasad et al., 2017b), and singular value decomposition (Chau and Wu, 2010)) could broaden the accuracy and scope of this study. A future work could apply copula functions (Nguyen-Huy et al., 2017) as statistical tools where joint behavior of multivariate data (e.g., yield and corresponding predictors) can be modeled for any target station.

6. Conclusion

This chapter has developed an ANN model using the wheat yield of neighboring stations as the predictors for the model estimation of regional wheat yield in a station that shares a common boundary. The wheat yield data from 1981 to 2013 for a total of eight stations were used in such a way that these stations were divided into two groups, each comprising of four stations. Then, three out of four stations were used as predictor variables to predict in the fourth (objective) station. The data were divided into training, validation, and testing periods in the traditional way. A large number of training algorithms and hidden transfer functions have been trialed to achieve a high level of accuracy of the ANN model. Levenberg–Marquardt (*trainlm*) algorithm with logarithmic sigmoid as hidden transfer and tangent sigmoid output function was adopted for the best prediction of wheat yield in Group 1 station. Similarly, the Levenberg–Marquardt (*trainlm*) algorithm with pure linear as a hidden transfer function and tangent sigmoid as the output function was found to be the most appropriate for prediction in Group 2 station.

The ANN model was compared with MPMR and GP. To attain an accurate model, the predictor data set was partitioned into 65% training, 15% validation, and 20% for testing. A total of 2700 models applying ANN architecture were tested with 90 of the best models being sorted for both station groups. Evidently, the performance of the ANN was found to be very much better than that of MPMR and the GP models. The better performance of the ANN model is evidenced in terms of low relative forecasting errors and high performance metrics. The prediction errors of all three ANN, MPMR, and GP models for the Group 1 objective station were $RMSE \approx 192.02$ kg/ha, 614.46 kg/ha, and 209.25 kg/ha, respectively, whereas the MAE was 162.75 kg/ha, 431.29 kg/ha, and 182.84 kg/ha, respectively. Similarly, for the Group 2 objective station, the prediction errors were $RMSE \approx 65.19$ kg/ha, 134.58 kg/ha, and 137.56 kg/ha, whereas the MAE was 41.31 kg/ha, 102.51 kg/ha, and 116.43 kg/ha, for each of the ANN, MPMR, and GP models, respectively. In terms of normalized performance metrics, the values of these metrics for the Group 1 station of all three models were $r \approx 0.983$ (ANN), 0.957 (MPMR), and 0.982 (GP), while $WI \approx 0.984$ (ANN), 0.544 (MPMR), and 0.980 (GP) and $EV \approx 0.962$ (ANN), 0.527 (MPMR), and 0.955 (GP), respectively. Similarly, for the Group 2 objective station, they were $r \approx 0.945$ (ANN), 0.845 (MPMR), and 0.752 (GP), while $WI \approx 0.954$ (ANN), 0.731 (MPMR), and 0.694 (GP) and $EV \approx 0.888$ (ANN), 0.438 (MPMR), and 0.503 (GP), respectively. By assessing the performance of ANN in relation to MPMR and the GP model using the most advanced normalized metrics of LM, the ANN was again found to have the highest agreement. The obtained LM agreement values between the predicted wheat yield and observed wheat yield for the Group 1 study site were $LM \approx 0.817$ (ANN), 0.427 (MPMR), and 0.794 (GP), respectively, whereas the relative percentage errors $RRMSE$ and $RMAE$ were only 8.23%, 7.29%

(ANN) compared with 24.78%, 34.03% (MPMR) and 8.97%, 8.90% (GP). Similarly, better performance of ANN was found over MPMR and the GP models for Group 2 stations in both the LM and relative percentage errors. The values of *LM, RRMSE,* and *RMAE* were 0.742%, 2.83%, 1.79% (ANN), 0.310%, 5.87%, 4.54% (MPMR), and 0.274%, 5.97%, 5.22% (GP) respectively. A reasonable degree of geographic variability was evident on the basis of relative percentage errors *RRMSE* and *RMAE* with the primal performance acquired for Group 1 and Group 2 in the accuracy of the ANN model being followed up by the MPMR and GP models.

In respect to the practical implication of the proposed technique, we aver that the models developed in this study are very important for farming precision and agricultural policymaking, given the importance of the nearest neighbor-based strategy to predict the crop yield at an objective station. It is important to note that an accurate and reliable model developed at a particular test site (trained with the data from nearest neighbor sites) is likely to be adopted in studying the agricultural practices and the detailed farming practices in neighboring site where the same a model is perhaps found to be less accurate. Therefore, this information is likely to enable farmers and policymakers to revisit their specific implementation of farm management strategies at a given site, for example, through a "peer-to-peer learning" process. The present models can therefore be extremely useful tools to enable the farmers to learn from each other the set of newer, more effective (and the most relevant) techniques of planting, and crop management strategies through the modeling of the neighboring station data (and hence their practices) if the neighboring sites have exceptionally effective practice. Moreover, the application of nearest neighbor data to predict crop yield at an objective site is also important from the perspective of modeling crop yield at that site without the use of other exogenous data that are difficulty to acquire especially in remote sites and more importantly, in developing nations such as Pakistan where reliable records of climate and associated data are not available. In these sorts of situations, only the crop yield data for the predictor site(s) can be used that, off course, will incorporate the influence of exogenous factors that govern the yield at an objective site. It is therefore envisaged that the present approach is extremely useful from a farming management point of view.

In a nutshell, this study advocates the possibility of using time series yield data to predict wheat to extend the scope to the other crops. To enhance the predictive accuracy, the ANN can be optimized and tuned with advanced learning techniques such as ensemble models, PSO, GAs, and fuzzy logics. Nonetheless, this study provides a strong baseline relevance of the ANN model, with models such as ANFIS, SVM, ELM, copulas, etc. being utilized to predict wheat and other crop yield more accurately in follow-up, future studies.

Acknowledgments

This chapter has utilized wheat yield acquired from Pakistan's Bureau of Statistics, Government of Pakistan: Islamabad, Pakistan that is duly acknowledged. The authors are duly acknowledged that this research project has been supported by The University of Southern Queensland's Postgraduate Research Scholarship (2017−19) awarded to the first author, managed by Office of Research and Graduate Studies Division.

References

Abbot, J., Marohasy, J., 2012. Application of artificial neural networks to rainfall forecasting in Queensland, Australia. Advances in Atmospheric Sciences 29, 717−730.

Abdulshahed, A.M., Longstaff, A.P., Fletcher, S., Myers, A., 2015. Thermal error modelling of machine tools based on ANFIS with fuzzy c-means clustering using a thermal imaging camera. Applied Mathematical Modelling 39, 1837−1852.

Adamowski, J., Fung Chan, H., Prasher, S.O., Ozga-Zielinski, B., Sliusarieva, A., 2012. Comparison of multiple linear and nonlinear regression, autoregressive integrated moving average, artificial neural network, and wavelet artificial neural network methods for urban water demand forecasting in Montreal, Canada. Water Resources Research 48.

Akhtar, I.U.H., 2014. Pakistan Needs a New Crop Forecasting System.

Ali, S., Smith, K.A., 2006. On learning algorithm selection for classification. Applied Soft Computing 6, 119−138.

Ali, S., Liu, Y., Ishaq, M., Shah, T., Ilyas, A., Din, I.U., 2017. Climate change and its impact on the yield of major food crops: evidence from Pakistan. Foods 6, 39.

Alvarez, R., 2009. Predicting average regional yield and production of wheat in the Argentine Pampas by an artificial neural network approach. European Journal of Agronomy 30, 70−77.

Amin, M., A, M., Akbar, A., 2014. Time series modelling for forecasting wheat production of Pakistan. The Journal of Animal and Plant Sciences 24, 7.

Anusree, K., Binu, G., 2011. Analysis of training functions in a biometric system. International Journal on Recent and Innovation Trends in Computing and Communication 2, 150−154.

Avriel, M., 2003. Nonlinear Programming: Analysis and Methods. Courier Corporation.

Azhar, B., Chaudhry, M.G., Shafique, M., 1973. A model for forecasting wheat production in the Punjab. Pakistan Development Review 12, 407−415.

Azhar, B., Chaudhry, M.G., Shafique, M., 1974. A forecast of wheat production in the Punjab for 1973-74. Pakistan Development Review 13, 106−112.

Balakrishnan, N., Muthukumarasamy, G., 2016. Crop production-ensemble machine learning model for prediction. International Journal of Computer Systems Science and Engineering 5, 148−153.

Battiti, R., 1992. First-and second-order methods for learning: between steepest descent and Newton's method. Neural Computation 4, 141−166.

Bauer, M.E., 1975. The role of remote sensing in determining the distribution and yield of crops. Advances in Agronomy 27, 271−304.

Bertsimas, D., Sethuraman, J., 2000. Moment problems and semidefinite optimization. Handbook of Semidefinite Programming 469−509.

Bokhari, A., 2013. Wheat Crisis in the Making. Dawn Newspaper Report. Dawn newspaper, Pakistan.

Briscoe, J.Q.,U., 2005. Pakistan'sWater Economy: Running Dry. Oxford University Press, Karachi, Pakistan, p. 2005.

Chau, K., Wu, C., 2010. A hybrid model coupled with singular spectrum analysis for daily rainfall prediction. Journal of Hydroinformatics 12, 458–473.

Chaudhry, M.G., Kemal, A.R., 1974. Wheat production under alternative production functions. Pakistan Development Review 13, 222–226.

Chen, C., McNairn, H., 2006. A neural network integrated approach for rice crop monitoring. International Journal of Remote Sensing 27, 1367–1393.

Co, H.C., Boosarawongse, R., 2007. Forecasting Thailand's rice export: statistical techniques vs. artificial neural networks. Computers and Industrial Engineering 53, 610–627.

Coopersmith, E.J., Minsker, B.S., Wenzel, C.E., Gilmore, B.J., 2014. Machine learning assessments of soil drying for agricultural planning. Computers and Electronics in Agriculture 104, 93–104.

Dawson, C.W., Abrahart, R.J., See, L.M., 2007. HydroTest: a web-based toolbox of evaluation metrics for the standardised assessment of hydrological forecasts. Environmental Modelling and Software 22, 1034–1052.

Dempewolf, J., Adusei, B., Becker-Reshef, I., Hansen, M., Potapov, P., Khan, A., Barker, B., 2014. Wheat yield forecasting for Punjab Province from vegetation index time series and historic crop statistics. Remote Sensing 6, 9653–9675.

Dennis Jr., J.E., Schnabel, R.B., 1996. Numerical Methods for Unconstrained Optimization and Nonlinear Equations. SIAM.

Deo, R.,C., Sahin, M., 2015a. Application of the artificial neural network model for prediction of monthly standardized precipitation and evapotranspiration index using hydrometeorological parameters and climate indices in Eastern Australia. Atmospheric Research 161–162, 65–81.

Deo, R.,C., Sahin, M., 2015b. Application of the extreme learning machine algorithm for the prediction of monthly effective drought index in Eastern Australia. Atmospheric Research 153, 512–525.

Deo, R.,C., Sahin, M., 2016. An extreme learning machine model for the simulation of monthly mean streamflow water level in eastern Queensland. Environmental Monitoring and Assessment 188, 1. https://doi.org/10.1007/s10661-016-5094-9.

Deo, R.C., Sahin, M., 2017. Forecasting long-term global solar radiation with an ANN algorithm coupled with satellite-derived (MODIS) land surface temperature (LST) for regional locations in Queensland. Renewable and Sustainable Energy Reviews 72, 828–848.

Deo, R.,C., Samui, P., 2017. Forecasting evaporative loss by least-square support-vector regression and evaluation with genetic programming, Gaussian process, and minimax probability machine regression: case study of Brisbane City. Journal of Hydrologic Engineering 22, 05017003.

Deo, R.C., Wen, X., Qi, F., 2016. A wavelet-coupled support vector machine model for forecasting global incident solar radiation using limited meteorological dataset. Applied Energy 168, 568–593.

Deo, R.C., Tiwari, M.K., Adamowski, J.F., Quilty, M.J., 2017. Forecasting effective drought index using a wavelet extreme learning machine (W-ELM) model. Stochastic Environmental Research and Risk Assessment 31, 1211–1240.

Dimitriadis, S., Goumopoulos, C., 2008. Applying machine learning to extract new knowledge in precision agriculture applications. In: Informatics, 2008. PCI'08. Panhellenic Conference on, IEEE, pp. 100–104.

Districts, 2008. Crops Area and Production by Districts 1981-2008, Food and Cash Crops. Federal Bureau of Statistics (Economic wing), Islamabad, Pakistan.

Dorosh, P., Salam, A., 2008. Wheat Markets and Price Stabilisation in Pakistan: An Analysis of Policy Options. The Pakistan Development Review, pp. 71–87.

Drummond, S.T., Sudduth, K.A., Joshi, A., Birrell, S.J., Kitchen, N.R., 2003. Statistical and neural methods for site–specific yield prediction. Transactions of the ASAE 46, 5.

Ertekin, C., Yaldiz, O., 2000. Comparison of some existing models for estimating global solar radiation for Antalya (Turkey). Energy Conversion and Management 41, 311–330.

FAO, 2013. Review of the Sector and Grain Storage Issue.

Faramarzi, M., Yang, H., Schulin, R., Abbaspour, K.C., 2010. Modeling wheat yield and crop water productivity in Iran: implications of agricultural water management for wheat production. Agricultural Water Management 97, 1861–1875.

Fletcher, R., Reeves, C.M., 1964. Function minimization by conjugate gradients. The Computer Journal 7, 149–154.

Foresee, F.D., Hagan, M.T., 1997. Gauss-Newton approximation to Bayesian learning. In: Neural networks, 1997, International Conference on, IEEE, pp. 1930–1935.

Ghodsi, R., Yani, R.M., Jalali, R., Ruzbahman, M., 2012. Predicting wheat production in Iran using an artificial neural networks approach. International Journal of Academic Research in Business and Social Sciences 2, 34.

Ghorbani, M.A., Deo Ravinesh, C., Zaher Mundher, Y., Mahsa H, K., Babak, M., 2018. Pan evaporation prediction using a hybrid multilayer perceptron-firefly algorithm (MLP-FFA) model: case study in North Iran. Theoretical and Applied Climatology 133 (3-4), 1119–1131.

Gilles, J., 2013. Empirical wavelet transform. IEEE Transactions on Signal Processing 61, 3999–4010.

Gonzalez-Sanchez, A., Frausto-Solis, J., Ojeda-Bustamante, W., 2014. Attribute selection impact on linear and nonlinear regression models for crop yield prediction. Science World Journal 2014.

Goyal, M.K., Bharti, B., Quilty, J., Adamowski, J., Pandey, A., 2014. Modeling of daily pan evaporation in sub tropical climates using ANN, LS-SVR, fuzzy logic, and ANFIS. Expert Systems with Applications 41, 5267–5276.

Guo, W., Li, L., Whymark, G., 2010. Simulating wheat yield in New South Wales of Australia using interpolation and neural networks. In: Neural Information Processing. Models and Applications, pp. 708–715.

Hamid, N.P., Alberto, T.V., Suzanne, G., 1987. The wheat economy of Pakistan setting and prospects. In: International Food Policy Research Institute, Ministry of Food and Agriculture, Government of Pakistan, Islamabad, Pakistan.

HariKumar, R., Vasanthi, N., Balasubramani, M., 2009. Performance analysis of artificial neural networks and statistical methods in classification of oral and breast cancer stages. International Journal of Soft Computing and Engineering 2.

Hestenes, M.R., Stiefel, E., 1952. Methods of Conjugate Gradients for Solving Linear Systems. NBS.

Hoang, N.-D., Pham, A.-D., Cao, M.-T., 2014. A novel time series prediction approach based on a hybridization of least squares support vector regression and swarm intelligence. Applied Computational Intelligence and Soft Computing 2014, 15.

Hsu, C.-W., Chang, C.-C., Lin, C.-J., 2003. A Practical Guide to Support Vector Classification.

Huang, N.E., Shen, Z., Long, S.R., Wu, M.C., Shih, H.H., Zheng, Q., Yen, N.-C., Tung, C.C., Liu, H.H., 1998. The empirical mode decomposition and the Hilbert spectrum for nonlinear and non-stationary time series analysis. In: Proceedings of the Royal Society of London A: Mathematical, Physical and Engineering Sciences. The Royal Society, pp. 903–995.

Huang, G.-B., Zhu, Q.-Y., Siew, C.-K., 2006. Extreme learning machine: theory and applications. Neurocomputing 70, 489−501.

Iqbal, N., Bakhsh, K., Maqbool, A., Ahmad, A.S., 2005. Use of the ARIMA model for forecasting wheat area and production in Pakistan. Journal of Agriculture and Social Sciences 1, 120−122.

Irmak, A., Jones, J., Batchelor, W., Irmak, S., Boote, K., Paz, J., 2006. Artificial neural network model as a data analysis tool in precision farming. Transactions of the ASABE 49, 2027−2037.

Karthika, B., Deka, P.C., 2015. Prediction of air temperature by hybridized model (Wavelet-ANFIS) using wavelet decomposed data. Aquatic Procedia 4, 1155−1161.

Kaul, M., Hill, R.L., Walthall, C., 2005. Artificial neural networks for corn and soybean yield prediction. Agricultural Systems 85, 1−18.

Kayarvizhy, N., Kanmani, S., Uthariaraj, R., 2014. ANN models optimized using swarm intelligence algorithms. WSEAS Transactions on Computers 13, 501−519.

Keskin, M.E., Terzi, Ö., 2006. Artificial neural network models of daily pan evaporation. Journal of Hydrologic Engineering 11, 65−70.

Khamis, A., Ismail, Z., Haron, K., Tarmizi Mohammed, A., 2006. Neural network model for oil palm yield modeling. Journal of Applied Sciences 6, 391−399.

Khan, S.B., Hafeezullah, K.A., Rehman, A., 2012. Agricultural Statistics of Pakistan Pakistan Bureau of Statistics. Government of Pakistan, Islamabad, Pakistan.

Kim, T.-W., Valdés, J.B., 2003. Nonlinear model for drought forecasting based on a conjunction of wavelet transforms and neural networks. Journal of Hydrologic Engineering 8, 319−328.

Koza, J.R., 1992. Genetic Programming: On the Programming of Computers by Means of Natural Selection. MIT press.

Kumar, D., Prasad, R.K., Mathur, S., 2013a. Optimal design of an in-situ bioremediation system using support vector machine and particle swarm optimization. Journal of Contaminant Hydrology 151, 105−116.

Kumar, M., Mittal, M., Samui, P., 2013b. Performance assessment of genetic programming (GP) and minimax probability machine regression (MPMR) for prediction of seismic ultrasonic attenuation. Earthquake Science 26, 147−150.

Kumar, P., Gupta, D.K., Mishra, V.N., Prasad, R., 2015a. Comparison of support vector machine, artificial neural network, and spectral angle mapper algorithms for crop classification using LISS IV data. International Journal of Remote Sensing 36, 1604−1617.

Kumar, R., Singh, M., Kumar, P., Singh, J., 2015b. Crop selection method to maximize crop yield rate using machine learning technique. In: Smart Technologies and Management for Computing, Communication, Controls, Energy and Materials (ICSTM), 2015 International Conference on, IEEE, pp. 138−145.

Lanckriet, G.R., Ghaoui, L.E., Bhattacharyya, C., Jordan, M.I., 2002. A robust minimax approach to classification. Journal of Machine Learning Research 3, 555−582.

Legates, D.R., McCabe, G.J., 1999. Evaluating the use of "goodness-of-fit" measures in hydrologic and hydroclimatic model validation. Water Resources Research 35, 233−241.

Legates, D.R., McCabe, G.J., 2013. A refined index of model performance: a rejoinder. International Journal of Climatology 33, 1053−1056.

Marquardt, D.W., 1963. An algorithm for least-squares estimation of nonlinear parameters. Journal of the Society for Industrial and Applied Mathematics 11, 431−441.

McCulloch, W.S., Pitts, W., 1943. A logical calculus of the ideas immanent in nervous activity. Bulletin of Mathematical Biophysics 5, 115−133.

McPhee, N.F., Poli, R., Langdon, W.B., 2008. Field Guide to Genetic Programming.

McQueen, R.J., Garner, S.R., Nevill-Manning, C.G., Witten, I.H., 1995. Applying machine learning to agricultural data. Computers and Electronics in Agriculture 12, 275−293.

Mehr, A.D., Kahya, E., Olyaie, E., 2013. Streamflow prediction using linear genetic programming in comparison with a neuro-wavelet technique. Journal of Hydrology 505, 240−249.

Mekanik, F., Imteaz, M., Gato-Trinidad, S., Elmahdi, A., 2013. Multiple regression and artificial neural network for long-term rainfall forecasting using large scale climate modes. Journal of Hydrology 503, 11−21.

Memon, M.I.N., Noonari, S., Laghari, M.A., Pathan, M., Pathan, A., Sial, S.A., 2015. Energy consumption pattern in wheat production in Sindh Pakistan. Energy 5.

Mohammadi, K., Shamshirband, S., Tong, C.W., Arif, M., Petković, D., Ch, S., 2015. A new hybrid support vector machine−wavelet transform approach for estimation of horizontal global solar radiation. Energy Conversion and Management 92, 162−171.

Møller, M.F., 1993. A scaled conjugate gradient algorithm for fast supervised learning. Neural Networks 6, 525−533.

Muhammad, K., 1989. Description of the Historical Background of Wheat Improvement in Baluchistan. Agriculture Research Institute (Sariab, Quetta, Baluchistan, Pakistan), Pakistan.

Muhammed, F., Siddique, M., Bashir, M., Ahamed, S., 1992. Forecasting rice production in Pakistan using ARIMA models. Journal of Animal Plant Sciences 2, 27−31.

Nash, J.E., Sutcliffe, J.V., 1970. River flow forecasting through conceptual models part I—a discussion of principles. Journal of Hydrology 10, 282−290.

Nguyen-Huy, T., Deo, R.C., An-Vo, D.-A., Mushtaq, S., Khan, S., 2017. Copula-statistical precipitation forecasting model in Australia's agro-ecological zones. Agricultural Water Management 191, 153−172.

Niaz, M.S., 2014. Wheat Policy—A Success or Failure. DAWN Newspaper.

Noh, H., Zhang, Q., Shin, B., Han, S., Feng, L., 2006. A neural network model of maize crop nitrogen stress assessment for a multi-spectral imaging sensor. Biosystems Engineering 94, 477−485.

Pal, S.K., Rai, C., Singh, A.P., 2012. Comparative study of firefly algorithm and particle swarm optimization for noisy non-linear optimization problems. International Journal of Intelligent Systems and Applications 4, 50.

Paswan, R.P., Begum, S.A., 2013. Regression and Neural Networks Models for Prediction of Crop Production, vol. 1.

Pham, D.T., Sagiroglu, S., 2001. Training multilayered perceptrons for pattern recognition: a comparative study of four training algorithms. International Journal of Machine Tools and Manufacture 41, 419−430.

Powell, M.J.D., 1977. Restart procedures for the conjugate gradient method. Mathematical Programming 12, 241−254.

Prasad, R., Deo, R.C., Li, Y., Maraseni, T., 2017a. Input selection and performance optimization of ANN-based streamflow forecasts in a drought-prone Murray Darling Basin using IIS and MODWT algorithm. Atmospheric Research 197.

Prasad, R., Deo, R.C., Li, Y., Maraseni, T., 2017b. Input selection and performance optimization of ANN-based streamflow forecasts in a drought-prone Murray Darling basin using IIS and MODWT algorithm. Atmospheric Research 197, 42−63.

Punjab, P., 2015. Population.

Qureshi, S.K., 1963. Rainfall, acreage and wheat production in west Pakistan: a statistical analysis. Pakistan Development Review 3, 566−593.

Qureshi, K.A., Buland Akhtar, M.A., Ullah, A., Hussain, A., 1992. An Analysis of the relative contribution of area and yield to total production of wheat and maize in Pakistan. Pakistan Journal of Agricultural Sciences 29.

Raftery, A.E., Madigan, D., Hoeting, J.A., 1997. Bayesian model averaging for linear regression models. 92 (437), 179−191.

Raheli, B., Aalami, M.T., El-Shafie, A., Ghorbani, M.A., Deo, R.C., 2017. Uncertainty assessment of the multilayer perceptron (MLP) neural network model with implementation of the novel hybrid MLP-FFA method for prediction of biochemical oxygen demand and dissolved oxygen: a case study of Langat River. Environmental Earth Sciences 76, 503.

Rahman, M.M., Haq, N., Rahman, R.M., 2014. Machine learning facilitated rice prediction in Bangladesh. In: Information and Computer Technology (GOCICT), 2014 Annual Global Online Conference on, IEEE, pp. 1−4.

Raorane, A., Kulkarni, R., 2012. Data Mining: an effective tool for yield estimation in the agricultural sector. International Journal of Emerging Trends of Technology in Computer Science 1, 75−79.

Sabir, H., Tahir, S., 2012. Supply and demand projection of wheat in Punjab for the year 2011−12. Interdisciplinary Journal Contemporary Research Business 3, 800−808.

Sabur, S.A., Haque, M.E., 1993. An analysis of rice price in Mymensing town market: pattern and forecasting. Bangladesh Journal of Agricultural Economics 16.

Saeed, N., Saeed, A., Zakria, M., Bajwa, T.M., 2000. Forecasting of wheat production in Pakistan using ARIMA models. International Journal of Agricultural Biology 2, 352−353.

Safa, M., Samarasinghe, S., 2011. Determination and modelling of energy consumption in wheat production using neural networks:"A case study in Canterbury province, New Zealand". Energy 36, 5140−5147.

Safa, M., Samarasinghe, S., Nejat, M., 2015. Prediction of Wheat Production Using Artificial Neural Networks and Investigating Indirect Factors Affecting it: Case Study in Canterbury Province. New Zealand.

Sajjad, S.A., 2017. Story of Pakistan's Elite Wheat. Published in The Express Tribune. THE EXPRESS TRIBUNE Pakistan.

Sarwar, U., 2014. Agriculture in Pakistan − an Overview.

Sedki, A., Ouazar, D., 2010. Hybrid particle swarm and neural network approach for streamflow forecasting. Mathematical Modelling of Natural Phenomena 5, 132−138.

Service, A.M.I., 2012. District-Wise Area of Wheat Crop. Directorate of Agriculture (Economics & Marketing) Punjab, Lahore.

Service, A.M.I., 2014. District-Wise Area of Wheat Crop. Directorate of Agriculture (Economics & Marketing) Punjab, Lahore Pakistan.

Sher, F., Ahmad, E., 2008. Forecasting Wheat Production in Pakistan.

Sohail, A., Sarwar, A., Kamran, M., 1994. Forecasting total food grains in Pakistan. Journal of Engineering and Applied Science 13, 140−146. Department of Mathematics and Statistics, University of Agriculture, Faisalabad.

Sreekanth, J., Datta, B., 2011. Coupled simulation-optimization model for coastal aquifer management using genetic programming-based ensemble surrogate models and multiple-realization optimization. Water Resources Research 47.

Stathakis, D., Savin, I., Nègre, T., 2006. Neuro-fuzzy modeling for crop yield prediction. The International Archives of the Photogrammetry, Remote Sensing and Spatial Information Sciences 34, p1−4.

Statistics, A., 2012. Federal Bureau of Statistics (Economic Wing). Federal Bureau of Statistics (Economic wing), Islamabad, Pakistan, Islamabad, Pakistan.

Statistics, P.D., 2015. Punjab Development Statistics. Bureau of Statistics Government of the Punjab Lahore Pakistan.

Strohmann, T., Grudic, G.Z., 2003. A formulation for minimax probability machine regression. Advances in Neural Information Processing Systems 785–792.

Survey, E., 2012. Pakistan Economic Survey 2011–2012. Ministry of Finance of the Government of Pakistan.

Taormina, R., Chau, K.-W., 2015. Data-driven input variable selection for rainfall-runoff modeling using binary-coded particle swarm optimization and extreme learning machines. Journal of Hydrology 529.

Taormina, R., Chau, K.-W., Sivakumar, B., 2015. Neural network river forecasting through baseflow separation and binary-coded swarm optimization. Journal of Hydrology 529, 1788–1797.

van Ittersum, M.K., Cassman, K.G., Grassini, P., Wolf, J., Tittonell, P., Hochman, Z., 2013. Yield gap analysis with local to global relevance—a review. Field Crops Research 143, 4–17.

Van Wart, J., Kersebaum, K.C., Peng, S., Milner, M., Cassman, K.G., 2013. Estimating crop yield potential at regional to national scales. Field Crops Research 143, 34–43.

Vogl, T.P., Mangis, J., Rigler, A., Zink, W., Alkon, D., 1988. Accelerating the convergence of the back-propagation method. Biological Cybernetics 59, 257–263.

Willmott, C.J., 1981. On the validation of models. Physical Geography 2, 184–194.

Willmott, C.J., 1982. Some comments on the evaluation of model performance. Bulletin of the American Meteorological Society 63, 1309–1313.

Willmott, C.J., 1984. On the evaluation of model performance in physical geography. Spatial Statistics and Models 443–460. Springer.

Willmott, C.J., Robeson, S.M., Matsuura, K., 2012. A refined index of model performance. International Journal of Climatology 32, 2088–2094.

Yaseen, Z.M., Ghareb, M.I., Ebtehaj, I., Bonakdari, H., Siddique, R., Heddam, S., Yusif, A., C, D.R., 2018. Rainfall pattern forecasting using novel hybrid intelligent model based ANFIS-FFA. Water Resources Management 32, 105–122.

Ye, X., Sakai, K., Garciano, L.O., Asada, S.-I., Sasao, A., 2006. Estimation of citrus yield from airborne hyperspectral images using a neural network model. Ecological Modelling 198, 426–432.

Yen, B.C., 1995. Discussion and closure: criteria for evaluation of watershed models. Journal of Irrigation and Drainage Engineering 121, 130–132.

Chapter 3

Monthly rainfall forecasting with Markov Chain Monte Carlo simulations integrated with statistical bivariate copulas

Mumtaz Ali[1,2], Ravinesh C. Deo[1], Nathan J. Downs[1], Tek Maraseni[1]

[1]*School of Agricultural, Computational and Environmental Sciences, University of Southern Queensland, Springfield, QLD, Australia;* [2]*Deakin-SWU Joint Research Centre on Big Data, School of Information Technology, Deakin University, Burwood, VIC, Australia*

3.1 Introduction

Owing to global warming, the change in rainfall patterns directly affects the agriculture sector as the rainfall plays a vital role in both the growth and production of crops. The effect is not only restricted to agriculture sector but also brings other major disasters (Barredo, 2007) such as droughts (Palmer, 1965) which causes water scarcity (Langridge et al., 2006; Vörösmarty et al., 2010) while excessive amount of rainfall causes flooding (Bhalme and Mooley, 1980). Extreme rainfall in 2017 led to widespread socioeconomic damages around the globe (O'Neill et al., 2017). Pakistan's economy was also severely damaged because of severe flooding caused by monsoon rains in 2010, especially infrastructure and crops (News, 2010). The estimated cost of damage in 2010 to the infrastructure was approximately four billion dollars, whereas the damage in agriculture sector accounts for about 500 million dollars (Hicks and Burton, 2010). The total economic damage was approximately 43 billion dollars in 2010 (Mansoor, 2010; Tarakzai, 2010).

Data-intelligent models using past data can offer an accurate solution to project future trends in rainfall (Luk et al., 2001). Machine learning models, which are highly nonlinear, use data that have input features valued for the rainfall forecasting (Nasseri et al., 2008). Chiew et al. (1998) studied rainfall forecasting in Australia using the empirical method. Sharma (2000) developed a nonparametric probabilistic model to forecast seasonal to interannual rainfall

Handbook of Probabilistic Models. https://doi.org/10.1016/B978-0-12-816514-0.00003-5

in Australia. Burlando et al. (1993) used Auto Regressive Moving Average (ARMA) model for short-term rainfall forecasting in the United States. Hung et al. (2009) applied an artificial neural network model for rainfall forecasting in Thailand. Lin et al. (2009) forecasted hourly rainfall using support vector machines in Taiwan. Mason (1998) forecasted seasonal rainfall of South Africa using a nonlinear discriminant analysis model. Nguyen-Huy et al. (2017) developed a copula-statistical rainfall forecasting model in Australia's agro-ecological zones. Some more studies on rainfall forecasting using data-intelligent models can be seen (Baratta et al., 2003; Chau et al., 2005; Folland et al., 1991; Hong and Pai, 2007; Luk et al., 2001; Nasseri et al., 2008; Shukla and Paolino, 1983; Xiong et al., 2001).

Few studies have been carried out in Pakistan regarding rainfall forecasting. Salma et al. (2012) forecasted rainfall trends in different climate zones of Pakistan using the ARIMA model. Archer and Fowler (2008) used meteorological data to forecast seasonal runoff on the River Jhelum, Pakistan, on the basis of multiple linear regression models. Reale et al. (2012) forecasted an extreme rainfall event (Indus River Valley, Pakistan, 2010) with global data assimilation and forecast model. Faisal and Gaffar (2012) used the Thiessen polygon method of weighted rainfall forecast in Pakistan. Ahasan and Khan (2013) simulated flood-producing rainfall event in 2010 over north-west Pakistan using the Weather Research forecasting (WRF) model.

All the previous studies indicate that rainfall forecasting has been mostly based on regression models. In addition to that, all these studies have been conducted for a large area, either for a whole province or national region, but not for a small locality. Moreover, there is a limitation of applying advanced data-intelligent algorithms for more accurate forecasting models at a micro-scale which can help with decision-making for better management in future to reduce the risk. To address these issues, there is an apparent need for data-intelligent models to forecast rainfall more accurately and at a much finer scale, as attempted previously. In this study, for the first time, Markov Chain Monte Carlo (MCMC)-based bivariate statistical copula models have been developed for rainfall forecasting in Faisalabad, Multan, Jhelum, and Peshawar in Pakistan. The novelty of this study is to use, yet untested, accurate copula models for rainfall forecasting in Pakistan.

To advance the application of copula models, this study will (1) apply MCMC-based statistical bivariate 25 copula models to determine the most accurate model for rainfall forecasting in Pakistan; (2) incorporate the antecedent significant lag (t-1) to forecast rainfall effectively the rainfall of consequent month; (3) validate the forecasting ability of each bivariate copula model for rainfall forecasting. The rest of the chapter is structured in the following way. In the next section, the forecasting MCMC-based statistical bivariate copula models considered in this chapter are described. In Section 3, the description of the experimental data set is presented. Furthermore, the model's development and forecasting skills metrics are also presented in the same Section 3. The application and results are discussed in Section 4. Section 5 presents the discussion and limitation of this work. Finally, the conclusion and remarks are presented in the last section.

3.2 Theoretical framework

3.2.1 Markov Chain Monte Carlo—based statistical copula models

Basically, copula is a set of mathematical tools that have the ability to connect two or more time-independent variables (Nelsen, 2003). A copula function is basically a mathematical function that is defined from $I^2(F, G)$ to $I(H)$ such that $[F(x), G(y), H(x, y)]$ is a point in I^3 with $I \in [0, 1]$, and X, Y are continuous random variables with distribution functions $F(x) = P(X \leq x)$ and $G(y) = P(Y \leq y)$, and $H(x, y) = P(X \leq x, Y \leq y)$ is a function that describes their joint distribution. In this chapter, we will use 25 types of copulas that are shown in the following equations where

I. the Gaussian copula (Li et al., 2013) is expressed as:

$$\int_{-\infty}^{\phi^{-1}(a)} \int_{-\infty}^{\phi^{-1}(b)} \frac{1}{2\pi\sqrt{1-\theta^2}} \exp\left(\frac{2\theta xy - x^2 - y^2}{2(1-\theta^2)}\right) dxdy, \quad \theta \in [-1, 1] \qquad (3.1)$$

II. a student t-copula (Li et al., 2013) can be formulated as:

$$\int_{-\infty}^{t_{\theta_2}^{-1}(a)} \int_{-\infty}^{t_{\theta_2}^{-1}(b)} \frac{\Gamma\left((\theta_2 + 2)/2\right)}{\Gamma\left(\theta_2/2\right)\pi\theta_2\sqrt{1-\theta_1^2}} \left(1 + \frac{x^2 - 2\theta_1 xy + y^2}{\theta_2}\right)^{\frac{(\theta_2+2)}{2}} \qquad (3.2)$$

$$dxdy, \theta_1 \in [-1, 1], \theta_2 \in (0, \infty)$$

III. a Clayton copula (Clayton, 1978) can be written as:

$$\max\left(a^{-\theta} + b^{-\theta} - 1, 0\right)^{-1/\theta}, \quad \theta_2 \in [-1, \infty)/0 \qquad (3.3)$$

IV. a Frank copula (Li et al., 2013) has the following mathematical formulation:

$$-\frac{1}{\theta}\ln\left(1 + \frac{(\exp(-\theta a) - 1)(\exp(-\theta b) - 1)}{\exp(-\theta) - 1}\right), \quad \theta \in \mathbb{R}/0 \qquad (3.4)$$

V. a Gumble copula (Li et al., 2013) can be expressed as:

$$\exp\left(-\left(\left((-\ln(a))^\theta\right)+(-\ln(b))^\theta\right)^{\frac{1}{\theta}}\right), \quad \theta \in [1, \infty)$$ (3.5)

VI. a Fischer-Hinzmann copula (Fischer and Hinzmann, 2006) can be given as:

$$\left[\theta_1(\min(a,b))^{\theta_2} + (1-\theta_1)(ab)^{\theta_2}\right]^{\frac{1}{\theta_2}}, \quad \theta_1 \in [0,1], \theta_2 \in \mathbb{R}$$ (3.6)

In all types of copula-based models, an unknown process κ links observation $\widetilde{Y}$ to parameters θ^* in the modeling inference analysis (Mojtaba Sadegh, 2017) and can be given through the following equation:

$$\widetilde{Y} = \kappa(\theta^*) + \xi$$ (3.7)

where ξ indicates a vector of measurement errors. The vector $e = \widetilde{Y} - Y$ is called the error residual and e = {e1,e2,...,en} where n is the number observations that include the effects of model structural errors (Mojtaba Sadegh, 2017). Bayesian analysis is going to be carried for model inference and uncertainty quantification purposes because Bayesian analysis quantifies uncertainty with a probability distribution (Mojtaba Sadegh, 2017).

Bayes' law attributes all modeling uncertainties to the parameters and estimates the posterior distribution of model parameters by the following equations (Mojtaba Sadegh, 2017):

$$p(\theta|\widetilde{Y}) = \frac{p(\theta)p(\widetilde{Y}|\theta)}{p(\widetilde{Y})}$$ (3.8)

where $p(\theta)$ and $p(\theta|\widetilde{Y})$ defines prior and posterior distribution of parameters, respectively. Further, $p(\widetilde{Y}|\theta) \simeq L(\theta|\widetilde{Y})$ denotes the likelihood given as,

$$L(\theta|\widetilde{Y}) = \frac{n}{2}\ln\frac{\sum_{i=1}^{n}\left[\widetilde{Y}_i - y_i(\theta)\right]^2}{n}.$$ (3.9)

To solve Eq. (3.9) analytically and numerically, a MCMC simulation technique will be adopted to sample from the posterior distribution. For more details, readers are referred to the studies by Mojtaba Sadegh (2017), and Mumtaz Ali et al., (2018). Literature on MCMC algorithms includes the study by Andrieu and Thoms (2008), Duan et al. (1993), Gelman and Rubin (1992), Gilks et al. (1994), Haario et al. (1999), (2001), Roberts and Rosenthal (2009), Roberts and Sahu (1997), Storn and Price (1995), Storn and Price (1997), Ter Braak (2006), ter Braak and Vrugt (2008). The modelling strategy designed in this chapter can be seen in Figure 3.1.

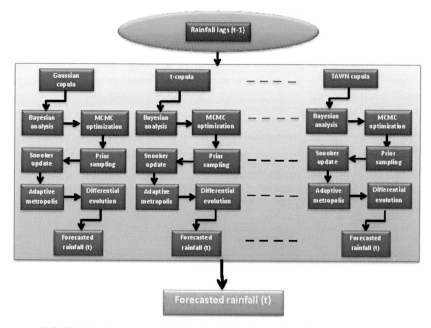

FIGURE 3.1   Flow chart of the Markov Chain Monte Carlo-based copula model.

3.3 Materials and method

3.3.1 Rainfall data set and study region

The rainfall data were obtained from the Pakistan Meteorological Department, Pakistan, for the year 1981−2013 (PMD, 2016). The missing values of monthly rainfall were substituted by average of the respective time-averaged value from the climatological period.

The selected region for this study is Peshawar district of Khyber Pakhtunkhwa (KP), Pakistan, which can be seen on the map in Fig. 3.2.

Peshawar (34.0151 degrees N, 71.5249 degrees E) is the capital of Khyber Pakhtunkhwa (KP) province, Pakistan. Peshawar features a hot semi-arid climate with very hot summers and mild winters. The highest amount of winter rainfall, measuring 236 mm, was seen in February 2007 (PMD, 2010, 2016), while the highest summer rainfall of 402 mm was observed in July 2010 (PMD, 2010). Based on a 30-year record, the average annual precipitation level was recorded as 400 mm, and the highest annual rainfall level of 904.5 mm was recorded in 2003 (PMD, 2016). Wheat, barley, millet, corn, cotton, peppers, and sugarcane are the primary crops in Peshawar.

Table 3.1 describes some statistics of the rainfall data for the designated region which have been used during the development of the forecasting models presented in this research. The rainfall at significant lag $(t-1)$ was used

FIGURE 3.2 Map of study regions.

as the significant input to develop the new MCMC-based statistical bivariate copula models.

3.3.2 Model performance criteria

To evaluate the performance of MCMC-based bivariate copula models, we used different types of statistical metrics. The mathematical formulations of all these assessment metrics are given as follows:

I. the Likelihood value (Max$_L$) (Thyer et al., 2009) is calculated as:

$$\max{}_L = -\frac{n}{2}\ln(2\pi) - \frac{n}{2}\ln\widetilde{\sigma}^2 - \frac{1}{2}\widetilde{\sigma}^{-2}\sum_{i=1}^{n}\left[R_{obs,i} - R_{for,i}\right]^2 \tag{3.10}$$

II. the Akaike information criterion (AIC) (Akaike, 1974) is given by:

$$AIC = 2D + n.\ln - \left(\frac{\sum_{i=1}^{n}\left[R_{obs,i} - R_{for,i}\right]^2}{n}\right) - 2CS \tag{3.11}$$

TABLE 3.1 Descriptive statistics of the study site.

Station	Geographic characteristics			Rainfall (mm)					
	Longitude	Latitude	Elevation (m)	Mean	Std.	Min	Max	Skewness	Kurtosis
Peshawar	34.01°	71.52°	331	44.97	48.71	1.0	409.0	2.60	10.67

III. the Bayesian information criterion (BIC) (Schwarz, 1978) is given by:

$$BIC = D.\ln + n.\ln - \left(\frac{\sum\limits_{i=1}^{n}\left[R_{obs,i} - R_{for,i}\right]^2}{n}\right) - 2CS \qquad (3.12)$$

where R_{obs} and R_{for} are the observed and simulated ith value of rainfall and n is the number of data points. The maximum likelihood (Max_L) minimizes the residuals between model simulations and observations. AIC, in contrast to the ad hoc likelihood value, takes into account both complexity of the model and minimization of error residuals and provides a more robust measure of quality of model predictions. A lower AIC value associates with a better model fit. Similar to AIC, a lower BIC value is associated with a better model fit.

3.4 Results

The results of MCMC-based statistical bivariate copula models have been evaluated based on the aforementioned criterion (Eqs. 3.10–3.12).

Fig. 3.3(A–D) displays a scatter plot showing the goodness of fit, and its correlation coefficient r is shown to depict the extent of agreement between forecasted and observed rainfall. The MCMC-based Farlie-Gumbel-Morgenstern (FGM), Gaussian, and Fischer-Kock copula models show better performance for Peshawar station than the rest of the tested MCMC-based copula models. Overall, all the 25 MCMC-based statistical bivariate copula models convincingly outperform in this study regions that can be confirmed by attaining the larger r-value.

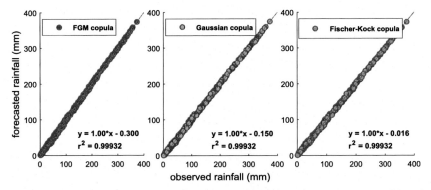

FIGURE 3.3 Scatter plot of forecasted and observed rainfall for the best three MCMC-based copula models using the significant lag at (t-1) with the coefficient of determination (r^2) inserted in each panel for the study zone Peshawar. *MCMC*, Markov Chain Monte Carlo.

TABLE 3.2 Local and MCMC parameters with 95% confidence of interval (CI) of MCMC-based copula models to forecast rainfall for the station Peshawar.

	Significant Lags (t-1)									
	MCMC based copula models									
Models	Copula	Local para 1	Local para 2	Local para 2	MCMC para 1	MCMC para 2	MCMC para 3	95% CI, Local MCMC para 1	95% CI, Local MCMC para 2	95% CI, Local MCMC para 3
	Peshawar									
M₁	Gaussian	0.10			0.13			[0.12 0.13]		
M₂	t	0.11	3001605.9		0.13	35.00		[0.12 0.14]	[21.15 34.86]	
M₃	Clayton	0.15			0.17			[0.15 0.18]		
M₄	Frank	0.66			0.74			[0.70 0.78]		
M₅	Gumbel	1.04			1.09			[1.08 1.09]		
M₆	AMH	0.33			0.33			[0.31 0.34]		
M₇	Joe	1.15			1.15			[1.14 1.16]		
M₈	FGM	0.37			0.37			[0.35 0.38]		
M₉	Gumbel-Barnet	0.00			0.00			[0.00 0.00]		
M₁₀	Plackett	1.45			1.45			[1.42 1.47]		
M₁₁	Cuadras-Auge	0.15			0.15			[0.13 0.15]		
M₁₂	Raftery	0.10			0.11			[0.01 0.99]		
M₁₃	Shih-Louis	0.11			0.11			[0.10 0.12]		
M₁₄	Linear-Spearman	0.11			0.11			[0.10 0.11]		
M₁₅	Cubic	-0.03			-0.03			[-0.32 0.24]		
M₁₆	Burr	5.65			5.65			[5.33 5.99]		
M₁₇	Nelsen	34.79			0.74			[0.70 0.77]		
M₁₈	Galambos	0.32			0.32			[0.30 0.32]		
M₁₉	M-O	5.30	0.06		0.20	0.11		[0.16 0.21]	[0.10 0.12]	
M₂₀	F-H	0.17	-0.75		0.17	-0.76		[0.15 0.18]	[-1.22 -0.31]	
M₂₁	Roch-Alegre	1.12	1.06		1.12	1.06		[1.07 1.19]	[1.03 1.07]	
M₂₂	Fischer-Kock	1.00	0.37		1.00	0.37		[1.00 1.09]	[0.35 0.39]	
M₂₃	BB1	0.04	1.07		0.04	1.07		[0.02 0.05]	[1.06 1.08]	
M₂₄	BB5	1.09	0.06		1.09	0.03		[1.041.09]	[0.00 0.23]	
M₂₅	TAWN	0.33	0.24	1.51	0.37	0.27	1.42	[0.22 0.61]	[0.17 0.59]	[1.17 1.92]

The parameters with CI of best mode are boldfaced (red). *MCMC*, Markov Chain Monte Carlo.

Table 3.2 shows the values of local and MCMC parameters with 95% confidence interval (CI) of MCMC-based bivariate copula models to forecast rainfall prediction for the study regions. The MCMC-based bivariate FGM-copula (M_8) appears to be the best on the basis of the MCMC optimization technique for Peshawar station, followed by Fischer-Kock (M_{22}) and Gaussian (M_1) copula models (Table 3.2). The values of local (parameter $1 \approx 0.37$) and MCMC (parameter $1 \approx 0.37$) with CI: 0.35 to 0.38 for the FGM-copula model are attained for Peshawar. Overall, all the MCMC-based bivariate copula models performed very well in forecasting rainfall for this selected study region.

Fig. 3.4 (A−C) plots a comparison boxplot of the best three MCMC-based bivariate copula models. The outliers specified by + in every boxplot represent the extreme magnitudes of the forecasting error (FE) within the testing months along with their upper quartile, median, and lower quartile values. To forecast

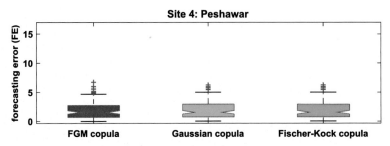

FIGURE 3.4 Boxplots of forecasted error (FE) of best three MCMC-based bivariate copula models of monthly forecasted rainfall in Peshawar. *MCMC*, Markov Chain Monte Carlo.

rainfall for the Peshawar station, the MCMC-based FGM, Gaussian, and Fischer-Kock copula models appeared to be best suitable than the rest of the models. By observing Fig. 3.4, the accuracy of MCMC-based copula models for Peshawar stations appeared to be good.

Table 3.3 presents the preciseness of different MCMC-based bivariate copula models, evaluated for Peshawar station in terms of *AIC, BIC*, and *Max$_L$*. The MCMC-based FGM-copula attained the highest values of *AIC $\approx$ -4167.1*, *BIC $\approx$ -4163.1*, and *Max$_L \approx$ 2084.5*, followed by MCMC-based Gaussian copula model where *AIC $\approx$ -4166.4*, *BIC $\approx$ -4162.4*, and *Max$_L \approx$ 2084.2* and MCMC-based Fischer-Kock copula model with *AIC $\approx$ -4165.0*, *BIC $\approx$ -4157.0*, and *Max$_L \approx$ 2084.5*.

Fig. 3.5 illustrates a polar-plot of the best three MCMC-based copula models FE between the forecasted and observed rainfall in the selected study sites. The distributed FE is justified by these polar-plots showing a much smaller value on x-axis and was achieved by MCMC-based FGM-copula model compared with MCMC-based Gaussian and Fischer-Kock copula models.

3.5 Discussion: limitations and opportunity for further research

To address the water scarcity and disruption of rainfall in the agricultural sector caused by climate change and increase in demand of water resources in Pakistan, there is a desire to develop strong predictive models for rainfall. This chapter is dedicated to modeling the monthly antecedent significant rainfall lags as a predictor to forecast the future rainfall in Pakistan using data from 1981 to 2015. This study aimed to validate MCMC-based copula models and evaluated its predictive ability. This study can be extended to other locations where rainfall data are available to provide the accurate estimation of future rainfall. In spite of the superior and accurate performance, it is to be noted that the MCMC-based copula models should be trialed on larger data sets of rainfall which would be of great interest to Government policy-makers and

TABLE 3.3 Evaluation of MCMC-based copulas models using AIC, BIC, and Max$_L$ to forecast rainfall for the station Peshawar.

Significant Lags (t-1)				
Models	Copula	AIC	BIC	Max$_L$
M$_1$	Gaussian	-4166.4	-4162.4	2084.2
M$_2$	t	-4156.7	-4148.7	2080.4
M$_3$	Clayton	-4000.7	-3996.7	2001.4
M$_4$	Frank	-4164.8	-4160.8	2083.4
M$_5$	Gumbel	-4144.6	-4140.6	2073.3
M$_6$	AMH	-4142.3	-4138.3	2072.1
M$_7$	Joe	-4053.2	-4049.2	2027.6
M$_8$	FGM	-4167.1	-4163.1	2084.5
M$_9$	Gumbel-Barnet	-3503.7	-3499.6	1752.8
M$_{10}$	Plackett	-4163.4	-4159.4	2082.7
M$_{11}$	Cuadras-Auge	-4074.5	-4070.5	2038.2
M$_{12}$	Raftery	-3901.7	-3897.7	1951.9
M$_{13}$	Shih-Louis	-4012.7	-4008.7	2007.3
M$_{14}$	Linear-Spearman	-4012.7	-4008.7	2007.3
M$_{15}$	Cubic	-3503.7	-3499.7	1752.9
M$_{16}$	Burr	-4112.8	-4108.8	2057.4
M$_{17}$	Nelsen	-4164.8	-4160.8	2083.4
M$_{18}$	Galambos	-4138.1	-4134.1	2070.1
M$_{19}$	Marshal-Olkin	-4104.2	-4096.2	2054.1
M$_{20}$	Fischer-Hinzmann	-4083.6	-4075.6	2043.8
M$_{21}$	Roch-Alegre	-4163.9	-4155.8	2083.9
M$_{22}$	Fischer-Kock	-4165.0	-4157.0	2084.5
M$_{23}$	BB1	-4157.2	-4149.2	2080.6
M$_{24}$	BB5	-4142.6	-4134.5	2073.3
M$_{25}$	TAWN	-4148.6	-4136.6	2077.3

The best model is boldfaced (red). *AIC*, Akaike information criterion; *BIC*, Bayesian information criterion; *Max$_L$*, maximum likelihood; *MCMC*, Markov Chain Monte Carlo.

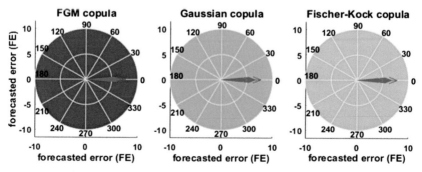

FIGURE 3.5 Polar-plots of forecasted error (FE) in testing period of best three MCMC-bivariate copula models of monthly forecasted rainfall for Peshawar. *MCMC*, Markov Chain Monte Carlo.

agricultural engineers to avoid the possibility of incorrect estimation in the future (Akhtar, 2014; Niaz, 2014). This model can be used to forecast rainfall in the future where the forecasting of rainfall will likely become even more important because of increasing demand for water resources and for agricultural growth in terms of irrigation.

Owing to the aforementioned qualities of the MCMC-based copula models, it is possible to apply the proposed MCMC-copula model for the forecasting of other climatological parameters such as temperature, humidity, wind speed, solar radiation, and soil moisture. As Pakistan is currently suffering from global warming, heatwave and drought forecasting can be another possible work in a follow-up study. Another possible study is to use several types of climatological parameters to forecast rainfall. The rainfall data of neighboring stations may be used to forecast the rainfall in a central station which shares a common boundary to the predictor stations.

In terms of model optimization for rainfall forecasting, it is believed that the hybridization of different models can generate better estimation than standalone models. Therefore, the MCMC-based copula models could be optimized with ensemble methods (Dietterich, 2002; Lei and Wan, 2012; Yun et al., 2008) to achieve more accurate results, improving on our 10% level of uncertainty. The Adaptive neuro-fuzzy inference system (ANFIS) that are very powerful can be another optimization method to be considered in this regard (Jang, 1993). Moreover, the more advanced models such as the ensemble method (Dietterich, 2002), particle swarm optimization (Chen and Yu, 2005; Zhisheng, 2010), genetic algorithms (Davis, 1991), chaos theory (Briggs and Peat, 1989) etc. coupled with copula (Nelsen, 2003) may generate good results. The regression models (Draper and Smith, 2014) can be coupled with copula in a way to optimize the forecasting ability of possible hybrid copula models. A more generalized framework of hybridizing copula with generalized mixed models (Draper and Smith, 2014) to develop MCMC-copula mixed models for rainfall is expected in the upcoming work. Autoregressive fractionally integrated moving average (Ling and Li, 1997)-based copula (ARFIMA copula) and least square support vector machine–based copula (LSSVM copula) models can be used to forecast rainfall. Support vector machine (Cortes and Vapnik 1995), extreme learning machine (Huang et al., 2003), etc. may also be very good options for rainfall forecasting in Pakistan. As the standard statistical approaches avoid the hurdle of model uncertainty that leads to overconfident inferences and more risky decisions, Bayesian model averaging (BMA) techniques (Hoeting et al., 1998; Raftery et al., 2005) have the ability to model uncertainty for accurate predictions. Therefore, BMA techniques provide yet another option to be used to model uncertainty in rainfall that is due to several factors such as missing climate data, extreme weather conditions, and the likely influence of climate change.

Multiresolution tools such as frequency resolution can also be applied in this area to broaden the scope of this study. In this regard, wavelet transformation (maximum overlap discrete wavelet) (Holschneider, 1988; Khalighi et al., 2011), empirical mode decomposition (Rilling et al., 2003), and singular value decomposition (De Lathauwer et al., 1994) can be used for prediction purposes. Other possible option could be the use of feature selection techniques (Guyon and Elisseeff, 2003; Jain and Zongker, 1997) to forecast rainfall.

3.6 Conclusion

This chapter developed MCMC-based statistical bivariate copula models using the significant antecedent lags of monthly rainfall as the predictor to forecast future rainfall for Peshawar in Pakistan. The rainfall data from 1981 to 2013 were used to develop MCMC-based 25 copula models to achieve a high level of accuracy. Furthermore, the best three MCMC-based copula models for each station were picked for comparison purposes.

To attain an accurate MCMC-based copula model, the MCMC algorithm adopted global optimization as well as local optimization technique to find the best copula parameters. Evidently, the performance of all the MCMC-based copula models was found to be very much better (Table 3.3). The better performance of the MCMC-copula models is evidenced in terms of low (AIC, BIC) and high (Max_L) performance metrics.

To enhance the forecasting accuracy, the MCMC-based copula models can be optimized and tuned with other advanced techniques such as ensemble methods, generalized mixed models to develop MCMC-copula mixed model for rainfall forecasting in upcoming work. ARFIMA copula and LSSVM copula models can be used to forecast rainfall. This study provides a baseline relevant with other models such as ANFIS, SVM, ELM, etc. being potentially used to forecast the rainfall and other climatological parameters more accurately in future studies.

Acknowledgments

This research used rainfall data that were acquired from Pakistan Meteorological Department, Pakistan, and duly acknowledged. This study was supported by the University Of Southern Queensland Office Of Graduate Studies Postgraduate Research Scholarship (2017−19).

References

Ahasan, M., Khan, A., 2013. Simulation of a flood producing rainfall event of 29 July 2010 over north-west Pakistan using WRF-ARW model. Natural Hazards 69 (1), 351−363.

Akaike, H., 1974. A new look at the statistical model identification. IEEE Transactions on Automatic Control 19, 716−723.

Akhtar, I.U.H., 2014. Pakistan Needs a New Crop Forecasting System.

Andrieu, C., Thoms, J., 2008. A tutorial on adaptive MCMC. Statistics and Computing 18, 343−373.

Archer, D.R., Fowler, H.J., 2008. Using meteorological data to forecast seasonal runoff on the river Jhelum, Pakistan. Journal of Hydrology 361, 10−23.

Baratta, D., Cicioni, G., Masulli, F., Studer, L., 2003. Application of an ensemble technique based on singular spectrum analysis to daily rainfall forecasting. Neural Network 16, 375−387.

Barredo, J.I., 2007. Major flood disasters in Europe: 1950−2005. Natural Hazards 42, 125−148.

Bhalme, H.N., Mooley, D.A., 1980. Large-scale droughts/floods and monsoon circulation. Monthly Weather Review 108, 1197−1211.

Briggs, J., Peat, F.D., 1989. Turbulent Mirror: An Illustrated Guide to Chaos Theory and the Science of Wholeness. HarperCollins Publishers.

Burlando, P., Rosso, R., Cadavid, L.G., Salas, J.D., 1993. Forecasting of short-term rainfall using ARMA models. Journal of Hydrology 144 (1−4), 193−211.

Chau, K., Wu, C., Li, Y., 2005. Comparison of several flood forecasting models in Yangtze river. Journal of Hydrologic Engineering 10, 485−491.

Chen, G.-C., Yu, J.-S., 2005. Particle swarm optimization algorithm. Information and Control-Shenyang 34 (3), 318.

Chiew, F.H., Piechota, T.C., Dracup, J.A., McMahon, T.A., 1998. El Nino/Southern oscillation and Australian rainfall, streamflow and drought: links and potential for forecasting. Journal of Hydrology 204, 138−149.

Clayton, D.G., 1978. A model for association in bivariate life tables and its application in epidemiological studies of familial tendency in chronic disease incidence. Biometrika 65, 141−151.

Cortes, C., Vapnik, V. J. M. l., 1995. Support-Vector Networks 20 (3), 273−297.

Davis, L., 1991. Handbook of Genetic Algorithms.

De Lathauwer, L., De Moor, B., Vandewalle, J., Higher-Order, B.S.S., 1994. Singular value decomposition. Proceedings of EUSIPCO-94, Edinburgh, Scotland, UK, pp. 175−178.

Dietterich, T.G., 2002. Ensemble learning. In: The Handbook of Brain Theory and Neural Networks, second ed., pp. 110−125.

Draper, N.R., Smith, H., 2014. Applied Regression Analysis. John Wiley & Sons.

Duan, Q., Gupta, V.K., Sorooshian, S., 1993. Shuffled complex evolution approach for effective and efficient global minimization. Journal of Optimization Theory and Applications 76, 501−521.

Faisal, N., Gaffar, A., 2012. Development of Pakistan's new area weighted rainfall using thiessen polygon method. Pakistan Journal of Meteorology 09 (07) (Pakistan).

Fischer, M.J., Hinzmann, G., 2006. A new class of copulas with tail dependence and a generalized tail dependence estimator. Diskussionspapiere//Friedrich-Alexander-Universität Erlangen-Nürnberg, Lehrstuhl für Statistik und Ökonometrie 77.

Folland, C., Owen, J., Ward, M.N., Colman, A., 1991. Prediction of seasonal rainfall in the sahel region using empirical and dynamical methods. Journal of Forecasting 10, 21−56.

Gelman, A., Rubin, D.B., 1992. Inference from iterative simulation using multiple sequences. Statistical Science 457−472.

Gilks, W.R., Roberts, G.O., George, E.I., 1994. Adaptive direction sampling. The Statistician 179−189.

Guyon, I., Elisseeff, A., 2003. An introduction to variable and feature selection. Journal of Machine Learning Research 3, 1157−1182.

Haario, H., Saksman, E., Tamminen, J., 1999. Adaptive proposal distribution for random walk Metropolis algorithm. Computational Statistics 14, 375−396.

Haario, H., Saksman, E., Tamminen, J., 2001. An adaptive Metropolis algorithm. Bernoulli 7, 223−242.

Hicks, M.J., Burton, M.L., 2010. Preliminary Damage Estimates for Pakistani Flood Events, 2010. Center for Business and Economic Research, Ball State University.

Hoeting, J.A., Madigan, D., Raftery, A.E., Volinsky, C.T., 1998. Bayesian model averaging. Proceedings of the AAAI Workshop on Integrating Multiple Learned Models 77−83.

Holschneider, M., 1988. On the wavelet transformation of fractal objects. Journal of Statistical Physics 50 (5), 963−993.

Hong, W.-C., Pai, P.-F., 2007. Potential assessment of the support vector regression technique in rainfall forecasting. Water Resources Management 21, 495−513.

Huang, G.-B., 2003. Learning capability and storage capacity of two-hidden-layer feedforward networks. IEEE Transactions on Neural Networks 14 (2), 274−281.

Hung, N.Q., Babel, M.S., Weesakul, S., Tripathi, N., 2009. An artificial neural network model for rainfall forecasting in Bangkok, Thailand. Hydrology and Earth System Sciences 13 (8), 1413−1425.

Jain, A., Zongker, D., 1997. Feature selection: evaluation, application, and small sample performance. IEEE Transactions on Pattern Analysis and Machine Intelligence 19 (2), 153−158.

Jang, J.-S., 1993. ANFIS: adaptive-network-based fuzzy inference system. IEEE Transactions on Systems, Man, and Cybernetics 23 (3), 665−685.

Khalighi, S., Sousa, T., Oliveira, D., Pires, G., Nunes, U., 2011. Efficient feature selection for sleep staging based on maximal overlap discrete wavelet transform and SVM. Annual International Conference of the IEEE Engineering in Medicine and Biology Society (EMBC), IEEE 3306−3309.

Langridge, R., Christian-Smith, J., Lohse, K., 2006. Access and resilience: analyzing the construction of social resilience to the threat of water scarcity. Ecology and Society 11.

Lei, K.S., Wan, F., 2012. Applying ensemble learning techniques to ANFIS for air pollution index prediction in Macau. International Symposium on Neural Networks, Springer 509−516.

Li, C., Singh, V.P., Mishra, A.K., 2013. A bivariate mixed distribution with a heavy−tailed component and its application to single−site daily rainfall simulation. Water Resources Research 49, 767−789.

Lin, G.F., Chen, G.R., Wu, M.C., Chou, Y.C., 2009. Effective forecasting of hourly typhoon rainfall using support vector machines. Water Resources Research 45 (8).

Ling, S., Li, W., 1997. On fractionally integrated autoregressive moving-average time series models with conditional heteroscedasticity. Journal of the American Statistical Association 92 (439), 1184−1194.

Luk, K.C., Ball, J.E., Sharma, A., 2001. An application of artificial neural networks for rainfall forecasting. Mathematical and Computer modelling 33, 683−693.

Mansoor, H., 2010. Pakistan Evacuates Thousands in Flooded South.

Mason, S., 1998. Seasonal forecasting of South African rainfall using a non−linear discriminant analysis model. International Journal of Climatology 18 (2), 147−164.

Mumtaz Ali, R.C.D., Downs, N.J., Maraseni, T., 2018. Multi-stage hybridized online sequential extreme learning machine integrated with Markov Chain Monte Carlo copula-Bat algorithm for rainfall forecasting. Atmospheric Research 218.

Nasseri, M., Asghari, K., Abedini, M., 2008. Optimized scenario for rainfall forecasting using genetic algorithm coupled with artificial neural network. Expert Systems with Applications 35, 1415−1421.

Nelsen, R.B., 2003. Properties and applications of copulas: a brief survey. In: Dhaene, J., Kolev, N., Morettin, P.A. (Eds.), Proceedings of the First Brazilian Conference on Statistical Modeling in Insurance and Finance. University Press USP: Sao Paulo, pp. 10−28.

News, D., 2010. Floods to Hit Economic Growth: Finance Ministry. Dawn News.

Nguyen-Huy, T., Deo, R.C., An-Vo, D.-A., Mushtaq, S., Khan, S., 2017. Copula-statistical precipitation forecasting model in Australia's agro-ecological zones. Agricultural Water Management 191, 153–172.

Niaz, M.S., 2014. Wheat Policy—A Success or Failure. DAWN Newspaper.

O'Neill, B.C., et al., 2017. IPCC Reasons for Concern Regarding Climate Change Risks 7 (1), 28.

Palmer, W.C., 1965. Meteorological drought. US Department of Commerce. Weather Bureau Washington. DC.

Palmer, W.C., 1965. Meteorological Drought. US Department of Commerce, Weather Bureau, Washington, DC.

PMD, 2010. Rainfall Statement. Pakistan Metoerological Department.

PMD, 2016. Pakistan Meteorological Department (Pakistan).

Raftery, A.E., Gneiting, T., Balabdaoui, F., Polakowski, M., 2005. Using Bayesian model averaging to calibrate forecast ensembles. Monthly Weather Review 133 (5), 1155–1174.

Reale, O., Lau, K., Susskind, J., Rosenberg, R., 2012. AIRS impact on analysis and forecast of an extreme rainfall event (Indus River Valley, Pakistan, 2010) with a global data assimilation and forecast system. Journal of Geophysical Research: Atmosphere 117 (D8).

Rilling, G., Flandrin, P., Goncalves, P., 2003. On empirical mode decomposition and its algorithms. In: IEEE-EURASIP Workshop on Nonlinear Signal and Image Processing, IEEER, Grado, Italy, pp. 8–11.

Roberts, G.O., Rosenthal, J.S., 2009. Examples of adaptive MCMC. Journal of Computational and Graphical Statistics 18, 349–367.

Roberts, G.O., Sahu, S.K., 1997. Updating schemes, correlation structure, blocking and parameterization for the Gibbs sampler. Journal of the Royal Statistical Society: Series B (Statistical Methodology) 59, 291–317.

Sadegh, M., Ragno, E., AghaKou, A., 2017. Multivariate copula analysis toolbox (MvCAT): describing dependence and underlying uncertainty using a bayesian framework. Water Resources Research 53, 17.

Salma, S., Shah, M., Rehman, S., 2012. Rainfall trends in different climate zones of Pakistan. Pakistan Journal of Meteorology 9.

Schwarz, G., 1978. Estimating the dimension of a model. The Annals of Statistics 6, 461–464.

Sharma, A., 2000. Seasonal to interannual rainfall probabilistic forecasts for improved water supply management: part 3—a nonparametric probabilistic forecast model. Journal of Optimization Theory and Applications 239, 249–258.

Shukla, J., Paolino, D.A., 1983. The Southern oscillation and long-range forecasting of the summer monsoon rainfall over India. Monthly Weather Review 111, 1830–1837.

Storn, R., Price, K., 1995. Differential Evolution—A Simple and Efficient Adaptive Scheme for Global Optimization over Continuous Spaces: Technical Report TR-95-012. International Computer Science, Berkeley, California.

Storn, R., Price, K., 1997. Differential evolution—a simple and efficient heuristic for global optimization over continuous spaces. Journal of Global Ooptimization 11, 341–359.

Tarakzai, S., 2010. Pakistan Battles Economic Pain of Floods. Jakarta Globe.

Ter Braak, C.J., 2006. A Markov chain Monte Carlo version of the genetic algorithm differential evolution: easy Bayesian computing for real parameter spaces. Statistics and Computing 16, 239–249.

ter Braak, C.J., Vrugt, J.A., 2008. Differential evolution Markov chain with snooker updater and fewer chains. Statistics and Computing 18, 435–446.

Thyer, M., Renard, B., Kavetski, D., Kuczera, G., Franks, S.W., Srikanthan, S., 2009. Critical evaluation of parameter consistency and predictive uncertainty in hydrological modeling: a case study using Bayesian total error analysis. Water Resources Research 45.

Vörösmarty, C.J., McIntyre, P.B., Gessner, M.O., Dudgeon, D., Prusevich, A., Green, P., Glidden, S., Bunn, S.E., Sullivan, C.A., Liermann, C.R., 2010. Global threats to human water security and river biodiversity. Nature 467, 555−561.

Xiong, L., Shamseldin, A.Y., O'connor, K.M., 2001. A non-linear combination of the forecasts of rainfall-runoff models by the first-order Takagi−Sugeno fuzzy system. Journal of Hydrology 245, 196−217.

Yun, Z., et al., 2008. RBF neural network and ANFIS-based short-term load forecasting approach in real-time price environment. IEEE Transactions on Power Systems 23 (3), 853−858.

Zhisheng, Z., 2010. Quantum-behaved particle swarm optimization algorithm for economic load dispatch of power system. Expert Systems with Applications 37 (2), 1800−1803.

Chapter 4

A model for quantitative fire risk assessment integrating agent-based model with automatic event tree analysis

Farid Wajdi Akashah[1], Rachid Ouache[2], Jianping Zhang[3], Michael Delichatsios[3]

[1]*Centre for Building, Construction and Tropical Architecture (BuCTA), Faculty of Built Environment, University of Malaya, Kuala Lumpur, Malaysia;* [2]*School of Engineering, Faculty of Applied Science, University of British Columbia, Okanagan campus, Kelowna, BC, Canada;* [3]*Fire Safety Engineering Research and Technology Centre (FireSERT), University of Ulster, Newtownabbey, United Kingdom*

1. Introduction

The world has seen tremendous increase in the growing population rate. Worldwide human populace development adds up to around 75 million every year, or 1.13% every year. The world population was 7.349 billion starting July 1, 2015, as indicated by the United Nations Department of Economic and Social Affairs, Population Division. It is relied upon to continue developing at 8.4 billion by mid-2030 and 9.6 billion by mid-2050. Therefore, the government of many countries around the world are establishing new residential buildings to prevent and protect the life of population. The building without taking into account the measures to prevent and protect life of population becomes among the priorities because of the catastrophic accidents that occurred in residential buildings around the world.

Malaysia is one of the countries that have seen catastrophic residential accidents in the last few years, such as fire. Fire accident is one of the catastrophic accidents that have occurred in the residential building, generally in the world and specifically in Malaysia with approximately 200,000 fire-related deaths occurring annually (Nilson et al., 2015). Owing to the increasing number of deaths that occur more globally due to fire in dwellings than other locations, many researchers have established studies for effective analysis of fire accident to protect people, environment, and properties (Harpur et al., 2013).

Handbook of Probabilistic Models. https://doi.org/10.1016/B978-0-12-816514-0.00004-7
107

Jennings (2013) reviewed literature on social, economic, and building stock characteristics as they relate to residential fire risk in urban neighborhoods. Liu and Chow (2014) made a literature review on design fires and fire load survey methods for buildings. Groner (2016) described an occupant movement decision model intended to assist the responsible way for managing a fire emergency by dividing building occupants into groups and recommend an appropriate protection strategy for each group. Ibrahim et al. (2011) reviewed the criteria and attributes of fire risks in building. Tancogne-Dejean and Laclémence (2016) studied fire risk perception and its influence on building evacuation to improve building evacuation processes.

Ran et al. (2014) analyzed the influences of an intelligent evacuation guidance system on crowd evacuation by simulating a fire scene on the experimental platform of the "black house". Li et al. (2016) presented a fire risk analysis in a case study of a six-storey apartment building of light-frame construction, aimed at showing the performance of wall barriers on building fire risk. Martinopoulos et al. (2016) compared the heating systems most commonly used in the residential sector, by using the equipment's efficiency and the lifecycle cost as criteria. Kang et al. (2016) focused on fire prevention and investigation and improvement of the system of full-time security staff. Shazmin et al. (2016) presented the incentives for development green building. Su and Bai (2016) used the background-oriented Schlieren technique to visualize the neutral plane when a fire whirl occurs in a vertical shaft with a single corner gap. Tian et al. (2016) relied on a full-scale in situ experiment and combined with uniform design experiment and nonparametric regression technique to create efficient response surface based on fire dynamic simulation of timber buildings. Ye et al. (2016) investigated different sprinkler system activation modes under different fire scenarios in large commercial buildings.

Crippa et al. (2009) have shown the benefits from the application of the fire risk assessment methodology. Tancogne-Dejean and Laclémence (2016) used qualitative research to study fire risk perception and to understand the attitudes and behaviors of individuals. Pasman and Reniers (2013) presented the story about the development of quantitative risk assessment (QRA) as the best approach compared with qualitative and semiquantitative ones in terms of precision, how it was conceptualized in the early 1970s, and what is its status today. In addition, Vinnem (1998) presented a brief summary of the development of quantified risk analysis in the offshore oil and gas industry for nearly 20 years. QRA is commonly used in the industry to support decision-making where risk is a function of frequency of events (probability) and associated consequences (negative outcomes) (Milazzo et al., 2015) and to demonstrate that risks are as low as reasonably practicable (Schofield 1998). Tong et al. (2016) analyzed the character of the risk of long-distance oil and gas pipeline, and then the fire risk of the long-distance oil and gas pipeline was processed based on QRA. Segui et al. (2014) developed a tool for the quantification of the consequences of toxic dispersions coming from fires in warehouses. Milazzo et al. (2015) presented the application of an extended risk analysis of loss of containments for a case study. Despite the developed models, still there are some points that challenge the researchers such as the following ones: (1) to determine and to calculate the evolving distribution of smoke, fire gases, and temperature throughout compartments of a building during a fire and to simulate the impact of potential fires and smoke in a specific building environment, (2) to predict a possible operational state for each agent for all possible scenarios, and (3) to coordinate interactions between all the agents and determine the risk of the scenarios.

In this study, a methodology for quantitative fire risk assessment using agent-based modeling and event tree analysis is developed to challenge the raised points. The expanded methodology is applied to a two-storey residential building. This chapter is organized as follows: the next section presents the methodology, including coupling deterministic fire modeling with uncertainty analysis, followed by a section that shows application of the developed methodology. Results from the case study are presented and discussed in the subsequent section. Conclusions are drawn in the final section, and suggestions are made for further development of the methodology.

2. Methodology

Fig. 4.1 presents the developed methodology for quantitative fire risk assessment. The methodology involves three main steps: (1) a deterministic model to determine the state of the fire using the Consolidated Model of Fire and Smoke Transport (CFAST) to calculate the evolving distribution of smoke, fire gases, and temperature throughout compartments of a building during a fire and to simulate the impact of potential fires and smoke in a specific building environment, (2) a probabilistic model to predict probability of a possible operational state for each agent for all possible scenarios, and (3) an agent-based model (ABM) to coordinate interactions between all the agents and determine the risk of the scenarios. The outputs from the ABM are processed through a computer algorithm to construct the event tree and subsequently its risk curve. The deterministic fire model produces heat and smoke development data for different fire scenarios, which are used to determine the response of agents and the consequences. The probabilistic model varies the state of each agent based on the output of the deterministic model and the probability of operation of each agent, and the ABM automatically generates all the scenarios based on the input provided by both the deterministic model and the probabilistic model. Specifically, the probabilistic model uses Monte Carlo simulation to process the reliability value of fire safety systems by randomly selecting values according to the probabilistic distributions specified by the user. These values are then passed on to the ABM to generate all scenarios together with their risks, namely their probability and consequence. The output is presented in the form of an event tree and its associated risk curve.

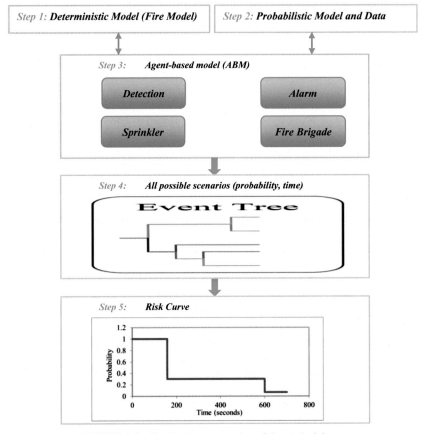

FIGURE 4.1 Graphical representation of the methodology.

The following subsections briefly explain the different components used in the aforementioned methodology starting with deterministic model, probabilistic model and data, the ABM with its agent's attributes and rules, and finally the presentation of results in terms of the event tree and its risk curve.

2.1 Deterministic model

To maintain the efficiency of the methodology in keeping the computational cost at a minimum, CFAST (Jones et al., 2005) was chosen as the deterministic model to model fire and smoke development. CFAST is a two-zone fire model that is based on ideal gas law and combustion modeling set into differential equation format to solve the homogeneous upper layer and lower layer of gas and temperature. Although CFAST was chosen in this study, other fire models including more advanced computational fluid dynamics (CFD) and models, such as fire dynamics simulator (FDS) (McGrattan et al., 2007), may be used.

In this study, the fire model produced the timeline of the temperature profiles of the fire and the activation of both detector and sprinkler systems. The output of the fire model is an input for ABM to be able to run dynamically using both time-driven and message-driven transitions for both interagent and intraagent interactions. The agents specified in ABM will be able to read the output supplied by fire model as both, the fire model and the ABM, were coupled together.

2.2 Probabilistic model and data—@RISK

In performance-based design, the reliability value of fire safety systems is an important input to derive engineering solution to satisfy fire safety objectives. Reliability includes both operational and performance reliability (Bukowski et al., 1999). Availability is a measure of operability of a system or component. Performance reliability is a measure of adequacy of a system or component during its operation. The reliability values used in this article refer to availability value based on selected values presented in.

The values in Table 4.1 are given based on a normal distribution. Other types of distribution can also be used as the Monte Carlo simulation used in this methodology supports more than 30 types of probabilistic distributions.

The Monte Carlo simulation used for this study is @RISK 5.5 (2009). @RISK is an "add-on" program installed in a spreadsheet. Its primary feature is to provide estimation to probability distribution of possible results for each selected output cell in the spreadsheets. In this study, @RISK performs two tasks:

1. It is used to vary the reliability values of fire safety systems.
2. After the ABM has produced all the event trees, it fits the distribution into a probability density function. The probability density functions produced show the effect of varying the reliability value of the fire safety.

TABLE 4.1 Selected values used for reliability value of fire safety systems (BSI, 2003b).

Fire safety systems	Mean, μ	Standard deviation, σ
Detector	0.86	0.035
Alarm	0.81	0.106
Sprinkler	0.90	0.043
Fire brigade	0.90	0.008

2.3 Agent-based model

ABM is one of the tools that are capable of simulating interactions between the components (or agents) of a system where multiple input states and outcomes are possible. It is ideally suited to processes such as simulating a battle of tanks (Fan, 2006), assessing risks for a nuclear plant (Hakobyan et al., 2008), or representing the fire safety engineering environment (e.g., architecture, regulations, search, and rescue) (Rueppel and Lange, 2009). The large and complex systems involved in these processes are deconstructed into interacting smaller components, each represented as an individual agent. By deconstructing the large system under review and providing an appropriate level of detail for the functioning of each agent, the user has the flexibility to customize his/her model according to the level of detail needed and/or available for each agent. This ability that ABM holds is the reason why it has been used across different industries for different applications from modeling a consumer market (Garifullin et al., 2007) to modeling a cooperation support in civil engineering (Rueppel and Lange, 2009) and people's evacuation (Guanquan and Jinhui, 2009; Tang and Ren, 2008).

ABM is a bottom-up modeling where the developer works at a level of an individual agent, sets its internal states, and then moves on to set rules that this agent needs to satisfy during both intraagent and interagent interactions. Some key concepts of ABM are presented as follows:

- Local view—the agent can receive sensory input from its environment, and it can perform actions which change the environment in some way,
- Autonomy—the system has control over its own actions and the internal state without direct intervention of humans, and
- Flexibility—the agents are capable of flexible actions to meet common objectives.

Based on these concepts, ABM has been widely applied in various domains, e.g., the game industry (Lysenko and D'Souza, 2008) and evacuation models (Tang and Ren, 2008). In this work, ABM is applied to perform risk assessment of fire safety systems that consist of an agent Detector, agent Alarm, agent Sprinkler, and agent Fire Brigade. It should be emphasized that there is no limit to the type of agents that can be modeled in this methodology, nor to the number of agents involved. In the study by Akashah et al. (2010), there were five types of agents, four agents mentioned earlier and agent Smoke Vent. In this work, the number of agents involved has increased from only five in a previous study (one for each type of agents) to 18, nine for agent Detector and agent Alarm, seven for agent Sprinkler, and one for agent Fire Brigade. The increased number of agents and the difference in the geometrical domain in which the methodology is applied resulted in the need to have more robust agent behavior rules to govern the process of interagent/intraagent communications that is presented in the next few subsections. The model being represented in this work was developed using AnyLogic (2007).

2.3.1 Attributes of agents involved

Every agent has its own unique behavior. In this methodology, this unique behavior is represented by an internal state of an agent. An example of an internal state that is unique to an agent is its reliability value and the time of its activation. In this application, each agent will have its own reliability value and time of activation. The reliability value of an agent in this application refers to availability value of a particular component of fire safety systems, e.g., availability value of a fire detector system while the time of activation of an agent is the time where a particular component of a fire safety system is activated in the event of fire, e.g., in the event where temperature near a detector reaches 47°C, the detector activates. In an ABM environment, the time of activation indicates the moment an agent becomes active and starts to perform actions within itself (intraagent communications) or with another agent (interagent communications) based on the agent behavior rules (which will be discussed in the next section) laid out by the user. Actions performed by the agents include obtaining the data from deterministic and probabilistic models into ABM, activating different agents and/or different state in an agent by means of passing the data to the subsequent agent, calculating the probability value of a particular scenario, and classifying the consequence of a series of scenarios in an event tree.

The probabilistic model provides the reliability value of each agent in the ABM. It is linked to the ABM through a database. The database contains the reliability values of the fire safety systems generated by the probabilistic model using Monte Carlo simulation. The database also stores the times of activation of fire safety systems involved sourced from the deterministic model. The reliability value and the time of activation of the fire safety systems are generated by the probabilistic model using Monte Carlo simulation. This process was incorporated into the methodology to enable uncertainty analysis to be performed.

2.3.2 Rules governing an agent's interactions for the present case

The interagent interactions are bounded by rules. These rules help to shape the structure and the sequence of events of the event tree that will be produced by the ABM. In AnyLogic, activation of an agent can be event-triggered or time-triggered. In this study, the rules that bound the interagent interactions are as follows:

- Initial event fire can interact directly with agent Detector and agent Sprinkler where both agents can operate if initial event fire reaches their respective activation time;
- Agent Alarm is dependent of agent Detector where agent Alarm will only operate if agent Detector operates;
- Agent Fire Brigade is dependent of agent Alarm where agent Fire Brigade will only operate if agent Alarm operates;

- "Fire" is extinguished once either agent Sprinkler or agent Fire Brigade operates;
- In the event of more than one agent Spinkler operate, the first agent Sprinkler to operate is assumed to extinguish fire. This rule also applies to agent Fire Brigade. As an example, if both agents Sprinkler in room A and in room B operate and agent Fire Brigade operates by signal given from both agents Alarm in room A and in room B, activation times of these agents (in this case four agents) are compared. The first agent to activate is assumed to extinguish the fire;
- If neither agent Sprinkler nor agent Fire Brigade operates, the fire will continue to burn until all the fuel load is consumed;
- Classification of the event consequence is made according to the agent and the location of the agent, i.e., which room in the house does the agent is placed in.

The rules governing agent's interactions will be different from case to case. For example, the rules for this case study are different to the one in the study by Akashah et al. (2010). These rules are different depending on the type of agents, the number of agents, and the characteristics of agents that the user has to deal with.

2.4 Generating automatic event trees—general rules

Once the agent behavior rules have been set up, the process of generating event trees automatically will be based on the following steps:

1. Generate an initial event, i.e., fire
2. Generate the agents involved in processing the initial event fire
3. Generate all possible scenarios as a result of intraagent interactions governed by the agent behavior rules, say, S_i, $i = 0,1, ..., m-1$
4. Each agent, A, is associated with its attributes: reliability value, R, and time of activation, t.
5. For each scenario, S_i,
   (a) Generate the actions performed by agents that lead to the scenario, S_i
   (b) Calculate the probability of S_i happening based on the product of reliability of agents involved.
   (c) Add the sequence of events that lead to scenario S_i into a collection file
   (d) Write the sequence of events into an output file
   (e) Repeat process 5.(a) to 5.(d) until no more events or actions are produced
   (f) If the same events are generated during the process, write only the first unique event produced.

The rules that have been set up are general rules to automatically generate an event tree. The same set of rules were used to produce event trees for the present case and another application in the study by Akashah et al. (2010).

2.5 Presenting the results

The ABM produces an event tree for each set of reliability values varied. The consequences of each event tree were then classified into three different categories: time of sprinkler to extinguish the fire, time of fire brigade to extinguish the fire, and time of fire burn out. For this to happen, a computer program based on formula translation (FORTRAN) was developed. The program is coupled with the ABM and is tasked to get the output from ABM, calculate the total value of probability value of scenarios produced according to the category of consequence being laid out by the user in rules governing agent's interactions, and finally produce risk curve as a two-dimensional plot which shows the probability of a consequence greater than a certain value. After the categorization of consequence is made and calculated, the risk curve is then plotted.

3. Case study

A building being assessed is a two-storey house with an open exterior door. The house has a floor dimension of 10.0×7.0 m and 2.4-m ceiling height for each floor. A 2-m^2-wide staircase connects the ground floor and the first floor of the house. The main door of the house has a dimension of 1.0 m width and 2.0 m height and is located at 0.15 m from the left of room hall. This door remains open throughout the simulation. Other doors throughout the house remain open throughout the simulation. The floor plan of the ground floor and first floor of the house are presented on the left- and right-hand side of Fig. 4.2. Fig. 4.2 also presents the location of the smoke detectors and sprinklers which are denoted by "+" symbol and "x" symbol, respectively. In total, there are 11 smoke detectors and eight sprinklers being installed throughout the house. The activation of the smoke detectors is at 28°C and that of the sprinklers is at 68°C. The sprinklers installed throughout the house are of fast-response type with an Relative temperature index (RTI) value of $35m^{1/2}s^{1/2}$ (Table 4.2).

In this study, a t-squared fire located at the center of the hall with a maximum heat release rate of 4 MW (MW). The simulation is run under a medium fire growth based on the design fire growth rates in the study by BSI, (2003a). During this run, the reliability of fire safety systems was fixed to just a single set of values, based on Table 4.1. Then, the effects of varying the reliability value of fire safety systems involved are examined. Probabilistic model through @RISK is used to randomly pick the reliability value of the detector, alarm, sprinkler, and fire brigade based on a normal distribution as explained earlier in section Probabilistic Model and Data—@RISK; for every set of reliability values selected by the probabilistic model, an event tree is generated. These steps are then repeated for the cases with different fire growth rate, i.e., slow, fast, and ultrafast. The effects of varying the reliability values

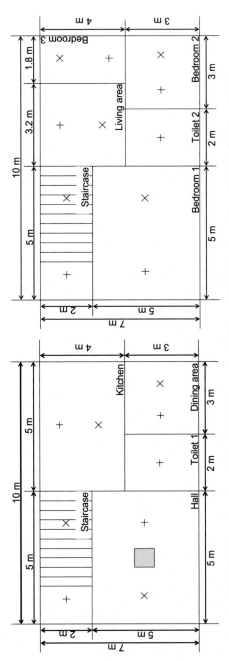

FIGURE 4.2   Floor plan with location of detectors, sprinklers, and fire at the ground floor (left) and first floor (right).

TABLE 4.2 Summary of the floor areas of the case study presented in Fig. 4.2.

Level	Room	Total (m^2)
Ground floor	Hall	25.0
	Kitchen	20.0
	Dining area	9.0
	Toilet 1	6.0
First floor	Living area	12.8
	Bedroom 1	25.0
	Toilet 2	6.0
	Bedroom 2	9.0
	Bedroom 3	7.2
Total		120.0

of fire safety systems and the fire growth rate are then analyzed and presented in the following sections.

4. Results and discussion

The event tree was first produced based on one set of reliability values assigned to fire safety systems. Fig. 4.3 shows the structure of the event tree produced by the ABM. The event tree presented shows the scenarios involving fire safety systems, i.e., detectors, alarms, and sprinklers that are installed in the room of fire origin and in the room adjacent to the room of fire origin as the probability of the fire safety systems in both of these rooms to fails is very small. Based on Fig. 4.3, the consequences of the event tree were then classified into six different categories: time of the sprinkler of the room of fire origin to extinguish the fire, t_{sp1}; time of the sprinkler of the room adjacent to the room of fire origin to extinguish the fire, t_{sp2}; time of the fire brigade came based on notification of an alarm in the room of fire origin to extinguish the fire, t_{fb1}; time of the fire brigade came based on notification of an alarm in the room adjacent to the room of fire origin to extinguish the fire, t_{fb2}; and, time of fire burn by itself, t_∞. The categorization was made based on the agent behavior rules. Based on the event tree produced (in Fig. 4.3), the summations of the probability values of scenarios with the same consequence are made.

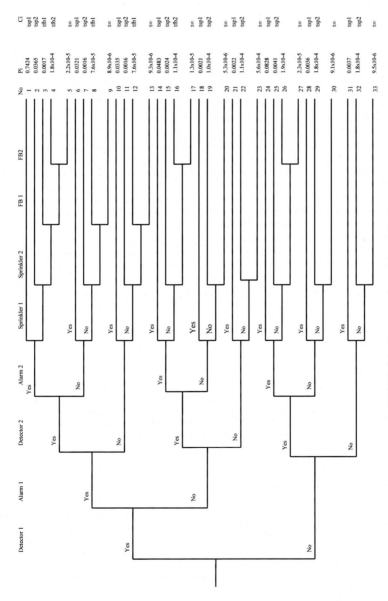

FIGURE 4.3 Event tree for present application.

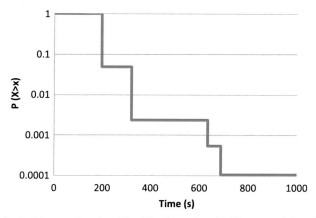

FIGURE 4.4 A risk curve based on Fig. 4.3 subjected to 4 MW t-squared fire with medium growth rate for fixed reliability value of fire safety systems' response.

The risk curve in Fig. 4.4 shows the probability of a consequence greater than a certain value which here is represented by a probability of fire to be extinguished after a fire suppression system, e.g., sprinkler system, suppresses the fire or burn out. Based on Fig. 4.4, the probability of the fire being extinguished after the sprinkler suppresses the fire, t_{sp1} or burn out is 0.048. The probability of the fire to be extinguished after the sprinkler in the roof adjacent to the room of fire origin suppresses the fire, t_{sp2} or burn out is 0.0023. The case being evaluated here has two different alarm activation times: one is of the alarm in the room of fire origin, and the other is of the alarm in the room adjacent to the room of fire origin. Subsequently, this has led to fire brigade being called upon to the scene at different times, hence another two categories of consequences. The probability of fire to be extinguished after time by the fire brigade that is called upon to the scene based on alarm being activated in the room of fire origin, t_{fb1} or burn out is 5.44×10^{-4}. The probability of fire to be extinguished after time by the fire brigade that is called upon to the scene based on an alarm being activated in the room adjacent to the room of fire origin, t_{fb2} or burn out is 1.04×10^{-4}.

4.1 Property damage due to fire

Damage of the property due to fire is defined as the ratio of the amount of fuel consumed to the amount of available fuel load.

$$\text{Property damage} = \frac{Q_{\text{consumed}}}{Q_{\text{available}}} \qquad (4.1)$$

The amount of fuel consumed is determined by the length of time a fire is sustained. It is assumed that the sprinkler system or the fire brigade will extinguish the fire. Assuming a t-squared growth fire, the amount of fuel involved in a fire can be calculated for each event

$$\dot{Q} = \alpha t^2 \tag{4.2}$$

$$Q_{consumed} = \int \alpha t^2 dt \tag{4.3}$$

$$Q_{consumed} = \frac{1}{3} \alpha t^3 + C \tag{4.4}$$

$Q_{consumed}$ is the amount of fuel consumed during the fire. α is the value of constant for fire growth rate. t is the duration of the fire and constant C value equals zero. The fire load density for the case considered is given as 780 MJ/ m^2 based on the value provided in PD7974-1 (BSI, 2003a). The area of the room of fire origin is 25 m^2, and from that the value of $Q_{available}$ which is 1.95×10^7 kJ is obtained. The property damage caused by different consequence is presented in Table 4.3.

From Table 4.3, scenarios with difference consequences are categorized into five different categories according to rules governing the agent's interactions. In the first row, consequence t_{sp1} where the sprinkler in the room of fire origin operates may damage the property and is calculated to be 14%. The damage caused by scenarios with consequence t_{sp2} is 0.62%. The percentage of damage is much higher as in t_{sp2}, the sprinkler adjacent to the room of fire origin is activated after the failure of the sprinkler in the room of fire origin to operate. It is assumed that when this happens, i.e., scenarios with consequence t_{sp2}, the fire is controlled within the room of fire origin from spreading to the room adjacent to the room of fire origin that has sprinkler in operation. The scenarios with consequence t_{fb1} result in a property damage of 4.78%, and scenarios with consequence t_{fb2} lead to a higher property damage of 5.91%. Different consequences for time of fire brigade t_{fb1} and t_{fb2} are a result of the assumption made in the subsection Agent behavior rules where the fire brigade is notified by an alarm. In the case of t_{fb1}, time of alarm in the room of fire origin and time of

TABLE 4.3 Damage percentage based on medium growth 4 MW t-squared fire.

Consequences		$Q_{consumed}$ [kJ]	Property damage [%]
t_{sp1}	196 s	2.79×10^4	0.14
t_{sp2}	319	1.20×10^5	0.62
t_{fb1}	633	9.32×10^5	4.78
t_{fb2}	688	1.15×10^6	5.91
t_∞	4873	1.95×10^7	100.00

alarm in the room adjacent to the room of fire origin is t_{fb2}. The scenarios with consequence t_∞ will cause 100% damage to the property.

4.2 Varying reliability values of fire safety systems

In the fire safety planning, most of the time, it is simply assumed that a system will function 100%. It is not the case in reality because the reliability of a system depends on factors such as installation and maintenance of the system. A risk curve similar to the one in Fig. 4.4 provides no information on the level of uncertainty. The probabilistic model, which incorporates Monte Carlo simulation within the methodology, is used to perform uncertainty analysis. The Monte Carlo simulation varies the reliability value data of the fire safety systems by randomly selecting the value based on the probability distribution specified by the user. For this work, the Monte Carlo simulation produced 30 different sets of values randomly according to probability distribution in Table 4.1. For every set of reliability values selected by the Monte Carlo simulation, event trees are produced using the methodology presented.

The effect of varying reliability values of fire safety systems is then represented in Fig. 4.5. Different curves in Fig. 4.5 correspond to different sets of reliability values being selected earlier by the probabilistic model. Four values on the Y-axis, probability of reliability that the sprinkler will extinguish the fire at time t_{sp1}, probability of reliability that the sprinkler will extinguish the fire at time t_{sp2}, probability of reliability that the fire brigade will extinguish the fire at time t_{fb1}, and the probability of reliability that the fire brigade will extinguish the fire at time t_{fb2}, are shown in Fig. 4.5 as having an uncertainty range of 0.3 ± 0.086, 0.09 ± 0.0052, 0.021 ± 0.0019, and 0.0048 ± 0.0006, respectively.

These four values can be better presented using probability density function to interpret the information. Therefore, the probabilistic model was used to fit the distribution of four different consequences from the event trees

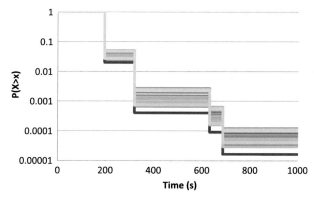

FIGURE 4.5 Risk curves based on reliability values with uncertainty of fire safety systems' response subjected to 4 MW t-squared fire with medium growth rate.

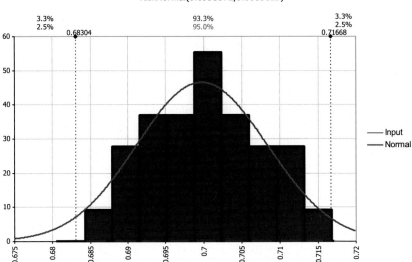

FIGURE 4.6 Probability density function (PDF) of the sprinkler in the room of fire origin to extinguish at $t_{sp1} = 196$ s.

produced, as shown in Fig. 4.6. Although only 30 sets of data were used, the resulted probability distribution is nearly a normal distribution. From Fig. 4.6, the probability density function of the sprinkler in the room of fire origin to activate and extinguish the fire at 196 s is 0.7 as the mean value with 0.009 as the standard deviation.

The summary of mean and standard deviation from the rest of the consequences are presented in Table 4.4. In the case where the sprinkler in the room adjacent to the room of fire origin to activate and control the fire at 319 s, the mean value and the standard deviation from the distribution fits by the probabilistic model is 0.21 and 0.003, respectively. The probability density

TABLE 4.4 Mean and standard deviation based on distribution fit of fire safety systems response to extinguish fire of 4 MW t-squared fire with different fire growth rate.

Consequences		Mean, μ	Standard deviation, σ
t_{sp1}	208 s	0.7	0.009
t_{sp2}	329 s	0.21	0.003
t_{fb1}	648 s	0.07	0.004
t_{fb2}	692 s	0.02	0.001

function of the fire brigade to extinguish the fire based on the alarm notification in the room of fire origin at 633 s is 0.07 as the mean value with 0.004 as the standard deviation. The probability density function of reliability of the fire brigade to extinguish the fire based on the alarm notification in the room next to the room of fire origin at 688 s shows the mean value of 0.02 and the standard deviation of 0.001.

4.3 The effects of different fire growth rate

In addition to varying the reliability value of the fire safety systems involved, different fire growth rate is also being considered. This step is another measure to include uncertainty into the risk assessment process. In this section, the property is subjected to 4 MW t-squared fire with slow, fast, and ultrafast growth, and the results are presented in the form of risk curves in Fig. 4.7 and the damage percentage in Table 4.3.

In the previous sections, when the reliability values of fire safety systems involved were a set of 30 values, the resulting risk curves in Fig. 4.5 show the probability of reliability of the sprinkler to extinguish fire at either t_{sp1} or t_{sp2} and probability of reliability of the fire brigade to extinguish the fire at either t_{fb1} or t_{fb2} were not single values. These values were presented in the form of probability density function (refer Fig. 4.6). This shows the uncertainty with regards to fire growth rate that was taken into consideration in the analysis.

Fig. 4.8 shows the risk curves produced for different fire growth rates for the property under review. In this section, when different fire growth rates are taken into consideration, the consequence value, that is the time of an event happening, is represented in the form of a range instead of just a single value. The time of the sprinkler in the room of fire origin to extinguish the fire is

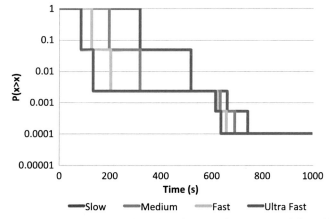

FIGURE 4.7 A risk curve based on Fig. 4.3 subjected to 4 MW t-squared fire with different growth rate.

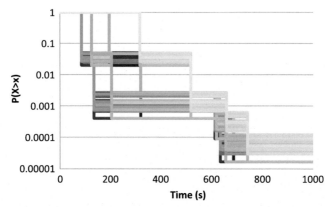

FIGURE 4.8 Risk curves with different fire growth rate and reliability value with uncertainty of fire safety systems.

between 84 and 318 s. If it is the sprinkler in the room adjacent to the room of fire origin, the time for it to extinguish fire is 133−518 s. The time range of the fire brigade to extinguish fire based on an alarm in the room of fire origin or an alarm in the room adjacent to the room of fire origin is 614−658 s and 634−742 s, respectively. The reliability values that were produced in the earlier section to assess the effect of reliability of fire safety systems were then used to plot risk curves for different fire growth rates as shown in Fig. 4.8. Fig. 4.8 is used to better represent the results with uncertainty of both reliability of fire safety systems and fire growth rate taken into consideration.

The percentages of damage to the room of fire origin caused by fire are represented in Table 4.4. Based on Table 4.5, events where the sprinkler system operates, whether it is in the room of fire origin, i.e., t_{sp1} or in the room adjacent to the room of fire origin i.e., t_{sp2}, have a property damage percentage between 0.14 and 0.18 and 0.62 and 0.72 respectively. Events with the consequence of the fire brigade extinguishing the fire result in much higher property damage percentage. The property damage percentage of event with the fire brigade extinguishing the fire based on an alarm in the room of fire origin, i.e., t_{fb1} as its consequence is between 1.47 and 10.54. The value is further increased if the fire brigade extinguished the fire based on the alarm in the room adjacent to the room of fire origin, i.e., t_{fb2} with property damage percentage between 2.11 and 10.95. The results indicate that the sprinkler system, if operated, is effective in minimizing the property damage.

The size and contents of the building play a significant role in deciding fire safety systems that are suitable to be installed in a building. The first priority of any decision in fire engineering will be based on its effect on life safety of the occupants. The property falls under the purpose group dwellings, and according to guidance documents for dwellings, medium fire growth is assumed. However, taking into account the uncertainty involved in assessing

TABLE 4.5 Effect of different fire growth on the fire safety system's response and damage percentage of the room of fire origin.

	Slow	Damage [%]	Medium	Damage [%]	Fast	Damage [%]	Ultrafast	Damage [%]
t_{sp1}	318 s	0.17	196 s	0.14	126 s	0.15	84 s	0.18
t_{sp2}	518 s	0.72	319 s	0.62	203 s	0.64	133 s	0.71
t_{fb1}	658 s	1.47	633 s	4.78	621 s	8.64	614 s	10.54
t_{fb1}	742 s	2.11	688 s	5.91	655 s	9.33	634 s	10.95
t_{∞}	4492 s	100.00	4873 s	100.00	4775 s	100	4825 s	100.00

fire risk assessment, slow, fast, and ultrafast fire growth rates were all considered and analyzed. Reliability values of fire safety systems involved were also varied using Monte Carlo simulation.

5. Conclusions

In this study, new methodology is developed based on the following steps: (1) CFAST is used as a deterministic model to determine the state of the fire, (2) @Risk is used as a probabilistic model to predict a possible operational state for each agent using MCS, and (3) ABM is used to coordinate interactions and determine the risk of all possible scenarios. The developed methodology demonstrates on a real case study how the methodology as presented by Akashah et al. (2010) is being extended and applied to a different domain, i.e., two-storey dwelling house. The methodology was successfully applied as a tool to aid decision-making process in deciding which fire safety systems are best suited for the two-storey dwelling house. There are two options being taken into consideration in this case study: (1) introduction of a detection system and (2) introduction of a detection system and a sprinkler system.

From the results, the introduction of a sprinkler does not help in maintaining a clear height of 2 m in the room of fire origin but is effective in minimizing the impact of fire by reducing the damage as presented in Table 4.4 which is determined by the ratio of the amount of fuel consumed to the amount of fuel, i.e., the length of time a fire is sustained.

This study also demonstrates, apart from being able to automate the process of generating event tree, the methodology is able to perform fire risk assessment in different geometry configurations. In this study, the uncertainty analysis that was incorporated in the methodology is highlighted. This study shows the advantages and flexibility that this methodology had and how QRA can be performed much faster and more comprehensive as opposed to perform a "snap shot" assessment, which is based on the "worst-case scenario."

In this case study, it was found that the location of the fire does not affect the response of the fire safety system but the size of the compartment in which the fire breaks. One of the reasons behind this observation is the deterministic fire model being used in this application, a zone model, CFAST. In CFAST, the fire enclosure is divided into zones with uniform fire characteristics. To be able to analyze the effect of different location of the fire in detail, the use of field models is recommended. In field models, instead of having uniform properties of upper and lower layers as in zone models, the enclosure is divided into a grid of cubes where equations of mass, momentum, and energy are solved at each element in within the cubes. These calculations account for physical changes generated within the cube and changes in the cube originating from surrounding cubes. The users are able to determine the conditions at any point, characterized by grid, in the compartment.

References

AnyLogic 6 User's Guide, 2007. XJ Technologies Company Ltd, p. 194.

Guide to Using @RISK: Risk Analysis and Simulation Add-In for Microsoft Excel, 2009. Palisade Corporation, Ithaca, NY, USA, p. 693.

Akashah, F., Zhang, J., Delichatsios, M., Wang, H., 2010. Agent Based Risk Assessment of Fire Safety Systems with Automated Event Tree Analysis, 8th International Conference on Performance-Based Codes and Fire Safety Design Methods. SFPE, Lund, Sweden.

BSI, 2003a. PD 7974-1:2003 Application of Fire Safety Engineering Principles to the Design of Buildings. BSI, London, p. 76.

BSI, 2003b. PD 7974-7:2003. Probabilistic risk assessment BSI, London, p. 88.

Bukowski, R.W., Budnick, E.K., Schemel, C.F., 1999. Estimates of the Operational Reliability of Fire Protection Systems, International Conference on Fire Research and Engineering (ICFRE3). Society of Fire Protection Engineers, Boston, MA, Chicago, IL, pp. 87−98.

Crippa, C., Fiorentini, L., Rossini, V., Stefanelli, R., Tafaro, S., Marchi, M., 2009. Fire risk management system for safe operation of large atmospheric storage tanks. Journal of Loss Prevention in the Process Industries 22, 574−581. https://doi.org/10.1016/j.jlp.2009.05.003.

Fan, C., 2006. DDSOS: A Dynamic Distributed Service-Oriented Modeling and Simulation Framework. Department of Computer Science and Engineering. Arizona State University, Tempe, AZ, p. 381.

Garifullin, M., Borshchev, A., Popkov, T., 2007. Using AnyLogic and Agent Based Approach to Model Consumer Market. Eurosim 2007, Ljubljana, Slovenia.

Groner, N.E., 2016. A decision model for recommending which building occupants should move where during fire emergencies. Fire Safety Journal 80, 20−29. https://doi.org/10.1016/j.firesaf.2015.11.002.

Guanquan, C., Jinhui, W., 2009. Stochastic Analysis on Probability of Fire Scenarios in Risk Assessment to Occupant Evacuation, pp. 2444−2448.

Hakobyan, A., Aldemir, T., Denning, R., Dunagan, S., Kunsman, D., Rutt, B., Catalyurek, U., 2008. Dynamic generation of accident progression event trees. Nuclear Engineering and Design 238, 3457−3467.

Harpur, A.P., Boyce, K.E., McConnell, N.C., 2013. An investigation into the circumstances surrounding fatal dwelling fires involving very young children. Fire Saf. J. 61, 72−82. https://doi.org/https://doi.org/10.1016/j.firesaf.2013.08.008.

Ibrahim, M.N., Ibrahim, M.S., Mohd-Din, A., Abdul-Hamid, K., Yunus, R.M., Yahya, M.R., 2011. Fire risk assessment of heritage building - perspectives of regulatory authority, restorer and building stakeholder. Procedia Engineering 20, 325 328. https://doi.org/10.1016/j.proeng.2011.11.173.

Jennings, C.R., 2013. Social and economic characteristics as determinants of residential fire risk in urban neighborhoods: a review of the literature. Fire Safety Journal 62, 13−19. https://doi.org/10.1016/j.firesaf.2013.07.002.

Jones, W.W., Peacock, R.D., Forney, G.P., Reneke, P.A., 2005. CFAST − Consolidated Model of Fire Growth and Smoke Transport (Version 6) Technical Reference Guide. National Institute of Standards and Technology, Washington, p. 126.

Kang, R., Fu, G., Yan, J., 2016. Analysis of the case of fire fighters casualties in the building collapse. Procedia English 135, 343−348. https://doi.org/10.1016/j.proeng.2016.01.140.

Li, X., Sun, X., Wong, C.-F., Hadjisophocleous, G., 2016. Effects of fire barriers on building fire risk - a case study using CUrisk. Procedia English 135, 444−453. https://doi.org/10.1016/j.proeng.2016.01.154.

Liu, J., Chow, K.W., 2014. Determination of fire load and heat release rate for high-rise residential buildings. Procedia English 84, 491−497. https://doi.org/10.1016/j.proeng.2014.10.460.

Lysenko, M., D'Souza, R.M., 2008. A framework for megascale agent based model simulations on graphics processing units. The Journal of Artificial Societies and Social Simulation 11, 10.

Martinopoulos, G., Papakostas, K.T., Papadopoulos, A.M., 2016. Comparative analysis of various heating systems for residential buildings in Mediterranean climate. Energy and Buildings. https://doi.org/10.1016/j.enbuild.2016.04.044.

McGrattan, K., Hostikka, S., Floyd, J., 2007. Fire Dynamics Simulator (Version 5) − User's Guide. BFRL, National Institue of Standards and Technology, Gaithersburg, MD.

Milazzo, M.F., Vianello, C., Maschio, G., 2015. Uncertainties in QRA: analysis of losses of containment from piping and implications on risk prevention and mitigation. Journal of Loss Prevention in the Process Industries 36, 98−107. https://doi.org/10.1016/j.jlp.2015.05.016.

Nilson, F., Bonander, C., Jonsson, A., 2015. Differences in determinants amongst individuals reporting residential fires in Sweden: results from a cross-sectional study. Fire Technology 51, 615−626. https://doi.org/10.1007/s10694-015-0459-0.

Pasman, H., Reniers, G., 2013. Past, present and future of quantitative risk assessment (QRA) and the incentive it obtained from land-use planning (LUP). Journal of Loss Prevention in the Process Industries 28, 2−9. https://doi.org/10.1016/j.jlp.2013.03.004.

Ran, H., Sun, L., Gao, X., 2014. Influences of intelligent evacuation guidance system on crowd evacuation in building fire. Automation in Construction 41, 78−82. https://doi.org/10.1016/j.autcon.2013.10.022.

Rueppel, U., Lange, M., 2009. An integrative process model for cooperation support in civil engineering. Journal of Information Technology in Construction 11, 509−528.

Schofield, S., 1998. Offshore QRA and the ALARP principle. Reliability Engineering and System Safety 61, 31−37. https://doi.org/10.1016/S0951-8320(97)00062-8.

Seguí, X., Darbra, R.M., Vílchez, J.A., Arnaldos, J., 2014. Methodology for the quantification of toxic dispersions originated in warehouse fires and its application to the QRA in Catalonia (Spain). Journal of Loss Prevention in the Process Industries 32, 404−414. https://doi.org/10.1016/j.jlp.2014.10.017.

Shazmin, S.A.A., Sipan, I., Sapri, M., 2016. Property tax assessment incentives for green building: a review. Renewable and Sustainable Energy Reviews 60, 536−548. https://doi.org/10.1016/j.rser.2016.01.081.

Su, C., Bai, J., 2016. Measurement of the neutral plane of an internal fire whirl using the background-oriented Schlieren technique for a vertical shaft model of a high-rise building. Measurement 78, 151−167. https://doi.org/10.1016/j.measurement.2015.10.004.

Tancogne-Dejean, M., Laclémence, P., 2016. Fire risk perception and building evacuation by vulnerable persons: points of view of laypersons, fire victims and experts. Fire Safety Journal 80, 9−19. https://doi.org/10.1016/j.firesaf.2015.11.009.

Tang, F., Ren, A., 2008. Agent-based evacuation model incorporating fire scene and building geometry. Tsinghua Science and Technology 13, 708−714.

Tian, D., Wu, X., Song, Z., Wang, H., 2016. Reverse analysis for fire pyrolysis parameters of timber buildings based on response surface method. Procedia Eng. 135, 19−24. https://doi.org/10.1016/j.proeng.2016.01.073.

Tong, S., Wu, Z., Wang, R., Wu, H., 2016. Fire risk study of long-distance oil and gas pipeline based on QRA. Procedia Engineering 135, 369−375. https://doi.org/10.1016/j.proeng.2016.01.144.

Vinnem, J.E., 1998. Evaluation of methodology for QRA in offshore operations. Reliability Engineering and System Safety 61, 39–52. https://doi.org/10.1016/S0951-8320(97)00063-X.

Ye, X., Ma, J., Shen, Y., Lin, L., 2016. Suppression effect of sprinkler system on fire spread in large commercial buildings. Procedia English 135, 455–462. https://doi.org/10.1016/j.proeng.2016. 01.155.

Further reading

Tancogne-Dejean, M., Laclémence, P., 2016. Fire risk perception and building evacuation by vulnerable persons: points of view of laypersons, fire victims and experts. Fire Safety Journal 80, 9–19. https://doi.org/10.1016/j.firesaf.2015.11.009.

Wang, S.-H., Wang, W.-C., Wang, K.-C., Shih, S.-Y., 2015. Applying building information modeling to support fire safety management. Automation in Construction 59, 158–167. https://doi.org/10.1016/j.autcon.2015.02.001.

Waterson, N.P., Castle, C.J.E., Bail, S.L., 2010. A Comparison of Evacuation Prediction Made Using Agent-Based Simulation and Code-Based Approaches, 8th International Conference on Performance-Based Codes and Fire Safety Design Methods. SFPE, Lund, Sweden.

Xin, J., Huang, C., 2013. Fire risk analysis of residential buildings based on scenario clusters and its application in fire risk management. Fire Safety Journal 62, 72–78. https://doi.org/10.1016/ j.firesaf.2013.09.022.

Xiong, L., Bruck, D., Ball, M., 2015. Comparative investigation of 'survival' and fatality factors in accidental residential fires. Fire Safety Journal 73, 37–47. https://doi.org/https://doi.org/10. 1016/j.firesaf.2015.02.003.

Chapter 5

Prediction capability of polynomial neural network for uncertain buckling behavior of sandwich plates

R.R. Kumar[1], Tanmoy Mukhopadhya[2], K.M. Pandey[1], S. Dey[1]

[1]*Department of Mechanical Engineering, National Institute of Technology Silchar, Silchar, Assam, India;* [2]*Department of Aerospace Engineering, Indian Institute of Technology Kanpur, Kanpur, India*

1. Introduction

Sandwich plate is commonly constructed with two facesheets (upper and lower) separated by a central soft core as shown in Fig. 5.1. The facesheets comprise a laminate of different orientations stacked one over the other in finite sequence to obtain the tailored properties. It portrays the superior properties in terms of weight sensitiveness, high resistance to environmental degradation, environmental friendly, and cost-effectiveness. Owing to these portfolio of merits, sandwich plates are exhaustively used in structural applications such as aerospace, marine, automobiles, civil construction, sports equipment's, and orthopedic medical devices. The weight sensitiveness becomes efficient by using the core of low strength material and

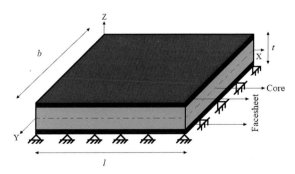

FIGURE 5.1 Laminated sandwich plate.

Handbook of Probabilistic Models. https://doi.org/10.1016/B978-0-12-816514-0.00005-9

facesheets of high strength (Chalak et al., 2012). The variation in properties of material throughout the thickness causes complex conduct of soft core laminated sandwich plates. In the present study, polynomial neural network (PNN) is used to construct the surrogate model for assessment of predictive capability in conjunction to buckling analysis of sandwich plates. For reliable design, the structural responses at different loading conditions are needed to be analyzed. The modeling of shear deformation in a refined manner is a major concern, and for that, several plate theories have been developed. On the basis of displacement field, the theory is widely divided into two sections namely single layer theory and layer-wise theory. In general, as per the single layer theory (also known as the first order shear deformation theory (FSDT) or Reissner-Mindlin's plate theory), the transverse shear strain is considered to be consistent across the plate thickness. An FSDT-based five-noded beam finite element (FE) model is developed by Goyal and Kapania (Goyal and Kapania, 2007). Ferreira (Ferreira, 2003) used a FSDT-based multiquadratic radial basis function method to analyze the thick rectangular laminated composite plate. In general, shear stress is varying parabolic, and shear correction factor is required in FSDT for the compensation of actual parabolic variation of shear stress. The use of shear correction factor in FSDT can be avoided by incorporating the actual cross-sectional warping and hence named as higher order shear deformation theories (HSDT). The development of HSDT provides the actual variation of transverse shear stresses and strains across the plate thickness (Reddy, 1984). Third order refined shear deformation theory is used to derive the complete set of equations for the analysis of thick elastic plates by Kant (Kant, 1982). Three dimensional Hook's law is used to obtain the quadratic and linear variation of transverse shear strain and transverse normal strain across the plate thickness (Kant, 1982). HSDT-based analytical solution was reported by Kant and Swaminathan (Kant and Swaminathan, 2002). Aagaah et al. (Aagaah et al., 2003) derived dynamic equations for modeling of laminated composite through Reddy's displacement field for third order shear deformation theory. Murthy et al. (Murthy et al., 2005) and Subramanian (Subramanian, 2006) also proposed an HSDT-based theory to analyze laminated beam. Laminated composite plate was statically analyzed by Pervez et al. (Pervez et al., 2005) using an HSDT-based two dimensional serendipity model. In static analysis of plates, 13 field variables per node are assumed for 3-noded triangular and 4-noded quadrilateral elements by Wu et al. (Wu et al., 2005) which satisfies weak continuity conditions. The limitation of zigzag theories is presented by Wu et al. (Wu et al., 2012), which leads to need of C_1 continuity condition. The value of shear rigidity is found to be different at the adjacent layer of laminated composite. As a result of which, HSDT shows continuous shear strain variation in transverse direction throughout the thickness, whereas discontinuous variation of shear stress is found to be at the layer interface. But, in actual, the composite laminate behaves in the opposite manner, wherein at layer interface, shear stress in transverse direction must be continuous, whereas strain, corresponding to transverse shear stress, may be intermittent (Sheikh and Chakrabarti, 2003). In present study, nine noded isoparametric quadratic plate elements with 11 field variables are used at each node. The displacement field is selected in such a way that it is not required to specify any new field variable (Kapuria and Kulkarni, 2007) in the proposed formulation, and it is also not required to apply any penalty stiffness (Pandit et al., 2008). In the past, many researchers studied stochastic behavior of composite structures (Karsh et al., 2018a,b; Kumar et al., 2019; Maharshi et al., 2018; Dey et al., 2015a,b,c,d,e, 2016a,b,c,d,e,f, 2017a, 2018; Mukhopadhyay et al., 2015, 2016, 2017; Mukhopadhyay and Adhikari, 2016). In the present study, prediction capability of PNN for uncertain first buckling load of sandwich plate is analyzed considering random variation in the input parameters of system. A PNN model is developed for reduction in computational time and cost without affecting the accuracy. Thereby, FE formulation in accordance with PNN model is used to layer-wise randomly varied input parameters for checking the predictability and error in result obtained through surrogate (PNN) model with respect to the Monte Carlo Simulation (MCS).

2. Governing equations

In the present study, FE formulation is applied by using nine-nodded isoparametric elements with each node having 11 degrees of freedom. The expression for elemental potential energy (Dey et al., 2017b) can be derived as

$$PE = U_s - U_{ext} = \frac{1}{2} \iint \{S\}^T [A(\varpi)]^T [E(\varpi)] [A(\varpi)]\{S\} dxdy$$

$$- \frac{1}{2} \iint \{S\}^T [A(\varpi)]^T [G(\varpi)] [A(\varpi)]\{S\} dxdy \qquad (5.1)$$

$$= \frac{1}{2}\{S\}^T [K_e(\varpi)]\{S\} - \frac{1}{2}\lambda\{S\}^T [K_G(\varpi)]\{S\}$$

where U_{ext} is the external in-plane load energy and U_s is the strain energy. $[K_e(\varpi)] = \int [A(\varpi)]^T [E(\varpi)] [A(\varpi)] dx$ and $[K_G(\varpi)] = \int [A(\varpi)]^T [G(\varpi)] [A(\varpi)] dx$. Random strain displacement matrix is denoted by $[A(\varpi)]$, while $[K_e(\varpi)]$ is stochastic elastic stiffness matrix and $[K_G(\varpi)]$ is geometric stiffness matrix. On minimizing PE equilibrium equation with respect to displacement vector $\{S\}$, the following equation can be obtained:

$$[K_e(\varpi)] \{S\} = \lambda(\varpi)[K_G(\varpi)] \{S\} \qquad (5.2)$$

where $\lambda(\varpi)$ is the stochastic buckling load factor. Global stiffness matrix is stored in a single array by using the skyline technique, and stochastic buckling Eq. (5.2) is solved by the simultaneous iteration technique.

3. Polynomial neural network

The input-output data in optimal PNN structure (Dey et al., 2017a) can be expressed as

$$(\mathbf{X}_i, \mathbf{Y}_i) = (x_{1i}, x_{2i}, x_{3i}, \ldots x_{ni}, y_i) \tag{5.3}$$

where $i = 1,2,3 \ldots n$.

Let Y be the output of the system which required to model. Polynomial regression equation is computed for each pair of input variable x_i and x_j and output Y

$$Y = A + Bx_i + Cx_j + Dx_i^2 + Ex_j^2 + Fx_i x_j \tag{5.4}$$

Where $i,j = 1,2,3 \ldots n$.

where coefficients of polynomial equation are represented as A, B, C, D, E, F.

The estimated output $\widehat{Y}_i$ can be expressed as

$$\widehat{Y} = \widehat{f}(x_1, x_2, x_3, \ldots x_n) = A_0 + \sum_{i=1}^{n} B_i x_i + \sum_{i=1}^{n}\sum_{j=1}^{n} C_{ij} x_i x_j$$
$$+ \sum_{i=1}^{n}\sum_{j=1}^{n}\sum_{k=1}^{n} D_{ijk} x_i x_j x_k + \cdots \tag{5.5}$$

where $i,j,k = 1,2,3 \ldots n$.

$X(x_1, x_2, \ldots, x_n) = $ input variable vector.

$P(A_0, B_i, C_{ij}, D_{ijk}, \ldots) = $ vector of coefficients or weight of ivakhnenko polynomials.

The components of the input vector X can be independent variables, functional forms, or finite difference terms. This algorithm provides model structure and model system output on the value of most noteworthy inputs of the system. The basic taxonomy for the architecture of PNN structure is given in Fig. 5.2. The way of probabilistic buckling analysis of sandwich plate through PNN as the surrogate model is indicated in Fig. 5.3.

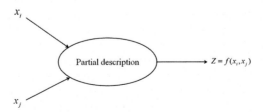

FIGURE 5.2 Taxonomy for architecture of PNN. *PNN*, polynomial neural network.

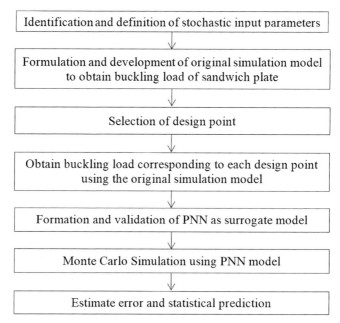

FIGURE 5.3 Flow diagram of stochastic analysis using PNN Model. *PNN*, polynomial neural network; *MCS*, Monte Carlo simulation.

4. Results and discussion

In the present study, a simply supported sandwich plate of length $(l) = 10$ cm, width $(b) = 10$ cm, total thickness $(t) = 1$ cm with core thickness $(t_c) = 0.8$ cm, and facesheet thickness $(t_f) = 0.2$ cm is considered for buckling analysis. Both the upper and lower half of the core are considered of $0°$ orientation angle, and facesheet is divided into eight laminates on both sides (upper and lower) having the orientation angle $(90°/0°/90°/0°/90°/0°/90°/0°)$. The FE code is validated with the results of Dey et al (Dey et al., 2017b). The percentage error of maximum value, minimum value, mean value, and standard deviation of first buckling load between MCS and PNN results with respect to different sample size is calculated. Based on the error analysis as depicted in Fig. 5.4, it is observed that with an increase in the number of sample size, mean percentage error is decreased, and a sample size of 256 is found to be optimum. The probability density function (PDF) (Fig. 5.5) and scatter plot (Fig. 5.6) corroborates the fact that PNN model is in good agreement with the original FE model. Scatter plot through the diagonal shows the applicability of the surrogate model instead of the original FE model. In the present analysis, combined variation of all random

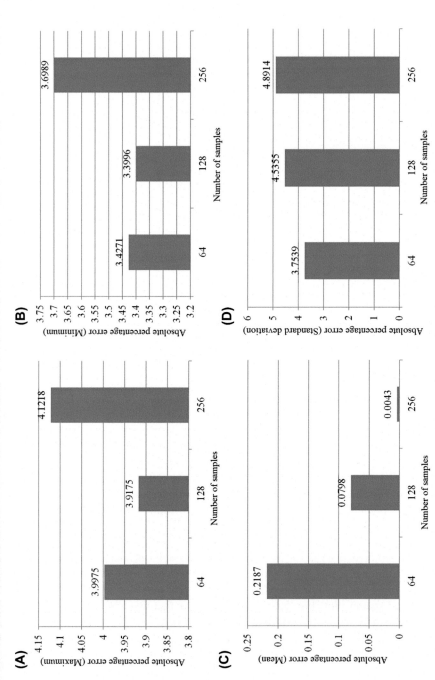

FIGURE 5.4 Percentage error of (A) maximum, (B) minimum, (C) mean, and (D) standard deviation of first buckling load between MCS and PNN results with respect to the sample size (n) considering combined variation. *MCS*, Monte Carlo simulation; *PNN*, polynomial neural network.

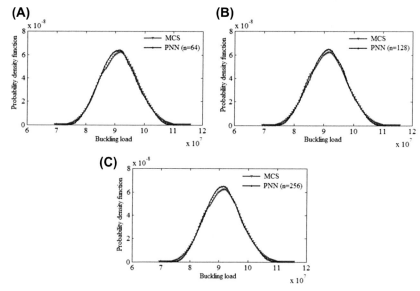

FIGURE 5.5 Probability density function of MCS and PNN for (A) 64, (B) 128, and (C) 256 sample run (n) of simply supported composite sandwich plates considering combined variation. *MCS*, Monte Carlo simulation; *PNN*, polynomial neural network.

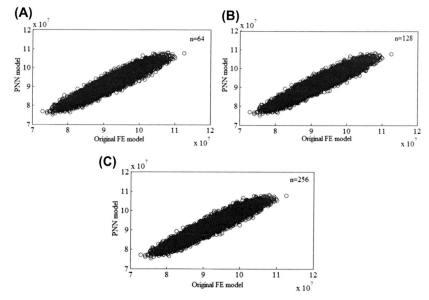

FIGURE 5.6 Scatter plot of PNN model and original FE model for (A) 64, (B) 128, and (C) 256 sample run (n) of simply supported composite sandwich plates considering combined variation. *FE*, finite element; *PNN*, polynomial neural network.

input parameters is considered. The PNN model is found efficient compared with the conventional MCS. The material properties (Kollar and Springer, 2003) considered for the present study are:

(a) For core: $E_1 = E_2 = E_3 = 0.5$ GPa, $G_{12} = G_{13} = 0.4$ GPa, $G_{23} = 0.2$ GPa, $v_{12} = v_{13} = v_{23} = v_{32} = 0.27$, $v_{21} = v_{31} = 0.006$, $\rho = 1000$ kg/m^3 and

(b) For facesheet: $E_1 = 38.6$ GPa, $E_2 = E_3 = 8.27$ GPa, $G_{12} = G_{13} = 4.14$ GPa, $G_{23} = 1.656$ GPa, $v_{12} = v_{13} = v_{23} = v_{32} = 0.26$, $v_{21} = v_{31} = 0.006$, $\rho = 2600$ kg/m^3.

5. Conclusion

The novelty of the present study includes the prediction ability of PNN for uncertain first buckling load of sandwich plate considering the combined variation of all random input parameters. To check the predictability of PNN as a surrogate model, error analysis is carried out in addition to PDF. The scatter plot of result is also obtained through PNN model with respect to the original FE model. PNN is used in spite of conventional MCS to obtain computational efficiency without affecting the accuracy. The predictability of the PNN model is reported as least error compared with conventional MCS. Hence, it is observed that computational time and cost are reduced compared with conventional MCS. It is observed that buckling load is remarkably deviating from its deterministic value. Hence, it is required to consider the effect of stochasticity of input parameters in the analysis of sandwich plate. The present model can be explored to assess the prediction capability for other complex structures.

References

Aagaah, M.R., Mahinfalah, M., Jazar, G.N., 2003. Linear static analysis and finite element modeling for laminated composite plates using third order shear deformation theory. Composite Structures 62, 27–39.

Chalak, H.D., Chakrabarti, A., Iqbal, M.A., Sheikh, A.H., 2012. An improved C$_0$ FE model for the analysis of laminated sandwich plate with soft core. Finite Elements in Analysis and Design 56, 20–31.

Dey, S., Mukhopadhyay, T., Khodaparast, H.H., Kerfriden, P., Adhikari, S., 2015. Rotational and ply-level uncertainty in response of composite shallow conical shells. Composite Structures 131, 594–605.

Dey, S., Mukhopadhyay, T., Sahu, S.K., Li, G., Rabitz, H., Adhikari, S., 2015. Thermal uncertainty quantification in frequency responses of laminated composite plates. Composites Part B: Engineering 80, 186–197.

Dey, S., Mukhopadhyay, T., Khodaparast, H.H., Adhikari, S., 2015. Stochastic natural frequency of composite conical shells. Acta Mechanica 226 (8), 2537–2553.

Dey, S., Mukhopadhyay, T., Adhikari, S., 2015. Stochastic free vibration analyses of composite shallow doubly curved shells—A Kriging model approach. Composites Part B: Engineering 99—112.

Dey, S., Mukhopadhyay, T., Adhikari, S., 2015. Stochastic free vibration analysis of angle-ply composite plates—A RS-HDMR approach. Composite Structures 122, 526—536.

Dey, S., Mukhopadhyay, T., Sahu, S.K., Adhikari, S., 2016. Effect of cutout on stochastic natural frequency of composite curved panels. Composites Part B: Engineering 105, 188—202.

Dey, S., Mukhopadhyay, T., Spickenheuer, A., Gohs, U., Adhikari, S., 2016. Uncertainty quantification in natural frequency of composite plates-an artificial neural network based approach. Advanced Composites Letters 25 (2), 43—48.

Dey, S., Mukhopadhyay, T., Khodaparast, H.H., Adhikari, S., 2016. Fuzzy uncertainty propagation in composites using Gram—Schmidt polynomial chaos expansion. Applied Mathematical Modelling 40 (7—8), 4412—4428.

Dey, S., Naskar, S., Mukhopadhyay, T., Gohs, U., Spickenheuer, A., Bittrich, L., Sriramula, S., Adhikari, S., Heinrich, G., 2016. Uncertain natural frequency analysis of composite plates including effect of noise—A polynomial neural network approach. Composite Structures 143, 130—142.

Dey, S., Mukhopadhyay, T., Spickenheuer, A., Adhikari, S., Heinrich, G., 2016. Bottom up surrogate based approach for stochastic frequency response analysis of laminated composite plates. Composite Structures 140, 712—727.

Dey, S., Mukhopadhyay, T., Khodaparast, H.H., Adhikari, S., 2016. A response surface modelling approach for resonance driven reliability based optimization of composite shells, periodica polytechnica. Civil Engineering 60 (1), 103.

Dey, S., Mukhopadhyay, T., Adhikari, S., 2017. Metamodel based high-fidelity stochastic analysis of composite laminates: a concise review with critical comparative assessment. Composite Structures 171, 227—250.

Dey, S., Mukhopadhyay, T., Naskar, S., Dey, T.K., Chalak, H.D., Adhikari, S., 2017. Probabilistic characterisation for dynamics and stability of laminated soft core sandwich plates. Journal of Sandwich Structures and Materials 01—32.

Dey, S., Mukhopadhyay, T., Sahu, S.K., Adhikari, S., 2018. Stochastic dynamic stability analysis of composite curved panels subjected to non-uniform partial edge loading. European Journal of Mechanics — A: Solids 67, 108—122.

Ferreira, A.J.M., 2003. A formulation of the multiquadratic radial basis function method for the analysis of laminated composite plates. Composite Structures 59, 385—392.

Goyal, V.K., Kapania, R.K., 2007. A shear deformeable beam element for analysis of laminated composites. Finite Elements in Analysis and Design 43, 463—477.

Kant, T., 1982. Numerical analysis of thick plates. Computer Methods in Applied Mechanics and Engineering 44 (4), 1—18.

Kant, T., Swaminathan, A., 2002. Analytical solutions for the static analysis of laminated composite and sandwich plates based on refined higher order shear deformation theory. Composite Structures 56, 329—344.

Kapuria, S., Kulkarni, S.D., 2007. An improved discrete Kirchhoff element based on third order zigzag theory for static analysis of composite and sandwich plates. International Journal for Numerical Methods in Engineering 69, 1948—1981.

Karsh, P.K., Mukhopadhyay, T., Dey, S., 2018. Stochastic dynamic analysis of twisted functionally graded plates. Composites Part B: Engineering 147, 259—278.

Karsh, P.K., Mukhopadhyay, T., Dey, S., 2018. Spatial vulnerability analysis for the first ply failure strength of composite laminates including effect of delamination. Composite Structures 184, 554—567.

Kollar, L.P., Springer, G.S., 2003. Mechanics of Composite Structures. Cambridge.

Kumar, R.R., Mukhopadhyay, T., Pandey, K.M., Dey, S, 2019. Stochastic buckling analysis of sandwich plates: The importance of higher order modes. International Journal of Mechanical Sciences 152, 630—643.

Maharshi, K., Mukhopadhyay, T., Roy, B., Roy, L., Dey, S., 2018. Stochastic dynamic behaviour of hydrodynamic journal bearings including the effect of surface roughness. International Journal of Mechanical Sciences 142, 370—383.

Mukhopadhyay, T., Adhikari, S., 2016. Free-vibration analysis of sandwich panels with randomly irregular honeycomb core. Journal of Engineering Mechanics 142 (11), 06016008.

Mukhopadhyay, T., Dey, T.K., Dey, S., Chakrabarti, A., 2015. Optimisation of fibre-reinforced polymer web core bridge deck—a hybrid approach. Structural Engineering International 25 (2), 173—183.

Mukhopadhyay, T., Naskar, S., Dey, S., Adhikari, S., 2016. On quantifying the effect of noise in surrogate based stochastic free vibration analysis of laminated composite shallow shells. Composite Structures 140, 798—805.

Mukhopadhyay, T., Chakraborty, S., Dey, S., Adhikari, S., Chowdhury, R., 2017. A critical assessment of kriging model variants for high-fidelity uncertainty quantification in dynamics of composite shells. Archives of Computational Methods in Engineering 24 (3), 495—518.

Murthy, M.V.V.S., Mahapatra, D.R., Badarinarayana, K., Gopalakrishnan, S., 2005. A refined higher order finite element for asymmetric composite beams. Composite Structures 67, 27—35.

Pandit, M.K., Sheikh, A.H., Singh, B.N., 2008. Buckling of laminated sandwich plates with soft core based on an improved higher order zigzag theory. Thin-Walled Structures 46, 1183—1191.

Pervez, T., Seibi, A.C., Jahwari, F.K.S.A., 2005. Analysis of thick orthotropic laminated composite plates based on higher order shear deformation theory. Composite Structures 71, 414—422.

Reddy, J.N., 1984. A simple higher order theory for laminated composite plates. Journal of Applied Mechanics 51, 745—752.

Sheikh, A.H., Chakrabarti, A., 2003. A new plate bending element based on higher order shear deformation theory for the analysis of composite plates. Finite Elements in Analysis and Design 39, 883—903.

Subramanian, P., 2006. Dynamic analysis of laminated composite beams using higher order theories and finite elements. Composite Structures 73, 342—353.

Wu, Z., Chen, R., Chen, W., 2005. Refined laminated plate element based on global local higher order shear deformation theory. Composite Structures 70, 135—152.

Wu, Z., Lo, S.H., Sze, K.Y., Chen, W., 2012. A higher order finite element including transverse normal strain including for linear elastic composite plates with general lamination configurations. Finite Elements in Analysis and Design 48, 1346—1357.

Chapter 6

Development of copula-statistical drought prediction model using the Standardized Precipitation-Evapotranspiration Index

Kavina S. Dayal[1], Ravinesh C. Deo[1], Armando A. Apan[2]
[1]*School of Agricultural Computational and Environmental Sciences, University of Southern Queensland, Springfield, QLD, Australia;* [2]*School of Civil Engineering and Surveying, University of Southern Queensland, Toowoomba, QLD, Australia*

1. Introduction

Modeling drought indicators via normalized metrics of rainfall deficiency and the resulting drought properties (Deo et al., 2016; Yevjevich, 1967a) is a topic of significant national interest to agricultural experts, hydrologists, and water resource managers who need to study the drought behavior and for making decisions in agriculture and other socioeconomic sectors. Generally, humans care more about socioeconomic drought events (i.e., one that impacts the supply and the demand of goods and services during the drought period); however, to address their impacts on the survival resources, scientists must first understand the leading meteorological drought event that is likely to explain the synoptic-scale drivers of precipitation deficiency.

The causes of drought in Australia and elsewhere are an ongoing debate (Ummenhofer et al., 2009; van Dijk et al., 2013; Verdon-Kidd and Kiem, 2009), but, as it has been noted in previous studies, drought properties can be influenced by large-scale atmospheric circulation patterns such as the El Niño—Southern Oscillation (ENSO), Pacific Decadal Oscillation (PDO), Southern Annular Mode (SAM), and many more (Bothe et al., 2010; van Dijk et al., 2013; Vicente;Serrano et al., 2011). Studies have discovered definite links of these oscillation systems, particularly with Australian drought events (van Dijk et al., 2013; Verdon-Kidd and Kiem, 2009). The basis for drought prediction is thus to better understand the association between atmospheric

Handbook of Probabilistic Models. https://doi.org/10.1016/B978-0-12-816514-0.00006-0

141

circulations and drought events. Wong et al. (2009) used the Southern Oscillation Index (SOI) to isolate the rainfall data into three states, El Niño, Neutral, and La Niña, to evaluate the trivariate joint behavior of D, S, and I using Gumbel–Hougaard and t-copula models for two locations, either side of the Great Dividing Range. Their findings show generally more frequent and severe droughts occurring during El Niño period. In another study, climate indices were used as predictors in a machine learning technique to predict the Standardized Precipitation Index (SPI), followed by the simulation of ensemble models of drought forecasts from a joint distribution of observed and predicted SPI using copula models in Rajasthan, India (Ganguli and Reddy, 2014). However, predicting droughts based on certain probabilities using a joint distribution of predictors and the predictand is crucially important, and this aspect is yet to be explored for managing the devastating impacts of drought and better and more informed decision-making in regard to agricultural water management.

With the theorem of Sklar (1959), several researchers have embraced copula models to study the dependence structure of different drought properties that are associated through their marginal distributions (González and Valdés, 2003; Kim et al., 2006; Kim et al., 2006; Reddy and Ganguli, 2012; Shiau and Modarres, 2009; Shiau and Shen, 2001). In hydrology, applications of copulas began after the work of De Michele and Salvadori (2003) who tested the Frank copula for a joint study of the negatively associated storm intensity and duration. Other research works (Evin and Favre, 2008; Salvadori and De Michele, 2006; Zhang and Singh, 2006; Zhang and Singh, 2007) incorporated copulae for an extreme analysis of rainfall and drought events. In an earlier study, Shiau (2006) assumed that the drought duration data exhibit an exponential function, while the corresponding severity of drought can be represented by the Gamma distribution function. However, every drought event possesses unique marginal distributions, as verified by Reddy and Ganguli (2012) and Ganguli and Reddy (2012), where the former study used severity–duration–frequency to show that the severity property was better represented by the log-normal, whereas the latter study found that the nonparametric kernel density function represented the drought severity property. This study first evaluates several marginal distributions and then select the best fitted model for each variable based on the maximum likelihood estimation method.

Copula applications for modeling drought properties in Australia have been limited (Rauf and Zeephongsekul, 2014; Wong, 2013; Wong et al., 2009; Wong et al., 2008). Copulae have been successfully applied for risk evaluation for several problems in hydrology and drought frequency analysis (Ganguli and Reddy, 2012; Reddy and Ganguli, 2012; Shiau, 2006; Shiau and Modarres, 2009). With copula, we can derive joint distribution functions of climate indices and drought index (and properties D, S and I) and develop probabilistic prediction model. The motivation for exploring and developing copula models for adequate estimate of drought risk is an interesting research endeavor in view of the plethora of hydrological applications. In a very recent study, Nguyen-Huy et al. (2017) used vine copula model for the first time in Australia's agro-ecological zones for probability-based seasonal rainfall predictions conditioned using the SOI and IPO Tripole Mode Polar Index (TPI). That study used vine copula for trivariate forecasting to yield a better accuracy than the bivariate model for the east and southeast agro-ecological zones. Importantly, the trivariate forecasting model was found to improve the forecasting of rainfall during the La Niña and negative TPI.

One popular index applied for copula-based modeling of drought is the SPI, calculated by the probabilistic analyses of rainfall (Ganguli and Reddy, 2012; McKee et al., 1993; Reddy and Ganguli, 2012; Wong et al., 2009). Despite its widespread applications in modeling and analysis of drought, the importance of the other related hydro-meteorological factors that can moderate the impact of a drought event is ignored. Recently, a modified version of SPI, Standardized Precipitation-Evapotranspiration Index (SPEI) was formulated with rainfall, temperature, and evapotranspiration data to examine the drought possibility (Vicente-Serrano et al., 2010; Vicente-Serrano et al., 2010). The SPEI can be considered advantageous in respect to the other indices (e.g., SPI) as it is able to incorporate the effects of evaporation on the demand for water and thus could allow decision-makers in agriculture and water resources to consider the drought event from a hydrologic perspective (Hanson, 1988). The utility of the SPEI for multiscale assessment of drought events has thus gained a notable degree of attention (Das et al., 2016; Liu et al., 2016; Zhang et al, 2017), but its application in Australia has remained less explored, with a couple of recent studies showing the great importance of the SPEI for drought prediction (Dayal et al., 2017a; Dayal et al., 2017b; Deo and Sahin, 2015). In particular, the earlier study of Dayal et al. (2017a) reaffirmed the utility of the SPEI for regional assessment of drought events. In this study, the capability of the SPEI for the detection of drought onsets and terminations, including drought ranking and drought recurrence evaluations, was considered, which are vital drought-risk statistics used in sustainable agricultural management and hydrologic systems engineering.

Following our earlier study on drought-risk monitoring (Dayal et al., 2017a), in this research article, we investigate the temporal behavior of the SPEI as an original contribution used for identifying and modeling the drought properties by means of developing bivariate and trivariate copula-based joint relationships. The study area is located in South East Queensland, which is a significant agricultural and socioeconomic hub of Australia's second largest state. The objectives of this chapter are as follows: (1) to compute the SPEI, including the duration (D), severity (S), and intensity (I) properties of iden-tified drought events and to statistically fit the marginal distributions based on goodness-of-fit tests; (2) to evaluate the nonparametric correlations between the SPEI (and D-S-I) in terms of 13 climatic mode indices and to screen the relevant indices for copula-statistical modeling; (3) to evaluate the potential utility of vine copulae and to deduce the optimum copula-statistical models for studying the bivariate and trivariate associations of the SPEI and climate mode indices, *D-S-I* and climate model indices, and between *D-S-I* properties; (4) to evaluate the utility of copula-based conditional models for probabilistic prediction of the SPEI and *D-S-I* parameters using the information of climate indices; and (5) to deduce conditional probabilities and joint return periods elucidating the importance of bivariate and trivariate copula models in drought-risk studies. It is especially noted that drought-risk studies through multivariate modeling and elucidation of its properties in respect to universal precursors (i.e., climate indices) is centrally important for agricultural water management and decision-making by farmers and the populations in South East Queensland where drought is considered a perpetuating risky phenomenon.

2. Materials and methods

2.1 Theoretical background

2.1.1 Standardized Precipitation-Evapotranspiration Index

In this study, point-based monthly time series of the SPEI over 1960—2016 period was computed. Based on the cumulative effects of precipitation (*P*) and reference evapotranspiration (*ETo*), the SPEI was generated where the *ETo* depicted the evaporative demand of the atmosphere, i.e., the evapotranspiration that would occur if sufficient water was available. To examine drought periods within the historical data, the *ETo* was subtracted from the total *P* (where $SDB_i = P_i - ETo_i$ and $i =$ the month) to deduce the surplus or deficit of water resources (*i.e.*, the computation of supply—demand balance, *SDB*) (Dayal et al., 2017a; Vicente-Serrano et al., 2010).

As the precipitation data generally exhibit seasonality, and the distribution of these data was especially pronounced in different climatic regimes, it was necessary to transform the *SDB* time series via an equal probability framework to a normal distribution with a mean of zero ($\mu = 0$) and standard deviation of one ($\sigma = 1$). This allows water deficit and surplus to be comparable in space and time and the SPEI to be standardized so that it is free from seasonality when assessing the different drought event. To achieve this, the *SDB* time series were fitted to the three-parameter log-logistic, Gamma, and Pearson III distributions based on goodness-of-fit tests, i.e., Kolmogorov—Smirnov (K—S) statistic.

With the null hypothesis that the *SDB* time series followed a specified distribution at significance level ($\alpha = 0.05$), the best (smallest) K—S statistics obtained were for log-logistic distribution, concurring with Vicente-Serrano et al. (2010). Thus, the transformation of the *SDB* time series used a probability density function of a three-parameter (α, β and γ) log-logistic distribution, $F(x)$ according to Vicente-Serrano et al., (2010a). This yielded the monthly SPEI with the reference (base) period as 1971−2000 (Dayal et al., 2017a; Deo et al., 2009). The full mathematical formulation of the SPEI can be found in the study by Vicente-Serrano et al. (2010a).

2.1.2 Copula theory

This work adopted copulae to model the bivariate and trivariate joint behavior of SPEI data and the identified drought properties with climate indices. The marginal distribution functions, $F_X(x)$ and $F_Y(y)$ of any two correlated variables, X and Y, are expressed by copula function C using Sklar's (1959) theorem as

$$F_{X,Y}(x,y) = C[F_X(x), F_Y(y)] = C(u,v) \qquad (6.1)$$

where $F_{X,Y}(x,y)$ is the joint cumulative distribution function (CDF) of X and Y. Copulae are scale-invariant under strictly increasing transformations of X and Y; hence, X and Y are transformed to [0,1]. This yields two uniformly distributed variables u and v, where $u = F_X(x)$ and $v = F_Y(y)$. To jointly model drought properties, a primal task is to construct a function $C(\bullet)$ as a bivariate distribution function with a mapping such that $C: [0,1]^2 \rightarrow [0,1]$. Similarly, three or more variables are formulated as

$$F(x_1, ..., x_n) = C[F_1(x_1), ..., F_n(x_n)] = C(u_1, ..., u_n) \qquad (6.2)$$

The u and v in any copula function, $C(u,v)$, must be monotonically increasing and therefore, satisfy (Sklar, 1959): $C(u,0) = 0$, $C(0,v) = 0$ and $C(u,1) = u$, $C(1,v) = v$.

A copula consists of a joint CDF by its definition, and their graphs are generally hard to interpret given that they are deduced as monotonically increasing functions. Because of this, plots of copula densities are typically used to illustrate distributions. Thus, if $C(\bullet)$ is a continuous function, the bivariate copula density is defined as the double derivative of C with respect to its marginal distributions, expressed as (Sklar, 1959):

$$c(u,v) = \frac{\partial^2 C(u,v)}{\partial u \partial v} \qquad (6.3)$$

where $c(\bullet)$ is the respective bivariate copula density constructed using the variables X and Y.

This work uses Archimedean copula family, namely Clayton, Gumbel, and Frank copulae. Table A1 in the Supplementary material lists the mathematical expressions of bivariate Clayton, Gumbel, and Frank copulae, where u and v are the uniform variables and θ is the copula parameter. Archimedean family of copulae is popular in hydrological and agricultural

TABLE A1 Mathematical expressions for bivariate copula functions.

Copula	Generator	Parameter	Bivariate Copula	Kendall's tau (τ)	Tail Dependence (lower, upper)
Clayton	$t^{-\theta} - 1$	$\theta > 0$	$\left(u^{-\theta} + v^{-\theta} - 1\right)^{-\frac{1}{\theta}}$	$\frac{\theta}{\theta+2}$	$\left(2^{-\frac{1}{\theta}}, 0\right)$
Gumbel	$(-\ln t)^{\theta}$	$\theta \geq 1$	$\exp\left\{-\left[(-\ln u)^{\theta} + (-\ln v)^{\theta}\right]^{\frac{1}{\theta}}\right\}$	$1 - \frac{1}{\theta}$	$\left(0, 2 - 2^{\frac{1}{\theta}}\right)$
Frank	$-\ln\frac{e^{-\theta t}-1}{e^{-\theta}-1}$	$-\infty < \theta < \infty,\ \theta \neq 0$	$-\frac{1}{\theta}\log\left[1 + \frac{(e^{-\theta u}-1)(e^{-\theta v}-1)}{e^{-\theta}-1}\right]$	$1 - \frac{4}{\theta} + 4\frac{D_1(\theta)}{\theta}$	$(0,0)$

Where: $D_1(\theta) = \int_0^\theta \frac{x/\theta}{\exp(x)-1}dx$ and $\Phi = \int_0^1 x \log(x)(1-x)^{\frac{2(1-\theta)}{\theta}}dx$

applications because it includes multivariate extreme distributions that exhibit tail dependence and reasonable empirical fit to hydrological data (Serinaldi et al., 2009). The Archimedean symmetric one-parameter copula has the form:

$$C(u_1, \dots, u_n) = \varphi^{-1}[\varphi(u_1) + \cdots + \varphi(u_n)] \tag{6.4}$$

where φ is the unique generator of the copula and $u_1, \dots, u_n$ is the uniform random variable. The symmetric copula is restrictive to two variables only because the correlations between any pair of variables is identical. However, this assumption is unrealistic for many hydrological variables. To overcome this, an asymmetric copula is constructed by nesting symmetric copulae (Joe, 1997), expressed as

$$\begin{aligned} C(u_1, \dots, u_n) &= C_1\{u_n, C_2[u_{n-1}, \dots, C_{n-1}(u_2, u_1)\cdots]\} \\ &= \varphi_1^{-1}\Big\lfloor \varphi_1(u_n) + \varphi_1\big(\varphi_2^{-1}\big(\varphi_2\{u_{n-1} + \cdots + \varphi_{n-1}^{-1}[\varphi_{n-1}(u_2) + \varphi_{n-1}(u_1)]\cdots\}\big)\big)\Big\rfloor \end{aligned} \tag{6.5}$$

The three-variable asymmetric Archimedean copula is given

$$C(u_1, u_2, u_3) = C_1[C_2(u_1, u_2), u_3] = \varphi_1^{-1}\big(\varphi_1\{\varphi_2^{-1}[\varphi_2(u_1) + \varphi_2(u_2)] + \varphi_1(u_3)\}\big) \tag{6.6}$$

where C_2 describes the dependence between variables u_1 and u_2 and the outer copula C_1 is a function of C_2 and variable u_3. This model assumes identical correlations between inner variables and outer variables, i.e., marginal copulae $C_1(u_1, u_3)$ and $C_2(u_2, u_3)$ are identical. Given the nonstationary nature of hydrological variables, any two pairs of variables will have different marginal copulae; therefore, a vine copula would be more suitable for constructing joint distributions of three or more variables.

2.1.3 Vine copula

The vine copula, introduced in the study by Joe (1996), is a graphical tool for describing multivariate, high-dimensional probability distributions through a cascade of bivariate copulae, so called *pair-copulae* (Brechmann and Schepsmeier, 2013). It uses the Markov trees to construct bivariate, pair-copulae. A vine copula decomposes a multivariate probability density into bivariate copulae where each pair-copula can be selected in an independent manner while allowing for an enormous flexibility in dependence modeling. The pair-copulae consider the asymmetries and tail dependence, as well as conditional independence, to build models that are more parsimonious. The "statistical breakthrough" of vines was because of Aas et al. (2009) who described statistical inference of vine into special classes of Canonical (C-) and D-vine functions. In the C-vine, the pair-copula for n variables can be constructed as (Bedford and Cooke, 2002)

$$
\begin{aligned}
(Tree\ 1) \quad & (u_1, u_2), (u_1, u_3), (u_1, u_4), \ldots, (u_1, u_n) \\
(Tree\ 1) \quad & (u_2, u_3|u_1), (u_2, u_4|u_1), \ldots, (u_2, u_n|u_1) \\
& , \ldots, \\
(Tree\ n-2) \quad & (u_{n-1}, u_n|u_1, \ldots, u_{n-2})
\end{aligned}
\tag{6.7}
$$

The vines arrange the $\frac{n(n-1)}{2}$ pair-copulae of an n-dimensional pair-copula construction in $n-1$ linked trees. In C-vine $Tree\,1$, the dependence with respect to one particular variable (first root node, i.e., conditional variable) is modeled using bivariate copulae for each pair. Conditioned on this variable, the pairwise dependency with respect to a second variable is modeled to obtain a C-vine $Tree\ 2$ (second root node). For n-dimensional copula, the decomposition of multivariate density with root nodes is written as

$$
f(x_1, \ldots, x_n) = \prod_{k=1}^{n} f_k(x_k) \times \prod_{i=1}^{n-1} \prod_{j=1}^{n-i} c_{i,j+1|1:(i-1)}[F(x_i|x_1, \ldots, x_{i-1}),
$$
$$
F(x_{i+j}|x_1, \ldots, x_{i-1})]
\tag{6.8}
$$

For three-dimensional copula model, this study adopts the recursive conditioning method, given as

$$
f(x_1, x_2|x_3) = f_1(x_1) \cdot f_2(x_2) \cdot f_3(x_3) \cdot c_{1,3}[F_1(x_1), F_2(x_2)]
$$
$$
\cdot c_{2,3}[F_2(x_2), F_3(x_3)] \cdot c_{1,2|3}[F_{1,3}(x_1|x_3), F_{2,3}(x_2|x_3)]
\tag{6.9}
$$

where the three-dimensional joint density is represented in terms of bivariate copulae $C_{1,3}, \cdot C_{2,3}$, and $C_{1,2|3}$ with densities $c_{1,3}, \cdot c_{2,3}$, and $c_{1,2|3}$ of pair-copulae, which can be independent of each other to achieve a wide range of dependence structures.

The pair-copula requires construction of conditional distribution function, $F(x|v)$ for an n-dimensional vector v. For a pair-copula term in tree $n+1$, the conditional distribution function can be established using pair-copula of previous trees $1, \ldots, n$ and by sequentially applying the relationship given as (Brechmann and Schepsmeier, 2013)

$$
h(x|v, \theta) := F(x|v) = \frac{\partial C_{xv_j|v_{-j}}[F(x|v_{-j}), F(v_j|v_{-j})|\theta]}{\partial F(v_j|v_{-j})}
\tag{6.10}
$$

where v_j is an arbitrary component of v and v_{-j} denotes $(n-1)$-dimensional vector v, excluding v_{-j} (Joe, 1996).

2.1.4 Conditional prediction model

The construction of prediction model uses the inverse form of the conditional distribution functions (Chen et al., 2009; Liu et al., 2015). Given two random variables (x_1, x_2) for a case of bivariate copula, the conditional distribution function, $h(u_1, u_2)$, can be used to obtain u_1 based on the information of u_2. For known probabilities, P on $(0,1)$, u_1 can be derived by the formula $u_1 = C_{1|2}^{-1}(P|u_2) = h^{-1}(P|u_2)$, where $C_{1|2}^{-1}$ is the inverse of copula $C_{1|2}$. The variable x_1 is then obtained by

$$x_1 = F^{-1}(u_1) = F^{-1}\left[C_{u_1|u_2}^{-1}(P, u_2)\right] = F^{-1}\left[h^{-1}(P|u_2)\right] \qquad (6.11)$$

where $F^{-1}(u_1)$ is the inverse marginal of u_1. Similarly, for a three-dimensional case, the random variable x_3 can be obtained based on the information of u_1 and u_2. Following Eq. 6.12, x_3 is computed as

$$x_3 = F^{-1}(u_3) = F^{-1}\left\{h^{-1}\left[h^{-1}(P|h(u_2, u_1))\right]u_1\right\} \qquad (6.12)$$

2.1.5 Joint return periods

A common approach undertaken for the proper design of hydrologic systems (e.g., water storage dams and agricultural irrigation systems) is the frequency analysis of drought, including the estimated recurrence interval or the return period of hydrologic drought events (Shiau and Shen, 2001). This is important to empower farmers and agricultural engineers in understanding the perpetuating risk of drought. The drought return periods, in particular, can provide useful information on the controlled usage of water under drought conditions (Serinaldi et al., 2009). The return period of a drought event can be defined in terms of the mean elapsed time (E_L) between the onsets of any two drought events (Shiau and Shen, 2001).

The univariate return period for D, S and I, according to Shiau and Shen (2001), for a severity, $S \geq s$; duration, $D \geq d$; or intensity $I \geq i$ using the mean clapsed time (F_L) can be calculated as

$$T_D = \frac{E_L}{1 - F_D(d)}; \quad T_S = \frac{E_L}{1 - F_S(s)}; \quad \frac{E_L}{1 - F_I(i)} \qquad (6.13)$$

The joint bivariate return period, according to Shiau (2006), can be calculated as either T_{AND} or T_{OR}, given as

$$T_{AND} = \frac{E_L}{1 - P(X_i \geq x_i, X_j \geq x_j)} = \frac{E_L}{1 - F(x_i) - F(x_j) + C[F(x_i), F(x_j)]} \qquad (6.14)$$

$$T_{OR} = \frac{E_L}{1 - P(X_i \geq x_i \; or \; X_j \geq x_j)} = \frac{E_L}{1 - C[F(x_i), F(x_j)]} \qquad (6.15)$$

where X_i and X_j are two random variables.

Similarly, the joint trivariate return period can be defined as

$$T_{AND} = \frac{E_L}{1 - P(X_i \geq x_i, X_j \geq x_j, X_k \geq x_k)}$$

$$= \frac{E_L}{1 - F(x_i) - F(x_j) - F(x_k) + C[F(x_i), F(x_j)] + C[F(x_i), F(x_k)]} \qquad (6.16)$$

$$+ C[F(x_j), F(x_j)] - C[F(x_i), F(x_j), F(x_k)]$$

$$T_{OR} = \frac{E_L}{1 - P(X_i \geq x_i \ or \ X_j \geq x_j \ or \ X_k \geq x_k)}$$

$$= \frac{E_L}{1 - C[F(x_i), F(x_j), F(x_k)]} \qquad (6.17)$$

In addition, drought return periods conditioned on certain variable threshold is also useful for management of water resources. Shiau (2006) defined bivariate conditional drought return period as

$$T_{x_i | x_j} = \frac{E_L}{\left(1 - F_{x_j}(x_j)\right)\left(1 - F_{x_i}(x_i) - F_{x_j}(x_j) + C(x_i, x_j)\right)} \qquad (6.18)$$

where $T_{x_i | x_j}$ is the conditional return period for X_i given $X_j \geq x_j$. The bivariate conditional return period of drought D, S, and I was calculated in this study where climate indices were the conditioned variables.

2.2 Study area and data

The case study area is a point-based location in the southeast Queensland (SEQ) region (Fig. 6.1), Australia, with the geographical coordinate (152.25°E, 28.25°S). This location has an elevation of 521 m above the sea level and falls within the Murray—Darling Basin, the hub for major agricultural activities. Millennium Drought (2001—09), the longest and most severe in the region, unveiled the vulnerability of SEQ's water supplies. With 2.6% population growth per annum (1985—2015), SEQ is expected to experience a significant increase in the demand for water and the need for water management strategies in terms of more efficient design of hydrologic systems (Seqwater, 2015). The annual evaporation rate is expected to increase by almost 16% in the next 60 years because of the increasing concentration of greenhouse gasses (Helfer et al., 2012). Considering the increase in evaporation rate and SEQ region being prone to frequent droughts, hydrologists and policy-makers must adopt statistical models that can provide probabilistic predictions of drought monitoring index and properties.

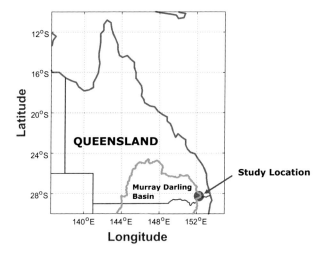

FIGURE 6.1 Map of the study location in drought-prone and economically active southeast Queensland (SEQ).

The monthly rainfall and reference evapotranspiration data were obtained from the Scientific Information for Land Owners (SILO) database for the period 1960 to 2016. The 13 different climate indices, i.e., Niño 3 sea surface temperature (SST), Niño 3.4 SST, Niño 4 SST, Southern Oscillation Index (SOI), Pacific Decadal Oscillation (PDO), Dipole Mode Index (DMI), El Niño Modoki Index (EMI), Southern Annular Mode (SAM), Trans Polar Index (TPI), quasi-biennial oscillation (QBO), Western Pacific Index (WPI), Oceanic Niño Index (ONI), and Multivariate ENSO Index (MEI), were obtained from various sources, such as the National Climate Prediction Center, British Antarctic Survey, JAMSTEC, and Bureau of Meteorology.

2.2.1 Characterization of drought properties

In this case study, we have used the SPEI for drought analysis that was also embraced recently for detecting drought onsets and terminations, drought ranking, and recurrence evaluation (Dayal et al., 2017a). Using total rainfall and the reference evapotranspiration data, the SPEI on a 3-month timescale was calculated for the present case study location. Because the SPEI was a standardized index, the value SPEI $= 0$ corresponded to the mean (normal) with respect to the base period 1971−2000 (Deo et al., 2009) and the SPEI $= \pm 1$ corresponded to the standard deviation where the negative (positive) SPEI indicated dry (wet) condition.

In accordance with the SPEI time series representing the deficits and surpluses of water resources relative to a well-defined base period, the drought onsets and terminations were then identified in periods when the SPEI declined to a value below zero (i.e., the standardized water deficit was below the normal value). The drought duration D, S and I properties were thus estimated from the SPEI time series via the widely adopted run-sum approach described in the study by Yevjevich (1967b):

$$D \equiv \sum_{j=1}^{j=n} n; \quad S = \sum_{j=1}^{D} -(SPEI_j); \quad I = \min(SPEI_j) \qquad (6.19)$$

where $j = 1$ is the start of a drought event when the SPEI drops below zero and its continuation as a negative value for at least 3 months to the length n, $D = $ total duration from the onset to the termination period, while the intensity, $I = $ the drought peak (i.e., minimum SPEI) for a given drought event.

We followed the rationale that a drought event that lasted for less than 3-month duration, which is generally regarded as insufficient to impact the available water resources, was ignored in this case study following the Australian Bureau of Meteorology's definition (*i.e.,* a drought condition is declared when precipitation is below normal for consecutive three or more months) (Mpelasoka et al., 2008). Therefore, only the drought events with $D \geq 3$ months were used in the analysis, following earlier studies (Deo et al., 2009).

2.2.2 Copula-statistical model development

Before constructing the copula-statistical models, we evaluated the significance of the correlation between the SPEI and climate mode indices (*CIs*) and the *D-S-I* and climate mode indices (averaged for the corresponding duration of drought events). The nonparametric Kendall's τ and corresponding *p*-values at 95% confidence interval for the SPEI, and 90% confidence interval for D, S and I drought properties, were obtained to test the hypothesis of no correlation against the alternative, that is, the *p*-values < 0.05 (*or 0.10*) indicated significant correlation between the SPEI (*or D, S and I*) and *CIs* as shown in Table A2 of Supplementary Material.

Evidently, all *CIs* except the PDO, QBO, and WPI exhibited significant correlation with the SPEI time series data. Those *CIs* having smaller correlations with the SPEI were disregarded, and consequently, the Niño 4 SST and SOI were selected for the drought analysis and prediction. The Niño 4 SST and SOI that had larger correlation values (i.e., -0.22 and 0.20, respectively) were used to assess ENSO independently whereby the Niño 4 SST was solely based on sea surface temperature, whereas SOI, on pressure (difference between pressure at Tahiti [149.6°W, 17.5°S] and Darwin [130.9°E, 12.4°S]). The warm phase of ENSO, i.e., El Niño, is known to influence (enhance) drought conditions in Australia (van Dijk et al., 2013; Verdon-Kidd and Kiem, 2009; Wong, 2013; Wong et al., 2008); therefore, probability-based statistical prediction of drought properties based on ENSO indicators is expected to provide useful information for water resource management.

In the case for drought properties, the Niño 4 SST for Duration and EMI for Severity and Intensity were selected based on the indicating *p*-values for

TABLE A2 Kendall's tau (τ) and associated *p*-value of the SPEI and drought severity, duration, and intensity with 13 climate mode indices. The statistically significant correlations are highlighted in bold italics, and selected for the study are highlighted in bold red italics.

Climate Mode Indices	SPEI		Severity		Duration		Intensity	
	τ	*p-value*	τ	*p-value*	τ	*p-value*	τ	*p-value*
Nino3	*-0.0760*	*0.0029*	0.0889	0.4565	-0.0795	0.5181	0.0000	1.0000
Nino3.4	*-0.1501*	*4.37E-09*	0.0476	0.6949	-0.0331	0.7939	0.0349	0.7764
Nino4	*-0.2247*	*1.57E-18*	-0.1714	0.1459	*0.2221*	*0.0675*	-0.1206	0.3094
SOI	*0.2050*	*1.07E-15*	0.0636	0.5952	-0.1046	0.3939	0.0541	0.6530
PDO	-0.0355	0.1648	-0.0095	0.9461	0.0630	0.6110	-0.0349	0.7764
DMI	*-0.0986*	*0.0002*	0.0317	0.7973	0.0331	0.7939	0.1841	0.1177
EMI	*-0.2145*	*4.70E-17*	*-0.1968*	*0.0940*	0.1723	0.1567	*-0.2032*	*0.0836*
SAM	*0.0991*	*0.0001*	0.0635	0.5978	-0.0166	0.9015	0.1905	0.1054
TPI	*0.0881*	*0.0007*	0.0487	0.6930	0.0070	0.9657	0.0017	1.0000
QBO	-0.0177	0.4897	0.0984	0.4088	-0.1061	0.3864	0.0032	0.9892
WPI	-0.0050	0.8451	0.0317	0.7973	-0.0762	0.5361	-0.0571	0.6359
ONI	*-0.1845*	*5.54E-13*	-0.0381	0.7558	0.0696	0.5730	-0.1079	0.3641
MEI	*-0.1818*	*1.15E-12*	-0.0857	0.4731	0.1094	0.3715	-0.0921	0.4403

Note: Significance test of SPEI at 95% confidence interval, while severity, duration, and intensity at 90% confidence interval.

significant correlations. The Niño 4 SST had a significant correlation with the drought duration (0.22), while the EMI had a significant correlation with the severity (-0.20) and intensity (-0.20) of drought events. The joint behavior of drought properties based on Niño 4 SST and EMI, which can aid in better management of agricultural systems (including irrigations and dams), is likely to be beneficial for the prediction of drought properties given their reasonably acceptable statistical dependencies.

2.2.3 Selection of marginal distributions

Because copula functions are able to join the marginal distributions of multivariate data to construct a joint distribution function, the foremost task was to fit appropriate marginal distribution to each drought-related variable. The suitable marginal distribution of the SPEI, Niño 4 SST and SOI for the monthly data from 1960 to 2016, as well as for the D, S, I, Niño 4 SST and EMI for all drought cases were determined. For an accurate estimation, several distributions defined by the extreme value, Gamma, generalized extreme value (GEV), logistic, log-logistic, log-normal, Nakagami, normal, Rician, exponential, and Weibull equations were evaluated based on the maximum likelihood method. The statistical significance was determined by p-values based on the Anderson–Darling (AD) and Kolmogorov–Smirnov (KS) test statistics. This considered the null hypothesis that variables were not from the assumed distribution against the alternate hypothesis that they were, that is, the larger the p-value, the more suitable fit the distribution is to the variable. Table A3 in Supplementary material lists the selected marginal distributions, parameters, and corresponding p values for the variables used in the analysis.

2.2.4 Selection of copulae

For any random variable with continuous marginal, the copula required the variable to be uniformly distributed. The pseudo-observation values were produced for each variable, transformed on [0,1] interval. Subsequently, the *BiCopSelect* function from "VineCopula" library in the statistical *"R"* software was applied to the transformed variables to construct the appropriate bivariate copula functions to model the joint relationships of the SPEI with *CIs* and drought properties with *CIs*. The *BiCopSelect* function investigates a rich variety of copulae and returns the most appropriate copula function based on the selection criteria.

During the selection of the copulae, statistical validation was undertaken using AIC, BIC, and log-likelihood, where the copula that yielded the minimum *(largest)* value of AIC and BIC *(log-likelihood)* at significance level $\alpha = 0.05$ was selected. Table 6.1(A–C) lists the selected bivariate and trivariate copula statistics where C-vine was used for trivariate conditional copula selection. Note that transforming the variables on [0, 1] interval does

TABLE A3 Marginal distribution parameters and *p*-values of observed variables.

Variable	Distribution	Parameters	KS p-value	AD p-value
For monthly data from 1960 to 2016				
SPEI	GEV	k= -0.23118, σ= 0.969683, μ= -0.52284	0.8288	0.8987
Nino4 SST	Weibull	a= 52.1350, b= 28.7875	0.5337	0.4622
SOI	Logistic	μ= -0.45166, σ= 5.25208	0.9528	0.8160
For drought case only				
Duration	Log Logistic	μ= 0.3145, σ= 1.7105	0.2338	0.2750
Severity	GEV	k= 0.3909, σ= 2.5901, μ= 3.1580	0.9695	0.9940
Intensity	GEV	k= -0.1927, σ= 0.5662, μ= 1.1355	0.9897	0.9989
EMI	GEV	k= -0.3997, σ= 0.4466, μ= -0.0798	0.3697	0.7620
Nino4 SST	Weibull	a= 55.9410, b= 28.8905	0.8495	0.9046

Mathematical equations of marginal distribution, KS and AD test:GEV:

$$f(x|k,\mu,\sigma) = \left(\tfrac{1}{\sigma}\right)\exp\left(-\left(1+k\tfrac{(x-\mu)}{\sigma}\right)^{\frac{-1}{k}}\right)\left(1+k\tfrac{(x-\mu)}{\sigma}\right)^{-1-\frac{1}{k}}, \text{ Weibull:}$$

$$f(x|a,b) = \tfrac{b}{a}\left(\tfrac{x}{a}\right)^{b-1}e^{-\left(\tfrac{x}{a}\right)^{b}}, \text{ Logistic: } f(x|\mu,\sigma) = \frac{\exp\left\{\tfrac{x-\mu}{\sigma}\right\}}{\sigma\left(1+\exp\left\{\tfrac{x-\mu}{\sigma}\right\}\right)^2}, \text{ Log Logistic:}$$

$$f(x|\mu,\sigma) = \tfrac{1}{\sigma}\tfrac{1}{x}\frac{e^{\frac{\log(x)-\mu}{\sigma}}}{\left(1+e^{\frac{\log(x)-\mu}{\sigma}}\right)^2}; \quad x\geq 0, \ AD^2 = -n-\tfrac{1}{n}\sum_{i=1}^{n}(2i-1).[\ln F(X_i)+\ln(1-F(X_{n-i+1}))],$$

$$KS = \max_{1\leq i\leq n}\left(F(x_i)-\tfrac{i-1}{n},\tfrac{1}{n}-F(x_i)\right)$$

not affect the correlation between variables, i.e., the observed and copula generated Kendall's τ values are similar.

$$AIC: = -2\sum_{i=1}^{N}\ln[C(u_i,v_i)|\theta]+2k \tag{6.20}$$

$$BIC: = -2\sum_{i=1}^{N}\ln[C(u_i,v_i)|\theta]+\ln(N)k \tag{6.21}$$

$$LogLik = \ln L(x,y;params,\theta)$$
$$= \ln L_C(F_X(x),F_Y(y);\theta)+\ln L_X(x;params)+\ln L_Y(y;params) \tag{6.22}$$

TABLE 6.1 Copula parameters and goodness-of-fit measures of the fitted copula models.

(a) For SPEI with CIs

	Copula	Par	ᵃKendall's tau	Kendall's tau	Upper TD	Lower TD	LogLik	AIC	BIC
Bivariate model									
$C(u_{speir}, u_{nino4})$	Frank	−2.34	−0.25	−0.25	N/A	N/A	47.36	−92.72	−88.19
$C(u_{speir}, u_{soi})$	Gumbel	1.29	0.24	0.23	0.29	N/A	51	−99.99	−95.46
$C(u_{nino4}, u_{soi})$	Frank	−7.85	−0.6	−0.6	N/A	N/A	331.07	−660.14	−655.61
Trivariate model									
$C(u_{nino4}, u_{soi} \mid u_{spei})$	Frank	−6.79	—	−0.55	N/A	N/A	372.49	−738.98	−725.39

(b) For drought properties with CIs

	Copula	Par	ᵃKendall's tau	Kendall's tau	Upper TD	Lower TD	LogLik	AIC	BIC
Bivariate model									
$C(u_1, u_2)$	Clayton	0.65	0.1723	0.25	N/A	0.35	2.53	−3.06	−1.48
$C(u_3, u_5)$	Frank	1.89	0.1968	0.2	N/A	N/A	1.56	−1.13	0.45
$C(u_4, u_5)$	Frank	2.02	0.2032	0.22	N/A	N/A	1.81	−1.62	−0.04

(c) For drought properties

	Copula	Par	[a]Kendall's tau	Kendall's tau	Upper TD	Lower TD	LogLik	AIC	BIC
Bivariate									
$C(u_d, u_s)$	Gumbel	4.36	0.81	0.77	0.83	N/A	35.81	−69.63	−68.04
$C(u_d, u_i)$	Frank	4.87	0.47	0.45	N/A	N/A	8.45	−14.89	−13.31
$C(u_s, u_i)$	Clayton	2.74	0.61	0.58	N/A	0.78	18.86	−35.72	−34.13
Trivariate									
$C(u_s, u_d, u_i)$	Frank	8.90	—	0.63	N/A	N/A	47.99	−89.99	−85.24

AIC, Akaike Information Criterion; BIC, Bayesian Information Criteria; LogLik, Log-likelihood; TD, tail dependence ; u_1, uniform Duration; u_2, uniform Nino4 SST; u_3, uniform Severity; u_4, uniform Intensity; u_5, uniform EMI.
[a]Kendall's tau of observed data.

The *params* are the parameters of marginal distribution of X and Y variables.

The three different sets of copula-drought analysis were carried out: (1) bivariate and the trivariate copula of SPEI with *CIs*, (2) bivariate copula of drought properties with *CIs*, and (3) bivariate and trivariate copula of drought properties. Table 6.1 lists the copula parameters and corresponding goodness-of-fit measures for each set of analysis. Because the SPEI, SOI, and EMI are all standardized indices that represent drought and synoptic-scale climate patterns where the values of these can range from being positive to negative, the Frank copula was found to be the most appropriate fit mainly because of its properties related to radial symmetry (Nelsen, 1999).

2.2.5 Dependence modeling

A potential avenue to validate the dependence structure of bivariate copula models is a Chi-plot proposed by Fisher and Switzer (1985). Chi-plot is based on two types of statistics: Chistatistic, χ_i and Lambda-statistic, λ_i viz (Genest and Favre, 2007):

$$\chi_i = \frac{\widehat{F}_{X,Y}(u_i, v_i) - \widehat{F}_X(u_i)\widehat{F}_Y(v_i)}{\sqrt{\widehat{F}_X(u_i)\left(1 - \widehat{F}_X(u_i)\right)\widehat{F}_Y(v_i)\left(1 - \widehat{F}_Y(v_i)\right)}} \qquad (6.23)$$

$$\lambda_i = 4\,\mathrm{sgn}\left(\overline{F}_X(u_i), \overline{F}_Y(v_i)\right) * \max\left(\overline{F}_X(u_i)^2, \overline{F}_Y(v_i)^2\right) \qquad (6.24)$$

where $i = 1, \ldots, N$ is the set of observations for (u_i, v_i), and $\widehat{F}_X, \widehat{F}_Y$, and $\widehat{F}_{X,Y}$ are the empirical distribution functions of the uniform random variables u and v, and $\widetilde{F}_X = \widehat{F}_X - 0.5$ and $\widetilde{F}_Y = \widehat{F}_Y - 0.5$.

In accordance with Eqs. 6.23 and 6.24, λ_i is adopted to estimate the distance between bivariate data points (u_i, v_i) and the median of the data set, while the χ_i corresponds to a correlation coefficient between dichotomized values of X and Y. Thus, a positive λ_i means that both u_i and v_i are large relative to their respective medians, or both small, whereas a negative corresponds to u_i and v_i being on opposite sides of their respective medians. Asymptotically, $\chi_i \cap \left(0, \frac{1}{N}\right)$ and $\lambda_i \sim \cup[-1, 1]$ under the condition of independence where a value of χ_i close to zero can indicate that the properties X and Y are independent of each other, i.e., $F_{X,Y} = F_X F_Y$. When there is a positive dependence margin between the properties, the pairs of (λ_i, χ_i) tend to be located above the confidence band and vice versa for negative dependent margins, while the points enclosed within the confidence band can indicate independence of the bivariate pair of the drought properties.

To further establish the statistical fitness of the bivariate joint dependence structure, the K-plot (also known as Lambda plot), which is a rank-based graphic tool for visualizing the dependence structure between two associated variables (Genest and Boies, 2003), was prepared. For any observation pair (u_i, v_i), where $i = 1, \ldots, N$, the K-plot is able to consider the two bivariate quantities that comprise first the ordered values of the empirical bivariate distribution function $H_i := \widehat{F}_{uv}(u_i, v_i)$ and second the quantity $W_{i:N}$ that shows the expected values of the order statistics from a random sample of size N of random variable, $W = C(X,Y)$. Under the null hypothesis of independence between the bivariate pairs X and Y, this is written as (Genest and Favre, 2007)

$$W_{in} = N \binom{N-1}{i-1} \int_0^1 \omega k_0(\omega)(K_0(\omega))^{i-1}(1 - K_0(\omega))^{N-i} d\omega \qquad (6.25)$$

where $K_0(\omega) = \omega - \omega \log(\omega)$ and $k_0(\cdot)$ is the corresponding copula density.

A plausible physical interpretation of the K-plot is that, if the datum points in the distribution lie on the diagonal, then the X and Y are independent of each other, whereas any deviation from this line is expected to indicate significant dependence. Importantly, when there is a positive dependence between bivariate properties, datum points are expected to lie above the diagonal, and vice versa for the negative dependence. Also, the degree of positive dependency between the drought-risk properties is likely to be strongest when the datum points $(W_{i:N}, H_i)$ are situated on the curve $K_0(\omega)$, above the diagonal $y = x$ line and the perfect negative dependence between X and Y exist when the points $(W_{i:N}, H_i)$ lie on x-axis (Schirmacher and Schirmacher, 2008).

The joint variables X and Y can be related in terms of their extremes (i.e., minimum and maximum) values, whereby the tail dependence notion is expected to relate to their amount of dependence in the upper-right or lower-left quadrant tail of bivariate joint distribution. The interpretation of upper (*lower*) tail dependence parameter is that the probability of one margin exceeds a high (*low*) threshold under the condition where the other margin exceeds a high (*low*) (Poulin et al., 2007). The formulae for the upper and lower tail dependence parameters are (Joe, 1997) as follows:

$$\lambda_U = \lim_{u \to 1^-} \frac{1 - 2u + (u,u)}{1-u}; \lambda_L \lim_{u \to 0^+} \frac{C(u,u)}{u} \qquad (6.26)$$

3. Results and discussion

3.1 Applications on the Standardized Precipitation-Evapotranspiration Index and climate indices

A random sample of 250 datum points was simulated to yield the Chi-plot of the SPEI versus the Niño 4 SST (Fig. 6.2A) and the SPEI versus the SOI data (Fig. 6.2C). There exists a strong positive dependence as most of the simulated datum points are situated outside the confidence band of 0.1.

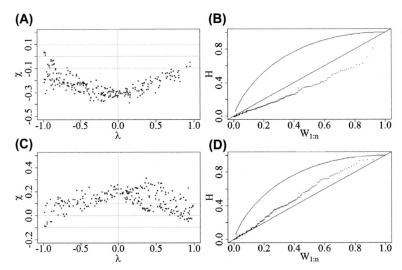

FIGURE 6.2 Chi-plots with "confidence band" at $\alpha = 0.1$ (*dashed lines*) for SPEI with Niño 4 SST (A) and SOI (C). *K*-plots with the straight line ($y = x$) corresponding to the case of independence and a smooth curve $K_0(\omega)$, associated with perfect positive dependence of SPEI and Niño 4 SST (B) and SOI (D) joint distribution generated from 250 random samples using Frank (for Niño 4 SST) and Gumbel (for SOI) copula. *SOI*, Southern Oscillation Index; *SPEI*, Standardized Precipitation-Evapotranspiration Index; *SST*, sea surface temperature.

Notwithstanding this, it is important that for the bivariate case of Niño 4 SST and SPEI joint behaviors, almost all of the simulated datum points are situated outside this confidence band and that the magnitude of χ_i is significantly large, indicating very strong joint dependence. In fact, the Chi-plot shows a bent course with the Chi-values concentrated around the zero mark of λ_i, confirming that the dependency is strong around the median of the distribution than around the tails.

In contrast, the bivariate combinations of the SPEI and SOI data (Fig. 6.2C) show strong dependence around the median and a weak dependence around the tail ends of the plot.

Fig. 6.2B,D shows the *K*-plot for bivariate joint distributions of Niño 4 SST and SPEI and SOI and SPEI using the simulated sample of 250 random points. It is evident that the lower tailed bivariate pair of data points appears to be independent, as the points lie closer to the diagonal line $y = x$, while the higher values indicate stronger negative (positive) dependence between the Niño 4 SST and SPEI (SPEI and SOI).

To address the limitations posed by the different marginal distributions, copula parameters from the selected bivariate and trivariate copula functions were applied to simulate margins of 2000 random pairs of (u_{nino4}, u_{spei}) and (u_{soi}, u_{spei}) that were back-transformed to their original units using their respective inverse marginal distributions. Fig 6.3A and B show the observed

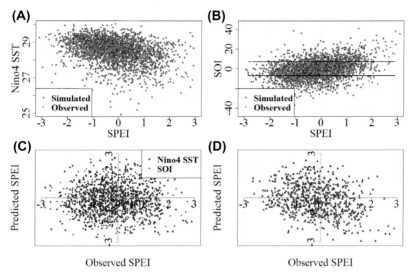

FIGURE 6.3 Comparison of observed SPEI with 2000 random samples (A, B) simulated from Frank (A) and Gumbel (B) copula. Scatterplot of observed versus predicted SPEI given information of Niño 4 SST and EMI using bivariate (C) Frank (for Niño 4 SST; *red*) and Gumbel (for SOI; *blue*) copula and using trivariate (D) Frank copula given combined information of Niño SST and SOI. *SOI*, Southern Oscillation Index; *SPEI*, Standardized Precipitation-Evapotranspiration Index; *SST*, sea surface temperature.

versus simulated SPEI and Niño 4 SST, and the SPEI and SOI, respectively. The black solid lines separate the El Niño (SOI < −7.0) and La Niña phases (SOI > 7.0) while everything in between corresponds to neutral ENSO conditions. The observed values in the scatterplot clearly overlap the simulated samples.

Fig 6.3C and D show the joint probabilistic predictions, expressed via the conditional bivariate and the conditional trivariate copula, respectively. To address the stochastic nature of drought and uncertainty in its predictions, the copula-statistical model was applied to generate an ensemble of 1000 predictions that were then averaged to obtain a single set of predictions with the same length (or number of datum points) as observed time series particularly for comparison purposes. The blue (*red*) scatters are the predicted SPEI for the random probabilities using features available from the SOI (*Niño 4 SST*) as shown in Fig. 6.3C based on their respective bivariate copula functions.

Using the trivariate Frank copula, the SPEI data were also predicted conditioned on the Niño 4 SST and SOI data shown in Fig. 6.3D. Notice that the observed and predicted values do not form a linear relationship (which are not expected to) as the predictions are to be made for any random probability of the event. Note that in the context of the present case study, the simulations and predictions differ from each other, that is, in the simulation process, all variables are randomly generated at the same time using the optimal copula parameters. Conversely, the predictions are conditional where one variable is then evaluated at a certain probability based on the information available from the other variable(s), which in fact may assist decision-makers in predicting the overall risk of the drought event. It is important to note that the copula-statistical models are able to predict the SPEI data directly based on the information derived from the Niño 4 SST and the SOI data, which act as synoptic-scale precursors of drought events.

Table 6.2 provides a comparison of the basic statistics of the data representing the observed and the predicted SPEI. Importantly, these results show relatively small differences between the observed and the predicted SPEI statistics. The absolute difference in the mean value of the trivariate copula-based prediction is slightly larger (≈ 0.0161) compared with the bivariate copula-based predictions (≈ 0.0045 and 0.0008). The root mean squared error (RMSE) computed between the observed and predicted SPEI is also slightly larger (≈ 1.5990) in the trivariate copula-based prediction case compared with approximately 1.3742 and 1.3545 from the bivariate copula model predictions. The prediction errors of SPEI using copula-statistical model are generally small, suggesting that the conditional probability bivariate or trivariate copula-based predictions have good accuracy and are potentially suitable for application in prediction modeling.

Figs 6.4 and 6.5 show conditional probability of the SPEI occurrence expressed via the joint probabilistic bivariate and trivariate copula models, respectively. The negative SPEI indicates the drought conditions given in terms of their standard deviations. In Fig. 6.4A, conditional on negative SOI values, the probability of obtaining a negative SPEI corresponding to drought conditions is higher in contrast to obtaining a lower probability with positive SOI values. For instance, to obtain an SPEI $= -2.0$ (i.e., when the normalized deficit in water resources is below two standard deviations relative to the comparative base period), the probability is found to be ~ 0.98 with conditional SOI $= -25.0$, whereas when the SOI $= 25.0$, the probability is ~ 0.57. Similarly, the probability of obtaining negative SPEI conditional on the Niño 4 SST data is higher with a larger Niño 4 SST value, as shown in Fig. 6.4B. For instance, to obtain an SPEI $= -1$, the probability is found to be ~ 0.98 (~ 0.88) with Niño 4 SST $= 30°C$ ($27°C$). Similarly, Fig. 6.5 shows probability distribution of SPEI conditional on Niño 4 SST and SOI simultaneously.

3.2 Applications on drought properties and climate mode indices

The drought properties represented as D, S and I were also modeled with climate mode indices using the optimal copulae developed in this case study. The selected bivariate copula for D and Niño 4 SST was Clayton, while for S and EMI and I and EMI, it was Frank. Using copula parameters, random values for D, S and I and their corresponding CI were simulated and predicted, as

TABLE 6.2 Comparison statistics for observed and predicted SPEI values for bivariate and trivariate joint copula models.

Conditional variable	Minimum	1st quartile	Median	Mean	3rd quartile	Maximum	AD_{mean}	AD_{median}	RMSE
Observed SPEI									
N/A	−2.8490	−0.8938	−0.2018	−0.1459	0.5683	2.8930	−	−	−
Bivariate model predictions									
Nino4	−2.5170	−0.7791	−0.1797	−0.1504	0.4576	2.7430	0.0045	0.0221	1.3732
SOI	−2.5230	−0.7492	−0.1399	−0.1451	0.4921	2.5130	0.0008	0.0619	1.3545
Trivariate Predictions									
[Nino4, SOI]	−2.9470	−0.8326	−0.1543	−0.1620	0.4869	2.4540	0.0161	0.0475	1.5990

AD, absolute difference; RMSE, root mean squared error between observed and predicted SPEI.

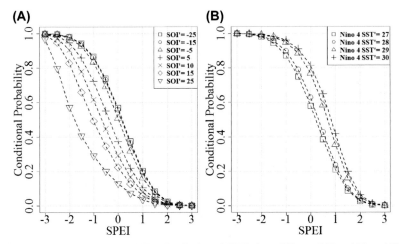

FIGURE 6.4 Conditional probability distribution of SPEI given different SOI and Niño 4 SST values using bivariate Gumbel (A) and Frank (B) copula. The Niño 4 SST' is in degrees Celsius. *SOI*, Southern Oscillation Index; *SPEI*, Standardized Precipitation-Evapotranspiration Index; *SST*, sea surface temperature.

shown in Fig. A1 in the Supplementary material. The comparison statistics between observed and predicted *D*, *S* and *I* are given in Table 6.3. The absolute difference in mean (AD$_{mean}$) is the smallest for the prediction of drought intensity, followed by the drought duration and drought severity. Because the information of EMI is used to predict both the intensity and the severity, and both use bivariate Frank copula, the smaller AD$_{mean}$ for intensity could be due to the larger Frank copula parameter ($\theta = 2.02$). The larger copula parameter indicates a higher dependence between the bivariate pair of variables. Importantly, the RMSE is also found to be the smallest for prediction of intensity.

An estimation of drought return period, which represents how often a drought is likely to occur (and hence, is closely linked to drought-risk on agriculture and water resources), is an important quantity in drought studies. Because every drought is different with the general notion that some events may persist as long drought episodes, an estimate of return period is crucial for designing hydraulic facilities (e.g., dams and irrigation systems). In this case study, the average interarrival time (i.e., *EL*) between drought events was found to be approximately 13.53 months. Fig. 6.6 shows the isopleths of the bivariate joint return period for the "AND" (a, c, e) and "OR" (b, d, f) cases of drought properties. The contours show the return periods (in years) for given drought properties and *CIs*. Notice that the isopleth patterns for the "AND" and the "OR" cases are unique because those for the various joint return periods defined by the "AND" case are bounded by horizontal (x) and vertical (y) axes, whereas there is no bound for isopleths for the specific return periods defined by the "OR" case.

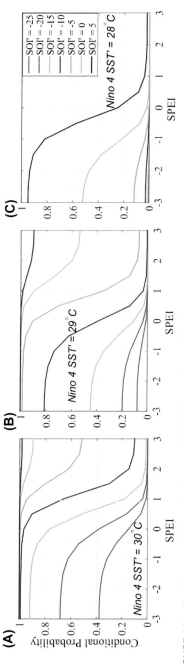

FIGURE 6.5 Conditional probability distribution of SPEI different Niño 4 SST (°C) and SOI values using trivariate Frank copula. Legend applies to all panels. *SOI*, Southern Oscillation Index; *SPEI*, Standardized Precipitation-Evapotranspiration Index; *SST*, sea surface temperature.

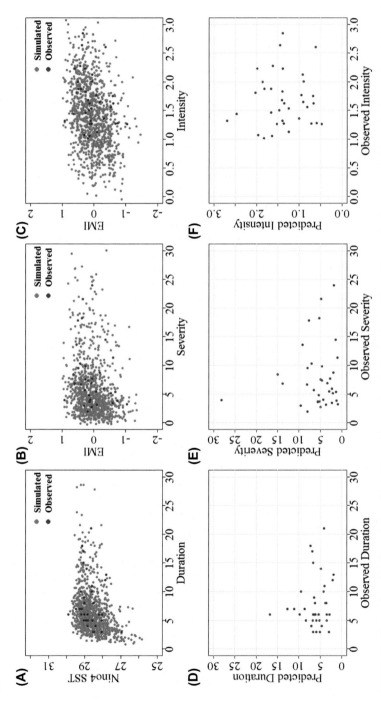

FIGURE A1 Comparison of observed data with 2,000 random samples (A, B, C) simulated from Clayton (A, B) and Frank (C) copula. Scatter plot of observed versus predicted duration (D), severity (E), and intensity (F) given information of Niño 4 SST (D) and EMI (E, F).

TABLE 6.3 Comparison statistics for observed and predicted duration (from Nino 4 SST), severity (from EMI), and intensity (from EMI) from the bivariate copula models.

	Minimum	1st quartile	Median	Mean	3rd quartile	Maximum	AD$_{mean}$	AD$_{median}$	RMSE
Duration									
Observed	3	5	6	8	10	21	–	–	–
Predicted	1.8380	4.0570	6.0170	6.1930	6.9100	16.7500	1.8073	0.0167	5.9194
Severity									
Observed	1.9450	3.7840	6.2840	7.8760	9.6080	24.0200	–	–	–
Predicted	0.9066	2.2670	4.8740	5.7090	7.2210	28.1900	2.1670	1.4091	8.0218
Intensity									
Observed	1.0200	1.3070	1.6230	1.6830	1.9090	2.8490	–	–	–
Predicted	0.4731	0.9193	1.3880	1.3620	1.6560	2.6970	0.3209	0.2354	0.7864

AD, absolute difference; RMSE, root mean squared error between observed and predicted values.

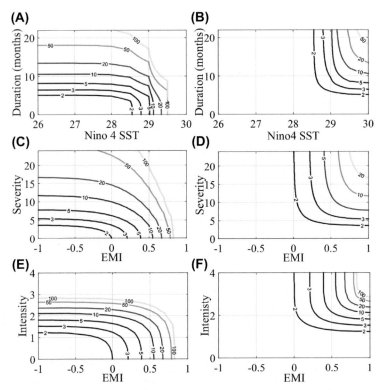

FIGURE 6.6   Bivariate drought return period for AND and OR case for duration (A, B) using Gumbel, severity (C, D) using Frank and intensity (E, F) using Clayton copula. *EMI*, El Niño Modoki Index; *SST*, sea surface temperature.

The return periods using two bivariate ("AND" and "OR") cases also differ markedly. Take for instance, for D versus Niño 4 SST joint pair (Fig. 6.6A,B), for any particular value of D and the corresponding Niño 4 SST, the estimated return period T_{OR} is less than the T_{AND} return period. In fact, for the "AND" case, the isopleths representing a given bivariate D and Niño 4 SST, I and EMI, and S and EMI, the joint pairs are situated over a much larger return period. For example, an estimated duration of 10 months and Niño 4 SST of 29°C, the "OR" case yields a return period of ∼3 years (Fig. 6.6B), whereas the respective bivariate equivalent return period, i.e., "AND" case, registers about ∼20 years (Fig. 6.6A). Similarly, the bivariate return periods of severity and intensity with EMI for "AND" and 'OR' cases are shown. The return period values using "AND" case suggest that any drought event with specific thresholds of D, S, I and CIs to occur simultaneously will have larger return periods as such events are relatively rare compared with when each variable is analyzed separately using "OR" case.

The bivariate conditional return periods of D, S and I were calculated (Fig. 6.7). The results show an increase in the return period as conditioned variable value increases. For instance, the return periods to obtain a drought event with a duration $D = 20$ months under Niño 4 SST conditions (27°C, 28°C, and 29°C) are 69.94, 85.33, and 544.82 years, respectively (Fig. 6.7A). Similarly, the return periods for severity $S = 10.0$ (*and Intensity, $I = 2.0$*) under EMI conditions (0, 0.2, 0.4, and 0.6) are 17.81 (17.46), 33.90 (33.01), 94.52 (91.46), and 493.18 (475.01) years, respectively (Fig. 6.7B,C). This shows that as the magnitude of ENSO indicators shifts toward their extreme values, in the direction of enhancing the drought conditions, the return period of drought properties increases as well, suggesting the rarity of such extreme events. The results thus reveal the importance of jointly modeling the drought properties to encapsulate the true associative behavior of drought properties with climate mode indices and how they manifest to produce an overall drought-risk in agricultural systems.

3.3 Applications on drought properties

Many previous studies have shown the interrelation between drought D, S and I properties. For agriculture and water management, estimating return period of a given drought event is crucial. As such, using the marginal and joint probabilities, the univariate, bivariate, and trivariate return periods were also calculated. Table 6.4 lists the marginal probabilities and univariate, bivariate, and trivariate return periods. The physical interpretation of this plot is greatly important for understanding drought-risk. Take for instance $D = 10.22$ months, the univariate return period is 5 years, whereas the joint bivariate return period with S *(or I)* for "AND" case is 5.87 (*or* 17.69) years and for "OR" case, it is 4.36 (*or* 2.91) years. Similarly, when all three variables are combined, the joint return period using the trivariate model for "AND" (*or* "OR") case is 31.25 (*or* 3.33) years. The dramatic difference between the bivariate (and trivariate) return period compared with the univariate counterpart clearly outlays the significance of considering the joint behavior of different properties to avoid an underestimation of the overall risk posed by a given drought event. Although univariate return period is somewhat limited in its ability to represent the joint behavior of more than one drought property, it is nonetheless important for drought-risk assessment when one random variable (e.g., duration) is considered relevant over the others.

The data in Table 6.4 show that the univariate return period is less than the joint return period T_{AND} and larger than T_{OR}. The trivariate T_{OR} is less than both univariate and bivariate ones. This is because adding one or more variables in the drought prediction model makes the exceedance probabilities $P(X_1 \geq x_1, X_2 \geq x_2, X_3 \geq x_3)$ smaller than two bivariate $P(X_1 \geq x_1, X_2 \geq x_2)$ or a univariate $P(X_1 \geq x_1)$ cases. This is evident across the rows where the marginal probabilities appear to be decreasing from the univariate to the trivariate models.

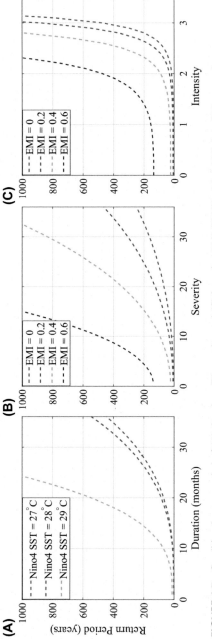

FIGURE 6.7 Conditional return period of (A) duration, (B) severity, and (C) intensity. *EMI*, El Niño Modoki Index; *SST*, sea surface temperature.

TABLE 6.4 Univariate and copula-based joint return periods (T_{AND} & T_{OR}; years) of drought duration, severity, and intensity with bivariate and trivariate joint copula models. $F(\bullet)$ is the marginal probability.

	Univariate model					Bivariate model									Trivariate model		
T	$F(x)$	D_d	S_s	I_i	$F(d,s)$	T_{AND}	T_{OR}	$F(d,i)$	T_{AND}	T_{OR}	$F(s,i)$	T_{AND}	T_{OR}	$F(d,s,i)$	T_{AND}	T_{OR}
5	0.77	10.22	9.96	2.00	0.74	5.87	4.36	0.61	9.25	2.91	0.67	9.52	3.39	0.66	11.97	3.33
10	0.89	14.54	16.27	2.26	0.87	11.91	8.62	0.79	27.69	5.41	0.81	30.95	5.96	0.80	29.49	5.51
20	0.94	17.34	19.40	2.62	0.93	24.00	17.14	0.89	91.77	10.41	0.90	109.42	11.01	0.88	50.83	9.26
50	0.98	20.34	23.50	2.80	0.97	60.24	42.73	0.96	501.06	25.41	0.96	629.47	26.03	0.94	90.13	19.64
100	0.99	21.00	24.02	2.85	0.99	120.65	85.38	0.98	1906.29	50.41	0.98	2444.97	51.04	0.97	150.67	36.48

4. Further discussion

This case study has explored and developed a probabilistic drought prediction model using Archimedean copulae (i.e., Clayton, Gumbel, and Frank) which have rarely been applied for drought-risk modeling in the drought-prone SEQ region. It therefore represents an important advancement in terms of potential applications for agriculture, water management, and related socioeconomic sectors where drought presents as a significant risk. The need for drought-risk assessment in a nonstationary climate has been emphasized in many studies (Mpelasoka et al., 2008; Verdon-Kidd and Kiem, 2010), and in recent studies, copula models have been adopted to illustrate the dependence structure of drought properties under different ENSO conditions (Ganguli and Reddy, 2014; Wong et al., 2009). The study by Wong et al. (2009) in New South Wales Australia provided strong evidence of dependence between drought duration, intensity, and severity under different ENSO conditions using Elliptical and Archimedean copulae. To further the investigation, we adopted three, popular in hydrology, Archimedean copulae to successfully derive joint distributions of the SPEI drought index and duration, severity, and intensity properties with ENSO climate mode indices, before developing the probabilistic prediction models. The probabilistic prediction is considered as an essential tenet for a number of end users such as water resource managers and farmers to develop management strategies for informed decision-making (Goddard et al., 2001).

It is imperative to mention that this is the first study to investigate a series of climate mode indices to produce conditional probabilistic prediction of the drought index SPEI and drought properties (duration, severity, and intensity) in Australia. The SPEI is a relatively new drought index that, in addition to precipitation, uses potential evapotranspiration in determining drought conditions. The incorporation of potential evapotranspiration in the equation allows the index to capture the increased temperature impact on water demand (Vicente-Serrano et al., 2010), and while Australia's drought is often influenced by warming and cooling phases of ENSO, the choice of SPEI justifies the drought-risk assessment in this case study. The application of copula and SPEI in recent investigations have demonstrated the significance of SPEI for drought-risk assessments in China (Chen et al., 2016; Fan et al., 2017), and the outcomes in our case study have highlighted the suitability of SPEI and copula-based drought modeling for Australian droughts. Furthermore, the forecasting of seasonal rainfall has been considered very important for sustainable agricultural management in Australia (Moeller et al., 2008; Nguyen-Huy et al., 2017; Stephens et al., 1994) and in the USA (Khedun et al., 2014; Mishra et al., 2015). In this case study, we took a further step and addressed the importance of drought forecasting using 3-month timescale SPEI that is important for providing information to agricultural sectors.

This case study has used the vine copulae (Brechmann and Schepsmeier, 2013) that have remained largely unexplored in the area of drought modeling and predictions in Australia. Unlike previous studies that have investigated joint relationships of drought properties only (e.g., duration with severity, duration with intensity, severity with intensity, or all three as trivariate), this case study has developed the copula-based probabilistic prediction models for SPEI and drought properties (D, S and I) using ENSO indicators (Niño 4 SST, SOI, and EMI) as predictors, providing significant novelty via probabilistic prediction of droughts with universal indicators of atmospheric circulation that modulates the supply of water and regulation of solar energy at the ground surface. The results are significant for activities such as agriculture because vine copulae allow the modeling of high-dimensional dependency between drought and its climate variables, and application of vines can be extended to both bivariate and trivariate copula-based modeling. Because application of vines in hydrological studies is relatively new (Gräler et al., 2013; Khedun et al., 2014; Nguyen-Huy et al., 2017; Vernieuwe et al., 2015), its application on drought modeling in Australia is a novel contribution.

In contrast to previous studies that have produced the simulations for assessing dependence structure for two or more variables simultaneously (Wong et al., 2008), this case study successfully predicted the core variables (SPEI, D, S and I) based on the information available from the climate indices (Niño 4 SST, SOI, and EMI). For better suitability in hydrology, the vine copula offer benefits of high-dimensional dependency between predictor and predictand, has better computational tractability (Joe, 1996), and is flexible in deducing multidimensional dependence structure using *tree* methods. Given the flexibility, vine copula offers a potential for multivariate dependence modeling by decomposing multidimensional multivariate density into bivariate copulae and its application in other climatic regions as well.

The initial stages of analysis involved evaluation of correlation of SPEI and its properties with 13 climate indices, and the subsequent analysis used only Niño 4 SST, SOI and EMI as these had statistically significant correlation and that are indicators of the ENSO event. For SPEI prediction, 10 indices had statistically significant correlations with SPEI; however, we selected only Niño 4 SST and SOI as they had highest correlations and are independently estimated, that is, the Niño 4 SST is solely based on temperature, while SOI is based on pressure. The analysis yielded satisfactory simulation and prediction by utilizing the most relevant, yet few climate indices. Using evaluation metrics, the results captured emphatically the joint dependence structure between predictors and predictand and highlights the importance of jointly predicting drought and its properties where correlation between predictors and predictand can be established. It is possible, however, that further investigation could incorporate other climate indices that have been found to influence Australian droughts in earlier studies (Ummenhofer et al., 2009; van Dijk et al., 2013; Verdon-Kidd and Kiem, 2009) for drought predictions.

In spite of the significant outcomes, this pilot study has shortcomings that create the opportunity for a follow-up work. While we used only the SPEI on a 3-month timescale and for only one case study location in the agricultural Murray—Darling Basin region, the results may be applicable for the study location considered owing to the regional nature of drought, yet the methodology can be easily adopted to any case. Therefore, the practical application of the methodology adopted in this case study requires site-specific SPEI for derivation of drought properties and selection of marginal distribution for predictors and predictand to be used in copula models. Also, to be consistent with the Australian Bureau of Meteorology definition of drought (and associated drought response) (Mpelasoka et al., 2008), we evaluated droughts with $D \geq 3$ months duration for modeling. The models were successful in yielding predictions for drought indicator and its properties conditioned on climatic indices that modulate droughts in Australia. As an alternative, the copula-based probabilistic predictive models can be applied to flash droughts that could be in the form of heat wave or precipitation deficit over relatively short period (Mo and Lettenmaier, 2016). In spite of the limited scope of this work, a follow-up study could implement vine copulae to droughts on different timescales and for study locations with varying climatic conditions, short- or long-term droughts, and flash droughts for a wide variety of statistical prediction modeling. This would provide appropriate information to agriculturists and farmers for their specific cropping needs and demand for water/soil moisture for different crops.

5. Summary

A study of the joint behavior of drought and its properties with climate indices is critical for water resource planning and drought-risk management. Drought is stochastic in nature where the duration, severity, and peak intensity property of various episodes can vary from one event to another, yet these properties can be strongly correlated with each other and therefore contribute differently to the overall risk of a given event. These properties are also influenced by large-scale oscillations in the atmosphere, such as the ENSO. In this work, we used vine copulae, as a new contribution to Australian drought-risk study, to model joint behavior of the SPEI and its drought properties (duration, severity, and intensity) with climate indices (Niño 4 SST, SOI and EMI). The Niño 4 SST, SOI and EMI are indicators of ENSO, which modulate drought conditions in Australia, where Niño 4 SST >26°C, SOI < −7, and EMI >0 represent El Niño, i.e., warm phase of ENSO that enhances the drought conditions.

Archimedean copulae (i.e., Clayton, Gumbel, and Frank) were found to be the most suitable models for modeling joint behavior of multivariable. The findings in this case study demonstrated the joint dependence structure of the SPEI and drought properties with climate indices, while the vine copulae illustrated the usefulness of conditional probability—based drought predictions. The comparison between predicted and observed values showed similarity, indicating satisfactory model performance. This technique can be applied to other study regions as well. Similarly, the prediction of D, S and I using Clayton and Frank copulae conditional on Niño 4 SST and EMI yielded small absolute difference in mean values from observed.

Overall, this case study demonstrated the feasibility of copula-statistical models in understanding the joint relationships between drought variables and climate indices in the present case study region. However, the practical relevance of these copula-statistical models can be further enhanced in a separate study by considering many other climate indices on other timescales (e.g., 1-, 6-, 12-, 24-month, seasonal, annual, decadal) using multivariate copula-statistical models, making them more applicable to a better assessment of the combined risk of drought events.

References

Aas, K., Czado, C., Frigessi, A., Bakken, H., 2009. Pair-copula constructions of multiple dependence. Insurance: Mathematics and Economics 44 (2), 182—198.

Bedford, T., Cooke, R.M., 2002. Vines: a new graphical model for dependent random variables. Annals of Statistics 1031—1068.

Bothe, O., Fraedrich, K., Zhu, X., 2010. The large-scale circulations and summer drought and wetness on the Tibetan plateau. International Journal of Climatology 30 (6), 844—855.

Brechmann, E.C., Schepsmeier, U., 2013. Modeling dependence with C-and D-vine copulas: the R-package CDVine. Journal of Statistical Software 52 (3), 1—27.

Chen, X., Koenker, R., Xiao, Z., 2009. Copula-based nonlinear quantile autoregression. The Econometrics Journal 12 (s1), S50—S67.

Chen, Y.D., Zhang, Q., Xiao, M., Singh, V.P., Zhang, S., 2016. Probabilistic forecasting of seasonal droughts in the Pearl River basin, China. Stochastic Environmental Research and Risk Assessment 30 (7), 2031—2040.

Das, P.K., Dutta, D., Sharma, J., Dadhwal, V., 2016. Trends and behaviour of meteorological drought (1901—2008) over Indian region using standardized precipitation—evapotranspiration index. International Journal of Climatology 36 (2), 909—916.

Dayal, KS., Deo, RC., Apan, AA., 2017a. Investigating drought duration-severity-intensity characteristics using the Standardized Precipitation-Evapotranspiration Index: case studies in drought-prone Southeast Queensland. J Hydrol Eng 23 (1), 05017029.

Dayal, K.S., Deo, R.C., Apan, A.A., 2017b. Drought modeling based on artificial intelligence and neural network algorithms: a case study in Queensland, Australia. In: Leal Filho W. (Ed.), Paper Presented at the Climate Change Adaptation in Pacific Countries: Fostering Resilience and Improving the Quality of Life, Climate Change Management. Springer, Cham. https://doi.org/10.1007/978-3-319-50094-2_11.

De Michele, C., Salvadori, G., 2003. A generalized Pareto intensity-duration model of storm rainfall exploiting 2-copulas. Journal of Geophysical Research: Atmosphere 108 (D2).

Deo, R., Sahin, M., 2015. Application of the artificial neural network model for prediction of monthly Standardized Precipitation and Evapotranspiration Index using hydrometeorological parameters and climate indices in Eastern Australia. Atmospheric Research 161, 65−81.

Deo, R.C., Byun, H.-R., Adamowski, J.F., Begum, K., 2016. Application of effective drought index for quantification of meteorological drought events: a case study in Australia. Theoretical and Applied Climatology 1−21.

Deo, R.C., Syktus, J., McAlpine, C., Lawrence, P., McGowan, H., Phinn, S.R., 2009. Impact of historical land cover change on daily indices of climate extremes including droughts in Eastern Australia. Geophysical Research Letters 36 (8).

Evin, G., Favre, A.C., 2008. A new rainfall model based on the Neyman-Scott process using cubic copulas. Water Resources Research 44 (3).

Fan, L., Wang, H., Wang, C., Lai, W., Zhao, Y., 2017. Exploration of use of copulas in analysing the relationship between precipitation and meteorological drought in Beijing, China. Advances in Meteorology 2017.

Fisher, N., Switzer, P., 1985. Chi-plots for assessing dependence. Biometrika 72 (2), 253−265.

Ganguli, P., Reddy, M.J., 2012. Risk assessment of droughts in Gujarat using bivariate copulas. Water Resources Management 26 (11), 3301−3327.

Ganguli, P., Reddy, M.J., 2014. Evaluation of trends and multivariate frequency analysis of droughts in three meteorological subdivisions of western India. International Journal of Climatology 34 (3), 911−928.

Genest, C., Boies, J.-C., 2003. Detecting dependence with Kendall plots. The American Statistician 57 (4), 275−284. https://doi.org/10.1198/0003130032431.

Genest, C., Favre, A.-C., 2007. Everything you always wanted to know about copula modeling but were afraid to ask. Journal of Hydrologic Engineering 12 (4), 347−368.

Goddard, L., Mason, S.J., Zebiak, S.E., Ropelewski, C.F., Basher, R., Cane, M.A., 2001. Current approaches to seasonal to interannual climate predictions. International Journal of Climatology 21 (9), 1111−1152.

González, J., Valdés, J.B., 2003. Bivariate drought recurrence analysis using tree ring reconstructions. Journal of Hydrologic Engineering 8 (5), 247−258.

Gräler, B., van den Berg, M., Vandenberghe, S., Petroselli, A., Grimaldi, S., De Baets, B., Verhoest, N., 2013. Multivariate return periods in hydrology: a critical and practical review focusing on synthetic design hydrograph estimation. Hydrology and Earth System Sciences 17 (4), 1281−1296.

Hanson, R.L., 1988. Evapotranspiration and droughts. In: Paulson, R.W., Chase, E.B., Roberts, R.S., Moody, D.W. (Eds.), Compilers, National Water Summary, pp. 99−104.

Helfer, F., Lemckert, C., Zhang, H., 2012. Impacts of climate change on temperature and evaporation from a large reservoir in Australia. Journal of Hydrology 475, 365−378.

Joe, H., 1996. Families of m-variate distributions with given margins and m (m-1)/2 bivariate dependence parameters. Lecture Notes-Monograph Series 120−141.

Joe, H., 1997. Multivariate Models and Multivariate Dependence Concepts. CRC Press.

Khedun, C.P., Mishra, A.K., Singh, V.P., Giardino, J.R., 2014. A copula-based precipitation forecasting model: investigating the interdecadal modulation of ENSO's impacts on monthly precipitation. Water Resources Research 50 (1), 580−600.

Kim, T.-W., Valdes, J.B., Aparicio, J., 2006a. Spatial characterization of droughts in the Conchos River Basin based on bivariate frequency analysis. Water International 31 (1), 50−58.

Kim, T.-W., Valdés, J.B., Yoo, C., 2006b. Nonparametric approach for bivariate drought characterization using Palmer drought index. Journal of Hydrologic Engineering 11 (2), 134−143.

Liu, S., Kang, W., Wang, T., 2016. Drought variability in inner Mongolia of Northern China during 1960−2013 based on standardized precipitation evapotranspiration index. Environmental Earth Sciences 75 (2), 1−14.

Liu, Z., Zhou, P., Chen, X., Guan, Y., 2015. A multivariate conditional model for streamflow prediction and spatial precipitation refinement. Journal of Geophysical Research: Atmosphere 120 (19).

McKee, T.B., Doesken, N.J., Kleist, J., 1993. The relationship of drought frequency and duration to time scales. In: Paper Presented at the Proceedings of the 8th Conference on Applied Climatology.

Mishra, A.K., Ines, A.V., Das, N.N., Khedun, C.P., Singh, V.P., Sivakumar, B., Hansen, J.W., 2015. Anatomy of a local-scale drought: application of assimilated remote sensing products, crop model, and statistical methods to an agricultural drought study. Journal of Hydrology 526, 15−29.

Mo, K.C., Lettenmaier, D.P., 2016. Precipitation deficit flash droughts over the United States. Journal of Hydrometeorology 17 (4), 1169−1184.

Moeller, C., Smith, I., Asseng, S., Ludwig, F., Telcik, N., 2008. The potential value of seasonal forecasts of rainfall categories—case studies from the Wheatbelt in Western Australia's Mediterranean region. Agricultural and Forest Meteorology 148 (4), 606−618.

Mpelasoka, F., Hennessy, K., Jones, R., Bates, B., 2008. Comparison of suitable drought indices for climate change impacts assessment over Australia towards resource management. International Journal of Climatology 28 (10), 1283−1292. https://doi.org/10.1002/joc.1649.

Nelsen, R.B., 1999. Introduction an Introduction to Copulas. Springer, pp. 1−4.

Nguyen-Huy, T., Deo, R.C., An-Vo, D.-A., Mushtaq, S., Khan, S., 2017. Copula-statistical precipitation forecasting model in Australia's agro-ecological zones. Agricultural Water Management 191, 153−172.

Poulin, A., Huard, D., Favre, A.-C., Pugin, S., 2007. Importance of tail dependence in bivariate frequency analysis. Journal of Hydrologic Engineering 12 (4), 394−403.

Rauf, U.F.A., Zeephongsekul, P., 2014. Analysis of rainfall severity and duration in Victoria, Australia using non-parametric copulas and marginal distributions. Water Resources Management 28 (13), 4835−4856.

Reddy, J.M., Ganguli, P., 2012. Application of copulas for derivation of drought severity−duration−frequency curves. Hydrological Processes 26 (11), 1672−1685.

Salvadori, G., De Michele, C., 2006. Statistical characterization of temporal structure of storms. Advances in Water Resources 29 (6), 827−842.

Schirmacher, D., Schirmacher, E., 2008. Multivariate dependence modeling using pair-copulas. Technical report.

South East Queensland Water Corporation (issuing body), 2015. Water for Life: your say on South East Queensland's water future 2015-2045. Ipswich. Qld SEQWater.

Serinaldi, F., Bonaccorso, B., Cancelliere, A., Grimaldi, S., 2009. Probabilistic characterization of drought properties through copulas. Physics and Chemistry of the Earth, Parts A/B/C 34 (10), 596−605.

Shiau, J., 2006. Fitting drought duration and severity with two-dimensional copulas. Water Resources Management 20 (5), 795−815.

Shiau, J.-T., Modarres, R., 2009. Copula-based drought severity-duration-frequency analysis in Iran. Meteorological Applications 16 (4), 481−489.

Shiau, J.-T., Shen, H.W., 2001. Recurrence analysis of hydrologic droughts of differing severity. Journal of Water Resources Planning and Management 127 (1), 30−40.

Sklar, M., 1959. Fonctions de répartition à n dimensions et leurs marges. Université Paris 8.

Stephens, D., Walker, G., Lyons, T., 1994. Forecasting Australian wheat yields with a weighted rainfall index. Agricultural and Forest Meteorology 71 (3−4), 247−263.

Ummenhofer, C.C., England, M.H., McIntosh, P.C., Meyers, G.A., Pook, M.J., Risbey, J.S., et al., 2009. What causes southeast Australia's worst droughts? Geophysical Research Letters 36 (4).

van Dijk, A.I.J.M., Beck, H.E., Crosbie, R.S., de Jeu, R.A.M., Liu, Y.Y., Podger, G.M., et al., 2013. The Millennium Drought in southeast Australia (2001−2009): natural and human causes and implications for water resources, ecosystems, economy, and society. Water Resources Research 49 (2), 1040−1057. https://doi.org/10.1002/wrcr.20123.

Verdon-Kidd, D.C., Kiem, A.S., 2009. Nature and causes of protracted droughts in southeast Australia: comparison between the federation, WWII, and big dry droughts. Geophysical Research Letters 36 (22). https://doi.org/10.1029/2009gl041067.

Verdon-Kidd, D.C., Kiem, A.S., 2010. Quantifying drought risk in a nonstationary climate. Journal of Hydrometeorology 11 (4), 1019−1031. https://doi.org/10.1175/2010jhm1215.1.

Vernieuwe, H., Vandenberghe, S., De Baets, B., Verhoest, N., 2015. A continuous rainfall model based on vine copulas. Hydrology and Earth System Sciences 19 (6), 2685−2699.

Vicente-Serrano, S.M., Beguería, S., López-Moreno, J.I., 2010a. A multiscalar drought index sensitive to global warming: the standardized precipitation evapotranspiration index. Journal of Climate 23 (7), 1696−1718.

Vicente-Serrano, S.M., Beguería, S., López-Moreno, J.I., Angulo, M., El Kenawy, A., 2010b. A new global 0.5 gridded dataset (1901−2006) of a multiscalar drought index: comparison with current drought index datasets based on the Palmer drought severity index. Journal of Hydrometeorology 11 (4), 1033−1043.

Vicente-Serrano, S.M., López-Moreno, J.I., Gimeno, L., Nieto, R., Morán-Tejeda, E., Lorenzo-Lacruz, J., et al., 2011. A multiscalar global evaluation of the impact of ENSO on droughts. Journal of Geophysical Research: Atmosphere 116 (D20).

Wong, G., 2013. A comparison between the Gumbel-Hougaard and distorted frank copulas for drought frequency analysis. International Journal of Horticultural Science and Technology 3 (1), 77−91.

Wong, G., Lambert, M., Leonard, M., Metcalfe, A., 2009. Drought analysis using trivariate copulas conditional on climatic states. Journal of Hydrologic Engineering 15 (2), 129−141.

Wong, G., Lambert, M.F., Metcalfe, A.V., 2008. Trivariate copulas for characterisation of droughts. ANZIAM Journal 49, 306−323.

Yevjevich, V.M., 1967a. An objective approach to definitions and investigations of continental hydrologic droughts. Hydrology Papers (Colorado State University) 23.

Yevjevich, V.M., 1967b. An objective approach to definitions and investigations of continental hydrologic droughts. Hydrology Papers (Colorado State University) 23.

Zhang, B., Wang, Z., Chen, G., 2017. A sensitivity study of applying a two-source potential evapotranspiration model in the standardized precipitation evapotranspiration index for drought monitoring. Land Degradation and Development 28 (1), 783−793. https://doi.org/10.1002/ldr.25486.

Zhang, L., Singh, V., 2006. Bivariate flood frequency analysis using the copula method. Journal of Hydrologic Engineering 11 (2), 150−164.

Zhang, L., Singh, V.P., 2007. Bivariate rainfall frequency distributions using Archimedean copulas. Journal of Hydrology 332 (1), 93−109.

Chapter 7

An efficient approximation-based robust design optimization framework for large-scale structural systems

Tanmoy Chatterjee[1,2], Rajib Chowdhury[2]
[1]*College of Engineering, Swansea University, Bay Campus, Swansea, United Kingdom;*
[2]*Department of Civil Engineering, Indian Institute of Technology Roorkee, Roorkee, Uttarakhand, India*

1. Introduction

The extensive developments in robust design optimization (RDO) reveal its profound utility in furnishing optimal solutions under stochastic environment (Taguchi, 1986; Zang et al., 2005; Park et al., 2006). A typical RDO problem is posed so as to minimize the effect of input uncertainties to the output quantities of interest, leading to least sensitive, i.e., *robust* solutions. In implementing this, the objective and constraints of the RDO formulation are represented as functions of statistical terms of the response quantities (Roy and Chakraborty, 2015; Chakraborty et al., 2012) or, first-order derivatives and other nonstatistical metrics to measure the variation of performance functions (Lee and Park, 2001; Han and Kwak, 2004). As a result, the solution of an RDO problem draws in uncertainty quantification of responses and thus entails significant computational effort (Schuëller and Jensen, 2008; Sudret, 2012). In this context, various approximation techniques referred to as metamodels have been developed over the years (Myers et al., 2009; Konakli and Sudret, 2016; Alis and Rabitz, 2001; Cortes and Vapnik, 1995; Adhikari et al., 2011) and used in RDO for computational viability (Jin et al., 2003; Bhattacharjya, 2010; Cheng et al., 2014).

However, it has been reported previously (Bhattacharjya, 2010; Chatterjee et al., 2017) that most of the metamodels perform vulnerably and often lead to false optima in the presence of uncertainties. This is mainly due to the fact that a marginal deviation in predicting the responses by the metamodel at any of the optimization iteration affects the final solutions significantly. To improve

Handbook of Probabilistic Models. https://doi.org/10.1016/B978-0-12-816514-0.00007-2
179

upon the aforementioned accuracy issue, metamodels have been built for approximating the response statistics within each optimization iteration. However, in doing so, the computational effort required in such metamodel-assisted RDO frameworks surges up, especially in dealing with large-scale engineering problems. Therefore, a trade-off between the desired accuracy level and computational feasibility is quite relevant in this scenario. For addressing the same, an adaptive metamodel-assisted RDO approach has been recently proposed (Chakraborty et al., 2017), which serves as a trade-off between the desired level of accuracy and affordable computational effort. However, still a large number of simulations have to be performed for evaluating the response statistics of the performance functions within the optimization cycle.

As a result, RDO solutions more often require a significantly high number of function evaluations to converge.

The primary motive of this chapter is to improve upon the present scenario of the metamodel-assisted RDO framework by taking into account the existing drawbacks as discussed previously. Specifically, the aim is to achieve accurate optimal solutions by using affordable computational effort. To this end, a simple but effective approximation-based approach has been developed as a part of this work. The novelty of the chapter lies in the following:

- The proposed framework approximates the response statistics outside the optimization cycle, i.e., globally, which allows a typical RDO formulation to be transformed into an equivalent deterministic optimization problem statement. By the aforementioned statement of transforming to an equivalent deterministic optimization, the authors mean that any simulation is not required for approximating the response statistics within the objective and constraints. This feature of the proposed framework attempts to improve the computational effort associated with a conventional metamodel-assisted RDO approach.
- Next improvement has been sought upon in context to the metamodels for enhancing the level of approximation accuracy. For approximating the response statistics, two compressive sensing—enabled hybrid Kriging models recently developed by the same authors group have been used (Chatterjee and Chowdhury, 2017, 2018). The proposed metamodels have been constructed so as to capture high degree of nonlinearity by using limited computational effort.

The rest of the chapter has been organized in the following sequence. Section 2 briefly discusses the overall framework of surrogate-assisted RDO. The novel metamodel-assisted response statistics—based RDO framework has been proposed in Section 3. Formulation of the proposed compressive sensing—enabled hybrid Kriging model has been illustrated in Section 4. Numerical examples have been illustrated in Section 5. Finally, the study has been summarized in Section 6.

2. Surrogate-assisted robust design optimization

2.1 Robust design optimization

The problem setup of RDO has been illustrated in this section. Let $\mathbf{x} := \{x_1, x_2, ..., x_N\}$ be $\mathbb{R}^N$ input vector defined in probability space (Ω, F, P) and $\mathbf{d}$ be the design variables. One of the possible ways in which an RDO problem can be stated is as follows:

$$
\begin{aligned}
&\min_{\mathbf{d} \subset D \in \mathbb{R}^N} & & c_f(\mathbf{d}) := f(E(y_0(\mathbf{x}, \mathbf{d})), \mathrm{var}(y_0(\mathbf{x}, \mathbf{d}))); \\
&\text{s.t.} & & c_g(\mathbf{d}) := g(E(y_c(\mathbf{x}, \mathbf{d})), \mathrm{var}(y_c(\mathbf{x}, \mathbf{d}))) \leq 0, c = 1, 2, ..., n_c; \quad (7.1) \\
& & & d_{i,l} \leq d_i \leq d_{i,u}, i = 1, 2, ..., n_v, \mathbf{d} \subset \mathbf{D} \in \mathbb{R}^{n_v}
\end{aligned}
$$

where $E(\bullet)$ and $\mathrm{var}(\bullet)$ represent mean and variance, respectively. It can be observed from Eq. (7.1) that the objective function c_f and constraints c_g in RDO framework are expressed as functions $f(\bullet)$, $g(\bullet)$ of mean and standard deviation of objective function y_0 and constraints y_c, in deterministic optimization framework, respectively, and n_c and n_v are the number of constraints and number of design variables, respectively. The ith design variable d_i varies from its lower bound $d_{i,l}$ to upper bound $d_{i,u}$.

Eq. (7.1) can be solved by using two different approaches. The first technique integrates the two objectives (i.e., mean and variance) by assigning proportional weights according to their importance (Chen et al., 1996) as

$$
\begin{aligned}
&\min_{\mathbf{d} \subset D \in \mathbb{R}^N} & & c_f(\mathbf{d}) := \alpha_w E(y_0(\mathbf{x}, \mathbf{d})) + (1 - \alpha_w)\sqrt{\mathrm{var}(y_0(\mathbf{x}, \mathbf{d}))}; \\
&\text{s.t.} & & c_g(\mathbf{d}) := E(y_c(\mathbf{x}, \mathbf{d})) + k_c\sqrt{\mathrm{var}(y_c(\mathbf{x}, \mathbf{d}))} \leq 0, c = 1, 2, ..., n_c; \\
& & & d_{i,l} \leq d_i \leq d_{i,u}, \quad i = 1, 2, ..., n_v, \mathbf{d} \subset \mathbf{D} \in \mathbb{R}^{n_v}
\end{aligned}
$$

$$(7.2)$$

where weightage factor $\alpha_w \in [0, 1]$. $k_c, c = 1, 2, ..., n_c$ are constant coefficients associated with the constraints. Eq. (7.2) is referred to as weighted sum method (WSM) and is quite popular because of its simplicity. Other related improved RDO approaches can be found in the studies by Chen et al. (1999, 2000), Messac (1996). The second class of approaches for solving Eq. (7.1) uses multiobjective optimization algorithms. Few prominent multiobjective optimization techniques can be found in the studies by Marano et al. (2010); Srinivas and Deb (1994); Sierra and Coello (2005); Fonseca and Fleming (1995); Zitzler et al. (2001); Zhang and Li (2007). In addition to the aforementioned category of approaches, there exist multiple variants of RDO formulation using other metrics of robustness (Huang and Du, 2007; Du et al., 2004), sensitivity-based approaches (Chakraborty et al., 2012; Lee and Park, 2001; Han and Kwak, 2004), surrogate-assisted frameworks (Ren and Rahman, 2013; Zhou et al., 2017; Shimoyama et al., 2009; Ray and Smith, 2006), adaptive strategies (Cheng et al., 2014; Zhou et al., 2017; Mortazavi et al., 2013; Shah et al., 2015), and so on.

2.2 Integration of surrogate models in robust design optimization

Primarily, there are two approaches for integrating surrogate models in RDO framework. The first one is a low-fidelity (LF) approach (Chakraborty et al., 2017) which involves a straightforward integration of a surrogate model into an optimization loop. A single surrogate model is constructed at the onset, and the same is used for the subsequent iterations within the optimization routine. As an obvious matter of fact, the computational effort involved is nominal. However, it has been observed that LF approach is often incapable of capturing the nonlinearity accurately. The second approach is a high-fidelity (HF) one (Chakraborty et al., 2016) which can adequately capture high degree of functional nonlinearity. Specifically, the HF approach constructs surrogate models for the objective function and constraints at each iteration. Consequently, it entails significant computational effort, making it unsuitable for large-scale problems. For further clarity, flowchart of the aforementioned existing frameworks have been presented in Fig. 7.1.

In this context, an adaptive improved version of the aforementioned two frameworks have been recently proposed in the study by Chakraborty et al. (2017), which serves as a trade-off between the desired level of accuracy and affordable computational effort. However, in all the aforementioned frameworks, a large number of simulations have to be performed for evaluating the response statistics of the performance functions within each optimization iteration. As a result, RDO solutions more often require a significantly high number of function evaluations to converge. Therefore, to avoid such computational issues within the optimization routine, a novel framework for surrogate-assisted RDO has been proposed in the following section.

3. Proposed surrogate-assisted robust design optimization framework

In most cases of RDO, as it is evident from Eq. (7.2), objective functions and constraints involve response statistical terms. Therefore, the fundamental concept underlying the proposed RDO framework is to approximate the response statistics, rather than the response quantities itself. The advantage associated with this principle is twofold, which is eliminating simulations to evaluate the response statistics and eventually transforming an RDO to an equivalent deterministic optimization. Thus, the efficient framework allows significant level of computations to be avoided and thus yields optimal solutions within much lesser number of function evaluations. To capture the variation of response statistics, two surrogate models have been used, one each for first and second moments of performance functions. The response statistics in objective function and constraints are approximated with the help of the global surrogate models at each design points. The steps of the proposed RDO framework have been presented in Algorithm 1.

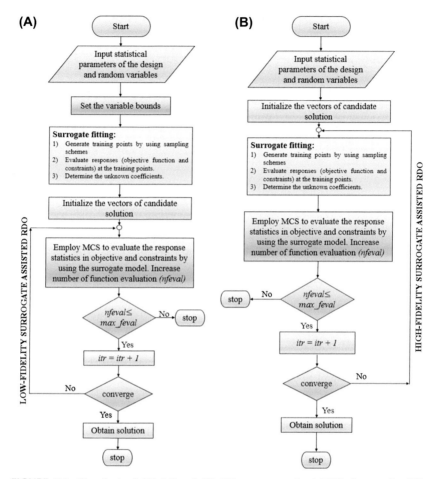

FIGURE 7.1 Flowchart of (A) LF and (B) HF surrogate-assisted RDO frameworks. *HF,* high-fidelity; *LF,* low-fidelity.

Algorithm 1 Stepwise illustration of the proposed RDO framework
- Global surrogate modeling for approximating the response statistics

1. Specify number of training points n_{s1}.
2. Generate n_{s1} training points for mean of each design variable.
3. Specify coefficient of variation for mean of each design variable.
4. Determine the upper and lower limits for mean of each design variable.
5. Generate n_{s2} random points according to the probability distribution of design variables and other stochastic parameters, if any.
6. Obtain responses corresponding to n_{s2} points in step 5, for each set of bound (step 4).
7. Evaluate n_{s1} number of mean and standard deviation (SD) of responses from each set of response vector of size n_{s2} obtained in step 6.

8. Construct surrogate models, each corresponding to map the mean of input variables (step 2) and output response statistics (step 7) (refer Section 4).

- Optimization loop

9. Approximate response statistics in objective function and constraints with the help of respective surrogate models constructed in step 8 at design points.

10. Increase iteration.

11. Set criterion for convergence. If criterion met, obtain optimal solution and stop, otherwise go to step 9 and repeat iterations.

It can be observed from steps 2, 7, and 8 of algorithm 1 that n_{s1} represents the number of training points for building the proposed surrogate model in approximating the response statistics (i.e., the mean and SD of response). It is also observed from steps 5 and 6 of algorithm 1 that n_{s2} represents the number of actual response evaluation in the proposed RDO framework. It should be noted that the parameters n_{s1} and n_{s2} have been selected on the basis of convergence study and do not involve any empirical formulae. More elaborately, convergence study has been carried out based on leave-one-out cross-validation (LOOCV) error in approximating the response statistics to determine n_{s1}. Furthermore, convergence study has been carried out to note the change in mean and SD of actual response. Accordingly, when the change in mean and SD of response is within the predefined tolerance, n_{s2} has been determined.

The lucid approximation framework as illustrated in steps 1–8 of algorithm 1 renders generality to the proposed RDO approach so that any surrogate model can be used; however, they should be robust enough to capture the variation of response statistics accurately to avoid convergence to a false optima. The efficiency of the proposed RDO framework has been clearly illustrated as there are no simulations within the optimization routine evident from step 9. Thus, the equivalent deterministic optimization of the RDO problem has been realized in steps 9–11. It should be further noted that the surrogate models have been constructed as global estimators (i.e., outside the optimization routine) for approximating the response statistics. In this scenario, a single surrogate may often be incapable of accurately approximating the response statistics at design points of subsequent iterations. Similar issue has also been observed in LF approach previously (Chakraborty et al., 2017). Therefore, ensuring accurate approximation of nonlinearly varying response statistics is the immediate motive for yielding desirable results by using the above proposed framework. For this purpose, two feature selection–enabled refined Kriging-based surrogate models developed recently by the same author group have been illustrated briefly in the next section.

4. Proposed surrogate models

The recently developed surrogate models proposed by the same author group in the study by Chatterjee and Chowdhury (2018) have been used in this work. The surrogate models are improvised so as to capture complex nonlinear and multimodal functional variation and deal with high-dimensional problems in a cost-effective way. Each of the models is based on Kriging (Sudret, 2012) with two-tier advantages, which are as follows:

- First, the models replace the polynomial trend portion of Kriging by the functional form of high-dimensional model representation (HDMR) (Chastaing et al., 2011). This allows to capture higher order of functional nonlinearity, ensuring convergence (Chatterjee and Chowdhury, 2017; Chatterjee et al., 2016).
- To enhance the computational complexity of the aforementioned model, a Bayesian formulation has been presented by utilizing relevance vector machine (RVM).
- Finally, an accelerated algorithm of RVM has been used to further improve the computational cost.

The formulation of the proposed surrogate models have been briefly outlined in the following subsections.

4.1 Integrating high-dimensional model representation and Polynomial Chaos Expansion (PCE) into Kriging trend

Kriging is widely acknowledged as a powerful surrogate modeling technique in which the interpolated values are modeled by Gaussian process governed by prior covariances (Sacks et al., 1989). The model output $\mathbf{M_K}(\mathbf{x})$ can be considered to be a realization of a Gaussian process as

$$\mathbf{M_K}(\mathbf{x}) = \boldsymbol{\beta}^T \mathbf{f}(\mathbf{x}) + \sigma^2 \mathbf{Z}(\mathbf{x}, \boldsymbol{\omega}) \tag{7.3}$$

where $\boldsymbol{\beta}^T \mathbf{f}(\mathbf{x}) = \sum_{j=1}^{P} \beta_j f_j(\mathbf{x})$ is the mean value of the Gaussian process and $\mathbf{Z}(\mathbf{x}, \boldsymbol{\omega})$ is a zero mean, unit variance Gaussian process. σ^2 denotes the process variance. The set of hyperparameters $\boldsymbol{\theta}$ defines the autocorrelation $R(\mathbf{x}, \mathbf{x}'; \boldsymbol{\theta})$ between two points $\mathbf{x}$ and $\mathbf{x}'$ (Dubourg et al., 2013). Also, R_{ij} is the $(i,j)^{\text{th}}$ element of the covariance matrix $\mathbf{R}$. $f_j(\mathbf{x}) \forall j = 1, \ldots, P$ are predefined polynomial functions, where P is the number of polynomials in the trend part of Kriging.

Now, substituting the functional form of HDMR in the trend part of Eq. (7.3), one obtains

$$\mathbf{M} = \left(g_0 + \sum_{k=1}^{M} \left\{ \sum_{i_2=1}^{N-k+1} \cdots \sum_{i_k=i_{k-1}}^{N} \sum_{r=1}^{k} \left[\sum_{m_1=1}^{s} \sum_{m_2=1}^{s} \cdots \sum_{m_r=1}^{s} \alpha_{m_1 m_2 \ldots m_r}^{(i_1 i_2 \ldots i_k) i_r} \psi_{m_1}^{i_1} \cdots \psi_{m_r}^{i_r} \right] \right\} \right)$$
$$+ \sigma^2 \mathbf{Z}(\mathbf{x}, \boldsymbol{\omega})$$

$$\tag{7.4}$$

where g_0 is a constant term representing the zeroth-order component function or the mean response of any response function $g(\mathbf{x})$. It has been observed that most real-life problems exhibit only the lower order cooperative effect (Li et al., 2001), and therefore, in Eq. (7.4), order of component function and bases ψ have been considered up to M and s, respectively. It is to be noted that the bases ψ satisfy the following relation

$$E(\psi_i, \psi_j) = m\delta_{ij}, \quad 0 \le |\mathbf{i}|, |\mathbf{j}| \le n, \; m \ge 0 \tag{7.5}$$

where

$$\delta_{\mathbf{ij}} = \prod_{k=1}^{n} \delta_{i_k j_k} = 1, \;\; \text{if } \mathbf{i} = \mathbf{j} \tag{7.6}$$
$$= 0, \;\; \text{elsewhere}$$

Based on the criteria illustrated in Eq. (7.5), the appropriate orthogonal bases are to be chosen. In this context, generalized polynomial chaos from the Askey scheme of hypergeometric orthogonal polynomials (Xiu and Karniadakis, 2002) has been used in Eq. (7.4). The determination of unknown coefficients α in Eq. (7.4) has been discussed in the next subsection.

4.2 Determination of unknown coefficients

The solution of Eq. (7.4) with respect to α may be represented as

$$(\boldsymbol{\psi}^T \mathbf{R}^{-1} \boldsymbol{\psi}) = \alpha = \boldsymbol{\psi}^T \mathbf{R}^{-1} \mathbf{d} \tag{7.7}$$

where $\mathbf{d}$ is the difference between response at the sample point and the mean response g_0. Alternatively, Eq. (7.7) may be stated as

$$\mathbf{B}\alpha = \mathbf{C} \tag{7.8}$$

where $\mathbf{B} = \boldsymbol{\psi}^T \mathbf{R}^{-1} \boldsymbol{\psi}, \mathbf{C} = \boldsymbol{\psi}^T \mathbf{R}^{-1} \mathbf{d}$. On obtaining the form as illustrated in Eq. (7.8), the unknown coefficients have been determined next.

4.2.1 Sparse recovery using Bayesian learning

Probabilistic sparse kernel model referred to as RVM has been used so as to recover the unknown coefficient vector. It is a Bayesian learning approach, where a prior governed set of hyperparameters is used and the most probable values are iteratively evaluated from the training data. Basically, sparsity is achieved as the posterior distribution of multiple weights are sharply peaked around zero. A multivariate Gaussian likelihood is defined as

$$p(\mathbf{C}|\alpha, \eta^2) = (2\pi)^{-P/2} \eta^{-P} \exp\left(-\frac{\|\mathbf{C} - \mathbf{B}\alpha\|^2}{2\eta^2}\right) \tag{7.9}$$

The prior governed by a set of hyperparameters can be represented as

$$p(\alpha|\gamma) = (2\pi)^{-M'/2} \prod_{m=1}^{M'} \gamma_m^{1/2} \exp\left(-\frac{\gamma_m \alpha_m^2}{2}\right) \tag{7.10}$$

$\gamma = \{\gamma_1, \gamma_2, ..., \gamma_{M'}\}^T$ is a vector of M' hyperparameters, which dominates the contribution of the prior over its associated coefficient. Note that the prior is responsible for rendering sparse features to the model. Subsequently, the posterior distribution based on the training set is obtained as the product of the likelihood and prior according to Bayes theorem as

$$p(\alpha|\mathbf{C}, \gamma, \eta^2) = p(\mathbf{C}|\alpha, \eta^2)p(\alpha|\gamma)/p(\mathbf{C}|\gamma, \eta^2) \qquad (7.11)$$

The posterior distribution is Gaussian $\mathbb{N}(\alpha, \delta)$ such that

$$\delta(\mathbf{D} + \eta^{-2}\mathbf{B}^T\mathbf{B})^{-1} \text{ and } \quad \alpha = \eta^{-2}\delta\mathbf{B}^T\mathbf{C} \qquad (7.12)$$

where $\mathbf{D} = diag(\gamma_1, \gamma_2, ... \gamma_{M'})$. Sparse Bayesian learning has been formulized as the maximization of the marginal likelihood with respect to α, and its logarithm $L(\gamma)$ is obtained (Tipping, 2001a) as

$$L(\gamma) = \log p(\mathbf{C}|\gamma, \eta^2)$$

$$= \log \int_{-\infty}^{\infty} p(\mathbf{C}|\alpha, \eta^2)p(\alpha|\gamma)d\alpha \qquad (7.13)$$

$$= -\frac{1}{2}\left(P \log 2\pi + \log|\Omega| + \mathbf{C}^T\Omega^{-1}\mathbf{C}\right)$$

where

$$\Omega = \eta^2\mathbf{I} + \mathbf{B}\mathbf{D}^{-1}\mathbf{B}^T \qquad (7.14)$$

A most probable point estimate γ_{MPE} has been obtained by type II maximum likelihood approach (Tipping, 2001b) as illustrated in Eqs. (7.13) and (7.14). Then, a point estimate μ_{MPE} can be evaluated by using Eq. (7.12) and upon substituting $\gamma = \gamma_{MPE}$, results to a posterior mean approximation. $\mathbf{B}\alpha_{MPE}$.

More details of the aforementioned approach can be found in the study by Chatterjee and Chowdhury (2017). The integrated approach as described previously been referred to as proposed model 1 (PM1), from now onward.

It is worth noting that the RVM scheme used previously suffer from the following drawbacks:

- The Bayesian learning is dependent on heuristic reestimation of the hyperparameter;, thus, iterative updating process is not convincing.
- The details of local maximization is not well explored (refer Eq. 7.13).
- PM1 is initiated with all P basis functions, and the hyperparameters γ are updated iteratively. As a result, the primary iterations remain computationally intensive.

Therefore, to address the aforementioned shortcomings, an improved algorithm has been used, which has been discussed next.

4.2.2 Sparse recovery using accelerated Bayesian learning

An improved framework of sparse Bayesian learning (Tipping and Faul, 2003) has been incorporated within the proposed model (Eq. 7.8) to further minimize the computational effort. In this algorithm, the basis functions are added sequentially so as to increase the marginal likelihood and the functions can also be deleted on becoming redundant. Specifically, this approach is a unique strategy for stimulating maximization of the marginal likelihood (Eq. 7.13). The steps have been illustrated in Algorithm 2.

Algorithm 2. Pseudo code for maximization of the marginal likelihood (Tipping and Faul, 2003)

1: Initialize. $\eta^2 \leftarrow \gamma\text{cov}(\mathbf{d}), \quad 0 < \gamma < 1$

2: Initialize

$$\gamma_i \leftarrow \frac{\|\mathbf{B}_i\|^2}{\|\mathbf{B}_i^T\mathbf{C}\|^2 \big/ \|\mathbf{B}_i\|^2 - \eta^2}$$

3: Set all other

$$\gamma_P \leftarrow \infty$$

4: $\delta \leftarrow \left(\mathbf{D} + \eta^{-2}\mathbf{B}^T\mathbf{B}\right)^{-1}$

5: $\alpha \leftarrow \eta^{-2}\delta\mathbf{B}^T\mathbf{C}$

6: **for** $i = 1 : P$

    $\xi_i \leftarrow l_i^2 - k_i$

    if $\xi_i > 0$ and $\gamma_i < \infty$ (i.e., $\mathbf{B}_i$ is within the model)

        re-evaluate. γ_i

    end if.

    if $\xi_i > 0$ **and**. $\gamma_i = \infty$

        add. $\mathbf{B}_i$

    end if.

    if $\xi_i \leq 0$ and $\gamma_i < \infty$

        remove $\mathbf{B}_i$ and assign. $\gamma_i \leftarrow \infty$

    end if.

    Update. $\eta^2 \leftarrow \|\mathbf{C} - \mathbf{B}\alpha\|^2 \Big/ \left(M' - P + \sum_{i=1}^{P} \gamma_i\delta_{ii}\right)$

    Update δ and α $k_P \leftarrow \frac{\alpha_P K_P}{\alpha_P - K_P}, l_P \leftarrow \frac{\alpha_P L_P}{\alpha_P - K_P}$

    $K_P \leftarrow \mathbf{B}_P^T\mathbf{H}\mathbf{B}_P - \mathbf{B}_P^T\mathbf{H}\mathbf{B}\delta\mathbf{B}^T\mathbf{H}\mathbf{B}_P$

    $L_P \leftarrow \mathbf{B}_P^T\mathbf{H}\mathbf{C} - \mathbf{B}_P^T\mathbf{H}\mathbf{B}\delta\mathbf{B}^T\mathbf{H}\mathbf{C}$, **where,** $\mathbf{H} \leftarrow \eta^2\mathbf{I}$

    if (change in $\log\gamma \leq$ tolerance and all other $\xi_i \leq 0$) (**Convergence criteria**)

        terminate.

    else.

        go to step 6.

    end if.

  end for.

7: End algorithm.

More details of the aforementioned algorithm can be found in the study by Chatterjee and Chowdhury (2018). From now onward, the approach illustrated in this section is referred to as proposed model 2 (PM2).

The proposed models (PM1 and PM2) have been used to approximate the response statistics within the efficient RDO framework (algorithm 1). Next, the proposed surrogate-assisted RDO framework has been implemented for numerical validation.

5. Numerical validation

To access the performance of the proposed surrogate-assisted RDO framework, it has been used to solve few numerical examples in this section. The proposed frameworks have been referred to as PM1-RDO and PM2-RDO from now onward. Latin-hypercube sampling (McKay et al., 1979) has been used for training the models in this chapter. Gaussian correlation function has been used in Kriging (Lophaven et al., 2002). The computational platform has been MATLAB® R2017b (n.d.). In case of the example involving finite element modeling, ANSYS Mechanical 18.1 (n.d.) has been integrated in MATLAB® environment. The in-house—developed computer codes have been simulated using Intel Xeon CPU E5645 @ 2.4 GHz. For single-objective and multi-objective optimization, MATLAB® toolbox *fmincon* and *gamultiobj* have been used, respectively. While solving the single-objective examples, the gradients of the objective function and constraints are evaluated by using finite-difference scheme.

5.1 Problem type 1: analytical test problems

The first problem constitutes single-objective function and multiple nonlinear constraints. The second problem is of multiobjective formulation.

5.1.1 Example 1: Welded beam design

The first example considered is that of a welded beam design as shown in fig. 7.2, adopted from the study by Deb (2001). The objective is to minimize the cost of the beam subject to constraints on shear stress, bending stress, buckling load, and end deflection. There are four continuous design variables, namely, beam thickness $t(= x_1)$, beam width $b(= x_2)$, weld length $l(= x_3)$, and weld thickness $h(= x_4)$ (Fig. 7.2).

The details of the problem formulation has been provided in Section A.1 of appendix A. Each of the design variables have been considered to be normally distributed with coefficient of variation as 5%. The RDO problem may be stated as

$$\boldsymbol{\mu_x} = \operatorname{argmin}[F := \mu(f(\mathbf{x})) + \sigma(f(\mathbf{x}))] \qquad (7.15)$$

where $\boldsymbol{\mu_x}$ denotes the mean of design variables at optimum F.

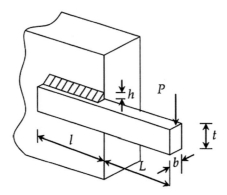

FIGURE 7.2   Schematic diagram of the welded beam.

The number of samples for MCS are adopted as 10^5 by carrying out the convergence of the response statistics and their confidence intervals (95%). For illustration purpose, convergence of the LOOCV error for determining n_{s1} in approximating the response statistics by using PM1 and PM2 has been illustrated in Fig. 7.3. Also, the convergence of objective function value by using PM1-RDO and PM2-RDO has been illustrated in Fig. 7.4 for determining n_{s2}.

The optimized design variables and objective function values obtained using the proposed RDO frameworks have been presented in Table 7.1. The adopted values for n_{s1}, n_{s2}, number of actual function evaluations n_a, and total CPU time required for yielding the optimal solutions have been presented in Table 7.2. The results illustrate excellent match between MCS and the proposed models. Comparison with other surrogate models such as, Kriging, radial basis function (RBF) (Deng, 2006) and multiadaptive regression splines (MARS) (Friedman, 1991) illustrate superior performance of PM1 and PM2 both in terms of approximation accuracy and CPU time.

5.1.2 Example 2: Multiobjective robust design optimization of a vibrating platform

The vibrating platform as illustrated in Fig. 7.5 has been adopted from the study by Bhattacharjya (2010) and involves simultaneous minimization of construction cost and maximizing the natural frequency of the platform. The details of the problem formulation has been provided in Section A.2 of xappendix A. The design variables are half thickness of the innermost layer $(t1)$ and the thicknesses of the other two layers $(t_2$ and $t_3)$, which have been assumed to be normally distributed with coefficient of variation as 5%. The design parameters (ρ_1, ρ_2, ρ_3), (E_1, E_2, E_3), and (c_1, c_2, c_3) represent the density, elastic modulus, and cost of the material corresponding to inner, middle, and outer layers of the platform, respectively. They have been considered as normally distributed with coefficient of variation as 10%. b and L are the width and length of the platform, respectively.

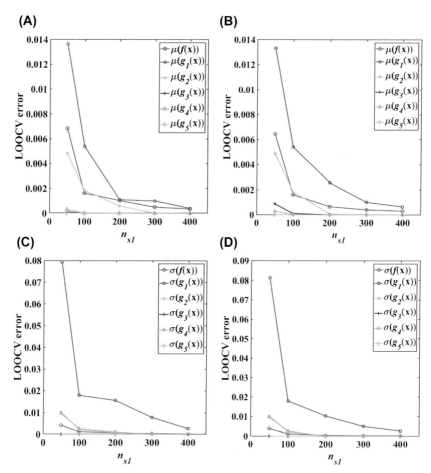

FIGURE 7.3 Convergence of the LOOCV error in approximating the mean (μ) of the performance functions by (A) PM1 (B) PM2 and standard deviation (σ) of the performance functions by (C) PM1 (D) PM2 to determine n_{s1}. *LOOCV*, leave-one-out cross-validation.

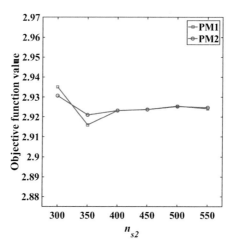

FIGURE 7.4 Convergence of the objective function value with increase in n_{s2}. *PM1*, proposed model 1; *PM2*, proposed model 2.

TABLE 7.1 Robust optimal solutions for example 1.

Design variables	MCS-RDO	PM1-RDO	PM2-RDO	Kriging-RDO	RBF-RDO	MARS-RDO
$\mu(x_1)$	0.2312	0.2282	0.2282	0.5852	0.3103	3.5638
$\mu(x_2)$	6.7505	6.8463	6.8458	1.3633	0.1	0.8112
$\mu(x_3)$	8.7636	8.6908	8.6915	0.4902	0.1	0.3011
$\mu(x_4)$	0.2299	0.2479	0.2479	6.143	10	3.5635
F^*	2.9059	2.9232	2.9233	2.9887	1.0259	12.9351

MARS, multiadaptive regression splines; PM1, proposed model 1; PM2, proposed model 2; RDO, robust design optimization;

TABLE 7.2 Number of functional evaluations and CPU time required for yielding optimal solutions in example 1.

	MCS-RDO	PM1-RDO	PM2-RDO	Kriging-RDO	RBF-RDO	MARS-RDO
n_{s1}	–	400	400	–	–	–
n_{s2}	–	400	400	–	–	–
n_a	10^5	–	–	128	128	128
Time (s)	8.5345	27.2006	7.8040	1009.9415	1378.6213	1550.8944

MARS, multiadaptive regression splines; PM1, proposed model 1; PM2, proposed model 2; RDO, robust design optimization

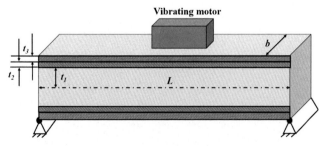

FIGURE 7.5 Schematic diagram of the vibrating platform.

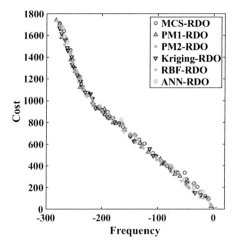

FIGURE 7.6 Pareto front depicting the trade-off between optimal cost and optimal frequency of the vibrating platform. Note that the negative sign in x-axis denote maximization of frequency. *ANN*, artificial neural network; *MARS*, multiadaptive regression splines; *MCS*, Monte Carlo simulation; *PM1*, proposed model 1; *PM2*, proposed model 2; *RBF*, radial basis function; *RDO*, robust design optimization.

TABLE 7.3 Number of functional evaluations and CPU time required for yielding optimal solutions in example 2.

	MCS-RDO	PM1-RDO	PM2-RDO	Kriging-RDO	RBF-RDO	MARS-RDO
n_{s1}	–	600	600	–	–	–
n_{s2}	–	600	600	–	–	–
n_a	1×10^4	–	–	500	500	500
Time (s)	98.3082	184.1751	27.3563	1004.0156	841.6614	6710.0481

RDO, robust design optimization; MARS, multiadaptive regression splines; PM1, proposed model 1; PM2, proposed model 2; RBF, radial basis function.

The robust optimal results have been represented in the form of Pareto front in Fig. 7.6. The close agreement of the points in Fig. 7.6 illustrates good approximation accuracy of PM1, PM2, and other surrogate models with that of benchmark solutions (MCS-RDO). The adopted values for n_{s1}, n_{s2}, number of actual function evaluations n_a, and total CPU time required for yielding the optimal solutions have been presented in Table 7.3. It can be observed from Table 7.3 that PM1 and PM2 prove to be significantly computationally efficient than that of other models.

In both test examples, excellent performance of the proposed RDO frameworks have been illustrated. The fact has been demonstrated both in terms of accurate optimal solutions and computational cost. The criticality of the response approximation can be realized from the fact that even after constructing surrogate models at each iteration (i.e., HF surrogate-assisted RDO approach), the other surrogate models (such as, Kriging, RBF, ANN, MARS) proved to be incapable of accurately capturing the functional nonlinearity. Considering the complexity of approximation, it is worth acknowledging the robustness of the proposed RDO framework which perform a single approximation of the response statistics and accurately map the overall functional behavior, eventually leading to the true optima.

Next, the proposed RDO framework has been used to solve a computationally expensive practical problem.

5.2 Problem type 2: finite element model of a building frame

The finite element model of a three-span five-storey building frame as shown in Fig. 7.7 has been used previously in applications, such as, uncertainty quantification and reliability analysis (Chakraborty and Chowdhury, 2015). In this work, it has been reformulated as an RDO problem, where the objective is to minimize the weight of the plane frame structure subjected to displacement constraints. The RDO formulation involves 18 stochastic variables, three applied lateral loads (P), two elastic moduli (E), density of concrete (ρ_c), four widths of the square columns (b_C), four widths of the beams (b_B), and four

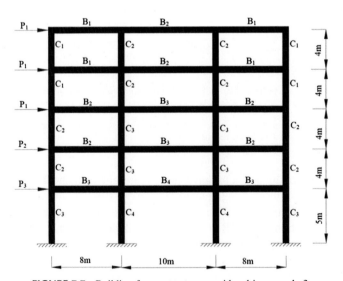

FIGURE 7.7 Building frame structure considered in example 3.

TABLE 7.4 Characteristics of frame elements in example 3.

Element	Young's modulus	Cross-sectional area
B_1	E_4	$b_{B_1} \times d_{B_1}$
B_2	E_4	$b_{B_2} \times d_{B_2}$
B_3	E_4	$b_{B_3} \times d_{B_3}$
B_4	E_4	$b_{B_4} \times d_{B_4}$
C_1	E_5	$b_{C_1} \times d_{C_1}$
C_2	E_5	$b_{C_2} \times d_{C_2}$
C_3	E_5	$b_{C_3} \times d_{C_3}$
C_4	E_5	$b_{C_4} \times d_{C_4}$

depths of the beams (d_B). The two elastic moduli (E) have been assumed to be correlated, with coefficient of correlation $= 0.7$. The mean of the 12 cross-sectional dimensions of beams and square columns are considered as design variables. The characteristics of the building frame elements have been presented in Table 7.4. The description of the random variables considered for this example have been provided in Table 7.5.

The RDO problem can be stated as

$$\text{Minimize} \quad f(\mathbf{x}, \mathbf{d_v}) = \alpha_w * \mu(W) + (1 - \alpha_w) * \sigma(W)$$

$$\text{subject to} \tag{7.16}$$

$$g_d(\mathbf{x}, \mathbf{d_v}) = [\mu(d) + 3\sigma(d)] - d_{\max} \leq 0$$

where $x = [P_1, P_2, P_3, E_4, E_5, \rho_c]$ and $d_v = [\mu(b_{C_1}), \mu(b_{C_2}), \mu(b_{C_3}), \mu(b_{C_4}), \mu(b_{B_1}), \mu(b_{B_2}), \dots \mu(b_{B_3}), \mu(b_{B_4}), \mu(b_{B_1}), \mu(b_{B_2}), \mu(b_{B_3}), \mu(b_{B_4})]$, α_w is the weightage factor, W is the overall weight of the frame structure, and μ and σ denote mean and standard deviation, respectively. Constraint g_d is such that maximum nodal displacement d should not exceed $d_{\max}$. Here, $d_{\max}$ is taken as 0.067 m.

It can be observed from Table 7.6 that results obtained by PM1 and PM2 are in close agreement as compared with MCS-RDO. This illustrates good approximation accuracy of the proposed models in a relatively high-dimensional RDO problem. The adopted values for n_{s1}, n_{s2}, number of actual function evaluations n_a, and total CPU time required for yielding the optimal solutions have been presented in Table 7.7. It can be observed from Table 7.7 that the proposed models yield the aforementioned optimal solutions by using 1.5% CPU time as compared with that of MCS-RDO.

TABLE 7.5 Description of the random variables in example 3.

Variable	Distribution	Mean	SD/cov[a]
P_1	Gumbel	135 kN	40 kN
P_2	Gumbel	90 kN	35 kN
P_3	Gumbel	70 kN	30 kN
$E_4^{**} E_4{}^d$	Lognormal	2.1×10^{10} N/m^2	1.9×10^9 N/m^2
$E_5^{**} E_5{}^d$	Lognormal	2.4×10^{10} N/m^2	1.9×10^9 N/m^2
ρ_C	Beta[b]	2450 kg/m^3	245 kg/m^3
b_{C_1}	Normal	DV[c]	0.1
b_{C_2}	Normal	DV	0.1
b_{C_3}	Normal	DV	0.1
b_{C_4}	Normal	DV	0.1
b_{B_1}	Normal	DV	0.1
b_{B_2}	Normal	DV	0.1
b_{B_3}	Normal	DV	0.1
b_{B_4}	Normal	DV	0.1
d_{B_1}	Normal	DV	0.1
d_{B_2}	Normal	DV	0.1
d_{B_3}	Normal	DV	0.1
d_{B_4}	Normal	DV	0.1

[a]SD/cov denote standard deviation and coefficient of variation, respectively.
[b]Shape parameters for beta distribution are considered as 5, 5.
[c]indicate design variable.
[d]E_4 and E_5 correlated random parameters with correlation coefficient = 0.7.

6. Summary and conclusions

A novel and efficient surrogate-assisted RDO framework has been proposed. Two refined feature selection—enabled surrogate models have been developed for this purpose. The study clearly indicates the excellent performance of the proposed models in terms of close agreement with that of MCS by using significantly less computational effort. Moreover, the proposed models have outperformed few existing surrogate models. The highlights in relation to the proposed RDO framework have been enumerated as follows:

- The proposed surrogate-assisted RDO methodology is a generalized framework, in which any of the existing and/or, improvised surrogate models can be integrated in a straightforward manner.

TABLE 7.6 Robust optimal solutions for example 3.

α_w	Response quantities/statistics	MCS-RDO	PM1-RDO	PM2-RDO
0	$f^*(\mathbf{x}, \mathbf{d}_v) \times 10^4$	1.8493	1.8729	1.8723
	$\mu(W) \times 10^5$	1.5580	1.5669	1.5666
	$\sigma(W) \times 10^4$	1.8599	1.8729	1.8723
0.25	$f^*(\mathbf{x}, \mathbf{d}_v) \times 10^4$	5.2773	5.3179	5.3177
	$\mu(W) \times 10^5$	1.5535	1.5642	1.5644
	$\sigma(W) \times 10^4$	1.8632	1.8764	1.8757
0.5	$f^*(\mathbf{x}, \mathbf{d}_v) \times 10^4$	8.7021	8.7587	8.7591
	$\mu(W) \times 10^5$	1.5533	1.5639	1.5641
	$\sigma(W) \times 10^4$	1.8644	1.8779	1.8769
0.75	$f^*(\mathbf{x}, \mathbf{d}_v) \times 10^4$	1.2126	1.2199	1.2200
	$\mu(W) \times 10^5$	1.5533	1.5639	1.5641
	$\sigma(W) \times 10^4$	1.8650	1.8787	1.8776

PM1, proposed model 1; PM2, proposed model 2; RDO, robust design optimization.

TABLE 7.7 Number of functional evaluations and CPU time required for yielding optimal solutions in example 3.

	MCS-RDO	PM1-RDO	PM2-RDO
n_{s1}	—	500	500
n_{s2}	-	500	500
n_a	1×10^5	—	—
Time (s)	1.3858×10^6	2.0810×10^4	2.0788×10^4

PM1, proposed model 1; PM2, proposed model 2; RDO, robust design optimization.

- The framework can also be easily extended to any other optimization algorithms so as to address typical features of problems.
- It is based on a simple yet effective principle, which is to directly approximate the response statistics allowing an RDO problem to be expressed as equivalent deterministic optimization. In other words, a large number of simulations can be avoided to approximate the response statistics, making the proposed RDO framework inherently suitable for large-scale problems.

The study has clearly paved way for the proposed surrogate-assisted RDO framework to be applied to further complex real-time applications.

Appendix A. Description of the test problems investigated in Section 5.1

A.1 Example 1: Welded beam design

$$\text{Minimize} \quad f(\mathbf{x}) = 1.10471 x_1^2 x_2 + 0.04811 x_3 x_4 (14 + x_2) \tag{A.1}$$

s.t.

$$
\begin{aligned}
g_1(\mathbf{x}) &= t - t_{\max} \le 0 \\
g_2(\mathbf{x}) &= s - s_{\max} \le 0 \\
g_3(\mathbf{x}) &= x_1 - x_4 \le 0 \\
g_4(\mathbf{x}) &= d - d_{\max} \le 0 \\
g_5(\mathbf{x}) &= P - P_c \le 0
\end{aligned}
\tag{A.2}
$$

$$M = P(L + x_2/2) \tag{A.3}$$

$$R = \sqrt{0.25\left(x_2^2 + (x_1 + x_3)^2\right)} \tag{A.4}$$

$$J = \sqrt{2} x_1 x_2 \left(x_2^2/12 + 0.25(x_1 + x_3)^2\right) \tag{A.5}$$

$$P_c = 64746.022(1 - 0.0282346 x_3) x_3 x_4^3 \tag{A.6}$$

$$t_1 = P\Big/\left(\sqrt{2} x_1 x_2\right) \tag{A.7}$$

$$t_2 = MR/J \tag{A.8}$$

$$t = \sqrt{t_1^2 + t_1 t_2 x_2/R + t_2^2} \tag{A.9}$$

$$S = 6PL\big/\left(x_4 x_3^2\right) \tag{A.10}$$

$$d = 2.1952\big/\left(x_4 x_3^3\right) \tag{A.11}$$

$$
\begin{aligned}
&P = 6000, L = 14, E = 30 \times 10^6, G = 12 \times 10^6, \\
&t_{\max} = 13600, s_{\max} = 30000, x_{\max} = 10, d_{\max} = 0.25 \\
&0.125 \le x_1 \le 10, \quad 0.1 \le x_i \le 10, \quad \text{for } i = 2, 3, 4.
\end{aligned}
\tag{A.12}
$$

A.2 Example 2: Vibrating platform

$$\text{minimize} \quad \text{cost} = 2bL(c_1 t_1 + c_2 t_2 + c_3 t_3)$$

and

$$\text{maximize} \quad \text{frequency} = \left(\frac{\pi}{2L^2}\right)\left(\frac{EI}{\mu}\right)^{1/2} \tag{A.13}$$

$$EI = \left[\frac{2b}{3}\right]\left[E_1 t_1^3 + E_2\left\{(t_1 + t_2)^3 - t_1^3\right\} + E_3\left\{(t_1 + t_2 + t_3)^3 - (t_1 + t_2)^3\right\}\right]$$
(A.14)

$$\mu = 2b(\rho_1 t_1 + \rho_2 t_2 + \rho_3 t_3)$$
(A.15)

$$0 \le t_1 \le 0.5, \quad 0 \le t_2 \le 0.15, \quad \text{and} \quad 0 \le t_3 \le 0.05; \quad b = 0.4 \text{ m}; \quad L = 4 \text{ m}$$
(A.16)

References

Adhikari, S., Chowdhury, R., Friswell, M.I., 2011. High dimensional model representation method for fuzzy structural dynamics. Journal of Sound and Vibration 330, 1516−1529.

Alis, Ö.F., Rabitz, H., 2001. Efficient implementation of high dimensional model representations. Journal of Mathematical Chemistry 29.

ANSYS ® Academic Research Mechanical, Release 18.1. (n.d.).

Bhattacharjya, S., 2010. Robust Optimization of Structures under Uncertainty. PhD Thesis. Department of Civil Engineering, Bengal Engineering and Science University, Shibpur.

Chakraborty, S., Chowdhury, R., 2015. A semi-analytical framework for structural reliability analysis. Computer Methods in Applied Mechanics and Engineering 289, 475−497.

Chakraborty, S., Bhattacharjya, S., Haldar, A., 2012. Sensitivity importance-based robust optimization of structures with incomplete probabilistic information. International Journal for Numerical Methods in Engineering 90, 1261−1277.

Chakraborty, S., Chatterjee, T., Chowdhury, R., Adhikari, S., 2016. Robust design optimization for crashworthiness of vehicle side impact. ASCE-ASME Journal of Risk and Uncertainty in Engineering Systems, Part B: Mechanical Engineering 1−9. https://doi.org/10.1115/1.4035439.

Chakraborty, S., Chatterjee, T., Chowdhury, R., Adhikari, S., 2017. A surrogate based multi-fidelity approach for robust design optimization. Applied Mathematical Modelling 47, 726−744.

Chastaing, G., Gamboa, F., Prieur, C., 2011. Generalized Hoeffding-Sobol decomposition for dependent variables -application to sensitivity analysis. Electronic Journal of Statistics.

Chatterjee, T., Chowdhury, R., 2017. An efficient sparse Bayesian learning framework for stochastic response analysis. Structural Safety 68, 1−14. https://doi.org/10.1016/j.strusafe.2017.05.003.

Chatterjee, T., Chowdhury, R., 2018. Refined sparse Bayesian learning configuration for stochastic response analysis. Probabilistic Engineering Mechanics 52, 15−27.

Chatterjee, T., Chakraborty, S., Chowdhury, R., 2016. A bi-level approximation tool for the computation of FRFs in stochastic dynamic systems. Mechanical Systems and Signal Processing 70−71, 484−505.

Chatterjee, T., Chakraborty, S., Chowdhury, R., 2017. A critical review of surrogate assisted robust design optimization. Archives of Computational Methods in Engineering. https://doi.org/10.1007/s11831-017-9240-5.

Chen, W., Allen, J., Tsui, K., Mistree, F., 1996. Procedure for robust design: minimizing variations caused by noise factors and control factors. Journal of Mechanical Design, Transactions of the ASME 118, 478−485.

Chen, W., Wiecek, M., Zhang, J., 1999. Quality utility — a compromise programming approach to robust design. Journal of Mechanical Design, Transactions of the ASME 121, 179–187.

Chen, W., Sahai, A., Messac, A., Sundararaj, G., 2000. Exploration of the effectiveness of physical programming in robust design. Journal of Mechanical Design, Transactions of the ASME 122, 155–163.

Cheng, J., Liu, Z., Wu, Z., Li, X., Tan, J., 2014. Robust optimization of structural dynamic characteristics based on adaptive Kriging model and CNSGA. Structural and Multidisciplinary Optimization 51, 423–437.

Cortes, C., Vapnik, V., 1995. Support-vector networks. Machine Learning 20, 273–297.

Deb, K., 2001. Multi-objective Optimization Using Evolutionary Algorithms. John Wiley & Sons, New York, USA.

Deng, J., 2006. Structural reliability analysis for implicit performance function using radial basis function network. International Journal of Solids and Structures 43, 3255–3291.

Du, X., Sudjianto, A., Chen, W., 2004. An integrated framework for optimization under uncertainty using inverse reliability strategy. Journal of Mechanical Design 126, 562–570.

Dubourg, V., Sudret, B., Deheeger, F., 2013. Metamodel-based importance sampling for structural reliability analysis. Probabilistic Engineering Mechanics 33, 47–57.

Fonseca, C., Fleming, P., 1995. Multiobjective genetic algorithms made easy: selection, sharing, and mating restriction. In: Proc. 1st Int. Conf. Genet. Algorithms Eng. Syst. Innov. Appl. IET, pp. 45–52.

Friedman, J.H., 1991. Multivariate adaptive regression splines. Annals of Statistics 19, 1–67.

Han, J.S., Kwak, B.M., 2004. Robust optimization using a gradient index: MEMS applications. Structural and Multidisciplinary Optimization 27, 469–478.

Huang, B., Du, X., 2007. Analytical robustness assessment for robust design. Structural and Multidisciplinary Optimization 34, 123–137.

Jin, R., Du, X., Chen, W., 2003. The use of metamodeling techniques for optimization under uncertainty. Structural and Multidisciplinary Optimization 25, 99–116.

Konakli, K., Sudret, B., 2016. Reliability analysis of high-dimensional models using low-rank tensor approximations. Probabilistic Engineering Mechanics 46, 18–36.

Lee, K.-H., Park, G.-J., 2001. Robust optimization considering tolerances of design variables. Computers and Structures 79, 77–86.

Li, G., Rosenthal, C., Rabitz, H., 2001. High dimensional model representation. The Journal of Physical Chemistry A 105, 7765–7777.

Lophaven, S., Nielson, H., Sondergaard, J., 2002. DACE A MATLAB Kriging Toolbox, IMM-TR-2002-12. Technical University of Denmark.

Marano, G.C., Greco, R., Sgobba, S., 2010. A comparison between different robust optimum design approaches: application to tuned mass dampers. Probabilistic Engineering Mechanics 25, 108–118. https://doi.org/10.1016/j.probengmech.2009.08.004.

MATLAB and Statistics Toolbox Release 2017b. (n.d.).

McKay, M.D., Beckman, R.J., Conover, W.J., 1979. A comparison of three methods for selecting values of input variables in the analysis of output from a computer code. Technometrics 2, 239–245.

Messac, A., 1996. Physical programming: effective optimization for computational design. AIAA Journal 34, 149–158.

Mortazavi, A., Azarm, S., Gabriel, S.A., 2013. Adaptive gradient-assisted robust design optimization under interval uncertainty. Engineering Optimization 45, 1287–1307.

Myers, R.H., Montgomery, D.C., Anderson-Cook, C.M., 2009. Response Surface Methodology: Process and Roduct Optimization Using Designed Experiments. Wiley, New York, U.S.A.

Park, G., Lee, T., Kwon, H., Hwang, K., 2006. Robust design: an overview. AIAA Journal 44, 181−191.

Ray, T., Smith, W., 2006. A surrogate assisted parallel multiobjective evolutionary algorithm for robust engineering design. Engineering Optimization 38, 997−1011.

Ren, X., Rahman, S., 2013. Robust design optimization by polynomial dimensional decomposition. Structural and Multidisciplinary Optimization 48, 127−148. https://doi.org/10.1007/s00158-013-0883-z.

Roy, B.K., Chakraborty, S., 2015. Robust optimum design of base isolation system in seismic vibration control of structures under random system parameters. Structural Safety 55, 49−59.

Sacks, J., Welch, W., Mitchell, T., Wynn, H., 1989. Design and analysis of computer experiments. Statistical Science 4, 409−423.

Schuëller, G.I., Jensen, H. a., 2008. Computational methods in optimization considering uncertainties − an overview. Computer Methods in Applied Mechanics and Engineering 198, 2−13.

Shah, H., Hosder, S., Koziel, S., Tesfahunegn, Y.A., Leifsson, L., 2015. Multi-fidelity robust aerodynamic design optimization under mixed uncertainty. Aerospace Science and Technology 45, 17−29.

Shimoyama, K., Lim, J.N., Jeong, S., Obayashi, S., Koishi, M., 2009. Practical implementation of robust design assisted by response Surface approximation and visual data-mining. Journal of Mechanical Design 131.

Sierra, M.R., Coello, C.A.C., 2005. Improving PSO-based multi-objective optimization using crowding, mutation and ∈-Dominance. In: Evol. Multi-Criterion Optim. Vol. 3410 Ser. Lect. Notes Comput. Sci. Springer Berlin Heidelberg, pp. 505−519.

Srinivas, N., Deb, K., 1994. Multiobjective optimization using nondominated sorting in genetic algorithms. Evolutionary Computation 2, 221−248.

Sudret, B., 2012. Meta-models for structural reliability and uncertainty quantification. In: Proc. 5th Asian-Pacific Symp. Stuctural Reliab. Its Appl. (APSSRA, 2012), Singapore, pp. 53−76.

Taguchi, G., 1986. Quality Engineering through Design Optimization. Krauss International Publications, White Plains, NY.

Tipping, M.E., 2001. Sparse kernel principal component analysis. In: Adv. Neural Inf. Process. Syst., vol. 13. MIT Press.

Tipping, M., 2001. Sparse Bayesian learning and the relevance vector machine. Journal of Machine Learning Research 1, 211−244.

Tipping, M.E., Faul, A., January 3-6, 2003. Fast marginal likelihood maximization for sparse Bayesian models. In: Proc. Ninth Int. Work. Artif. Intell. Stat. Key West, FL.

Xiu, D., Karniadakis, G.E., 2002. The Wiener−Askey polynomial chaos for stochastic differential equations. SIAM Journal on Scientific Computing 24, 619−644.

Zang, C., Friswell, M.I., Mottershead, J.E., 2005. A review of robust optimal design and its application in dynamics. Computers and Structures 83, 315−326.

Zhang, Q., Li, H., 2007. MOEA/D: a multiobjective evolutionary algorithm based on decomposition. IEEE Transactions on Evolutionary Computation 11, 712−731.

Zhou, H., Zhou, Q., Liu, C., Zhou, T., 2017. A kriging metamodel-assisted robust optimization method based on a reverse model. Engineering Optimization. https://doi.org/10.1080/0305215X.2017.1307355.

Zitzler, E., Laumanns, M., Thiele, L., 2001. SPEA2: Improving the Strength Pareto Evolutionary Algorithm, Tech. Rep. 103, Computer Engineering and Networks Laboratory (TIK). Department of Electrical Engineering, Swiss Federal Institute of Technology (ETH) Zurich.

Chapter 8

Probabilistic seasonal rainfall forecasts using semiparametric D-vine copula-based quantile regression

Thong Nguyen-Huy[1], Ravinesh C. Deo[2], Shahbaz Mushtaq[3], Shahjahan Khan[1]

[1]*School of Agricultural, Computational and Environmental Sciences, Centre for Applied Climate Sciences, University of Southern Queensland, Toowoomba, QLD, Australia;* [2]*School of Agricultural Computational and Environmental Sciences, University of Southern Queensland, Springfield, QLD, Australia;* [3]*Centre for Applied Climate Sciences, University of Southern Queensland, Toowoomba, QLD, Australia*

1. Introduction

Skillful probabilistic seasonal rainfall forecasting plays an important role in supporting water resource users, developing agricultural risk-management plans, and improving decision-making processes. The use of climate information in explaining rainfall variability, and its application to managing climate risks, has been well documented globally (Nicholson and Kim, 1997; Corte-Real et al., 1998; Enfield et al., 2001). However, traditional probabilistic forecasting approaches, focusing on the mean values, are unable to quantify the tail dependence when extreme events occur. In this context, the quantile regression method, as proposed by Koenker and Bassett (1978), is an essential tool for capturing the full dependency structure between the climate indices and seasonal rainfall. This approach measures the association of the predictor variables with a conditional quantile of a dependent variable without any specific assumption on the conditional distributions. Therefore, quantile regression models are useful for quantifying the dependencies between variables in the outer regions of the conditional distribution.

This chapter develops a novel copula-based quantile regression method for investigating the impacts of various climate indices on rainfall variability, particularly when extreme events occur. To provide a clear focus, the Australian Wheatbelt will be used as a case study. Australia is an agricultural nation with climate variability that is more spatially and temporally diverse than any other country (Nicholls et al., 1997; Best et al., 2007). The remote, synoptic-scale drivers, including El Niño—Southern Oscillation (ENSO) and Indian Ocean Dipole (IOD) modes, are the principal factors influencing the interannual and interseasonal rainfall variabilities. ENSO and IOD are representative of the synoptic-scale processes of the air—sea interaction over the tropical Pacific and the Indian Ocean regions, respectively. Although many local factors such as atmospheric blocking and the subtropical ridge also influence the variability of Australian rainfall (King et al., 2014), the relationship between the remote drivers and Australian rainfall variation is the primary subject of discussion in this chapter.

The effects of ENSO on Australian rainfall fluctuation have been extensively investigated since the early 1980s (McBride and Nicholls, 1983; Nicholls et al., 1996). It is well known that the oscillating phases of ENSO are the main factors explaining Australia's rainfall variability, in particular during the period July—March. Risbey et al. (2009) reported that ENSO has the strongest relationship to rainfall in the east of Australia, where generally La Niña phases bring more rainfall and El Niño phases are linked to decreased rainfall. McKeon et al. (2004) found that the El Niño phases were associated with drought events over much of the Australia continent. Furthermore, ENSO also has a significant influence on the rainfall patterns in north and northeast Australia (Holland, 1986; Brown et al., 2011). The ENSO—rainfall relationship varies across the Australian continent (Power et al., 2006; Nguyen-Huy et al., 2017), even within particular regions such as southeast Queensland (Cai et al., 2010) and southeast Australia (King et al., 2013). In general, the influence of ENSO on rainfall during the La Niña phase is stronger than that during the El Niño phase.

The IOD, similar to ENSO, is associated with the variability of Australian rainfall depending on seasons and times. IOD mainly modulates interannual rainfall in western and southern Australia during the winter and spring seasons (Risbey et al., 2009). The influence of IOD on the climatic conditions of southeast Australia has also been observed in some studies (Meyers et al., 2007) where it has been associated with drought events in this region (Ummenhofer et al. 2009, 2011). Moreover, it was also observed throughout the 20th century that the increased occurrences of the positive IOD phases were key drivers of major drought events in southeast Australia, where ENSO conditions are not usually assumed (Cai and Rensch, 2012).

Several studies have identified the relative roles of climate mode indices on Australian rainfall variability within either particular regions (Gallant et al., 2007; Klingaman et al., 2013) or a whole country (Risbey et al., 2009; Schepen et al., 2012; Min et al., 2013). However, there has been less attention paid to the relationship between joint climate drivers and extreme rainfall. In one of the first studies, Nguyen-Huy et al. (2017) found that the rainfall forecast was significantly improved in the upper and lower tails using the combination of

ENSO and Interdecadal Pacific Oscillation (IPO) Tripole Index (TPI). Therefore, further studies of the association between rainfall and multiple climate modes are required to better understand how these remote drivers are able to modulate extreme rainfall over a monthly and seasonal timescale.

In this chapter, we adopt the copula theorem (Sklar, 1959) as a way to provide a powerful approach for modeling the nonlinear dependencies among bivariate, trivariate, and multivariate random variables. In a copula-based joint distribution, the associations between the relevant variables are modeled independently with the individual marginal distribution of each variable. As a result, statistical copula-based models can overcome the issues of normal and symmetric assumptions in traditional forecast methods. Therefore, recent years have witnessed extensive applications of copula-based modeling in a wide range of fields such as economics and finance (de Melo Mendes et al., 2010; Nguyen and Bhatti, 2012), water resources and hydrology (Hao and Singh, 2012; Grimaldi et al., 2016), agriculture (Bokusheva, 2011; Nguyen-Huy et al., 2018), and environment (Kao and Govindaraju, 2010; Sraj et al., 2015).

This chapter aims to develop new understandings and applications of copula models by investigating the teleconnections related to climate variability between the different remote synoptic-scale climate drivers and extreme seasonal rainfall across the Australian Wheatbelt. Comparisons are made between the novel D-vine quantile regression and traditional quantile regression. Fivefold cross-validation is also applied to evaluate their out-of-sample performance and observe the sensitivity of the predictor set. The primary contribution of this chapter is to develop and validate the suitability of a copula-statistical methodology for the quantile-based forecasting of rainfall using large-scale climate mode influences and the implications of the model in agricultural risk-management and decision-making.

A brief description of the data used and methodologies applied is presented in Section 2. Results and analysis of climate−rainfall relationship and model performance are described in Section 3. Discussion of the results and future works and the conclusions are given in Sections 4 and 5, respectively.

2. Data and methodology

2.1 Cumulative rainfall index

The monthly and seasonal total precipitation data used in this chapter were obtained from the daily rainfall data covering the period from January 1, 1889, until December 31, 2012. These data sets are available from Scientific Information for Land Owners (SILO) and can be downloaded via the website of The Long Paddock, Queensland Government (https://legacy.longpaddock.qld.gov.au/silo/). These SILO databases are constructed from historical observational climate records provided by the Bureau of Meteorology (BOM). These time series are acquired for 16 weather stations that are spread over the Australian Wheatbelt and span different climate regimes (Fig. 8.1).

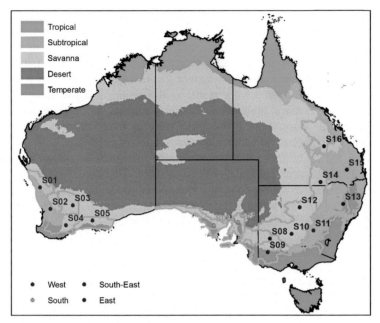

FIGURE 8.1 Selected weather stations across the Australian Wheatbelt (green border) spanning different climate conditions.

The cumulative rainfall index (CRI) is derived from measurements based on these daily observations. It is commonly used as the hypothetical underlying for agricultural weather insurance. In this study, CRI measures the rainfall within the two main vegetation periods of wheat crops that last from April 1 until June 30 and from July 1 until September 30. The index is calculated as follows (Xu et al., 2010):

$$CRI_{i,t} = \sum_{j=T_B}^{j=T_E} P_{j,t,i}, \tag{8.1}$$

where $P_{j,t,i}$ is the daily precipitation (mm) observed at day j in year $t = 1, \ldots,$ 124 and station $i = 1, \ldots,16$. T_B and T_E denote the beginning and the end of the considered period, respectively. In Australia, wheat is the main grain crop grown in the Wheatbelt where the sowing season is commonly from the mid-April to June, depending on the rainfall pattern. The harvest season is normally from mid-October until January of the next year. Therefore, two cumulative rainfall indices are derived for two periods, namely April–June (AMJ CRI) (sowing stage) and July–September (JAS CRI) (before harvesting). In the context of weather index–based insurance, this index addresses drought risk (Martin et al., 2001; Xu et al., 2010). The information and statistics of these selected weather stations are described in Table 8.1.

TABLE 8.1 Mean and standard deviation of cumulative rainfall index (CRI) in the period of April–June (AMJ) and July–September (JAS) for 16 weather stations.

Code	ID	Name	Coordinates	AMJ CRI		JAS CRI	
				Mean	Standard deviation	Mean	Standard deviation
West							
S01	08,088	Mingenew	115.44°E − 29.19°S	163.921	70.357	167.024	54.916
S02	10,111	Northam	116.66°E − 31.65°S	159.671	58.497	180.270	55.290
S03	12,074	Southern cross	119.33°E − 31.23°S	98.798	45.547	89.271	34.163
S04	10,627	Pingrup	118.51°E − 33.53°S	117.550	43.055	123.364	37.276
S05	12,070	Salmon Gums	121.64°E − 32.98°S	100.510	39.377	103.171	33.431
South							
S06	18,064	Lock	135.76°E − 33.57°S	116.136	47.562	145.624	45.207
S07	21,027	Jamestown	138.61°E − 33.20°S	129.480	56.715	166.443	52.743
South-east							
S08	76,047	Ouyen	142.32°E − 35.07°S	86.548	43.783	95.815	38.653
S09	79,023	Horsham Polkemmet	142.07°E − 36.66°S	127.763	53.566	144.585	47.872
S10	75,031	Hay	144.85°E − 34.52°S	98.559	51.546	95.123	41.671
S11	73,000	Barmedman	147.39°E − 34.14°S	115.492	59.153	116.589	48.070
East							
S12	48,030	Cobar	145.80°E − 31.50°S	84.715	53.988	76.996	37.743
S13	55,054	Tamworth	150.85°E − 31.09°S	130.139	67.544	137.860	55.555
S14	44,030	Dirranbandi	148.23°E − 28.58°S	96.633	71.250	77.356	50.829
S15	41,023	Dalby	151.26°E − 27.18°S	114.245	71.076	101.434	60.020
S16	35,059	Rolleston	148.63°E − 24.46°S	113.331	86.799	80.285	65.712

It is clear that weather and climate regimes vary over the Australian Wheatbelt. In particular, the eastern part of the Wheatbelt has a subtropical and savanna climate with the average rainfall during AMJ higher than during JAS months. Furthermore, the rainfall variability in this region is generally higher than that in other remaining sites in both summer and winter. The south, southwest, and west have a subtropical, savanna, and temperate climate, respectively, experiencing more rainfall in the winter season. In addition, the western part of the Wheatbelt receives the highest rainfall on average. Remote drivers influence these weather systems mentioned previously in a complicated way, resulting in rainfall variability in Australia.

2.2 Climate indices

Eight synoptic-scale climate indices are used for a comprehensive analysis of their influence on seasonal rainfall variability. These climate mode indices have been well documented in many studies investigating the climate—rainfall relationship in Australia (Risbey et al., 2009; Kirono et al., 2010; Schepen et al., 2012). ENSO is represented by several different indicators including Niño3.0, Niño3.4, and Niño4.0 (i.e., sea surface temperature [SST] representative), Southern Oscillation Index (SOI) (air pressure) and El Niño Modoki (EMI) (coupled ocean-atmospheric). The Dipole Mode Index (DMI) characterizes the intensity and Indonesian Index (II) the individual pole of IOD over the tropical Indian Ocean. The Tasman Sea Index (TSI) is included in this research to consider a potential link between extratropical SST and rainfall variability.

In terms of the origin of these data, the monthly SST anomalies for the period from January 1, 1889, to December 31, 2012, are derived from NOAA Extended Reconstructed Sea Surface Temperature Anomalies (SSTA) data, version 4, downloaded from the Asia-Pacific Data Research Center (APDRC). Monthly SOI data were acquired from the BOM, Australia. The seasonal climate indices are calculated as the average of 3-month values. Table 8.2 summarizes the description of these climate indices including the formula of EMI and DMI calculations and their components.

2.3 Methodology

2.3.1 Copula theorem

A copula, as explained by Sklar (1959), is a function used to link multiple univariate marginal distributions of random variables into a multivariate distribution. In brief, suppose a d-dimensional random vector $X = (x_1, ..., x_d)^T$ has its marginal cumulative distribution functions (CDFs) $F_1(x_1)$, $...,F_d(x_d)$

TABLE 8.2 Climate mode indices derived from NOAA Extended Reconstructed Sea Surface Temperature Anomalies (SSTA) data, version 4, and downloaded from Asia–Pacific Data Research Center (APDRC). SOI data acquired from Bureau of Meteorology, Australia (BOM).

Predictor variables	Description	Region
Niño3.0	Average SSTA over 150^0–90^0W and 5^0N–5^0S	Pacific
Niño3.4	Average SSTA over 170^0E–120^0W and 5^0N–5^0S	Pacific
Niño4.0	Average SSTA over 160^0E–150^0W and 5^0N–5^0S	Pacific
EMI	C $-$ 0.5 $\times$ (E + W) Where the components are average SSTA over C: 165^0E–140^0W and 10^0N–10^0S E: 110^0–70^0W and 5^0N–15^0S W: 125^0–145^0E and 20^0N–10^0S	Pacific
SOI	Pressure difference between Tahiti and Darwin as defined by Troup (1965)	Pacific
DMI	WPI $-$ EPI Where the components are average SSTA over WPI: 50^0–70^0E and 10^0N–10^0S EPI: 90^0–110^0E and 0^0N–10^0S	Indian
II	Average SSTA over 120^0–130^0E and 0^0N–10^0S	Indian
TSI	Average SSTA over 150^0–160^0E and 30^0S–40^0S	Extratropical

and probability density functions (PDFs) $f_1(x_1)$, ..., $f_d(x_d)$. Their joint CDF is expressed as (Sklar, 1959)

$$F(x_1, ..., x_d) = C[F_1(x_1), ..., F_d(x_d)] = C(u_1, ..., u_d) \qquad (8.2)$$

and the corresponding joint PDF

$$f(x_1, ..., x_d) = \left[\prod_{i=1}^{d} f_i(x_i) \right] c(u_1, ..., u_d), \qquad (8.3)$$

where C denotes the copula function and $c = \frac{\partial^d C}{\partial F_1, ..., \partial F_d}$ the corresponding copula density. If marginal distributions are continuous, then C is unique and $u_i = F_i(x_i), i = 1, ..., d$ is the univariate probability integral transformation (PIT). Copula families are generally distinguished as empirical, elliptical, Archimedean, extreme value, vine, and entropy copulas. This study focuses on the use of the D-vine copula approach, described in the following, which serves a quantile forecast purpose.

2.3.2 D-*vine copulas*

D-vine copula, a special form of vine family, was first proposed by Joe (1997) and further developed by Bedford and Cooke (2001, 2002). In short, the copula density in Eq. (8.3) is decomposed into the conditional and unconditional bivariate densities, so-called bivariate pair-copulas, as follows (Czado, 2010):

$$
c(u_1, \ldots, u_d) = \prod_{i=1}^{d-1} \prod_{j=i+1}^{d} c_{ij;i+1,\ldots,j-1} \big[C_{i|i+1,\ldots,j-1}(u_i | u_{i+1}, \ldots, u_{j-1}),
$$

$$
C_{j|i+1,\ldots,j-1}(u_j | u_{i+1}, \ldots, u_{j-1}) \big]
$$

(8.4)

In this construction, each pair-copula is selected independently from the others, allowing the flexible model of full dependence structure of high-dimensional random variables existing as the characteristics of asymmetric and tail dependences. Therefore, the D-vine approach can address the limitations of other copula families such as meta-elliptical or symmetric Archimedean copulas.

2.3.3 *Semiparametric* D-*vine quantile regression*

Eq. (8.2) reveals that the construction of copula-based models commonly including fitting the marginal distributions and fitting the copulas. In general, these both procedures can be fitted either parametrically or nonparametrically. In this study, the nonparametric approach is used to fit marginal distributions and the copulas are fitted parametrically, resulting in a semiparametric quantile regression model. Constructing the model in this way can minimize the bias and inconsistency issues often faced by the fully parametric model when one of the parametric components is misspecified (Noh et al., 2013).

As the first step, marginal distributions are fitted nonparametrically using the univariate local-polynomial likelihood density estimation method (Nagler, 2017). Given a sample $x^i, i = 1, \ldots, n$ with unknown PDF, the estimated kernel density is defined as (Geenens and Wang, 2018)

$$
\widehat{f}(x) = \frac{1}{nh} \sum_{i=1}^{n} K \left(\frac{x - x^i}{h} \right),
$$

(8.5)

where h is the bandwidth parameter and $K(x)$ the smoothing kernel function. In this study, the Gaussian kernel and the plug-in bandwidth are used as in the methodology developed by Sheather and Jones (1991). The degree of the polynomial is selected as the log-quadratic fitting (Nagler, 2017).

The estimated marginal distribution functions $\widehat{F}_Y$ and $\widehat{F}_j$ can be then obtained for the response variable Y and predictor variables $X_1, \ldots, X_d, j = 1, \ldots, d$, respectively. These functions are used to convert the observed data to pseudo-copula data, which are, $\widehat{v}^i = \widehat{F}_Y(y^i)$ and $\widehat{u}_j^i = \widehat{F}_j(x_j^i)$. These pseudo-copula data $\widehat{v} = (\widehat{v}^i)$ and $\widehat{v} = (\widehat{u}_j^i)$ approximate to an i.i.d sample from the PIT vector $(V, U_1, \ldots, U_d)^T$ and are therefore able to be used for the D-vine copula estimate in the next step (Kraus and Czado, 2017).

These pseudo-copula data are fitted to a D-vine model with an order $V - U_{l_1} - ... - U_{l_d}$, where $L = (l_1, ..., l_d)^T$ is the arbitrary ordering resulting in $d!$ possible models. Therefore, this study applies the new algorithm proposed by Kraus and Czado (2017) to automatically select the parsimonious D-vine model. In short, only the most influential predictors are added into the model in an order that minimizes the AIC-corrected conditional log-likelihood cll^{AIC}. As a result, the conditional quantile prediction model has the highest explanation for the response variable. Furthermore, this algorithm overcomes the common issue in terms of conventional regression, involving collinearity, transformation, and the inclusion and exclusion of covariates.

Finally, for quantile levels $\alpha \in (0,1)$, the quantile $\widehat{q}_\alpha$ of a response variable Y given predictor variables $X_1, ..., X_d$ can be obtained using the inverse forms of the marginal distribution function F_Y^{-1} and the conditional copula function $C_{V|U_1,...,U_d}^{-1}$ conditional on $\widehat{u}_1, ..., \widehat{u}_d$ which is defined as (Kraus and Czado, 2017):

$$\widehat{q}_\alpha(x_1, ..., x_d) = \widehat{F}_Y^{-1}\left[\widehat{C}_{V|U_1,...,U_d}^{-1}\left(\alpha\middle|\widehat{u}_1, ..., \widehat{u}_d\right)\right]. \tag{8.6}$$

More details of this approach can be found in the study by Kraus and Czado (2017) and Schallhorn et al. (2017).

2.3.4 Linear quantile regression

For the purpose of comparison, this study also utilizes the traditional linear quantile regression (LQR) model to predict rainfall with the same predictor sets for the D-vine copula model. The LQR approach, first introduced by Koenker and Bassett (1978), assumed the conditional quantile of the predicted variables to be linear in the predictors. This assumption can be expressed as (Schallhorn et al., 2017)

$$\widehat{q}_\alpha(x_1, ..., x_d) = \widehat{\beta}_0 + \sum_{j=1}^{d} \widehat{\beta}_j x_j, \tag{8.7}$$

where the estimates of the regression coefficients $\widehat{\beta}_j$ are acquired by solving the minimization problem:

$$\min_{\beta \in R^{d+1}}\left[\alpha\sum_{i=1}^{n}\left(y^i - \beta_0 - \sum_{j=1}^{d}\beta_j x_j^i\right)^+ + (1-\alpha)\sum_{i=1}^{n}\left(\beta_0 + \sum_{j=1}^{d}\beta_j x_j^i - y^i\right)^+\right]. \tag{8.8}$$

The LQR method has a number of limitations such as a very restrictive assumption of normal margins and the changeable slopes at different quantile levels (Bernard and Czado, 2015; Kraus and Czado, 2017; Schallhorn et al., 2017).

2.3.5 Evaluation of model performance

To assess the forecast performance of semiparametric D-vine and linear quantile regression, this study applies a fivefold cross-validation test to evaluate the out-of-sample performance. Therefore, the total 124 data points (1889–2012) are split into fivefolds where each fold will become an evaluation data set. The remaining data corresponding to each fold are used as training data sets. As a result, all data points are joined in training and testing processes.

Because in this out-of-sample test, the true regression is unknown (Kraus and Czado, 2017), only a realization for each seasonal rainfall can be obtained, and an averaged cross-validated tick-loss function $L_{\alpha,m}^{j}$ is used to evaluate the forecasted α-quantiles for $\alpha \in (0,1)$. The expression of this computation is expressed as follows (Komunjer, 2013; Kraus and Czado, 2017; Schallhorn et al., 2017):

$$L_{\alpha,m}^{j} = \frac{1}{n_{eval}} \sum_{i=1}^{n_{eval}} \rho_{\alpha}\left(y^{i} - \widehat{q}_{\alpha,m}^{i}\right), \tag{8.9}$$

where y^{i} and $\widehat{q}_{\alpha,m}^{i}$ denote the observation and forecast quantile using method m at the station j and a point i in a sample size n_{eval}. The function $\rho_{\alpha} = y$ $[\alpha - I(y < 0)]$ is the check or tick function. The lower values of the averaged cross-validated tick-loss function imply better performance of the forecast model.

3. Results

3.1 Climate–rainfall relationships

The influence of climate indices on rainfall variability is inspected as an initial analysis before the development of the probabilistic models. The Kendall statistic has been used to estimate the rank-based measure of association between climate indices and CRI at different lag times in the entire data set for different weather stations. The concurrent relationship of climate indices and CRI has been explored as well. Such simultaneous relationships may benefit the seasonal rainfall forecast models because the recent maturity of climate forecasting systems allows information of climate mode indices to be forecast with sufficient lead time and accuracy (Chen et al., 2004). The correlation coefficients between AMJ CRI and climate conditions during the period JFM and AMJ are illustrated in Table 8.3. All this climate information is used together with JAS climate indices to analyze JAS CRI, and the results are represented in Table 8.4.

It can be seen from Table 8.3 that all JFM climate indices provide very limited information for forecasting of AMJ CRI. This is to be expected because the impact of remote drivers on rainfall variability, as mentioned previously, is generally strong from July of the year being considered to March of the next year. The results agree with findings from a study undertaken by

TABLE 8.3 Kendall-tau correlation coefficients with significant *P*-values at 10% (bold) and 5% (underlined bold) significance levels between January–March (JFM) and April–June (AMJ) climate indices and AMJ cumulative rainfall index (CRI) in 16 weather stations.

	S01	S02	S03	S04	S05	S06	S07	S08	S09	S10	S11	S12	S13	S14	S15	S16
JFM Niño3.0	**-0.105**	**-0.132**	0.009	-0.084	-0.069	-0.075	-0.059	-0.025	-0.036	-0.028	-0.022	0.041	0.055	-0.007	-0.019	-0.055
JFM Niño3.4	-0.091	-0.099	0.030	-0.044	-0.017	-0.075	-0.026	-0.002	-0.020	-0.003	-0.007	0.021	0.035	-0.014	-0.037	-0.053
JFM Niño4.0	-0.090	-0.099	0.047	-0.039	0.002	-0.070	-0.025	-0.001	-0.016	0.007	0.001	0.006	0.023	-0.021	-0.049	-0.059
JFM SOI	0.042	0.056	-0.030	0.005	-0.020	0.013	-0.034	-0.061	-0.059	-0.090	-0.086	**-0.103**	-0.096	-0.035	-0.013	-0.006
JFM EMI	-0.047	-0.026	0.050	0.030	0.049	-0.028	0.020	0.007	-0.025	-0.017	-0.051	-0.085	-0.016	-0.072	**-0.100**	**-0.104**
JFM DMI	0.027	0.006	0.046	0.023	-0.047	-0.005	-0.012	0.023	-0.011	0.007	0.012	0.001	-0.097	0.000	-0.052	-0.094
JFM II	-0.092	**-0.136**	0.029	**-0.128**	-0.031	0.006	0.000	0.003	-0.006	0.010	-0.002	0.081	**0.118**	0.038	0.016	0.076
JFM TSI	**-0.137**	-0.090	0.026	**-0.136**	-0.047	0.075	0.078	0.007	-0.008	0.023	-0.037	0.024	0.000	-0.042	**-0.123**	0.024
AMJ Niño3.0	-0.038	-0.071	-0.007	-0.093	-0.099	**-0.107**	**-0.103**	-0.079	**-0.139**	-0.066	-0.033	-0.012	-0.044	-0.043	-0.067	-0.057
AMJ Niño3.4	-0.067	-0.089	0.002	**-0.122**	-0.075	-0.087	-0.082	-0.032	-0.064	-0.065	-0.018	-0.018	-0.033	-0.038	-0.062	-0.092
AMJ Niño4.0	-0.086	-0.096	0.012	-0.114	-0.077	-0.077	-0.049	-0.026	-0.054	-0.086	-0.018	-0.041	-0.033	-0.047	-0.072	-0.098
AMJ SOI	**0.164**	**0.127**	0.037	**0.135**	**0.135**	**0.201**	**0.163**	**0.141**	**0.175**	**0.101**	**0.131**	0.071	**0.132**	**0.118**	**0.129**	**0.149**
AMJ EMI	-0.076	-0.054	0.003	-0.099	-0.026	0.027	0.041	0.060	0.037	-0.062	-0.029	-0.069	-0.007	-0.062	-0.055	-0.095
AMJ DMI	-0.098	-0.009	-0.014	-0.039	-0.072	**-0.109**	**-0.151**	-0.078	**-0.114**	**-0.102**	**-0.127**	-0.096	**-0.157**	**-0.123**	**-0.126**	**-0.196**
AMJ II	-0.035	-0.065	0.053	**-0.126**	0.046	0.070	0.094	0.061	0.045	0.084	0.057	**0.111**	**0.143**	0.081	-0.006	0.092
AMJ TSI	-0.062	-0.078	0.068	**-0.109**	0.060	0.068	0.083	0.069	0.039	0.081	0.068	**0.107**	0.085	**0.121**	0.034	**0.126**

TABLE 8.4 Kendall-tau correlation coefficients with significant *P*-values at 10% (bold) and 5% (underlined bold) significance levels between January–March (JFM), April–June (AMJ), and July–September (JAS) climate indices and JAS cumulative rainfall index (CRI) in 16 weather stations.

	S01	S02	S03	S04	S05	S06	S07	S08	S09	S10	S11	S12	S13	S14	S15	S16
JFM Niño3.0	−0.013	−0.071	−0.001	0.015	−0.035	−0.056	−0.072	−0.010	−0.046	0.036	−0.016	−0.078	−0.065	−0.065	−0.068	**−0.100**
JFM Niño3.4	−0.001	−0.057	0.007	0.000	−0.034	−0.042	−0.054	0.023	−0.016	0.046	0.019	−0.036	−0.027	−0.027	−0.034	−0.080
JFM Niño4.0	−0.004	−0.051	0.013	−0.017	−0.036	−0.036	−0.055	0.020	−0.017	0.019	0.009	−0.035	−0.026	−0.016	−0.052	−0.087
JFM SOI	−0.041	0.024	−0.051	−0.039	0.018	−0.030	−0.072	**−0.107**	−0.060	**−0.123**	−0.091	−0.027	−0.093	−0.035	−0.049	−0.022
JFM EMI	−0.023	−0.068	−0.016	−0.045	−0.035	−0.029	−0.065	0.054	0.022	−0.004	0.017	−0.013	−0.018	−0.006	−0.012	−0.082
JFM DMI	**−0.113**	−0.051	−0.019	−0.085	−0.081	−0.031	−0.008	−0.019	0.016	−0.015	−0.049	0.012	−0.026	0.004	−0.068	−0.034
JFM II	0.002	−0.001	0.067	−0.039	0.063	−0.028	0.029	0.088	0.027	0.108	0.069	−0.010	0.047	0.071	0.007	0.027
JFM TSI	−0.048	0.018	0.065	−0.082	0.000	−0.028	−0.003	0.039	0.034	0.077	0.058	−0.002	0.005	0.035	−0.066	−0.031
AMJ Niño3.0	**−0.106**	**−0.119**	−0.078	−0.091	**−0.128**	−0.046	−0.084	**−0.163**	**−0.145**	−0.063	**−0.137**	**−0.172**	**−0.186**	**−0.240**	**−0.242**	**−0.274**
AMJ Niño3.4	−0.080	−0.090	−0.041	−0.055	−0.080	−0.029	−0.079	−0.100	**−0.111**	−0.060	**−0.109**	**−0.138**	**−0.154**	**−0.188**	**−0.210**	**−0.240**
AMJ Niño4.0	−0.085	−0.094	−0.032	−0.070	−0.051	−0.026	−0.051	−0.065	−0.080	−0.059	−0.077	**−0.115**	**−0.119**	**−0.138**	**−0.192**	**−0.214**
AMJ SOI	0.082	**0.106**	0.071	0.044	0.048	**0.163**	**0.127**	**0.198**	**0.152**	**0.172**	**0.199**	**0.193**	**0.222**	**0.275**	**0.250**	**0.223**

AMJ EMI	−0.024	−0.008	0.030	0.007	0.010	0.003	−0.034	0.066	0.023	−0.055	−0.026	−0.042	−0.030	−0.022	−0.057	−0.068
AMJ DMI	−0.091	−0.071	−0.036	−0.136	−0.090	−0.084	−0.141	−0.132	−0.117	−0.177	−0.118	−0.113	−0.127	−0.100	−0.101	−0.072
AMJ II	−0.054	−0.060	0.028	−0.148	0.033	−0.013	0.067	0.127	0.067	0.155	0.107	0.018	0.054	0.046	−0.024	0.039
AMJ TSI	−0.130	−0.071	−0.018	−0.169	0.004	−0.004	0.023	0.059	0.033	0.133	0.108	0.024	0.035	0.053	−0.052	−0.012
JAS Niño3.0	−0.139	−0.139	−0.080	−0.126	−0.075	−0.081	−0.132	−0.212	−0.149	−0.117	−0.185	−0.156	−0.231	−0.222	−0.283	−0.289
JAS Niño3.4	−0.127	−0.141	−0.060	−0.100	−0.074	−0.084	−0.112	−0.204	−0.144	−0.131	−0.206	−0.191	−0.279	−0.262	−0.342	−0.320
JAS Niño4.0	−0.102	−0.140	−0.018	−0.083	−0.052	−0.108	−0.097	−0.188	−0.151	−0.131	−0.186	−0.209	−0.274	−0.259	−0.352	−0.335
JAS SOI	0.192	0.195	0.154	0.118	0.124	0.225	0.208	0.299	0.271	0.249	0.285	0.290	0.317	0.310	0.318	0.351
JAS EMI	−0.081	−0.081	0.019	−0.030	−0.024	−0.077	−0.034	−0.009	−0.037	−0.089	−0.106	−0.144	−0.143	−0.124	−0.225	−0.187
JAS DMI	−0.158	−0.140	−0.093	−0.132	−0.083	−0.228	−0.257	−0.276	−0.249	−0.215	−0.173	−0.162	−0.196	−0.201	−0.191	−0.221
JAS II	0.088	0.053	0.159	−0.010	0.109	0.219	0.246	0.298	0.265	0.321	0.293	0.198	0.254	0.221	0.154	0.237
JAS TSJ	−0.111	−0.100	0.065	−0.148	0.002	−0.067	0.023	−0.011	0.003	0.093	0.081	−0.029	−0.007	−0.002	−0.075	−0.088

Schepen et al. (2012). The concurrent relationship between a range of climate indices and rainfall is stronger in AMJ. It is worth pointing out that the ENSO plays an important role in the variations of AMJ rainfall over the Australian Wheatbelt, as indicated by the significant coefficients between SOI and rainfall across most of the weather stations. According to Lo et al. (2007), SST-based indices are more useful as the predictors or drivers of Australian rainfall at longer times ahead than SOI. However, our observations show that the SOI can be potentially used to forecast rainfall at the same lead timescale with SST-based indices depending on locations and seasons (Tables 8.3 and 8.4). The mechanism for this relationship may be explained by the fact that the SOI is related to the large-scale surface pressure, and therefore, its variability is more closely associated with the rainfall process. Furthermore, the collection of SOI data is based on consistent pressure values observed from two stations providing more confidence in the early record than SST-based indices that are interpolated from the observations of sparse stations (Risbey et al., 2009).

The IOD, similar to ENSO, is an index representing a coupled interaction of ocean—atmosphere phenomena in the equatorial Indian Ocean (Saji et al., 1999). In this study, we found that the DMI and II anomalies derived from the Indian Ocean have a similar impact on Australian rainfall to that of the Pacific region. There is minimum evidence encouraging the use of any lagged climate drivers (i.e., JFM DMI and JFM II) to forecast AMJ CRI. However, variations of AMJ DMI are related to AMJ CRI along the east to the southeast and south Australia, but excluding the western region. The impact of II and extratropical index TSI can be useful as a predictor to forecast AMJ rainfall in some stations, but are very limited in general using either lagged or simultaneous information.

Table 8.4 represents the usefulness of lag ENSO information in forecasting seasonal rainfall, in particular in the east and southeast regions, during the period of July—August. However, it is interesting that the influence of different ENSO indicators on JAS rainfall varies at different locations. Niño4.0 affects the rainfall in the east region only, while there is no significant correlation between EMI and rainfall at all weather stations. On the other hand, Niño3.4, SOI, and Niño3.0 extend their effects on JAS CRI to the southeast, south and west regions, respectively. The influence from the Indian Ocean anomalies on rainfall patterns is similar to ENSO, where the lag information of AMJ DMI and AMJ II is potentially useful for forecasting JAS rainfall from the east to the south and southwest regions. The influence from the extratropical region on rainfall can be observed in the southeast, southwest, and western Australia where AMJ TSI has a significant correlation with S10—11, S04, and S01, respectively.

As expected, simultaneous correlation coefficients between climate drivers and rainfall of the JAS period are stronger than the lag coefficients. In regard to ENSO phenomena, while EMI influences rainfall in the eastern region only (i.e., S11-16) and the impact from all Niño indices does not cover some regions in the south and westeast (i.e., S03 and S05-06), the information of

JAS SOI can be used to skillfully forecast JAS rainfall over much of the Australian Wheatbelt. Taking into consideration the impact from the Indian Ocean, DMI affects most of the weather stations except S03 and S05. In addition, II can be used to compensate for the lack of forecasting information in these weather stations. These results agree with a study undertaken by Risbey et al. (2009) where IOD generally peaks in spring (September–October) but can be observed from May to November. Furthermore, there is no evidence supporting the use of TSI as a predictor for rainfall forecast over the Wheatbelt regions except for the western region where it is useful for explaining JAS rainfall in three out of five stations.

The results indicate that probabilistic seasonal rainfall forecasts can be performed efficiently in all regions with sufficient lead time using multiple climate drivers. For these reasons, it is obvious that a robust rainfall-forecasting model should take multiple climate drivers and lag information into account to achieve better performance, at least in terms of time sufficiency and spatial coverage. In addition, if the IOD and ENSO events occur together, they can reinforce each other (Kirono et al., 2010), although this need not necessarily happen (Meyers et al., 2007; Risbey et al., 2009). The question of whether these combinations can improve the accuracy of rainfall forecasts will be addressed in the following section.

3.2 Rainfall quantile forecast

We now present the forecast of seasonal rainfall and the evaluation of D-vine and benchmark these results with an LQR model performance. The rainfall forecast is made at three quantile levels: 0.05 (lower tail), 0.50 (median), and 0.95 (upper tail) for two periods of AMJ and JAS using various combinations of climate drivers as predictor variables. In particular, three predictor sets including climate indices observed in JFM (i.e., eight predictors), AMJ (i.e., eight predictors), and JFM + AMJ (i.e., 16 predictors) are used to forecast AMJ CRI. Similarly, JAS CRI at 16 weather stations is forecast using six predictor sets consisting of the former sets and three predictor sets of JAS, AMJ + JAS, and JFM + AMJ + JAS (i.e., 24 predictors).

Figure 8.2 displays the results of the AMJ CRI forecast at the three alpha levels derived from all predictor sets and both models. The first visual inspection indicates that the results of rainfall forecast and model performance vary across the study regions depending on alpha levels, predictor sets, and models used. In general, the D-vine copula model provides better accuracy than LQR for the west-west region (S01-02) at the lower tail and for the south (S06-07) and southeast (S09-11) regions at the median and upper tail for all predictor sets. These findings imply that the impact of climate indices on these stations are more scattered and nonlinear at the median and upper extreme events which cannot be captured by the traditional LQR method. Furthermore, both models reveal that the use of simultaneous information or its combination

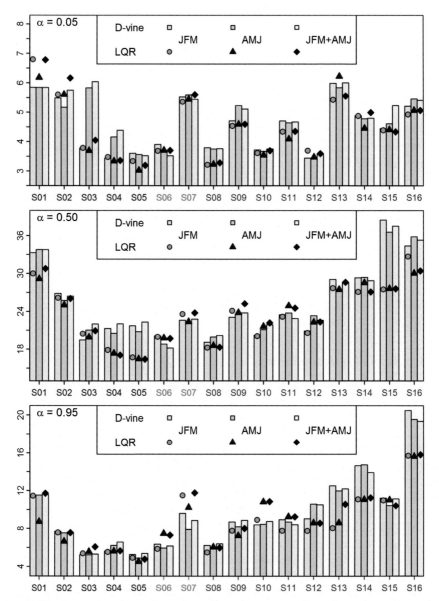

FIGURE 8.2 Averaged fivefold cross-validated tick-loss of April−June cumulative rainfall index (AMJ CRI) forecast at 16 stations using different sets of predictors January−March (JFM), AMJ, and JFM + AMJ and using D-vine (bar charts) and linear (symbols) quantile regression (LQR) for different quantile levels.

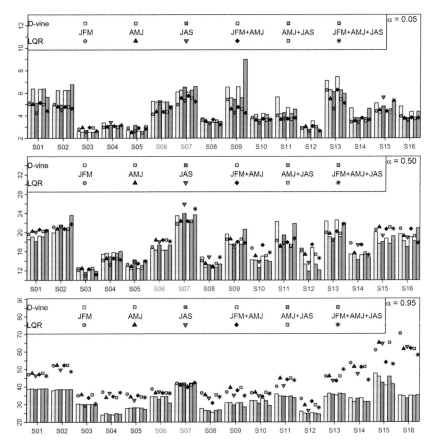

FIGURE 8.3 Averaged fivefold cross-validated tick-loss function of July–September cumulative rainfall index (JAS CRI) forecast at 16 stations using different sets of predictors January–March (JFM), April–June (AMJ), JAS, JFM + AMJ, AMJ + JAS, and JFM + AMJ + JAS and using D-vine (bar charts) and linear (symbols) quantile regression (LQR) for different quantile levels.

with lag information does not always improve the forecasting performance. These outcomes reflect the spatiotemporal characteristic of influences of climate indices on Australian rainfall.

This spatiotemporal variability affecting rainfall of climate drivers is especially emphasized in Fig. 8.3 where the D-vine copula model outperforms the LQR approach in most cases at the median and upper extreme levels. This highlights the usefulness of the copula-based model in forecasting JAS rainfall above the median level. Furthermore, it is clear that the impact of climate on Australian seasonal rainfall is asymmetric where the upper tail is more scattered

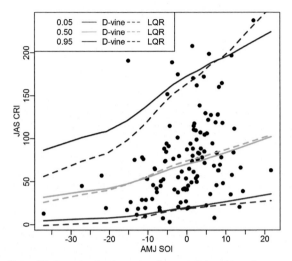

FIGURE 8.4 Exemplified scatterplot between observed July−September cumulative rainfall index (JAS CRI) and April−June SOI (AMJ SOI) (dotted points) overlaid with locally weighted regression lines of forecasted JAS CRI from fivefold cross-validation in three quantile levels (0.05, 0.50, and 0.95).

and nonlinear. In regard to the use of different predictor sets, the results again show an inconsistent pattern of using information from lag, concurrent and their combination. For example, JAS climate information yields the best performance of JAS CRI forecast in the S16 below the median levels. However, to forecast JAS CRI in the upper quantiles, the climate information of lagged and concurrent times and their combination are seen to provide almost the same results.

The differences in the forecasting performance of D-vine and LQR are illustrated as an example in Fig. 8.4 for S16. In this figure, the observations of AMJ SOI and JAS CRI are represented by dotted points overlaid with locally weighted regression lines of forecasted JAS CRI from fivefold cross-validation in three quantile levels (0.05, 0.50, and 0.95). It is clear that the relationship in the upper tail between observed AMJ SOI and JAS CRI is scattered and nonlinear. This empirical pattern of dependence may explain the reason for the outperformance of the D-vine copula-based model in the quantile level of 0.95 where the LQR method is inadequate. Both models yield a small difference in the forecast lines of the lower and median quantile levels. However, there is large divergence between forecast lines in the upper quantile which is in agreement with the results derived from the average cross-validated tick-loss function (Fig. 8.3).

Finally, the results (not shown here) also spell out the fact that the influence of climate indices on seasonal rainfall can vary over a decadal timescale in the present study region. This was indicated by the change of predictor sets selected for the training model in each fold of the cross-validation process.

Furthermore, the maximum number of selected predictors was four in all considered cases. These findings question whether it is possible to build up a certain predictor set of climate indices for rainfall forecast in each region. However, the answer to this problem, although will it bring interesting insights, is out of the scope of this chapter.

4. Discussion

This chapter has explored the association between a number of climate mode indices and rainfall observed in many weather stations across Australia's Wheatbelt regions. The stations selected were those that geographically distribute over the Australian Wheatbelt and experience different climate conditions. The lagged and concurrent information derived from the Pacific and Indian Oceans were found to be useful for rainfall forecast systems. Based on this analysis, the chapter also developed a quantile rainfall forecast model using the vine copula approach. The semiparametric D-vine copula-based model used in this chapter showed better performance for rainfall forecasting in the median and upper levels. To minimize the model misspecification further (Noh et al., 2013), future research may apply the fully nonparametric copula-based approach (Schallhorn et al., 2017) meaning that both estimates of marginal distributions and copulas are nonparametric. Furthermore, the results from this chapter also showed that the LQR provides better agreement of rainfall forecast in the lower tail. Therefore, other several quantile regression models such as boosting additive (Koenker, 2011) or nonparametric quantile regression (Li et al., 2013) may be used in future work for the purpose of comparison.

It is worth noting that the predictor data sets chosen for the training models in each fold of the cross-validation method and each location differed from each other. These changes reflected the spatiotemporal characteristics of the impact of various climate mode indices on Australian rainfall. Therefore, further research could assist in building up a certain predictor data set of climate indices for rainfall forecast corresponding to each study site. This work can be carried out by using a comparison of model performance between a fixed predictor set and exchangeable predictor sets using a k-fold cross-validation approach. To provide a more comprehensive analysis, the rainfall forecast may be conducted at more timescale points.

The Australian climate is also affected by many local factors such as atmospheric blocking and the subtropical ridge, which were not considered in this study. Atmospheric blocking to the southeast of Australia has been examined as the driver of rainfall increase across a large part of Australia while the position and intensity of the subtropical ridge affect rainfall in the east of Australia (Risbey et al., 2009; Cai et al., 2011; Schepen et al., 2012). According to Zscheischler et al. (2018), extreme events are often the result of the processes that many drivers interact together and have spatiotemporal dependencies. As a result, the risk assessment is potentially underestimated. Therefore, investigating the joint influence of large-scale and local drivers to improve risk management is an important consideration for future research.

In practice, probabilistic seasonal rainfall forecasts can be derived from both empirical and dynamic climate forecasting models up to a year ahead (Goddard et al., 2001; Schepen et al., 2012). However, they all have their own advantages and limitations. The empirical models might be categorized into statistically based (Rajeevan et al., 2007; Nguyen-Huy et al., 2017) or machine learning methods (Ramirez et al., 2005). Empirical models use the empirical relationships between historically observed variables and therefore depend on the availability of recorded data length and assume stationary relationships between variables. On the other hand, dynamic forecasting models (Druce, 2001; Vieux et al., 2004) rely on numerical simulations directly modeling physical processes; however, they often cost more than statistical models in terms of implementation and operation. Therefore, a hybrid integrated forecasting system is preferred to provide greater accuracy and precision of rainfall forecasts while being more economically viable.

5. Conclusions

This chapter has demonstrated that the information derived from large-scale oceanic-atmospheric processes is potentially useful for seasonal rainfall forecasting in Australia. In general, the simultaneous relationships between climate mode indices and rainfall are strongest. This finding agrees with results from studies undertaken by Risbey et al. (2009). With the development of climate forecast systems, climate information such as prominent El Niño events could be successfully predicted up to 2 years ahead (Chen et al., 2004). Therefore, it is possible to achieve seasonal rainfall forecasts with high accuracy and longer lagged time using statistical models and climate drivers. In addition, lagged climate indices also expose significant evidence to support the seasonal rainfall forecast.

Climate drivers have an asymmetric influence on seasonal rainfall in Australia. Climate indices derived from oceanic and atmospheric variability in the Pacific region such as SOI exhibit strong evidence for forecasting seasonal rainfall over much of the Australian Wheatbelt. The impact of the Indian Ocean on the seasonal rainfall represents similar evidence in supporting the forecast of all weather stations. The extratropical region also shows a significant relationship with rainfall in some regions and during some seasons. The strongest and most spatially widespread evidence supporting JAS rainfall forecast comes from the JAS climate indices. Furthermore, the joint occurrence of extreme climate events may reinforce rainfall fluctuation (Nguyen-Huy et al., 2017) and may subsequently affect crop yield (Nguyen-Huy et al., 2018). Therefore, using a copula-based model with multiple climate indices as predictors could improve the forecast of seasonal rainfall ahead.

The copula-based joint probability modeling method was applied to forecast the seasonal cumulative rainfall across Australian Wheatbelt using different predictor sets of lagged and concurrent climate indices. In addition, the traditional linear quantile regression is simultaneously implemented for a comparison. The fivefold cross-validation was used to evaluate the out-of-sample performance of both models. Furthermore, the most influential predictors were selected based on the AIC-corrected conditional log-likelihood to form the parsimonious model. In general, the D-vine copula-based model shows greater potential for forecasting rainfall above the median level. The results imply that the impact of climate indices on rainfall is nonlinear in the upper quantiles where they may be unable to be measured by the traditional LQR.

The usefulness of lagged, concurrent, or combined climate information for seasonal rainfall forecast varies with locations and times. The performance of seasonal rainfall quantile forecasts using the information of lagged climate indices may be higher than that of using simultaneous predictor sets. Furthermore, the selected parsimonious predictor sets are different from each other for each training model in each fold of the cross-validation method. Therefore, a potential study is to test the performance of seasonal rainfall using certain predictor sets for each location. In addition, future research may be conducted with more climate indices at more timescales using fully nonparametric models.

Probabilistic seasonal rainfall forecasts derived from statistical models can provide important information to a variety of users related to water resource in regard to planning and decision-making processes. For example, seasonal rainfall forecasts may assist water managers to make operational decisions on water allocation for rival users (Kirono et al., 2010). In addition, seasonal rainfall forecasting is one of the most effective means to adapt to and diminish the vagaries of adverse weather and support the development of risk-management strategies. For example, skillful quantification of seasonal rainfall in extreme cases with a sufficient time lag can support agricultural producers geographically diversifying farming systems to minimize climate risk and optimize profitability (Larsen et al., 2015). We are currently studying the use of a copula-based approach for evaluating the weather (rainfall) systemic risk (Xu et al., 2010; Okhrin et al., 2013) in Australia. Because the spread of the rainfall extremes is modeled through a joint distribution using a robust vine copula approach, these results will potentially improve risk management and crop insurance.

Acknowledgments

The project was financed by the University of Southern Queensland Post Graduate Research Scholarship (USQPRS 2015–18), School of Agricultural, Computational and Environmental Sciences and Drought and Climate Adaptation Program (DCAP) (Producing Enhanced Crop Insurance Systems and Associated Financial Decision Support Tools). The authors would like to acknowledge Dr. Louis Kouadio for his support in collecting rainfall data.

References

Bedford, T., Cooke, R.M., 2001. Probability density decomposition for conditionally dependent random variables modeled by vines. Annals of Mathematics and Artificial Intelligence 32 (1–4), 245–268.

Bedford, T., Cooke, R.M., 2002. Vines: a new graphical model for dependent random variables. Annals of Statistics 1031–1068.

Bernard, C., Czado, C., 2015. Conditional quantiles and tail dependence. Journal of Multivariate Analysis 138, 104–126.

Best, P., Stone, R., Sosenko, O., 2007. Climate risk management based on climate modes and indices-the potential in Australian agribusinesses. In: 101st Seminar, July 5–6, 2007. European Association of Agricultural Economists, Berlin Germany.

Bokusheva, R., 2011. Measuring dependence in joint distributions of yield and weather variables. Agricultural Finance Review 71 (1), 120–141.

Brown, J.R., Power, S.B., Delage, F.P., Colman, R.A., Moise, A.F., Murphy, B.F., 2011. Evaluation of the South Pacific Convergence Zone in IPCC AR4 climate model simulations of the twentieth century. Journal of Climate 24 (6), 1565–1582.

Cai, W., Rensch, P., 2012. The 2011 southeast Queensland extreme summer rainfall: a confirmation of a negative Pacific Decadal Oscillation phase? Geophysical Research Letters 39 (8).

Cai, W., Van Rensch, P., Cowan, T., 2011. Influence of global-scale variability on the subtropical ridge over southeast Australia. Journal of Climate 24 (23), 6035–6053.

Cai, W., van Rensch, P., Cowan, T., Sullivan, A., 2010. Asymmetry in ENSO teleconnection with regional rainfall, its multidecadal variability, and impact. Journal of Climate 23 (18), 4944–4955.

Chen, D., Cane, M.A., Kaplan, A., Zebiak, S.E., Huang, D., 2004. Predictability of El Niño over the past 148 years. Nature 428 (6984), 733.

Corte-Real, J., Qian, B., Xu, H., 1998. Regional climate change in Portugal: precipitation variability associated with large-scale atmospheric circulation. International Journal of Climatology 18 (6), 619–635.

Czado, C., 2010. Pair-copula constructions of multivariate copulas. In: Copula Theory and its Applications. Springer, pp. 93–109.

de Melo Mendes, B.V., Semeraro, M.M., Leal, R.P.C., 2010. Pair-copulas modeling in finance. Financial Markets and Portfolio Management 24 (2), 193–213.

Druce, D.J., 2001. Insights from a history of seasonal inflow forecasting with a conceptual hydrologic model. Journal of Hydrology 249 (1–4), 102–112.

Enfield, D.B., Mestas-Nuñez, A.M., Trimble, P.J., 2001. The Atlantic multidecadal oscillation and its relation to rainfall and river flows in the continental US. Geophysical Research Letters 28 (10), 2077–2080.

Gallant, A.J., Hennessy, K.J., Risbey, J., 2007. Trends in rainfall indices for six Australian regions: 1910-2005. Australian Meteorological Magazine 56 (4), 223–241.

Geenens, G., Wang, C., 2018. Local-likelihood transformation kernel density estimation for positive random variables. Journal of Computational and Graphical Statistics 27 (4), 822−835.

Goddard, L., Mason, S.J., Zebiak, S.E., Ropelewski, C.F., Basher, R., Cane, M.A., 2001. Current approaches to seasonal to interannual climate predictions. International Journal of Climatology 21 (9), 1111−1152.

Grimaldi, S., Petroselli, A., Salvadori, G., De Michele, C., 2016. Catchment compatibility via copulas: a non-parametric study of the dependence structures of hydrological responses. Advances in Water Resources 90, 116−133.

Hao, Z., Singh, V.P., 2012. Entropy-copula method for single-site monthly streamflow simulation. Water Resources Research 48 (6).

Holland, G.J., 1986. Interannual variability of the Australian summer monsoon at Darwin: 1952−82. Monthly Weather Review 114 (3), 594−604.

Joe, H., 1997. Multivariate Models and Multivariate Dependence Concepts. CRC Press.

Kao, S.-C., Govindaraju, R.S., 2010. A copula-based joint deficit index for droughts. Journal of Hydrology 380 (1), 121−134.

King, A.D., Alexander, L.V., Donat, M.G., 2013. Asymmetry in the response of eastern Australia extreme rainfall to low-frequency Pacific variability. Geophysical Research Letters 40 (10), 2271−2277.

King, A.D., Klingaman, N.P., Alexander, L.V., Donat, M.G., Jourdain, N.C., Maher, P., 2014. Extreme rainfall variability in Australia: patterns, drivers, and predictability. Journal of Climate 27 (15), 6035−6050.

Kirono, D.G., Chiew, F.H., Kent, D.M., 2010. Identification of best predictors for forecasting seasonal rainfall and runoff in Australia. Hydrological Processes 24 (10), 1237−1247.

Klingaman, N.P., Woolnough, S., Syktus, J., 2013. On the drivers of inter-annual and decadal rainfall variability in Queensland, Australia. International Journal of Climatology 33 (10), 2413−2430.

Koenker, R., 2011. Additive models for quantile regression: model selection and confidence bandaids. Brazilian Journal of Probability and Statistics 25 (3), 239−262.

Koenker, R., Bassett, G., 1978. Regression quantiles. Econometrica 46 (1), 33−50.

Komunjer, I., 2013. Quantile prediction. In: Handbook of Economic Forecasting, vol. 2. Elsevier, pp. 961−994.

Kraus, D., Czado, C., 2017. D-vine copula based quantile regression. Computational Statistics & Data Analysis 110, 1−18.

Larsen, R., Leatham, D., Sukcharoen, K., 2015. Geographical diversification in wheat farming: a copula-based CVaR framework. Agricultural Finance Review 75 (3), 368−384.

Li, Q., Lin, J., Racine, J.S., 2013. Optimal bandwidth selection for nonparametric conditional distribution and quantile functions. Journal of Business & Economic Statistics 31 (1), 57−65.

Lo, F., Wheeler, M.C., Meinke, H., Donald, A., 2007. Probabilistic forecasts of the onset of the north Australian wet season. Monthly Weather Review 135 (10), 3506−3520.

Martin, S.W., Barnett, B.J., Coble, K.H., 2001. Developing and pricing precipitation insurance. Journal of Agricultural and Resource Economics 261−274.

McBride, J.L., Nicholls, N., 1983. Seasonal relationships between Australian rainfall and the southern oscillation. Monthly Weather Review 111 (10), 1998−2004.

McKeon, G., Hall, W., Henry, B., Stone, G., Watson, I., 2004. Pasture Degradation and Recovery in Australia's Rangelands: Learning from History.

Meyers, G., McIntosh, P., Pigot, L., Pook, M., 2007. The years of El Niño, La Niña, and interactions with the tropical Indian ocean. Journal of Climate 20 (13), 2872−2880.

Min, S.K., Cai, W., Whetton, P., 2013. Influence of climate variability on seasonal extremes over Australia. Journal of Geophysical Research Atmospheres 118 (2), 643–654.

Nagler, T., 2017. A Generic Approach to Nonparametric Function Estimation with Mixed Data arXiv preprint arXiv:1704.07457.

Nguyen-Huy, T., Deo, R.C., An-Vo, D.A., Mushtaq, S., Khan, S., 2017. Copula-statistical precipitation forecasting model in Australia's agro-ecological zones. Agricultural Water Management 191.

Nguyen-Huy, T., Deo, R.C., Mushtaq, S., An-Vo, D.-A., Khan, S., 2018. Modeling the joint influence of multiple synoptic-scale, climate mode indices on Australian wheat yield using a vine copula-based approach. European Journal of Agronomy 98, 65–81.

Nguyen, C.C., Bhatti, M.I., 2012. Copula model dependency between oil prices and stock markets: evidence from China and Vietnam'. Journal of International Financial Markets, Institutions and Money 22 (4), 758–773.

Nicholls, N., Drosdowsky, W., Lavery, B., 1997. Australian rainfall variability and change. Weather 52 (3), 66–72.

Nicholls, N., Lavery, B., Frederiksen, C., Drosdowsky, W., Torok, S., 1996. Recent apparent changes in relationships between the El Niño-Southern Oscillation and Australian rainfall and temperature. Geophysical Research Letters 23 (23), 3357–3360.

Nicholson, S.E., Kim, J., 1997. The relationship of the El Niño–southern oscillation to African rainfall. International Journal of Climatology 17 (2), 117–135.

Noh, H., Ghouch, A.E., Bouezmarni, T., 2013. Copula-based regression estimation and inference. Journal of the American Statistical Association 108 (502), 676–688.

Okhrin, O., Odening, M., Xu, W., 2013. Systemic weather risk and crop insurance: the case of China. Journal of Risk & Insurance 80 (2), 351–372.

Power, S., Haylock, M., Colman, R., Wang, X., 2006. The predictability of interdecadal changes in ENSO activity and ENSO teleconnections. Journal of Climate 19 (19), 4755–4771.

Rajeevan, M., Pai, D., Kumar, R.A., Lal, B., 2007. New statistical models for long-range forecasting of southwest monsoon rainfall over India. Climate Dynamics 28 (7–8), 813–828.

Ramirez, M.C.V., de Campos Velho, H.F., Ferreira, N.J., 2005. Artificial neural network technique for rainfall forecasting applied to the Sao Paulo region. Journal of Hydrology 301 (1–4), 146–162.

Risbey, J.S., Pook, M.J., McIntosh, P.C., Wheeler, M.C., Hendon, H.H., 2009. On the remote drivers of rainfall variability in Australia. Monthly Weather Review 137 (10), 3233–3253.

Saji, N., Goswami, B., Vinayachandran, P., Yamagata, T., 1999. A dipole mode in the tropical Indian Ocean. Nature 401 (6751), 360–363.

Schallhorn, N., Kraus, D., Nagler, T., Czado, C., 2017. D-vine Quantile Regression with Discrete Variables arXiv preprint arXiv:1705.08310.

Schepen, A., Wang, Q.J., Robertson, D., 2012. Evidence for using lagged climate indices to forecast Australian seasonal rainfall. Journal of Climate 25 (4), 1230–1246.

Sheather, S.J., Jones, M.C., 1991. A reliable data-based bandwidth selection method for kernel density estimation. Journal of the Royal Statistical Society: Series B 683–690.

Sklar, M., 1959. Fonctions de répartition à n dimensions et leurs marges. Université Paris 8.

Sraj, M., Bezak, N., Brilly, M., 2015. Bivariate flood frequency analysis using the copula function: a case study of the Litija station on the Sava River. Hydrological Processes 29 (2), 225–238.

Troup, A., 1965. 'The 'southern oscillation''. Quarterly Journal of the Royal Meteorological Society 91 (390), 490–506.

Ummenhofer, C.C., England, M.H., McIntosh, P.C., Meyers, G.A., Pook, M.J., Risbey, J.S., Gupta, A.S., Taschetto, A.S., 2009. What causes southeast Australia's worst droughts? Geophysical Research Letters 36 (4).

Ummenhofer, C.C., Sen Gupta, A., Briggs, P.R., England, M.H., McIntosh, P.C., Meyers, G.A., Pook, M.J., Raupach, M.R., Risbey, J.S., 2011. Indian and Pacific Ocean influences on southeast Australian drought and soil moisture. Journal of Climate 24 (5), 1313−1336.

Vieux, B.E., Cui, Z., Gaur, A., 2004. Evaluation of a physics-based distributed hydrologic model for flood forecasting. Journal of Hydrology 298 (1−4), 155−177.

Xu, W., Filler, G., Odening, M., Okhrin, O., 2010. On the systemic nature of weather risk. Agricultural Finance Review 70 (2), 267−284.

Zscheischler, J., Westra, S., Hurk, B.J., Seneviratne, S.I., Ward, P.J., Pitman, A., AghaKouchak, A., Bresch, D.N., Leonard, M., Wahl, T., 2018. Future climate risk from compound events. Nature Climate Change 1.

Chapter 9

Geostatistics: principles and methods

Saman Maroufpoor[1], Omid Bozorg-Haddad[1], Xuefeng Chu[2]
[1]Department of Irrigation and Reclamation Engineering, Faculty of Agricultural Engineering and Technology, College of Agriculture and Natural Resources, University of Tehran, Tehran, Iran; [2]Department of Civil and Environmental Engineering, North Dakota State University, Fargo, ND, United States

1. Introduction

Although the motivations of using geostatistics were raised in the 1960s by miners, its ideas have already existed in other fields. The first documentary report was published in 1911 (Mercer and Hall, 1911). It examined the product yield changes in small plots, and the results showed that the variance from one plot to another, by a small increase in plot size, was reduced. In addition, the products of adjacent plots were very similar to each other, and the reason that was offered for it was self-correlation and randomness. That article showed the basic characteristics of geostatistics. Krige (1960) found that if the degree of a rock was calculated in the neighboring blocks, the estimate could be improved. The basis of this method was self-correlation, which became an important method in gold mines. In the early 20th century, the tendency to predict in the branches of environmental sciences was seen only in climate and soil studies. Geostatistics was applied to mining and oil engineering because of financial incentives and resources. In this regard, mine engineers sought to estimate the value of metals in the body of rock, and oil engineers sought to identify the locations and the volume of oil reservoirs. These needs form the main stimulus of geostatistics because more accurate estimates mean more profit and less loss. Currently, geostatistics has been widely used in many different areas.

2. Difference between geostatistics and classical statistics methods

The methods of classical estimation are based on random sampling of a set of data, all of which carry the same weight. In other words, in classical statistical surveys, the samples obtained from the community are mostly considered

Handbook of Probabilistic Models. https://doi.org/10.1016/B978-0-12-816514-0.00009-6
229

random, and the value of measurement of a given quantity in a particular sample gives no information about the value of same quantity in other samples with a determined distance. In geostatistics, however, it is possible to communicate among the values of a quantity, its distance, and direction of its placement. In addition, in classical statistics, variability is assumed to be random. But in geostatistics, first, the existence or absence of a locative structure among the data is examined, and then, if there is a locative structure, data analysis is carried out. The classical statistical methods are only able to model structural changes and are incapable of processing random changes. In addition, the classical statistical methods are based on a set of assumptions that may not be valid in many conditions in relation to the characteristic of the parameter under consideration.

2.1 Geostatistics

It is the science of sample collection, organizing observations and processing them, interpreting the results of processing, and generalizing the results from sample to target. In other definitions, it is introduced as the statistics of changes science. Most data and information used by environmental specialists are in some way related to their locations and distribution patterns. In recent decades, a branch of statistics science has been developed under the name of locative statistics. The locative statistics includes methods, algorithms, and various tools for analyzing locative data and modeling their locative distribution in sampling points. In fact, locative statistics has made it possible for specialists to understand the distribution pattern of locative data by combining numerical information about variables and their geographic information and using mathematical and statistical algorithms. One of the most important subsets of locative statistics is geostatistics. Geostatistics is a branch of applied statistics science, in which the desired characteristics at unsampled points are estimated by using the information obtained from the sampling points. One of the geostatistics tools is a variogram, which allows for the analysis of the structure of the scale and extent of the locative changes of the regional variables. If the variogram is correctly determined, it can be used not only for statistical estimation but also for designing and modifying the sampling network. The development of various interpolation techniques has led to the flexibility and wide-scope use of geostatistics to analyze many of the existing issues in environmental sciences. Therefore, in geostatistics, we examine the variables that represent the locative structure. The regional variable is a sample of these variables.

3. Regionalized variables

The term "regionalized variables" was first introduced by Matheron (1989). In geostatistics, a regional variable depends on the location and time (i.e., locative dependence) (Matérn, 1960). To better explain the regional variable, we first need to define the random variable.

4. Random variable

A random variable gains any value within its yield; it would have a certain probability of occurrence. For each random variable at any point in space, independent of the coordinates of that point, two states occur:

1. The difference of the values of a random variable at two different points in space is independent of the distance between the two points. In other words, this difference does not have a distance format.
2. The difference of the values of a random variable at two points in space depends on the distance between those two points, and this difference has a distance format. If the random variable has such values in space that state two occurs, that random variable is called regional variable (Cressie, 1993). The value of the regional variable of $Z(x)$ at any point in space can be decomposed into random and definite components [(Eq. (9.1)]:

$$Z(x) = N(x) + S(x) \tag{9.1}$$

where $Z(x)$ is the regional variable at coordinate x, $N(x)$ is the definite component of the regional variable, and $S(x)$ is the random component of the regional variable. The difference in the values of a regional variable in space depends on the distance of those values, which is referred to as a spatial structure. It is obvious that such a structure, although independent of coordinates, is dependent on distance. The magnitude of the difference of the values is proportional to the distance.

5. Variogram and semivariogram

The variogram function is a key tool in the theory of regional variables and geostatistics estimation methods. But determining whether a calculated variogram is an appropriate and reliable estimation of the actual variogram of the studied regional variable is a difficult task. On the other hand, the calculation of the appropriate variogram and the provision of accurate estimation require a significant number of samples. A variogram is used to determine the locative continuity of a variable. Locative continuity means that adjacent samples are interdependent to a certain distance, and it is assumed that this dependence can be presented as a mathematical model called a variogram. The variance of the values among the points to the distance L from each other can express the mutual correlation of the values of two points to the distance L. In the case of existence of a locative structure, it is natural that the dependence of the values of the nearest points is more than the dependence of the values of the distant points. Therefore, this type of variance can be used to display the impact or influence of a sample value on the values of its adjacent environment. This distance-dependent variance is called a variogram, and it is shown with $2\gamma(L)$ (Goovaerts, 1997). $\gamma(L)$ is a semivariogram. The value of the variogram is calculated by Eq. (9.2):

$$2\gamma(L) = \frac{1}{N(L)} \sum_{i=1}^{N(L)} [Z(x_i + L) - Z(x_i)]^2 \tag{9.2}$$

where L is the separation distance in the specified direction between points $(x_i + L)$ and (x_i), $N(L)$ is the number of pairs of samples with distance L from each other, $2\gamma(L)$ is the variogram for distance L, $Z(x_i)$ is the sample value at point (x), and $Z(x_i + L)$ is the sample value at point $(x_i + L)$.

6. Variogram specifications

To plot the variogram, first calculate $2\gamma(L)$ for different values of L, and then the variogram values for different separation distances of L are plotted in a diagram, which shows the spatial dispersion of the regional variables. Fig. 9.1 shows the variogram plot, including the range, sill, and nugget.

7. Range

For a regional variable with a spatial structure, the points close to each other exhibit strong correlations, compared to the distant points, within a certain distance, beyond which the variogram becomes constant. This means that the samples affect each other to a certain distance and have locative continuity. There is no locative continuity beyond this distance, and the samples act independently. The distance at which the variogram reaches a fixed limit is called range. This domain specifies a range, within which the data can be used to estimate the values for any unknown points. It is obvious that a larger impact domain implies a wider locative continuity.

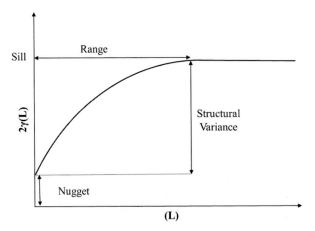

FIGURE 9.1 Variogram, range, sill, and nugget.

8. Sill

The fixed value at which the variogram in the domain of impact reaches constant is called sill. The sill value is equal to the variance of all the samples used in the calculation of the variogram. In the Kriging method, variograms that reach a specific sill are more important. In some cases, there are variograms that do not tend to approach a fixed limit within the range of considered intervals. These variograms may indicate a trend in data or instability of data.

9. Nugget

The value of a variogram at the origin of coordinates (i.e., for $L = 0$) is nugget (C_0). Ideally, the value C_0 must be zero. But in most cases, it is larger than zero. Nugget effect represents an unstructured variance. The structured component (C) is equal to the difference between sill and nugget. C depends on the location of the data (i.e., the distance and direction of their placement) and also indicates the changes that their cause can be found in the characteristics of a regional variable. Usually, two factors create the nugget effect in the variogram: (1) the existence of random components in the variable distribution that actually returns to the randomness of the process and (2) the existence of a sampling error, preparation, and laboratory analysis.

The ratio of the sill effect to the nugget effect $\left(\frac{C}{C_0}\right)$ is the locative structure of the variogram. The locative structure of 75% and more indicates a strong structure, a value between 25% and 75% indicates a moderate structure, and a value less than 25% indicates a weak locative structure for the studied variable. Another ratio $\frac{C_0}{C_0+C}$ represents how much of the total variability is justified by the nugget effect (Cambardella et al., 1994). The most important application of the variogram is the use of its information in geostatistics algorithms (Isaaks and Srivastava, 1989).

10. Model fitting to empirical variogram

After calculating the empirical variogram, a mathematical model is fitted to it. The main and basic models of variograms are divided into two categories: models that have a fixed limit (models containing sill) and models that do not have such a condition (no sill). The variogram models with no sill do not reach a fixed value and continue to increase with increasing L.

11. Models with sill

In these models, with increasing L to a certain value, the value of $2\gamma(L)$ increases and then it approaches to a fixed limit (i.e., sill). The models with a sill include the following ones:

12. Spherical model

A spherical model is one of the most commonly used models in geostatistics (Armstrong, 1989). This model starts from the origin of coordinates and has a linear behavior near it. As L increases, $\gamma(L)$ increases rapidly and eventually reaches the sill. Eq. (9.3) shows the spherical model (Jean-Paul and Pierre, 1999):

$$\gamma(L) = \begin{cases} 0 & if(L=0) \\ c\left[1.5\dfrac{L}{r}0.5\left(\dfrac{L}{r}\right)^3\right] & if(0 < L \leq r) \\ c & if(L > 0) \end{cases} \qquad (9.3)$$

where r is the range and c is the sill.

13. Exponential model

The exponential model, similar to the spherical model, starts from the origin of the coordinates and has a linear behavior near it. But the ascending slope of the curve is less than that of the spherical model. The variogram of an exponential model never reaches a sill, so its range influence is uncertain. Eq. (9.4) shows the exponential model (Jean-Paul and Pierre, 1999):

$$\gamma(L) = \begin{cases} 0 & \text{if } (L=0) \\ c\left[1 - \exp\left(-\dfrac{(L)}{r}\right)\right] & \text{if } (L \neq 0) \end{cases} \qquad (9.4)$$

where r is the range and c is the sill.

14. Gaussian model

The Gaussian model has a parabolic behavior near the origin of coordinates. The slope of this model is initially zero and gradually increases up to the turning point and then quickly climbs to the sill. This model represents the high continuity degree of the regional variable. One of the applications of this model is the use in meteorological issues (Delfiner, 1973; Schlatter, 1975; Chauvet et al., 1976). Eq. (9.5) shows the Gaussian model (Jean-Paul and Pierre, 1999):

$$\gamma(L) = \begin{cases} 0 & \text{if } (L=0) \\ c\left[1 - \exp\left(-\dfrac{L^2}{r^2}\right)\right] & \text{if } (L \neq 0) \end{cases} \qquad (9.5)$$

15. Models without sill

In no-sill models, the variogram increases with increasing L in such a way that it does not reach a fixed limit, and hence, the variogram does not have a sill. Typical no-sill models include the following ones:

16. Linear model

The linear model is widely used for characterizing nonsill variogram using a simple straight line, as expressed in Eq. (9.6):

$$\gamma(L) = aL + C_0 \qquad (9.6)$$

where a is the slope and C_0 is the nugget.

17. Parabolic model

Eq. (9.7) shows a parabolic model, in which a^2 is a fixed value.

$$\gamma(L) = \frac{1}{2}a^2 L^2 \qquad (9.7)$$

18. The DeWijsian model

Eq. (9.8) shows the DeWijsian model, in which H is equal to three times the absolute dispersion coefficient and K is related to the sampling length. Because the DeWijsian model is logarithmic, it cannot be applied to values such as $L < 1$ (DeWijs, 1976).

$$\gamma(L) = H \ln(L) + K \qquad (9.8)$$

19. Selection of the theoretical variogram models

The choice of the theoretical variogram models depends on the motion pattern of the function of variogram near the source. If the considered phenomenon is completely connected, the variogram will have a parabolic motion near the source. In such a situation, the Gaussian model is appropriate. If the variogram has a linear motion near the source, both the exponential and spherical models will be appropriate. The sill is easily selected from the peak of the empirical variogram. Having the sill, we can determine other parameters through trial and error. In addition, the nugget effect can be determined based on the initial behavior of the empirical variogram near the vertical axis.

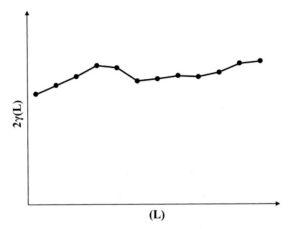

FIGURE 9.2   Variogram without proper locative continuity.

20. Locative continuity analysis of variogram

With an increase of the separation distance, the variogram increases regularly and reaches the sill value, and we have locative continuity. Fig. 9.2 shows the variogram without the direction of a variable. Although a relative increase of $2\gamma(L)$ can be observed with increasing distance, this increase is negligible and irregular. In addition, the low slope of the variogram near the source indicates the locative continuity. A higher variogram slope at the source coordinates indicates more discontinuity of the regional variable.

21. Anisotropy in a variogram

In addition to the distinction of locative continuity, more information such as anisotropy can be extracted from the variogram. If there is a difference between the range and the sill in different directions, the variogram possesses anisotropic properties. What causes these changes is the heterogeneity that exists in different directions. In general, there are two types: geometric anisotropy and zonal anisotropy (Cressie, 1993).

22. Geometric anisotropy

If the variograms plotted for a given variable in different directions have the same sill, but different range values, that variable thus exhibits geometric anisotropy. Geometric anisotropy means that although variability is the same in different directions, the magnitude of the range values along different directions is different. The fitted ellipse to the anisotropy diagram shows the range influence in different directions (Fig. 9.3). This ellipse is called anisotropy ellipse. Each radial of this ellipse shows the range influence in the desired direction (Fig. 9.3).

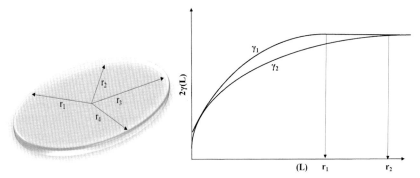

FIGURE 9.3 Anisotropy ellipse for variograms in different directions.

23. Zonal anisotropy

In this type of anisotropy, the sill of the variogram varies in different directions, while the range is the same for all directions. This inconsistency may exist in the nature of the desired quantity, so it cannot be overcome. For example, the variograms of the hydraulic conductivity in a laminate and substrate soil in horizontal and vertical directions exhibit zonal anisotropy (Hsieh, 2015).

24. Geostatistics methods

Geostatistics methods are often used for estimating the values of variables at nonsampled points (Maroufpoor et al., 2017). The basis of the methods is the calculation of the weighted average of the values in the neighborhood of the points without statistics.

25. Kriging Interpolation method

Kriging was first introduced by Krige (1981), a mining engineer in South African gold lands. Kriging is a weighted linear estimator; therefore, the purpose of Kriging is to find statistical weights of samples so that, besides the unbiasedness of the estimation, the estimation variance is also minimized.

26. Kriging equations

Suppose n is a measure of variable Z in locative coordinates $x_1, x_2, x_3, ..., x_n$, the value of $\widehat{Z}$ at point x_0 is estimated by Eq. (9.9) (Matheron, 1963).

$$\widehat{Z}(x_0) = \sum_{i=1}^{n} \lambda_i Z(x_i) \tag{9.9}$$

In the aforementioned equation, the unknown values at the right side of the equation are a set of coefficients (λ_1 to λ_n). The difference between the actual value of $Z(x_0)$ and the estimated value of $\widehat{Z}(x_0)$ is obtained from Eq. (9.10).

$$\widehat{Z}(x_0) - Z(x_0) = \left(\sum_{i=1}^{n} \lambda_i Z(x_i) \right) - Z(x_0) \tag{9.10}$$

As shown in Eq. (9.10), the Kriging system attempts to calculate an estimator that has the following properties:

26.1 Unbiasedness

This property means that the average error of the system should be zero.

$$E\left[\widehat{Z}(x_0) - Z(x_0)\right] = \sum_{i=1}^{n} \lambda_i m - m = \left(\sum_{i=1}^{n} \lambda_i - 1 \right) m = 0 \tag{9.11}$$

To establish this property, Eq. (9.12) is needed.

$$\sum_{i=1}^{n} \lambda_i = 1 \tag{9.12}$$

26.2 Minimum variance

This property means that the mean square error must be minimized.

Given the two properties, the Kriging problem is converted into an optimization one, in which the coefficients λ_1 to λ_n must be selected (Deutsch and Journel, 1998).

27. Kriging types

The Kriging system is divided into different types based on output or computational structure. Some of the most widely used types of Kriging are listed as follows (Webster and Oliver, 2007):

28. Ordinary Kriging

The most common type of Kriging is ordinary Kriging, in which the average value of the unknown element is assumed to be independent of the coordinates of the area under study. In this system, the average is part of the answer to the problem.

29. Simple Kriging

In this type of Kriging, the average of the data is assumed to be known and independent of the coordinates. The variable at an unknown point is estimated by using Eq. (9.13):

$$\widehat{Z}(x_0) = m + \sum_{i=1}^{n} \lambda_i (Z(x_i) - m) \tag{9.13}$$

where m is the mean of the variable being investigated.

30. Lognormal Kriging

One of the most common conditions that may occur is the existence of data with high skewness and abnormality. The variograms show a high sensitivity to positive skewness. Skewness can be eliminated by converting the data distribution, for example, taking logarithm and reducing the variance. If we obtain a normal distribution by converting the initial data to logarithmic values, the distribution of the data is lognormal, and lognormal Kriging can be applied. Note that the lognormal Kriging cannot provide unbiased results.

31. Universal Kriging

This method is used when there is a trend in the data. This means that the variogram plotted for the regional variable does not reach a specified sill and that with the increase of distance, it increases too. In this case, nonlinear estimators such as universal Kriging are used. In this way, the existing process in the data is identified and removed. Estimation takes place based on the remainder of the data that are detrended. Finally, the values of the deducted trend are added to the estimated values.

32. Indicator Kriging

The indicator Kriging is a nonlinear and nonparametric type of Kriging in which continuous variables are converted binary. This method is most commonly used because it can handle any types of distribution. This method can also modify qualitative information for better prediction.

33. Disjunctive Kriging

Disjunctive Kriging is a nonlinear Kriging method but a parametric one. This method is very valuable for decision-making because it determines both estimation and probabilities more and less than a certain limit.

34. Co-Kriging

Sometimes a variable may not be sufficiently sampled because of various reasons such as the difficulty of sampling or high costs. Based on this variable, it is not possible to carry out statistical estimation with proper accuracy. In such cases, we can correct the estimate and increase the accuracy by considering the locative correlation between this variable and other well-sampled variable(s). This can be performed by using Co-Kriging method (Myers, 1982):

$$\widehat{Z}(x_0) = \left(\sum_{z=1}^{n} \lambda_{z,i} Z(x_i) \right) \left(\sum_{k=1}^{n} \lambda_{k,y} T(x_k) \right) \qquad (9.14)$$

where $\widehat{Z}_0$ is the estimated value of the variable at the unknown point, $\lambda_{z,i}$ is the weight of the variable Z, $\lambda_{k,y}$ is the weight of the auxiliary variable T, $Z(x_i)$ is the observed value of the main variable, and $T(x_k)$ is the observed value of the auxiliary variable.

35. Kriging parameters

The implementation of a precise geostatistics method requires the determination of the parameters regarding the distribution of data and the locative structure. In other words, the estimates under the conditions of a strong/weak spatial structure and a wide/narrow range are different. The most important parameters include the weights, neighborhood, and search radius.

36. Weights

In general, more weights are allocated to the nearest points than the farther points, and the cluster points have smaller weights than the single points in the same distance. In the Kriging method, the weight of each known sample in the estimation for an unknown point depends entirely on the locative structure of the regional variables, whereas in other methods, the weights only depend on a geometric characteristic such as distance. As the spatial structure weakens, the position of the samples decreases so that in the nugget effect mode, the weights of all the samples will be equal. Finally, with the increase of the range, higher weights will be assigned to the samples far from the unknown point.

37. Neighborhood

The neighborhood is one of the important factors in the Kriging estimation, which provides benefits for this method. For example, the closest points to the target point have significant weights. Although there are no restrictive rules for the definition of the neighborhood, some situations need to be considered. If the variogram is bounded and the data are dense, it is possible to set the neighborhood close to the range. Therefore, any data points away from the range will have small weights. If the variance is large, there is the possibility of carrying important weights by the points. As a solution, the user may choose the closest points for calculation, which will limit the choice of the number of neighboring points. If the sampling structure is irregular, the number of neighborhood points will be more or less varied as much as the target point is displaced. If we consider a maximum radius for the neighborhood, we must also consider a minimum value for the number of neighborhood points. If the dispersion of the points is very heterogeneous, it is better to divide the space around the target point into one-eighths and select the two nearest points to the target point.

38. Search radius

One of the reasons that lead to a weakened spatial structure and ultimately cause to disappear is the increase in the distance of the points from each other. Therefore, whatever the distance between the measured points and the estimated points gets larger, it is practically ineffective for the estimation and there is no need to enter them in the estimation process. This distance is called range influence of the variogram, and the maximum distance within which the points will be included in estimation is called search radius.

The search radius is sometimes equivalent to the range of the variogram and sometimes equivalent to two-third of it. In the case of anisotropy, the value of search radius in different directions will be different. In this case, the search radius in different directions will be determined based on the anisotropy ellipse. Therefore, an anisotropy ellipse is determined for each estimation point, and all data points within the ellipse will be used in the estimation.

39. Conclusions

This chapter described principles and concepts of Geostatistics. The chapter presented a brief literature review of the variogram specifications, selection of the theoretical variogram models, and Kriging interpolation method. At the end, types of Kriging methods were investigated.

References

Armstrong, M., 1989. Geostatistics, vol. 2. Kluwer Academic Publishers, Dordrecht.

Cambardella, C., Moorman, T., Parkin, T., et al., 1994. Field-scale variability of soil properties in central Iowa soils. Soil Science Society of America Journal 58 (5), 1501−1511.

Chauvet, P., Pailleux, J., Chiles, J., 1976. Analyse objective des champs météorologiques par cokrigeage. La Météorologie-Sciences et Techniques 6 (4), 37−54.

Cressie, N., 1993. Statistics for Spatial Data. John Wiely.

Delfiner, P., 1973. Analyse objective du géopotentiel et du vent géostrophique par krigeage universel. La Météorologie 25, 1−57.

Deutsch, C.V., Journel, A.G., 1998. Geostatistical Software Library and User's Guide. Oxford University Press, New York.

DeWijs, H., 1976. Models for the estimation of world ore reserves. Geologie en Mijnbouw 55 (1−2), 46−50.

Goovaerts, P., 1997. Geostatistics for Natural Resources Evaluation. Oxford University Press, Oxford.

Hsieh, A.I.-J., 2015. Geostatistical modeling and upscaling permeability for reservoir scale modeling in bioturbated, heterogeneous tight reservoir rock: Viking Fm, Provost Field, Alberta. Science: Department of Earth Sciences.

Isaaks, E.H., Srivastava, R.M., 1989. An Introduction to Applied Geostatistics. Oxford university press.

Jean-Paul, C., Pierre, D., 1999. Geostatistics: Modeling Spatial Uncertainty. Jhon Wiley & Sons Inc., New York, p. 695.

Krige, D., 1960. On the departure of ore value distributions from the lognormal model in South African gold mines. Journal of the South African Institute of Mining and Metallurgy 61 (4), 231−244.

Krige, D.G., 1981. Lognormal-de Wijsian geostatistics for ore evaluation. South African Institute of mining and metallurgy Johannesburg.

Maroufpoor, S., Fakheri-Fard, A., Shiri, J., 2017. Study of the spatial distribution of groundwater quality using soft computing and geostatistical models. ISH Journal of Hydraulic Engineering 1−7.

Matérn, B., 1960. Spatial variation. Stochastic models and their application to some problems in forest surveys and other sampling investigations. Meddelanden Fran Statens Skogsforskningsinstitut 49.

Matheron, G., 1963. Principles of geostatistics. Economic Geology 58 (8), 1246−1266.

Matheron, G., 1989. Estimating and Choosing. An Essay on Probability in Practice. Springer-Verlag. Translated from the French and with a preface by AM Hasofer.

Mercer, W., Hall, A., 1911. The experimental error of field trials. The Journal of Agricultural Science 4 (2), 107−132.

Myers, D.E., 1982. Matrix formulation of co-kriging. Journal of the International Association for Mathematical Geology 14 (3), 249−257.

Schlatter, T.W., 1975. Some experiments with a multivariate statistical objective analysis scheme. Monthly Weather Review 103 (3), 246−257.

Webster, R., Oliver, M.A., 2007. Geostatistics for Environmental Scientists. John Wiley & Sons.

Chapter 10

Adaptive H_∞ Kalman filtering for stochastic systems with nonlinear uncertainties

Yuankai Li
University of Electronic Science and Technology of China, Chengdu, China

1. Introduction

For most real systems in stochastic environments, nonlinear uncertainties, composed of inner modeling errors and outer disturbances primarily, are necessary to be considered for high-quality control purpose, including for state estimation and parameter identification. For example, in the problem of maneuvering target tracking, nonlinear uncertainties are presented by a combination of the errors of tracking models, unknown target maneuvers, and stochastic noises during system process and measurements.

For such a dynamic system with nonlinear uncertainties, an effective way to estimate the system state accurately is to use the robust filtering method (Bishop et al., 2007), and a typical form of that is the H_∞ filter. In essence, the filter is an estimator to system states via output measurements, guaranteeing an H_∞ norm of error propagation from nonlinear uncertainties to estimation error to be minimized (Shen and Deng, 1997; Seo et al., 2006). This norm value is called H_∞ criterion, making the filter get the best estimation results of the worst-possible case. Such an H_∞ filter does not need to know the exact statistical properties of system uncertainties. If only the energy of uncertainties such as noises and disturbances is bounded, the filter can work effectively. For the sake of the simple assumption, H_∞ filter has been applied in lots of areas (Soken and Hajiyev, 2010; Li et al., 2013; Zhong et al., 2010) for solving estimation problems of systems with uncertainties.

However, traditional H_∞ filter that has only single filtering mode appears conservative estimation precision. To acquire system state with high quality, it is necessary to have a better filter with more than one filtering mode to adapt to real situation of uncertainties and make the best decision for state estimation. Therefore, in this chapter, an adaptive H_∞ filtering method is

Handbook of Probabilistic Models. https://doi.org/10.1016/B978-0-12-816514-0.00010-2
243

presented. It aims to realize state estimation with both high accuracy and high precision for stochastic systems with nonlinear uncertainties.

2. Problem: H_∞ state estimation

A typical dynamic model of stochastic systems can be expressed by

$$X_{k+1} = F_k X_k + G_k w_k + D \tag{10.1}$$

$$Z_k = H_k X_k v_k \tag{10.2}$$

where matrices F, G, and H are unrelated with X, the system states. Z is the measurement vector. w and v are process noise and measurement noise, respectively, which are white with their covariance W and V.

We aim to find a filter that can estimate X_k from Z_k such that the estimation error $\widetilde{X}_{k|k}$ can minimize an H_∞ norm criterion, which is essentially an error propagation function as

$$T_\infty = \left\| \frac{\widehat{X}_{k|k} - \widehat{X}_k}{X_k^Z - \overline{X}_k} \right\|_\infty \tag{10.3}$$

where X^Z is the pseudo—real-system states yielded by measurements. In this chapter, those variables with sharp hat denote their estimation value, with tilde for estimation error and bar for true value.

Clearly, the deviation of X^Z from $\overline{X}_k$ is caused by the input system uncertainties, so minimizing (10.3) is equivalent to satisfying such an inequity

$$max \frac{\|\widetilde{X}_{k|k}\|_2^2}{\|w_k\|_2^2 + \|v_k\|_2^2 + \|D\|_2^2} \leq \gamma^2 \tag{10.4}$$

for some minimum $\gamma \in R$. The γ shows the maximum efficiency of energy propagation from total uncertainties to estimation error, giving the worst-possible case performance.

2.1 Standard H_∞ filter

A typical solution of state estimation for (10.1) and (10.2) constrained by (10.4) is the H_∞ filter (H_∞F), which is built based on game theory (Banavar, 1992; Einicke and White, 1999). The filter uses the H_∞ criterion to generate filtering gain, resembling structure of usual Kalman filter (KF) and performed with the following standard steps.

Step 1: *State Prediction.*

One-step prediction state and covariance:

$$\widehat{X}_{k+1|k} = F_k \widehat{X}_{k|k} \tag{10.5}$$

$$P_{k+1|k} = F_k P_{k|k} F_k^T + G_k W_k G_k^T \tag{10.6}$$

Step 2: *Gain Configuration.*

Configure H_∞ prediction covariance with

$$\sum\nolimits_{k+1|k} = \left(P_{k+1|k}^{-1} - \gamma^{-2} I \right)^{-1} \tag{10.7}$$

where γ satisfies (10.4). I is a unit matrix, the dimension of which is the same with P.

Step 3: *Measurement Innovation.*

Measurement residue and covariance:

$$\widetilde{Z}_{k+1} = Z_{k+1} - H_{k+1} \widehat{X}_{k+1|k} \tag{10.8}$$

$$P_{k+1}^Z = H_{k+1} \sum\nolimits_{k+1|k} H_{k+1}^T + V_{k+1} \tag{10.9}$$

Step 4: *Estimate Update.*

Filtering gain, state estimate, and covariance:

$$K_{k+1} = \sum\nolimits_{k+1|k} H_{k+1}^T \left(P_{k+1}^Z \right)^{-1} \tag{10.10}$$

$$\widehat{X}_{k+1|k+1} = \widehat{X}_{k+1|k} + K_{k+1} \widetilde{Z}_{k+1} \tag{10.11}$$

$$P_{k+1|k+1} = \left(\sum\nolimits_{k+1|k}^{-1} + H_{k+1}^T V_{k+1}^{-1} H_{k+1} \right)^{-1} \tag{10.12}$$

It is clear from (10.4), γ shows the level of filtering robustness. Decreasing γ may increase robustness. Nevertheless, (10.12) implies that decreasing γ also makes P, the mean square error of the state estimates, increase and reduce the filtering optimality. On the contrary, if $\gamma \to \infty$, $\Sigma_{k+1|k}$ equals to $P_{k+1|k}$ and the $H_\infty F$ reverts to the usual KF. Hence, finding γ is to seek for the balance between robustness and optimality.

2.2 Conservativeness and optimization

An intrinsic problem of the H_∞ filter is that the γ needs to be given a priori value and mismatching of the value will lead to excessive robustness, which means excessive filtering optimality is sacrificed to exchange robustness, reducing the filtering precision. In short, the filter solution is conservative.

To improve the conservativeness of the H_∞ filter, many works are focused on adaptive H_∞ filtering.

A typical form is with the name switched H_∞ filter (SH$_\infty$F), which is presented in Xiong et al., 2008; Li et al., 2010, 2014. The filter alternates two filtering modes that are H_∞ and optimal by introducing a test condition related with real measurement covariance, so that filtering robustness can be finite horizon and get much more flexibility. With the capability of resetting filtering gain, the SH$_\infty$F can improve the sacrifice of optimality and precision and increase the estimation accuracy under time-variant uncertainties.

For the SH$_\infty$F, the introduced switching mechanism replaces the bound of uncertainties with a test condition. While an auxiliary parameter is designed to control the switching threshold (Li et al., 2010, 2014), the parameter also needs to be given a priori value and the value usually depends on the tolerance level of system uncertainties. Setting the parameter is simpler than to find the γ in H$_\infty$F, but mismatch of the parameter value will also reduce estimation precision remarkably. Therefore, it is expected that the switching parameter can be on-line given and optimized, and the SH$_\infty$F can be independent with system itself.

To achieve that objective, an optimal-switched H_∞ filter (OSH$_\infty$F) is presented in Li et al., 2016. The filter is established on the combination of the standard H$_\infty$F given by Theodor et al. (Theodor et al., 1994) and the switched structure by Li et al (Li et al., 2010, 2014), embedded an extra optimization mechanism to on-line determine the best switching condition at each iteration. The extra mechanism is built by optimizing a quadratic optimality-robustness cost function (ORCF) that is designed as the weighted sum of measurement estimation square error and measurement prediction covariance. These two parts evaluate robustness and optimality, representing estimation accuracy and precision, respectively, and their balance controlled by a nondimensional weight factor (WF).

To use the filter, the WF is the only parameter that needs set. It is to determine that how much filtering optimality and robustness is contained in the estimation result. That avoids worsening of filtering performance caused by potential mismatch of switching parameter, and the filtering result can be regarded as optimal with the given WF. Without independence of system uncertainty, the OSH$_\infty$F is a unified form of H$_\infty$F and has been verified that the OSH$_\infty$F can achieve much better filtering performances than many other typical filters.

3. Approach: adaptive H_∞ filtering

For the problem of H_∞ state estimation, in this section, we are going to discuss the adaptive H_∞ filtering method with presenting the aforementioned two typical filter forms: SH$_\infty$F and OSH$_\infty$F.

3.1 Switched H_∞ Kalman filter

For the H_∞ prediction covariance with expression as (10.7), a given γ makes the H_∞ filter work at a constant robustness level and expenses filtering optimality on infinite horizon.

That results in excessive loss of estimation precision. Therefore, we replace (10.7) with such a dual-mode structure

$$\Sigma_{k+1|k} = \begin{cases} \left(P_{k+1|k}^{-1} - \gamma^{-2}L_k^T\right)^{-1}, & \overline{P}_{k+1}^Z > \alpha P_{k+1}^Z \\ P_{k+1|k}, & \overline{P}_{k+1}^Z \leq \alpha P_{k+1}^Z \end{cases} \tag{10.13}$$

where $L_k \in R^{n \times n}$ is tunable. α is a diagonal matrix, called error-redundant factor, and

$$\overline{P}_{k+1}^Z = \frac{\rho P_k^Z + \widetilde{Z}_{k+1} \widetilde{Z}_{k+1}^T}{\rho + 1} \tag{10.14}$$

where ρ is a forgetting factor. The dual-mode structure shows that the filter works in the H_∞ robust mode only when the innovation covariance is less than its real value. By using the test condition, filtering mode can switch between optimal and robust adaptively.

With the limitation from filtering stability, γ should be at a bounded range. The range is identified as follows.

Write the estimation error as $\widetilde{X}_k$ so that the real prediction error and covariance are

$$\widetilde{X}_{k|k-1} = \beta_k F_{k-1} \widetilde{X}_{k-1} + w_k \tag{10.15}$$

and

$$\overline{\Sigma}_{k|k-1} = \beta_k F_{k-1} P_{k-1} F_{k-1}^T \beta_k + W_k \tag{10.16}$$

in which β is called nonlinearity factor that is a diagonal matrix, evaluating the system nonlinearity caused by inner modeling errors and outer disturbances. Then, γ needs to satisfy

$$\left\{\max eig\left[P_{k|k-1}\right]\right\}^{1/2} < \gamma < \left\{\max eig\left[P_{k|k-1}^{-1} - \overline{\Sigma}_{k|k-1}^{-1}\right]\right\}^{-1/2} \tag{10.17}$$

such that $\Sigma_{k|k-1}$ is positive definite and

$$\Sigma_{k|k-1} \geq \overline{\Sigma}_{k|k-1} \tag{10.18}$$

In (10.17), $\max ei()$ shows the maximum eigenvalue of a matrix, and (10.17) identifies the upper bound of γ to guarantee estimation divergence, and (10.18) provides a sufficient condition (Xiong et al., 2008) for filtering stability. The condition can be transformed into a more applicable form by taking L_k as

$$L_k = \gamma\left(P_{k|k-1}^{-1} - \overline{\varepsilon}_{max}^{-2}I\right)^{1/2} \tag{10.19}$$

where $\overline{\varepsilon}_{max}^{-2} I$ is a compensation function satisfying

$$\overline{\varepsilon}_{max}^{-2}I \leq \overline{\Sigma}_{k|k-1}^{-1} \tag{10.20}$$

to guarantee (10.18).

Practically, it is hard to get the exact state equation. It is necessary to replace (10.1) with

$$\widehat{X}_{k|k-1} = F_{k-1}\widehat{X}_{k-1|k-1} \qquad (10.21)$$

which results in the linearization error. Hence, we need, let $\alpha \geq 1$, to introduce a certain error redundancy in the switching condition, assuming that part of disturbance and model errors are tolerable.

With introduction of error redundancy, the stability criterion (10.18) can be loosened by applying Lemma 1 (Reif et al., 1999). Denote $\|\cdot\|$ as 2 norm of real vectors and spectral norm of real matrices.

Lemma 1. There are stochastic processes ζ_k and $\Gamma(\zeta_k)$ and positive real $\tau_{min,max}$, μ and $\lambda \in (0,1]$ satisfying

$$\tau_{min}\|\zeta_k\| \leq \|\Gamma(\zeta_k)\| \leq \tau_{max}\|\zeta_k\| \qquad (10.22)$$

$$E\{\Gamma(\zeta_k)|\zeta_{k-1}\}\Gamma(\zeta_{k-1}) \leq \mu - \lambda\Gamma(\zeta_{k-1}) \qquad (10.23)$$

formed with probability one. Then, ζ_k is bounded in mean square, that is

$$E\left\{\|\zeta_k\|^2\right\} \leq \frac{\tau_{max}}{\tau_{min}}E\left\{\|\zeta_k\|^2\right\}(1-\lambda)^k + \frac{\mu}{\tau_{min}}\sum_{i=1}^{k-1}(1-\lambda)^i \qquad (10.24)$$

In the Lemma, (10.23) is to guarantee the bounded conditional expectation. The reason is if

$$E\{\Gamma(\zeta_k)|\zeta_{k-1}\} \leq \varepsilon_{max} \qquad (10.25)$$

there must exist $\eta \in (0,1]$ such that

$$E\{\Gamma(\zeta_k)|\zeta_{k-1}\} - \Gamma(\zeta_{k-1}) \leq \eta[\varepsilon_{max} - \Gamma(\zeta_{k-1})] \qquad (10.26)$$

It is exactly (10.23) if $\lambda = \eta$ and $\mu = \eta\varepsilon_{max}$. For filtering stability, Lemma 1 states that the ζ_k converges gradually if the Lyapunov function $\Gamma(\zeta_k)$ and its one-step conditional expectation are both bounded.

Based on Lemma 1, the loosened stability criterion can be given by Theorem 1.

Theorem 1. For stochastic system as (10.1) and (10.2) and the filtering algorithm presented by (10.5) to (10.12), there exist real $f_{min,max}$, $w_{min,max}$, $h_{min,max}$, $v_{min,max}$, $\beta_{min,max}$, and $p_{min,max}$ satisfying

$$f_{min}I \leq F_kF_k^T \leq f_{max}I, w_{min}I \leq W_k \leq w_{max}I \qquad (10.27)$$

$$h_{min}I \leq H_kH_k^T \leq h_{max}I, v_{min}I \leq V_k \leq v_{max}I \qquad (10.28)$$

$$\beta_{min}I \leq \beta_k\beta_k^T \leq \beta_{max}I, p_{min}I \leq P_{k|k} \leq p_{max}I \qquad (10.29)$$

Then, note $\beta_m = \left\|I - F_k^TH_k^TV_k^{-1}\left(V_k - H_kP_kH_k^T\right)V_k^{-1}H_kF_kP_{k-1}\right\|^{-1/2}$ if

$$\beta_{max} \leq \beta_m \qquad (10.30)$$

$$\Sigma_{k|k-1} \geq \beta_k^{-2}\overline{\Sigma}_{k|k-1} \tag{10.31}$$

there exist real $\mu_{max} > 0$ and $0 < \lambda_{min} \leq 1$ forming

$$E\left\{\|\widetilde{X}_k\|^2\right\} \leq \frac{p_{max}}{p_{min}}E\left\{\|\widetilde{X}_{k_0}\|^2\right\}(1-\lambda_{min})^k + \frac{\mu_{max}}{p_{min}}\sum_{i=1}^{k-1}(1-\lambda_{min})^i \tag{10.32}$$

In Theorem 1, it can be obtained that $\beta_m \in \left[1, \left(1 - \frac{1}{4}f_{max}^2 p_{max} p_{min}^{-1}\right)^{-1/2}\right]$, μ_{max} and λ_{min} determined by (10.23). The proof on that is given in Appendix of this chapter.

Obviously, (10.27−10.29) are formed naturally. Theorem 1 indicates that the estimation error of the filtering algorithm is bounded if (10.30) and (10.31) form. Compared with (10.18), (10.31) is constrained by (10.30) and extends the range of $\Sigma_{k|k-1}$. With introduction of the coefficient β_k^{-2}, the lower bound of $\Sigma_{k|k-1}$ is loosened to its linear part, yielding the most possible redundancy for filtering stability.

To specify the entire filtering process, design the compensation function in (10.19) as

$$\overline{\varepsilon}_{max}^{-2}I = f_\alpha^{-1}P_{k+1|k}^{-1} \tag{10.33}$$

where

$$f_\alpha = \alpha^{-1}diag\frac{\overline{P}_{k+1}^Z}{P_{k+1}^Z}\otimes I_{exp} \tag{10.34}$$

called compensation factor, in which α satisfies

$$I_{exp}\otimes\alpha \leq \beta_k^2 \tag{10.35}$$

where $\otimes$ refers to Kronecker multiplication, expanding α with I_{exp} to match dimension. For example, use a second-order unit matrix for three-dimensional Z to match a six-dimensional X. The f_α in (10.34) is determined by the ratio of real measurement covariance to the switching threshold shown in (10.13). That is to say, the level of filtering robustness can vary with real situations of system uncertainties, rather than at a constant as $H_\infty F$. Clearly, $f_\alpha > I$ when the filter switches to the H_∞ robust filtering mode.

Then, the $SH_\infty F$ can be expressed as (10.13), (10.14), (10.19), and (10.33)−(10.35), whereas stability of the $SH_\infty F$ can be guaranteed by Theorem 2 below.

Theorem 2. Consider the filter expressed by (10.5), (10.6), (10.13), (10.19), (10.33)−(10.35) and (10.8)−(10.12), and the assumptions of (10.1), (10.2), (10.27)−(10.29) that are the same to Theorem 1, and take $I_{exp} = I_2$. There exist real $\mu_{max} > 0$ and $0 < \lambda_{min} \leq 1$ such that (10.32) forms.

Proof of Theorem 2.

Based on the proof of Theorem 1, we only need to verify (10.32) under (10.13) and (10.33)–(10.35).

According to (10.2) and (10.16), we have

$$\overline{P}_{Y_k} = H_k \overline{\Sigma}_{k|k-1} H_k^T + V_k \tag{10.36}$$

Subtracting α times (9), (36) becomes

$$\overline{P}_{Y_k} - \alpha P_{Y_k} = H_k \left[\overline{\Sigma}_{k|k-1} - (I_2 \otimes \alpha) \Sigma_{k|k-1} \right] H_k^T \tag{10.37}$$

If $P_{Yk} \leq \alpha P_{Yk}$, noticing that $I_2 \otimes \alpha \leq \beta_k^2$ from (10.35), we may get (10.31) by

$$\Sigma_{k|k-1} \geq (I_2 \otimes \alpha)^{-1} \overline{\Sigma}_{k|k-1} \geq \beta_k^{-2} \overline{\Sigma}_{k|k-1} \tag{10.38}$$

Otherwise, if $\overline{P}_{Y_k} > \alpha P_{Y_k}$, substituting (10.33) and (10.34) into (10.13) forms

$$\Sigma_{k|k-1} - \beta_k^{-2} \overline{\Sigma}_{k|k-1} \geq \left[(I_2 \otimes \alpha)^{-1} - \beta_k^{-2} \right] \overline{\Sigma}_{k|k-1} \geq 0 \tag{10.39}$$

which can be guaranteed by (10.35). Therefore, (10.31) is verified.

Applied with Theorem 1, Theorem 2 can be proven.

Because of introduction of the switching logic, the presented $SH_\infty F$ can alternate between robust filtering mode as an $H_\infty F$ and optimal filtering mode as a KF by test if the strength of uncertainties exceeds the tolerable threshold, and the threshold is tunable by α, a switching parameter. Clearly, the $SH_\infty F$ is independent with γ and the only parameter is α. Compared with γ, α is a normalized index to evaluate uncertainties with clear physical interpretation, so it is easier to provide accurately.

3.2 Optimal-switched H_∞ Kalman filter

In the $SH_\infty F$, γ is transformed into α by the switching mechanism. With various α, the filter can get filtering results with different weight of optimality or robustness. However, the relationship between α and the weight is not clear. It does not identify that, for given α, the filtering result is based on how much optimality and how much robustness, and for given weight of optimality and robustness, what is the best α. As a result, it is hard to evaluate α and achieve the best filtering performance. To optimize the filter, we need to identify the relationship of them and optimize the α in real time with given weight.

We adopt a weighted sum function, named ORCF, to quantify the overall performances of optimality and robustness at each iteration. It has the form as

$$\Gamma = \lambda E \left[\left(Z_k - \widehat{Z}_k \right) \left(Z_k - \widehat{Z}_k \right)^T \right] + P_k^Z \tag{10.40}$$

where $\widehat{Z}_k = H_k \widehat{X}_{k|k}$ and λ is called the weight factor (WF). It has two parts, evaluating robustness and optimality, respectively. The weight λ controls their balance. We should minimize the ORCF to optimize the filtering performance.

First of all, take (10.11) into $\widehat{Z}_k$ and notice that (10.9) forms. It has

$$Z_k - \widehat{Z}_k = V_k \left(P_k^Z \right)^{-1} \widetilde{Z}_k \tag{10.41}$$

Substitute (10.41) into (10.40) and notice that $P_k^Z = E\left(\widetilde{Z}_k \widetilde{Z}_k^Z \right)$. The ORCF can be reformed as

$$\Gamma = \lambda \left[V_k \left(P_k^Z \right)^{-1} \right] P_k^Z \left[V_k \left(P_k^Z \right)^{-1} \right]^T + \left[V_k \left(P_k^Z \right)^{-1} \right]^{-1} V_k \tag{10.42}$$

For simple expression, note $\lambda I_m = \lambda' V_k$. Take derivative of (10.42) to $V \left(P_k^Z \right)^{-1}$ and notice that the second derivative is positive definite. The minimum Γ can be obtained if

$$P_k^Z = \left(2\lambda' \overline{P}_k^Z \right)^{1/3} V_k \triangleq \left(P_k^Z \right)^* \tag{10.43}$$

Superscript * represents optimum of a variable.

Furthermore, define diagonal matrices μ_1 and μ_2 satisfying

$$H_k \Sigma_{k|k-1} H_k^T = \mu_1 H_k P_{k|k-1} H_k^T \tag{10.44}$$

$$P_k^Z = \mu_2 \left(H_k P_{k|k-1} H_k^T + V_k \right) \tag{10.45}$$

showing amplification ratio of the prediction and measurement covariance, respectively.

They have

$$\mu_1 = \left(I_m + \frac{V_k}{H_k P_{k|k-1} H_k^T} \right) \left(I_m - \frac{V_k}{P_k^Z} \right) \mu_2 \triangleq k_k^\mu \mu_2 \tag{10.46}$$

where k_k^μ is a transformation matrix. Let P_k^Z take $\left(P_k^Z \right)^*$ as (10.43), then (10.45) is rewritten as

$$\mu_2 = \left(P_k^Z \right)^* T_k^{-1} \tag{10.47}$$

where $T_k = H_k P_{k|k-1} H_k^T + V_k$.

To match the dimension of measurements with system states, according to (10.33), expand (10.44) by keeping

$$\Sigma_{k|k-1} = f_\alpha P_{k|k-1} \tag{10.48}$$

where $f_\alpha = \mu_1 \otimes I_2$. Then, it has

$$(\alpha^*)^{-1} diag \frac{\overline{P}_k^Z}{P_k^Z} = \mu_1 \tag{10.49}$$

Substitute (10.46), (10.47), and (10.43) into (10.49), then α^* can be calculated by

$$\alpha_k^* = T_k V_k^{-1} \left(2\lambda V_k^{-1} \overline{P}_k^Z \right)^{-1/3} \left(k_k^\mu \right)^{-1} diag \frac{\overline{P}_k^Z}{P_k^Z} \tag{10.50}$$

It indicates that the α is optimized by the ORCF under given λ.

Take α in (10.13) and (10.34) as α^* in (10.50), then the SH$_\infty$F can be transformed to the OSH$_\infty$F, and the OSH$_\infty$F can be expressed by (10.5), (10.6), (10.13), (10.50), and (10.8) to (10.12). Considering that the switching threshold is determined by α, the OSH$_\infty$F can get the optimal threshold in real time in the sense of λ. In other words, for given λ, the filter can optimize filtering robustness and achieve worst-case optimal estimation.

4. Example: maneuvering target tracking in space

In this section, we use an example to verify the filtering approach for the problem of H_∞ state estimation. Consider a scenario of space target tracking. There are two spacecraft called Chaser and Target. The Target has potential maneuver and the Chaser flies freely. The Chaser is required to estimate the relative motional states with the Target. We first build the orbital relative motion model for the two spacecraft, and then formulate the tracking algorithm based on the adaptive H_∞ filtering approach.

4.1 Orbital relative motion model

For the assumed Chaser and Target, orbital relative motion model is expressed as (Psiaki, 1999)

$$\ddot{R} = -2\omega_C \times \dot{R} - \dot{\omega}_C \times R - \omega_T \times (\omega_T \times R) + \frac{\mu}{r_C^3} \times \left(\rho_C - \frac{r_C^3}{r_T^3} \rho_T \right) + D^R \tag{10.51}$$

where ρ denotes the distance vector in the Earth-centered frame. $R = [R_x, R_y, R_z]$ is the relative range in the radial, in-track, and cross-track (RIC) frame of the Chaser. μ, r, and ω are geocentric gravitational constant, geocentric distance, and orbital angular rate, respectively. D^R shows disturbances, composed of J_2, the oblateness perturbation of the Earth, and U_T, the target maneuver, with bounded energy. Subscripts C and T represent Chaser and Target, respectively.

Notice that relative range is usually far less than geocentric distance of the chaser, i.e., $\|R\| \ll r_C$. Then, (10.51) can approximate to a discrete form as (Kawase, 1990)

$$X_{k+1} = \phi_{k+1} \varphi_{k+1,k} \phi_k^{-1} X_k + D \tag{10.52}$$

in which $X = [R, \dot{R}]^T$ is the relative state and D expresses the uncertainties resulted from D^R. ϕ and φ satisfy

$$X_k = \boldsymbol{\phi}_k \delta\sigma_k \tag{10.53}$$

$$\delta\sigma_{k+1} = \boldsymbol{\varphi}_{k+1} \delta\sigma_k \tag{10.54}$$

where $\delta\sigma$ is the difference of orbital elements of the two spacecrafts, i.e., $\delta\sigma = \sigma_T - \sigma_C$. Define orbital element set as $\sigma = [a, u+f, i, e\cos u, e\sin u, \Omega]^T$ where a, e, i, u, f, and Ω are semimajor axis, eccentricity, inclination, argument of periapsis, true anomaly, and longitude of the ascending node, respectively. The expressions of $\boldsymbol{\phi}$ and $\boldsymbol{\varphi}$ (Hablani, 2003) are

$$\varphi(t, t_0) = \begin{bmatrix} 1 & 0 & 0 & 0 & 0 & 0 \\ -\dfrac{3n(t_0)t - t_0}{2a(t_0)G_\theta(t)} & \dfrac{G_\theta(t_0)}{G_\theta(t)} & 0 & \dfrac{G_{q_1}(t) - G_{q_1}(t_0)}{G_\theta(t)} & \dfrac{G_{q_2}(t) - G_{q_2}(t_0)}{G_\theta(t)} & 0 \\ 0 & 0 & 1 & 0 & 0 & 0 \\ & O_{3\times3} & & & I_{3\times3} & \end{bmatrix}$$

in which

$$G_\theta = \frac{r\eta}{a(1+f_1)}, \eta = \sqrt{1 - q_1^2 - q_2^2}, f_1 =$$

$$q_1 \cos\theta + q_2 \sin\theta, f_2 = q_1 \sin\theta - q_2 \cos\theta$$

$$G_{q_1} = \frac{q_2}{\eta(1+\eta)} + \frac{q_1 f_2}{\eta(1+f_1)} - \frac{r(a+r)}{\eta a^2}(q_2 + \sin\theta)$$

$$G_{q_2} = -\frac{q_1}{\eta(1+\eta)} + \frac{q_2 f_2}{\eta(1+f_1)} + \frac{r(a+r)}{\eta a^2}(q_1 + \cos\theta)$$

and

$$\phi(t) = \begin{bmatrix} \phi_{11}(t) & \phi_{12}(t) \\ \phi_{21}(t) & \phi_{22}(t) \end{bmatrix}$$

where

$$\phi_{11} = \begin{bmatrix} r/a & rf_2/(1+f_1) & 0 \\ 0 & r & 0 \\ 0 & 0 & r\sin\theta \end{bmatrix}$$

$$\phi_{21} = \frac{n}{\eta} \begin{bmatrix} -f_2/2 & a - r/\eta & 0 \\ -3(1+f_1)/2 & -af_2 & 0 \\ 0 & 0 & a(q_1 + \cos\theta) \end{bmatrix}$$

$$\phi_{12} = \frac{1}{\eta^2} \begin{bmatrix} -2raq_1 - r^2\cos\theta/a & -2raq_2 - r^2\sin\theta/a & 0 \\ 0 & 0 & \eta^2 r \cos i \\ 0 & 0 & -\eta^2 r \sin i \cos\theta \end{bmatrix}$$

$$\phi_{22} = \frac{n}{\eta^3} \begin{bmatrix} aq_1 f_2 + r\sin\theta(1+f_1) & aq_1 f_2 + r\sin\theta(1+f_1) & 0 \\ (1+f_1)(3aq_1 + 2r\cos\theta) & (1+f_1)(3aq_2 + 2r\cos\theta) & \eta^2 af_2 \cos i \\ 0 & 0 & \eta^2 a(q_2 + \sin\theta)\sin i \end{bmatrix}$$

Denote radar measurements of the Chaser to the Target as $\mathbf{Y} = [\rho, \theta_a, \theta_e]$ with noise covariance $\mathbf{V}^Y = \left[V_\rho^Y, V_a^Y, V_e^Y \right]^T$, where the components represent relative range, azimuth, and elevation, respectively. The measurement equation can be written as

$$\mathbf{C}\mathbf{Y}_k = \mathbf{R}_k + \mathbf{v}_k \qquad (10.55)$$

where $\mathbf{C}$ is an unbiased conversion matrix that is derived by using the unbiased converted measurement technique (Hablani, 2009). $\mathbf{v}_k$ is the converted noise with covariance $\mathbf{V}$. $\mathbf{C}$ and $\mathbf{V}$ are determined by $\mathbf{Y}$ and $\mathbf{V}^Y$, expressed as (Gurfil and Mishne, 2007)

$$\mathbf{C} = \begin{bmatrix} \lambda_a \lambda_e & 0 & 0 \\ 0 & \lambda_a \lambda_e & 0 \\ 0 & 0 & \lambda_a \end{bmatrix} \qquad (10.56)$$

and $\mathbf{V}(t) = \left[V_{i,j}(t) \right]_{i,j=1,2,3}$ where

$$V_{11}(t) = -\lambda_a^2 \lambda_e^2 \rho^2(t) \cos^2\theta_a(t) \cos^2\theta_e(t)$$
$$+ \frac{1}{4}\left(\rho^2(t) + V_\rho^{Y2}\right)\left(1 + \lambda_a^4 \cos 2\theta_a(t)\right)\left(1 + \lambda_e^4 \cos 2\theta_e(t)\right)$$

$$V_{22}(t) = -\lambda_a^2 \lambda_e^2 \rho^2(t) \cos^2\theta_a(t) \sin^2\theta_e(t)$$
$$+ \frac{1}{4}\left(\rho^2(t) + V_\rho^{Y2}\right)\left(1 + \lambda_a^4 \cos 2\theta_a(t)\right)\left(1 + \lambda_e^4 \cos 2\theta_e(t)\right)$$

$$V_{33}(t) = -\lambda_a^2 \rho^2(t) \sin^2\theta_a(t) + \frac{1}{2}\left(\rho^2(t) + V_\rho^{Y2}\right)\left(1 - \lambda_a^4 \cos 2\theta_a(t)\right)$$

$$V_{12}(t) = V_{21}(t) = -\frac{1}{2}\lambda_a^2 \lambda_e^2 \rho^2(t) \cos^2\theta_a(t) \sin 2\theta_e(t)$$
$$+ \frac{1}{4}\left(\rho^2(t) + V_\rho^{Y2}\right)\left(1 + \lambda_a^4 \cos 2\theta_a(t)\right)\lambda_e^4 \sin 2\theta_e(t)$$

$$V_{13}(t) = V_{31}(t) = -\frac{1}{2}\left(\rho^2(t)\left(1 - \lambda_a^2\right) - V_\rho^{Y2}\right)\lambda_a^2 \lambda_e \sin 2\theta_a(t) \cos\theta_e(t)$$

$$V_{23}(t) = V_{32}(t) = \frac{1}{2}\left(\rho^2(t)\left(1 - \lambda_a^2\right) - V_\rho^{Y2}\right)\lambda_a^2\lambda_e \sin 2\theta_a(t)\sin \theta_e(t)$$

in which $\lambda_a = e^{-V_a^{Y2}/2}$ and $\lambda_e = e^{-V_e^{Y2}/2}$.

Then, the relative motion model for space target tracking can be given by (10.52) and (10.55).

4.2 Adaptive H_∞ tracking algorithm

Applying the model into the $\mathrm{SH}_\infty\mathrm{F}$ and the $\mathrm{OSH}_\infty\mathrm{F}$ discussed before, i.e., considering the space relative motion model (10.53) and (10.55) as the system model (1) and (2) into the filters, we may obtain adaptive H_∞ tracking algorithms for target state estimation.

5. Numerical simulations

We use two cases to verify the tracking algorithm and the presented filtering approach. For one case, assume that the Target is flying freely, and for the other case, the Target has potential maneuver of continuous finite thrust. The Chaser is at an in-track formation with the Target so that their relative motion is only along the Chaser in-track (Lane and Axelred, 2006).

The initial orbits of the two spacecrafts are the same with orbital elements as $a = 6578.137$ km, $e = 0$, $i = 28.5$ degrees, $f = 180$ degrees, and $\Omega = 45$ degrees, except u for which $u_T = 0.05$ degrees and $u_C = 0°$. The variances of the Chaser measurements are assumed $(1m)^2$ for range and $(0.1 rad)^2$ for elevation and azimuth. The variances of initial states are $(1m)^2$ for position and $(0.1\ m/s)^2$ for velocity. We take KF, $H_\infty\mathrm{F}$, $\mathrm{SH}_\infty\mathrm{F}$, and $\mathrm{OSH}_\infty\mathrm{F}$ as the tracking algorithm, respectively.

Figs. 10.1 and 10.2 give the result of the $\mathrm{SH}_\infty\mathrm{F}$ compared with the $H_\infty\mathrm{F}$ and the KF for the first case. At the $\mathrm{SH}_\infty\mathrm{F}$, $\alpha = diag(3.5, 2.5, 10)$. Fig. 10.1b indicates that the $\mathrm{SH}_\infty\mathrm{F}$ needs less gain compensation than the $H_\infty\mathrm{F}$, improving the filtering conservativeness and increasing the tracking precision as shown in Fig. 10.1a. Compared with the KF, the $\mathrm{SH}_\infty\mathrm{F}$ has higher level of robustness to the nonlinear modeling error and achieves improved tracking accuracy, as Fig. 10.2 shows.

Now assume that the Target maneuvers from 10,300 s and lasts for 200 s, along both radial and in-track with acceleration 0.02m/s^2. For such case, the tracking results are compared in Fig. 10.3 and 10.4.

Clearly, the KF is divergent after target maneuvers. The $\mathrm{SH}_\infty\mathrm{F}$ obtains the best results, showing higher precision than the traditional $H_\infty\mathrm{F}$. We can conclude that the H_∞ filtering criterion improves the estimation accuracy, while the switching mechanism increases the estimation precision, generating the capability of fast tracking.

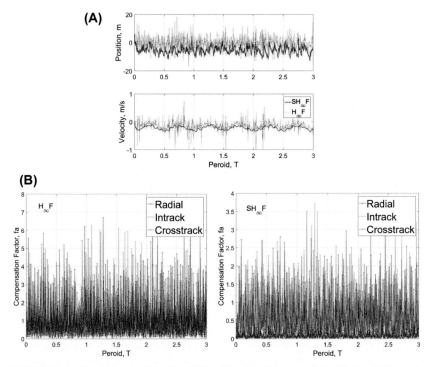

FIGURE 10.1 Comparisons of H∞F and SH∞F for free-flying target. (A) Radial relative state estimation error, (B) Compensation factor.

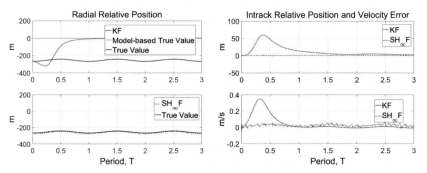

FIGURE 10.2 Comparison of Kalman filter (KF) and SH∞F for free-flying target.

In addition, we suppose that the Target has constant thrust of 2000 s from 1000 s along radial and in-track. The first half of the thrust is negative and the second half is positive, both of which have the same amplitude 0.001 m/s^2. This case is used to verify the OSH∞F.

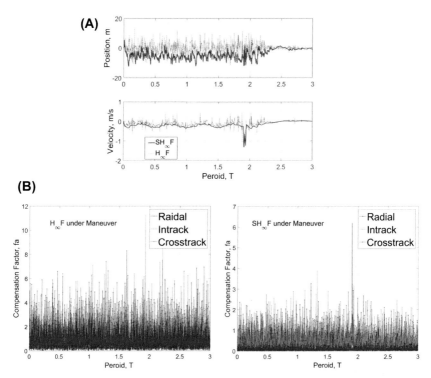

FIGURE 10.3 Comparisons of $H_\infty F$ and $SH_\infty F$ for maneuvering target. (A) Radial relative state estimation error, (B) Compensation factor.

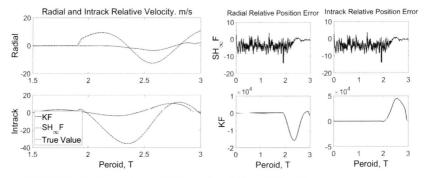

FIGURE 10.4 Comparisons of Kalman filter (KF) and $SH_\infty F$ for maneuvering target.

The target maneuver and target trajectory relative to the Chaser are as in Fig. 10.5. Fig. 10.6 shows in-track estimation errors of the $OSH_\infty F$ compared with KF, $H_\infty F$, and the $SH_\infty F$.

Clearly, the KF has serious deviation due to lack of robustness so that the filter requires long time to recover the convergence. The $H_\infty F$ achieves better

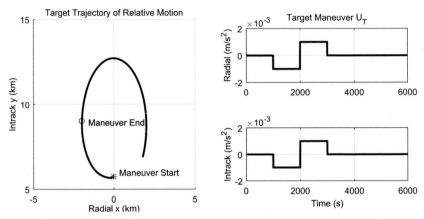

FIGURE 10.5   The trajectory of the target maneuvering with constant thrust.

estimation accuracy, but the infinite-horizon robustness results in filtering conservativeness and the precision is low. With introduction of a dual-mode switching mechanism, the $SH_\infty F$ improves the conservativeness and has a higher precision, as presented in subfigure (c) where $\alpha = 1$.

The $OSH_\infty F$ on-line optimizes the $SH_\infty F$ so that it obtains the best estimation performance among all the compared filters. It can be seen from subfigure (d) where $\lambda = 0.5$.

The tracking errors of the $OSH_\infty F$ with various λ are demonstrated in Fig. 10.7. It indicates that filtering robustness is enhanced with λ. It is clear that when $\lambda = 0.05$, the filter has smoother but larger deviation than when $\lambda = 5$, whereas $\lambda = 0.5$ gets the best balance on accuracy and precision, among the three cases.

6. Conclusions

In this chapter, an adaptive H_∞ filtering approach is discussed with application to target tracking in space. For the approach, two typical filters, $SH_\infty F$ and $OSH_\infty F$, are presented sequentially. With introduction of a dual-mode switching structure, the $SH_\infty F$ can control the weight of filtering robustness and optimality to improve the design conservativeness of the traditional $H_\infty F$, achieving higher filtering performance. Furthermore, the $OSH_\infty F$ is an optimal filter that can on-line optimize the switching control parameter of the $SH_\infty F$ by minimizing a unified cost function in real time, achieving worst-case optimal estimation, i.e., conservativeness of the traditional $H_\infty F$ is optimized. Compared with KF and $H_\infty F$, the $SH_\infty F$ and the $OSH_\infty F$ can achieve superior estimation performance for stochastic systems with nonlinear uncertainties.

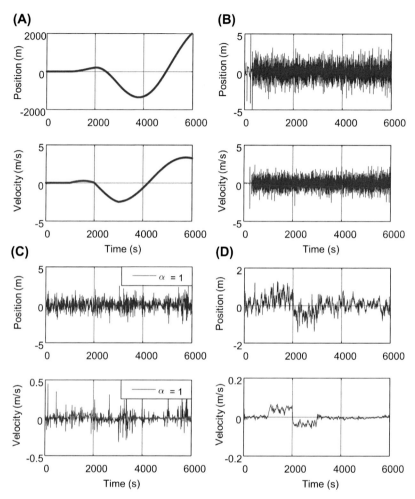

FIGURE 10.6 Comparison of the tracking error for small constant maneuvering target. (A) In-track Error of KF, (B) In-track error of $H_\infty F$, (C) In-track Error of $SH_\infty F$, (D) In-track error of $OSH_\infty F$.

7. Appendix

Proof of Theorem 1

  Choose Lyapunov function as

$$\boldsymbol{\Gamma}\left(\widetilde{\boldsymbol{X}}_k\right)\widetilde{\boldsymbol{X}}_k^T\boldsymbol{P}_{k|k}^{-1}\widetilde{\boldsymbol{X}}_k \tag{10.57}$$

which is bounded if considering (10.29), so

$$p_{\max}^{-1}\left\|\widetilde{\boldsymbol{X}}_k\right\|^2 \le \boldsymbol{\Gamma}\left(\widetilde{\boldsymbol{X}}_k\right) \le p_{\max}^{-1}\left\|\widetilde{\boldsymbol{X}}_k\right\|^2 \tag{10.58}$$

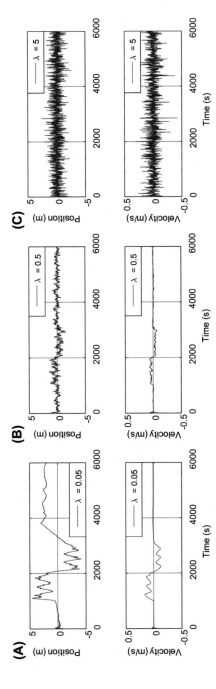

FIGURE 10.7 The in-track estimation error of the OSH∞F with various λ. (A) In-track error when λ=0.05, (B) In-track Error when λ=0.5, (C) In-track error when λ=5.

According to (10.2), (10.11), and (10.15), the estimation error of the SH$_\infty$F can be represented as

$$= (I - K_k H_k)\beta_k F_{k-1}\widetilde{X}_{k-1} + (I - K_k H_k)w_k - K_k v_k \tag{10.59}$$

Substituting (10.59) into (10.57) and taking conditional expectation, we have

$$E\{\Gamma(\widetilde{X}_k)|\widetilde{X}_{k-1}\} = \tau_k + \mu_k \tag{10.60}$$

where

$$\tau_k = \widetilde{X}_{k-1}^T F_{k-1}^T \beta_k \left[\Sigma_{k|k-1}^{-1} - H_k^T\left(V_k^{-1} - V_k^{-1}H_k P_k H_k^T V_k^{-1}\right)H_k\right]\beta_k F_{k-1}\widetilde{X}_{k-1} \tag{10.61}$$

$$\mu_k = E\left\{W_k^{-1}(I - K_k H_k)^T P_k^{-1}(I - K_k H_k)W_k + v_k^T K_k^T P_k^{-1}K_k V_k\right\} \tag{10.62}$$

Based on (10.16) and (10.31), it has

$$\Sigma_{k|k-1} \geq \beta_k^{-2}\left(\beta_k F_{k-1}P_{k-1}F_{k-1}^T\beta_k + W_k\right) \geq F_{k-1}P_{k-1}F_{k-1}^T \tag{10.63}$$

According to matrix theory, there is

$$V_k^{-1} - V_k^{-1}H_k P_k H_k^T V_k^{-1} = \left(H_k \Sigma_{k|k-1}H_k^T + V_k^{-1}\right)^{-1} \tag{10.64}$$

Substituting (10.63) and (10.64) into (10.61) forms

$$E\{\Gamma(\widetilde{X}_k)|\widetilde{X}_{k-1}\}$$
$$\leq \widetilde{X}_{k-1}^T\beta_k\left[P_{k-1}^{-1} - F_{k-1}^T H_k^T\left(H_k\Sigma_{k|k-1}H_k^T + V_k^{-1}\right)^{-1}H_k F_{k-1}\right]\beta_k\widetilde{X}_{k-1} + \mu_k \tag{10.65}$$

which can be reshaped into the form similar to (10.23), expressed as

$$E\{\Gamma(\widetilde{X}_k)|\widetilde{X}_{k-1}\} - \Gamma(\widetilde{X}_{k-1}) \leq \mu_k - \lambda_k\Gamma(\widetilde{X}_{k-1}) \tag{10.66}$$

where

$$\lambda_k = 1 - \widetilde{X}_{k-1}^T\beta_k\left[P_{k-1}^{-1} - F_{k-1}^T H_k^T\left(H_k\Sigma_{k|k-1}H_k^T + V_k^{-1}\right)^{-1}H_k F_{k-1}\right] \times \beta_k\widetilde{X}_{k-1}\Gamma^{-1}(\widetilde{X}_{k-1}) \tag{10.67}$$

According to Lemma 1, the rest objective is to find a $\mu > 0$ and a $\lambda \in (0,1]$. At first, from (10.62), it is obvious that $\mu_k > 0$ and

$$\mu_k = \leq tr\left[\left(P_k^{-1} + H_k^T V_k^{-1}H_k P_k^{-1}H_k^T V_k^{-1}H_k\right)W_k\right] + tr\left(K_k^T P_k^{-1}K_k V_k\right) \tag{10.68}$$

so, we may get the upper bound as

$$\mu_{max} = \left(p_{min}^{-1} + h_{max}^4 v_{min}^{-2}p_{max}\right)q_{max}n + h_{max}^2 p_{max}v_{min}^{-1}m \tag{10.69}$$

Furthermore, it is clear that

$$
\begin{aligned}
&\overset{\lambda_k}{=1-\widetilde{X}_{k-1}^T\beta_k\left[P_{k-1}^{-1}-F_{k-1}^T H_k^T\left(H_k\Sigma_{k|k-1}H_k^T V_k^{-1}\right)^{-1}H_k F_{k-1}\right]\beta_k\widetilde{X}_{k-1}\Gamma^{-1}\left(\widetilde{X}_{k-1}\right)}\\
&\leq 1-\widetilde{X}_{k-1}^T\beta_k\left[P_{k-1}^{-1}-F_{k-1}^T H_k^T\left(H_k P_{k|k-1}H_k^T V_k^{-1}\right)^{-1}H_k F_{k-1}\right]\beta_k\widetilde{X}_{k-1}\Gamma^{-1}\left(\widetilde{X}_{k-1}\right)\\
&=1-\widetilde{X}_{k-1}^T\beta_k\left[P_{k-1}+P_{k-1}F_{k-1}^T H_k^T\left(H_k W_k H_k^T V_k\right)^{-1}H_k F_{k-1}\right]\beta_k\widetilde{X}_{k-1}\Gamma^{-1}\left(\widetilde{X}_{k-1}\right)\\
&\qquad\qquad\qquad\qquad\leq 1
\end{aligned}
$$

$$(10.70)$$

After that, we can learn from (10.67) that $\lambda_k > 0$ if only

$$
\beta_k < \left[I-F_{k-1}^T H_k^T\left(H_k\Sigma_{k|k-1}H_k^T V_k^{-1}\right)^{-1}H_k F_{k-1}P_{k-1}\right]^{-1/2}
\tag{10.71}
$$

Taking (10.64) into consideration, it may become

$$
\beta_k < \left[I-F_{k-1}^T H_k^T V_k^{-1}\left(V_k-H_k P_k H_k^T\right)V_k^{-1}H_k F_{k-1}P_{k-1}\right]^{-1/2}
\tag{10.72}
$$

Taking spectral norm to (10.72) obtains

$$
\|\beta_k\| < \left\|I-F_{k-1}^T H_k^T V_k^{-1}\left(V_k-H_k P_k H_k^T\right)V_k^{-1}H_k F_{k-1}P_{k-1}\right\|^{-1/2}\triangleq\beta_m
\tag{10.73}
$$

Considering (10.30), we can see $\beta_{\max}\leq\beta_m$. Then, (10.71) to (10.73) form, that is to say, $\lambda_k > 0$. Therefore, $\lambda_k\in(0,1]$ and there must exist a minimum $\lambda_{\min}\in(0,1]$. With $\mu_{\max}$ in (10.69) and based on Lemma 1, we can conclude that (10.32) forms. That means, X_k is bounded.

Range determination of βm:

From (10.71), we can learn $\beta_m=\geq 1$. Note $\delta_k=H_k^T V_k^{-1}H_k$ and reshape βm as

$$
\beta_m=\left\|I-F_{k-1}^T(\delta_k-\delta_k P_k\delta_k)F_{k-1}P_{k-1}\right\|^{-1/2}
\tag{10.74}
$$

Clearly, when $\delta_k=0$ or $\delta_k=P_k^{-1}$, β_m reaches its inferior, whereas when $\delta_k=P_k^{-1}/2$, β_m is at the superior, as

$$
\beta_m=\left\|I-\frac{1}{4}F_{k-1}^T P_{k-1}^{-1}F_{k-1}P_{k-1}\right\|^{-1/2}
\tag{10.75}
$$

so the superior of β_m is $\left(1-\frac{1}{4}f_{\max}^2 p_{\max}p_{\max}^{-1}\right)^{-1/2}$ if satisfying $f_{\max}^2 < 4p_{\min}p_{\max}^{-1}$.

Therefore, $\beta_m\in\left[1,\left(1-\frac{1}{4}f_{\max}^2 p_{\max}p_{\max}^{-1}\right)^{-1/2}\right]$.

Acknowledgments

The work was supported in part by China Natural Science Foundation (60775022 and 60674107), the Project Sponsored by SRF for ROCS, SEM, and Fundamental Research Funds for the Central Universities (ZYGX-2014J098).

Thanks for the permissions of the Institution of Engineering & Technology (IET) and the International Federation of Automatic Control (IFAC) to reproduce the figures in the following original works: 1) Y.K. Li, Z.L. Jing and S.Q. Hu, Redundant Adaptive Robust Tracking of Active Satellite and Error Evaluation, IET Control Theory and Applications, vol. 4, no. 11, pp. 2539−2553, Nov. 2010; 2) Y.K. Li, S. Zhang, L. Ding and Z.G. Shi, Optimal-Switched H_∞ Robust Tracking for Maneuvering Space Target, Proceedings of the 20th IFAC Symposium on Automatic Control in Aerospace, Sherbrooke, Canada, IFAC-Papers Online vol. 49, no. 17, pp. 415−419, Aug. 2016.

References

Banavar, R., 1992. A Game Theoretic Approach to Linear Dynamic Estimation. Doctoral Dissertation, University of Texas at Austin.

Bishop, A.N., Pathirana, P.N., Savkin, A.V., 2007. Radar target tracking via robust linear filtering. IEEE Signal Processing Letters 14 (12), 1028−1031.

Einicke, G.A., White, L.B., 1999. Robust extended Kalman filtering. IEEE Transactions on Signal Processing 47 (9), 2596−2599.

Gurfil, P., Mishne, D., 2007. Cyclic spacecraft formations: relative motion control using line-of-sight measurement only. AIAA Journal of Guidance, Control and Dynamics 30 (1), 214−226.

Hablani, H.B., 2003. Autonomous navigation, guidance, attitude determination and control for spacecraft rendezvous. In: Proceeding of AIAA Conference and Exhibit on Guidance, Navigation and Control, Austin, Texas.

Hablani, H.B., 2009. Autonomous inertial relative navigation with sight-line-stabilized sensors for spacecraft rendezvous. AIAA Journal of Guidance, Control and Dynamics 32 (1), 172−183.

Kawase, S., 1990. Intersatellite tracking methods for clustered geostationary satellites. IEEE Transactions on Aerospace and Electronic Systems 26 (3), 469−474.

Lane, C., Axelred, P., 2006. Formation design in eccentric orbits using linearized equations of relative motion. AIAA Journal of Guidance, Control and Dynamics 29 (1), 146−160.

Li, Y.K., Jing, Z.L., Hu, S.Q., 2010. Redundant adaptive robust tracking of active satellite and error evaluation. IET Control Theory & Applications 4 (11), 2539−2553.

Li, W., Gong, D.R., Liu, M.H., Chen, J.A., Duan, D.P., 2013. Adaptive robust Kalman filter for relative navigation using global position system. IET Radar, Sonar & Navigation 7 (5), 471−479.

Li, Y.K., Jing, Z.L., Liu, G.J., 2014. Maneuver-aided active satellite tracking using six-DOF optimal dynamic inversion control. IEEE Transactions on Aerospace and Electronic Systems 50 (1), 704−719.

Li, Y.K., Zhang, S., Ding, L., Shi, Z.G., 2016. Optimal-switched H∞ robust tracking for maneuvering space target, Proceedings of the 20th IFAC Symposium on automatic control in Aerospace, Sherbrooke, Canada. IFAC-Papers Online 49 (17), 415−419.

Psiaki, M.L., 1999. Autonomous orbit determination for two spacecraft from relative position measurements. AIAA Journal of Guidance, Control and Dynamics 22 (2), 305−312.

Reif, K., Gunther, S., Yaz, E., et al., 1999. Stochastic stability of the discrete-time extended Kalman filter. IEEE Transactions on Automatic Control 44 (4), 714−728.

Seo, J., Yu, M.J., Park, C.G., Lee, J.G., 2006. An extended robust $H\infty$ filter for nonlinear constrained uncertain systems. IEEE Transactions on Signal Processing 54 (11), 4471−4475.

Shen, X., Deng, L., 1997. Game theory approach to discrete $H\infty$ filter design. IEEE Transactions on Signal Processing 45 (4), 1092−1095.

Soken, H.E., Hajiyev, C., 2010. Pico satellite attitude estimation via robust unscented Kalman filter in the presence of measurement faults. ISA Transactions 49 (3), 249−256.

Theodor, Y., Shaked, U., de Souza, C.E., 1994. A game theory approach to robust discrete-time $H\infty$-estimation. IEEE Transactions on Signal Processing 42 (6), 1486−1495.

Xiong, K., Zhang, H., Liu, L.D., 2008. Adaptive robust extended Kalman filter for nonlinear stochastic systems. IET Control Theory & Applications 2 (3), 239−250.

Zhong, M.Y., Zhou, D.H., Ding, S.X., 2010. On designing H∞ fault detection filter for linear discrete time-varying systems. IEEE Transactions on Automatic Control 55 (7), 1689−1695.

Chapter 11

R for lifetime data modeling via probability distributions

Vikas Kumar Sharma
Institute of Infrastructute Technology Research and Management (IITRAM), Department of Mathematics, Ahmedabad, Gujarat, India

1. Introduction

R is an open-source software that provides an emphatic programming environment to perform statistical computing. It has received huge popularity among students, academicians, industries, and researchers during last decade. Its invention in 1993 is because of the pioneer efforts of Robert Gentleman and Ross Ihaka (University of Auckland) who introduced R as a reduced version of language S. The S language was discovered at Bell Laboratories by Becker and Chambers (1984); see also Chambers (1998). In 1995, R was made freely available worldwide under General Public License. We may refer the journal article of Gentleman and Ihaka (Ross & Robert G, 1996) to the readers for basic information and invention of R. After that in 1997, an R Development Core Team (R Core Team, 1996) consisting S programmers (John Chambers is one of the members), statistician, and social scientists was made and an online network called CRAN (Comprehensive R Archive Network) was launched, which stores up-to-date versions of R codes and documentations. Since then, we have seen the increasing participation by the researchers in its developments taken place around the world.

Statistical computing is an integral part of any scientific study and involves a wide variety of computing tasks such as manipulation and display of data and computing statistical quantities, etc. This chapter introduces the statistical modeling using R with special reference to lifetime data. Mainly three components are discussed: the basic introduction of R, construction of graphs, and implementation of statistical tools for modeling real-life data using probability distributions. R can be effectively used in such statistical computing. R covers almost all statistical techniques ranging from descriptive measures to modeling

Handbook of Probabilistic Models. https://doi.org/10.1016/B978-0-12-816514-0.00011-4

265

of complex multivariate data. As R also provides a programming environment, it can be extended to any extent. For more acquaintance of R and its uses to various areas, readers may follow the monographs by Dalgaard (2008), Rizzo (2008), Matloff (2011), Ohri (2012), Crawley (2013), and Peng (2015).

In this chapter, it is assumed that random observations follow the generalized inverse Lindley distribution (GILD). The codes of R will be constructed for analyzing the lifetime data sets using GILD. The probability density function (pdf) is given by

$$f(x; \alpha, \theta) = \frac{\alpha \theta^2}{(1+\theta)} \left[\frac{1+x^\alpha}{x^{2\alpha+1}} \right] e^{-\frac{\theta}{x^\alpha}}; (x, \alpha, \theta) > 0, \qquad (11.1)$$

where θ and α are scale and shape parameters, respectively. $\text{GILD}(\alpha, \theta)$ can be used to denote the GILD in (11.1) henceforth.

Sharma et al. (2016) introduced this distribution and discussed its properties. They have shown an application of distribution for fitting the maximum flood levels data over many standard competing distributions. The distribution is flexible enough for modeling practical data that are positively skewed and show upside-down bathtub failure rates (see Sharma et al., 2014). Sharma 2018 developed estimation methods for estimating stress—strength reliability derived from the GILD for survival times of head and neck cancer (HNC) patients.

In this chapter, analyses of maximum flood levels and survivals of HNC patients are presented along with extensive discussion on R codes used for analyzing these data. The data set of maximum flood levels is completely observed, while the lifetimes of patients suffering from HNC are censored observations. Therefore, the discussion on how to estimate the parameters for both complete and censored data is given. Before proceeding to the analysis, the discussion on how to start with R for data handling and curve sketching are given in first few sections.

2. R installations, help and advantages

R application and packages can be obtained from the website called Comprehensive R Archive Network (CRAN) at https://cran.r-project.org/, see Fig. 11.1. At this website, R software is available for Linux, Mac, and window operating systems, and it can be easily downloaded and installed as we do for other software packages.

We see an R logo [R] on the desktop after successful installation of the application. To start R program, we click the R icon and as R session starts, an introductory message appears on the R console window, see Fig. 11.2.

The easiest way of getting help when working in the R environment is to click the help button on the toolbar. Alternatively one can type `help(name of function)`. Another way of getting help is to type the name of the function you require followed by a question mark, e.g., ?mean will open an HTML file that provides details of the `mean()` function.

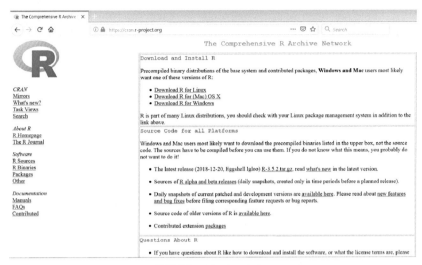

FIGURE 11.1   R website.

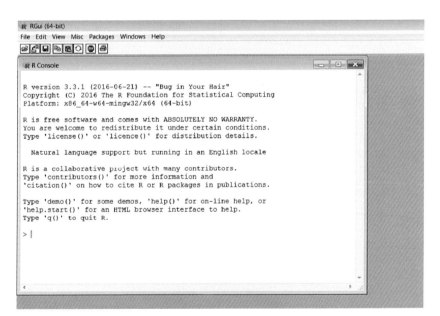

FIGURE 11.2   R console window.

R being an open-source computing package received huge growth in popularity in the last few years.

- R is freely available as it is published under general public license. Its users can see how the codes are written and can modify it. Being an open-source package, it is easily obtainable by students and economical to install in our computing laboratories.
- It is easy to learn. Those who are familiar with some programming concepts can easily switch to R programming.
- It has a good collection of tools that can be effectively used in data manipulation.
- Besides built-in packages, it also provides an extensible programming environment that allows users to develop their own programs.
- As it is object-oriented language, it provides a faster way to program the real-life problems at lower development cost.
- It has command-based outstanding capability of producing high-quality graphs with various formats such as eps, jpg, png and pdf, etc.
- It has very nice import/export facility that makes R compatible with other computer applications such as csv, text, SAS and SPSS, etc.
- It has a very rich collection of packages. It crossed over 10,000 packages for statistical data analysis.
- Researcher, scientists, and programmers are welcomed to contribute to the development of the R. One can develop the R packages and share them worldwide. This is one of the reasons for fast growth and popularity of R.

3. Operators in R

Once R is opened with a prompts symbol ">" for input, the console window screen is ready for direct arithmetic and algebraic calculations without caring of hard and fast programming concepts. Thus R can be used as a calculator. Similar to other programming languages, R also has well-defined arithmetic, logical, and relational operators. The operators are illustrated here.

Arithmetic operators:

Operator	Symbol	Task	Code/Output
+	Plus	Addition	>2+3 is 5
–	Dash	Subtraction	>3-2 is 1
*	Asterisk	Multiplication	>3/2 is 1.5
/	Slash	Division	>2*3 is 6
^	Cap	Exponentiation (2^3)	>2**3 is 8
**	Double asterisk	Exponentiation (2^3)	>2^3 is 8
%%	Double percent	Modulus	>3%%2 is 1
%/%	Percent slash percent	Integer division	>7%/%3 is 2

Logical operators: The logical operator returns either TRUE or FALSE as the output.

Operator	Symbol	Code/Output
<	Less than	>x=5; y=10; >(x<y) is TRUE
<=	Less than equal to	>(x<=y) is TRUE
>	Greater than	>(x>y) is FALSE
>=	Greater than equal to	>(x>=y) is FALSE
==	Equal to	>(x==y) is FALSE
!=	Not equal to	>(x!=y) is TRUE
!	Logical NOT	>!(x>y) is TRUE
||	Logical OR	>(x > 5 || y < 15) is TRUE
|	Logical OR	Used for vectors
&&	Logical AND	>(x == 3 && x <= 10) is FALSE
&	Logical AND	Used for vectors
isTRUE(expression)	Test if expression is true	isTRUE(x == 3) returns FALSE

Also note the following,

- semicolon (;) is used to terminate the statement,
- hash (#) at the beginning of each line serves as the comment line operator,
- symbols "=" and "<-" are used as the assignment operator.

3.1 Data entry in R

There are various built-in functions in R that can be used for generating the sequences of numbers and entering data points into R workspace. The general syntaxes of the commands are described as follows.

1. `a:b`: it generates a sequence of integers starting from a to b, e.g., `1:5` gives 1, 2, 3, 4, 5.
2. `c(arg)`: it is the most basic command to enter data into R and creates a vector of `arg`, a list of elements separated by commas, for example, `c(2,5,3,1,9)` returns a vector consisting 2, 5, 3, 1, 9 as of its elements.
3. `seq(from=a, to=b, by=c)` or `seq(from=a, to=b, length=d)`: it generates a sequence of real numbers from a to b with consecutive distance c or length d.
4. `data.frame(arg)`: it helps to arrange the data in a table form having variables in columns and respective entries in rows. Here `arg` is a list of vectors/variables separated by commas. It is worth noting that each variable must have same length of entries. Otherwise, the entries of the variable having lesser numbers are recycled. In such cases, we recommend to use `list()` command that coerce to store the variable of different lengths and types in a single object.
5. `matrix(x,ncol=n,nrow=m,byrow=TRUE)`: it is used to enter the data in the $m \times n$ matrix form. The matrix transpose, multiplication, and inverse can be performed by the following built-in functions, `t()`, `%*%`, and `solve()`, respectively.

For more details, readers are advised to read the help page of each commends. We can extract the data, from data structures discussed above, in the following ways:

- x[i] extracts the ith element of vector x.
- Dframe$x extracts the column named x of the data frame, Dframe.
- Dframe[,i] extracts the ith column of the data frame, Dframe.
- Mat[i,j] extracts the element of the (i, j)th cell of the matrix, Mat.
- Mat[,j] extracts the jth column of the matrix, Mat.
- Mat[i,] extracts the ith row of the matrix, Mat.

4. Programming with user-defined functions

In R language, we can create our own functions like the built-in functions such as mean(), matrix(), and dexp(), etc., supplied as part of the R system. The users can define function on their own using function() command and the function thus obtained is referred as user-defined function. The general syntax of function() is given by

```
>func_name=function(arg){statements
return(statement)
}
```

Here func_name is the function name to be given by the programmer, arg is the list of arguments separated by comma, and statements include the R expressions to be evaluated at arg. The function can be called anywhere in R environment as

```
>func_name(arg)
```

and it returns the value of statement written within return() command.

4.1 Remarks

- The arguments are dummy/local and outside the function they have no meaning and values. The order in which they are written is important while evaluating the functions for given values of arguments.
- There is no need of using {} and return() command if the function has a single statement.
- The arg may be a vector valued.

Let us define the density function of the exponential distribution $f(x; \theta, \mu) = \theta e^{-\theta(x-\mu)}, \theta > 0, x \geq \mu > 0$. There are many ways to define R function depending on how the arguments are supplied. Three possible programs to define the exponential density are given below.

```
# All arguments separated by commas.
>f=function(x,th,mu) th * exp(-th*(x-mu))
>f(3,1.5,2)
[1] 0.3346952    # Value of the density function as output

# Both parameters as vector and x is supplied separately.
>f=function(x,TH){TH[1] * exp(-TH[1]*(x-TH[2]))}
>f(3,c(1.5,2))
[1] 0.3346952
```

```
# Arguments (parameters) are provided as a single vector.
>f=function(par){
x=par[1];th=par[2];mu=par[3]
d=th * exp(-th*(x-mu))
return(d)
}
>x=3;th=1.5;mu=2    # Values of the parameters and x.
>f(c(x,th,mu))
[1] 0.3346952
```

The built-in function `dexp()` is tailored in R to compute the exponential density and can be utilized as follows.

```
> x=3;mu=2;th=1.5
> dexp(x-mu,rate=th)
[1] 0.3346952
```

which gives the output same as produced by the user-defined functions.

5. Loops and if/else statements in R

In R programming, the order of statements in which they are written is very important. A program is executed statement by statement (line after line from up to down) in a sequence. This flow of execution may be considered as the normal flow. But in some cases, such execution flow can be restricted because of some conditions. Similar to other programming languages, R has strong tools to control the logical/conditional flow of the program. The `for()` and `while()` are two important loops in R.

The general syntaxes of `for()` and `while()` loops are given by

```
for(index in vector){
statement(s) evaluated at index
}
```

```
while(logical expression){
statement(s)
}
```

The `for()` loop executes the statement(s) (within braces) repeatedly for all possible values of index in vector. For example,

```
>for(k in 1:5) print(k)
```

returns 1, 2, 3, 4, and 5 as output in the respective iterations for k in Becker and Chambers (1984); Zuur et al. (2009). We can compute the sum of a series of numbers using the `for()` loop as follows.

```
SUM=function(y){
     ADD=0; n=length(y)
     for(i in 1:n){
     ADD = ADD + y[i]
     }
  return(ADD)}
>y=1:10
>SUM(y)          # user defined function
[1] 55
>sum(y)          # Built-in function
[1] 55
```

In the `while()` loop, the executions of the statement(s) (within braces) continue repeatedly till the condition (`logical expression`) results as TRUE. One can see the program for computing factorial using the `while()` loop for the understanding of its functioning.

```
FACT=function(N){
     j=N; P=1
     while(j>= 1){
     P = P *j
     j=j-1
     }
return(P) }
>FACT(10)               # user defined function
[1] 3628800
>factorial(10)          # Built-in function
[1] 3628800
```

R features the implementation of selection/conditional structures that enable us to introduce decision point in a program at which decision is made during the program execution to follow one of the several actions. The general syntaxes of if and if/else statements are given by

```
1.  if(logical_condition-1){Statement(s)}
2.  if(logical_condition-2){Statement-1
                }else {Statement-2}
```

The if() statement executes the statement(s) within the braces only if logical_condition-1 returns TRUE. In if/else statement, statement-1 is executed only if logical_condition-2 is TRUE, otherwise statement-2 is executed. Program (named as MAX) to find a maximum from three numbers.

```
MAX=function(x,y,z){
max=0
if(x>y){if(x>z){max=x}else{max=z}
}else{ if(y>z){max=y}else{max=z}}
return(max)
}
>MAX(30,10,20)
[1] 30
```

6. Curve plotting in R

Graphical study of the data is also an important component of statistical analysis. R is capable of producing wide variety of graphs and flexible enough to create customized plots. The graphical commands are categorized into three groups.

High-level plotting functions: These functions are used to create a new plot on R graphics window and automatically generate appropriate axes, titles, and labels that we can customize later. The main function tailored for this purposed is plot(), which produces a wide varieties of plots (scattered plot, curve, bar diagram, etc.) depending on the input parameters and data types. One may also use curve() for sketching the plot of a function. Other functions such as hist(), pie(), contour (), and persp() are also useful to draw histogram, pie chart, contour plot, and 3D plots, respectively.

Low-level plotting functions: These commands are used to add more information to an existing graph generated from high-level plotting functions, such as extra points, lines, text, legend, and mathematical annotation, etc. Some important functions for low-level plotting are as follows: lines(), points(), text(), mtext(), axis(), and legend().

Interactive graphics functions: These codes allow us to interactively add information or extract information using mouse, a pointing device. It is very limited in R. See the use of `locator()` function.

In addition to that, the `par()` function includes a list of graphical parameters such as margin of the graph, line width, number of plots in a figure, colors of plots, etc., and is very useful to customize the whole graphical device. Along with some important parameters, the `par()` function is written as

```
>par(mfrow, mar, lwd, col, las=1)
```

where `mfrow` is a vector of length 2 that represents the number of row and column of graphical window, respectively, `mar` is a vector of length 4 representing the margins of bottom, left, top, and right sides of the graph, respectively, `lwd` is a line width, `col` is a string that defines the color of the figure, and `las` in 0,1,2,3 sets the style of axis labels.

To demonstrate R graphics, we plot the density function of the GILD and histogram of the simulated samples from the distribution. Sharma et al. (2016) proposed two methods, based on mixture distributions and inverse cumulative distribution function (cdf), for generating random numbers from the GILD. They also provided R codes for the inverse cdf method where quantile equation is solved using Newton's iterative procedure. Note from Sharma et al. (2016) that the GILD is a mixture of power-transformed inverse gamma and inverse Weibull distributions. Here we provide R codes for the random number generation from the GILD using the definition of the mixture distributions. R codes for this method are given below.

```
set.seed(123)
n=10;alpha=2;theta=2              # Sample size and parameters values
U=runif(n);                       # Uniform random number
p=theta/(1+theta)                 # Mixing proportion
Y=rgamma(n,shape=2,scale=1/theta) # Gamma random sample
V=(1/Y)^(1/alpha)            # Power transformed inverse gamma random sample
W=(-theta/(log(U)))^(1/alpha)     # Inverse Weibull random sample
x=c()
for(i in 1:n){                    # Loop for each sample with mixing proportion
if(runif(1)<=p){x[i]=W[i]
}else {x[i]=V[i]}
}
>print(x)                         # GILD random sample of size 10
 [1] 0.6791668 2.8996470 1.4956252 3.9163006 5.7083049 0.8046740 1.7698987
 [8] 4.1918594 1.8330408 1.5972777
```

The above methods are now tailored in "LindleyR" package of Mazucheli et al. (Mazucheli et al., 2016). One can use the built-in codes in "LindleyR" for generating the random numbers from various families of distributions. The codes `rgenilindley()` and `dgenilindley()` are defined to generate random numbers and

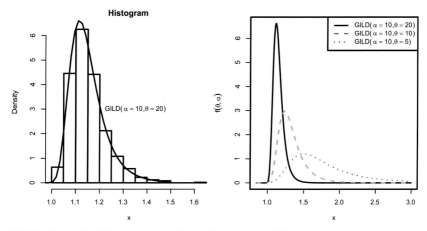

FIGURE 11.3 (A) Histogram of the simulated sample and (B) pdf of the generalized inverse Lindley distribution (GILD).

define the pdf of the GILD, respectively. We use these built-in functions to draw the density curves of the distribution. The R codes are given by

```
library(LindleyR)
set.seed(123)
x <- rgenilindley(n = 1000, theta = 10, alpha = 20) # Generate random sample
par(mfrow=c(1,2),mar=c(4,4,1,1),lwd=2,cex=.7)   # Set graphical parameters
hist(x, prob = TRUE, main = 'Histogram',ylim=c(0,6.5))  # Plot histogram
curve(dgenilindley(x, theta = 10, alpha = 20),0,1.5,add=TRUE)  # Plot pdf
    curve in existing plot
text(1.35, 3, expression(paste("GILD(",alpha==10,",",theta==20,")"))) # Add
    text
S=seq(0,3,length=10^3)   # A sequence of numbers from 0 to 3
                          # Create a new plot of the GILD pdf
plot(S, dgenilindley(S, theta = 10, alpha = 20), type="l", xlab = 'x',
                ylab = expression(paste(f(theta,alpha))))
lines(S, dgenilindley(S, theta = 10, alpha = 10), lty=2,col="green")
lines(S, dgenilindley(S, theta = 10, alpha = 5), lty=3,col="purple")
legend("topright",          # To add a legend in existing plot
c(expression(paste("GILD(",alpha==10,",",theta==20,")")),
expression(paste("GILD(",alpha==10,",",theta==10,")")),
expression(paste("GILD(",alpha==10,",",theta==5,")"))
),lty=1:3,col=c("black", "green", "purple"))
```

Fig. 11.3 is obtained as an output of the codes above.

7. Maximum likelihood estimation

Once an appropriate model, indexed by some parameters, is selected for a particular inferential problem, the next objective is to analyze the model and draw relevant inferences. To specify the assumed model completely, one has to estimate the unknown parameters based on observed data and this is known as problem of point estimation. Information about θ is gathered from the randomness of the sample observations and it is obtained through the joint probability distribution, say $f_X(\underline{x}; \theta)$, where $\underline{x} = \{x_1, x_2, \ldots, x_n\}$. In practice, we are provided with a sample and wish to draw inferences about the unknown parameter. Thus we treat $f_X(\underline{x}; \theta)$ as function of the parameter for given sample x. To utilize the above idea, we define a function of θ for given observed data $\underline{x}$ as

$$l(\theta|\underline{x}) = f_X(\underline{x}; \theta) = \prod_{i=1}^{n} f_X(x_i; \theta) \qquad (11.2)$$

which is called the *likelihood function* for n independent and identically distributed (iid) observations. The likelihood function can easily be used to compare the plausibility of various parameter values. For example, $l\left(\widehat{\theta}_1\Big|\underline{x}\right) = 2l\left(\widehat{\theta}_2\Big|\underline{x}\right)$, then, in some sense, $\widehat{\theta}_1$ is twice as plausible as $\widehat{\theta}_2$. Therefore, the desired estimate $\widehat{\theta}$ of θ must satisfy

$$\begin{cases} l\left(\widehat{\theta}\Big|\underline{x}\right) \geq l(\theta|\underline{x}), \text{for all } \theta \in \Theta, \\ i.e., \ l\left(\widehat{\theta}\Big|\underline{x}\right) = \sup_{\theta \in \Theta} \{l(\theta|\underline{x})\} \end{cases}$$

Such a procedure, of obtaining the point estimate of the model parameter(s), is called the maximum likelihood estimation and the estimate thus obtained is called the maximum likelihood estimate (MLE). The MLE is the most popular and widely used method because it holds the desired properties of an estimator such as consistency, invariance, and asymptotic properties.

Besides point estimation, one may be interested in finding a set of values, say $A(\theta)$, such that $A(\theta)$ contains the true value of the parameter θ with a certain high probability $(1 - \phi), \phi \in (0, 1)$. For statistics, $T_1 = t_1(X_1, X_2, \ldots, X_n)$ and $T_2 = t_2(X_1, X_2, \ldots X_n), T_1 \leq T_2$, define

$$P_\theta[T_1 \leq \theta \leq T_2] = 1 - \phi \ \forall \ \theta \in \Theta \qquad (11.3)$$

where $(1 - \phi)$ is called coverage probability. Then the random interval (T_1, T_2) is called the $100 \times (1 - \phi)\%$ confidence interval (CI) for θ. Exact CIs are generally not available for some distributions, especially in case of multiparameter model. In such case, the asymptotic distribution property of MLE is very useful in constructing asymptotic CIs and this is one of the reasons for the popularity of likelihood principle. The asymptotic CIs are defined as

$$\left\{ \widehat{\theta} - Z_{\phi/2}\sqrt{var\left(\widehat{\theta}\right)}, \widehat{\theta} + Z_{\phi/2}\sqrt{var\left(\widehat{\theta}\right)} \right\}$$

where $Z_{\phi/2}$ is the $(\phi/2)$th upper percentile of a standard normal variables, and $var\left(\widehat{\theta}\right)$ is the asymptotic variance of θ and is obtained as diagonal elements of inverse Fisher's information matrix. Suppose there are k parameters $\theta = (\theta_1, \theta_2, ..., \theta_k)$, then variance−covariance matrix, say Σ, is defined by

$$\Sigma(\theta) = \begin{bmatrix} \delta_{11} & \delta_{12} & \cdots \delta_{1k} \\ \delta_{21} & \delta_{22} & \cdots \delta_{2k} \\ \vdots & \vdots & \vdots \\ \delta_{k1} & \delta_{k2} & \cdots \delta_{kk} \end{bmatrix}^{-1}_{\theta = \hat{\theta}}$$

where $\delta_{ij} = -\frac{\partial^2}{\partial \theta_i, \partial \theta_j} \log l(i, j) = 1, 2, ..., k$.

7.1 Maximum likelihood estimate for censored data

In the previous section, we discussed the maximum likelihood estimation method for the case of complete sample of observations. In some situations, we are unable to observe the lifetimes of all items under study, rather for some individuals it is only known that the lifetime (T, say) is greater or lesser than some value. The observations thus obtained are termed as censored observations. For instance, suppose in a clinical study, a patient moves to another clinic/hospital for treatment and can no longer be treated under the same study. The only information about the survival of that patient is the last date on which the patient reported to a clinic for a regular checkup and was known to be alive. An actual survival time of patient can also be regarded as censored when death is from a cause that is known to be unrelated to the treatment. For example, some patients suffering from cancer may die from other causes such as heart attack, high blood pressure, and failure of some other parts of the body. The concept of censoring leads to the most important developments in survival analysis; for details see the books Lawless (1982); Nelson (1982), which adduced the techniques for survival data analysis. There are various categories of right censoring, such as time censoring (type-I censoring), failure censoring (type-II censoring), and progressive censoring; see Balakrishnan and Aggrawalla (2000) and Balakrishnan and Cramer (2014).

Consider a general situation of the censoring such that $T_1, T_2, ..., T_n$ denote the lifetimes of n individuals. It is further assumed that the lifetimes are iid with pdf, $T_i \sim f_X(\mathbf{x}; \theta), i = 1, 2, ..., n$. Define an indicator function $\Delta i = I(T_i = t_i)$ such that $\Delta_i = 1$ when T_i is completely observed, $T_i = t_i$ and $\Delta_i = 0$ if T_i is censored, $T_i > t_i$. Thus the data are observed of the form (t_i, Δ_i), $i = 1, 2, ..., n$, where t_i is a censored time or lifetime of ith item. The likelihood function under such data takes the following form

$$l(\theta | \underline{\mathbf{x}}) = \prod_{i=1}^{n} (f_X(x_i; \theta))^{\Delta_i} (1 - F_X(x_i; \theta))^{1-\Delta_i} \tag{11.4}$$

The likelihood function (11.4) includes various cases of right censoring as one of the special cases, e.g., Type-I and Type-II censoring. Similarly, the construction of the likelihood under other censoring schemes is straightforward and left to exercise to the readers.

7.2 The GILD likelihoods and survival estimates

GILD likelihood under complete case: Let $X_1, X_2, \ldots, X_n$ be n iid random observations that represent the lifetimes of individuals and follow the GILD. Based on observed lifetimes, say $\underline{x} = \{x_1, x_2, \ldots, x_n\}$, the log-likelihood function (11.2) is given by

$$\log l(\theta, \alpha | \underline{x}) = n \ln(\alpha) + 2n \ln(\theta) - n \ln(1 + \theta) + \sum_{i=1}^{n} \ln\left(1 + x_i^\alpha\right)$$

$$-(2\alpha + 1) \sum_{i=1}^{n} \ln(x_i) - \theta \sum_{i=1}^{n} x_i^{-\alpha} \tag{11.5}$$

GILD likelihood under censored case: Suppose the life of r items is completely observed and the lifetimes of the remaining $(n - r)$ are censored. Let Δ be an indicator function and ith observation is censored if $\Delta_i = 0$, otherwise $\Delta_i = 1$. Assuming lifetimes follow the GILD in (11.1), the log-likelihood function (11.4) under censoring can be readily obtained as

$$\log l(\theta, \alpha | \underline{x}_r) = r \ln(\alpha) + 2r \ln(\theta) - r \ln(1 + \theta)$$

$$+ \sum_{i=1}^{n} \Delta_i \ln\left(1 + x_i^\alpha\right) - \theta \sum_{i=1}^{n} \Delta_i x_i^{-\alpha}$$

$$-(2\alpha + 1) \sum_{i=1}^{n} \Delta_i \ln(x_i) + \sum_{i=1}^{n} (1 - \Delta_i) \ln\left(1 - \left(1 + \frac{\theta x_i^{-\alpha}}{1 + \theta}\right) e^{-\theta x_i^{-\alpha}}\right)$$

$$\tag{11.6}$$

Eqs. (11.5) and (11.6) are solved numerically to obtain the estimates of the parameters. Suppose $\widehat{\theta}$ and $\widehat{\alpha}$ denote the MLEs of θ and α, respectively. The MLEs of the pdf and distribution function of the GILD are

$$\widehat{f}(x) = \frac{\widehat{\alpha}\widehat{\theta}^2}{(1+\widehat{\theta})} \left[\frac{1 + x^{\widehat{\alpha}}}{x^{2\widehat{\alpha}+1}}\right] e^{-\frac{\widehat{\theta}}{x^{\widehat{\alpha}}}} \text{ and } \widehat{F}(x) = \left[1 + \frac{\widehat{\theta} x^{-\widehat{\alpha}}}{(1+\widehat{\theta})}\right] e^{-\frac{\widehat{\theta}}{x^{\widehat{\alpha}}}}, \text{ respectively. The MLE}$$

of the reliability/survival is obtained as $\widehat{S}(x) = 1 - \widehat{F}(x)$. Once we have these estimated/fitted functions for the data, it is obvious to compare them with their nonparametric versions. Nonparametric estimate of the distribution function, in case of complete sample of ordered observations, $\left\{x_{(1)}, x_{(2)}, \dots x_{(n)}\right\}$, is defined by

$$\widehat{F}(x) = \left\{\frac{\text{Number of observation} \leq x}{n}\right\} = \begin{cases} 0 & \text{if } x < x_{(1)}, \\ \dfrac{k}{n} & \text{if } x_{(k)} \leq x < x_{(k+1)}, \\ 1 & \text{if } x \geq x_{(n)} \end{cases} \quad (11.7)$$

where $k = 1, 2, \dots, (n-1)$. If all observations are distinct, this is a step function that increases by $1/n$ just after each observed lifetime.

We now present nonparametric estimate of survival investigated by (Kaplan and Meier 1958) that allows the censored observations in study. The Kaplan and Meier estimate $\widehat{S}_{KP}(x)$ of survival function is given by

$$\widehat{S}_{KP}(x) = \prod_{j:x_{(j)} < x} \left(1 - \frac{d_j}{n_j}\right), \quad (11.8)$$

where n_j is the number of individuals at risk at $x_{(j)}$ and $d_{(j)}$ is the number of deaths at $x_{(j)}$.

8. Lifetime data modeling

8.1 Complete case: maximum flood levels data

In this subsection, we present R codes for fitting the complete sample of observations representing maximum flood levels (in millions of cubic feet per second) for the Susquehanna River, Pennsylvania, from 1890 to 1969. The data set consists of the maximum flood level. The data set was first discussed by Dumonceaux and Antle (1973) and they have proposed the use of lognormal over the Weibull distribution for this data set.

Upadhyay and Peshwani (2003) performed discrimination analysis between lognormal and Weibull models under Bayesian setup and showed that lognormal distribution gives a better fitting for the data set than the Weibull distribution while stating that the data set has unimodel failure rate function. Estimation of the parameters of inverse Weibull distribution is discussed by Maswadah (2010) for this data set. Singh et al. (2013) also accessed the goodness of fit of inverse Weibull distribution for the data set and compare the fitting results with lognormal, Weibull, gamma, and flexible Weibull distributions. Recently, Sharma et al. (2016) proposed the use of the GILD for modeling this data set. They compared the performance of the GILD with generalized inverse exponential, inverse Gaussian, inverse gamma, and inverse Weibull distributions. According to KS and Akaike information criterion

(AIC), the GILD was found to be a better model among others. We show how to do such modeling in R. The summary(x) function is used for computing descriptive measures of the given data, x. To obtain coefficients of kurtosis and skewness, we need to load Time Series Analysis (TSA) package available at https://CRAN.R-project.org/package=TSA. The descriptive statistics along with respective R codes are given by

```
# Enter the observations using c()
x=c(0.654, 0.613, 0.315, 0.449, 0.297, 0.402, 0.379, 0.423, 0.379, 0.324,
    0.269, 0.740, 0.418, 0.412, 0.494, 0.416, 0.338, 0.392, 0.484, 0.265)
> n=length(x); n          # Number of observations
[1] 20
> summary(x)
   Min. 1st Qu.  Median    Mean 3rd Qu.    Max.
 0.2650  0.3345  0.4070  0.4232  0.4578  0.7400
> sd(x)
[1] 0.1252582
> library(TSA)            # Load TSA package
> skewness(x)            # Coefficient of skewness
[1] 1.067651
> 3-kurtosis(x)          # Coefficient of kurtosis
[1] 2.400018
```

Under complete case, the MLE for the GILD can be performed by using the following R codes.

```
LogLike=function(th){                # Define Negative Log-likelihood function
alpha=th[1];theta=th[2]
zz=n*log(alpha)+2*n*log(theta)-n*log(1+theta)+sum(log(1+x^alpha))-(2*alpha+1)*
    sum(log(x))-theta*sum(x^(-alpha))
return(-zz)
}
MLE=nlm(LogLike,c(1,.01),hessian=T) # nlm() for getting MLEs
```

The nlm() returns the following objects as an output.

```
print(MLE)
$minimum                          # Optimum values of the negative log-likelihood.
[1] -16.14752
$estimate                         # Estimates of the parameters
[1] 3.0766661 0.0898681
$gradient                         # Gradient score
[1] -9.930664e-08 -1.634248e-07
$hessian                          # Hessian Matrix
            [,1]       [,2]
[1,]    47.67513   460.1236
[2,]   460.12361  4924.9421
$code                             # Convergence code of the nlm
[1] 1
$iterations                       # Number of iterations to reach optimum value
[1] 28
```

The MLEs of the GILD parameters are $\widehat{\alpha} = 3.0766661$ and $\widehat{\theta} = 0.0898681$. The 95% asymptotic CIs are obtained as follows.

```
Z=diag(solve(MLE$hessian))  # diagonal elements of variance co-variance matrix.
Z
[1] 0.213353028 0.002065331
MLE$estimate[1]+c(-1,1)*1.965*sqrt(Z[1])    # CI of shape parameter
[1] 2.169030 3.984303
MLE$estimate[2]+c(-1,1)*1.965*sqrt(Z[2])    # CI of scale parameter
[1] 0.0005668925 0.1791693164
```

The performance of the two models can be accessed and compared using the likelihood ratio (LR) test. Comparing a model with its subclass is of importance as it may prove the significance of producing generalized case. The hypothesis can be stated as $H_0 : \alpha = 1$ (data follow ILD) versus $H_1 : \alpha \neq 1$ (data follow GILD), which can access the performance of the shape parameter added to the GILD over the inverse Lindley distribution. The test statistic,

$\xi = -2\left(\frac{\log(L_0)}{\log(L_1)}\right)$, where L_1 and L_0 denote the likelihood functions under H_1 and

H_0, respectively, can be used to test H_0 against H_1. The asymptotic distribution of ξ is χ_k^2, where k is the degree of freedom and denotes the number of parameters to be involved in H_0. We reject H_0 if $\xi > \chi_k^2(\gamma)$, 100% quantile of the χ_k^2 distribution. The MLE of the inverse Lindley distribution (ILD) parameter is obtained by

```
LogLike0=function(th){          # Negative Log-likelihood of ILD
alpha=1;theta=th[1]
zz=n*log(alpha)+2*n*log(theta)-n*log(1+theta)+sum(log(1+x^alpha))-(2*alpha+1)*
    sum(log(x))-theta*sum(x^(-alpha))
return(-zz)
}
MLE0=nlm(LogLike0,c(1),hessian=T)
MLE0$estimate                   # MLE of ILD parameter
[1] 0.6344513
```

The LR statistic along with corresponding *P*-value (PV) is computed as follows.

```
L0=-MLE0$minimum                # ILD log-likelihood
L1=-MLE$minimum                 # GILD log-likelihood
Del=-2*(L0-L1);print(Del)       # LR test statistic
[1] 31.12417
PV=1-pchisq(Del, df=1);print(PV)  # P-value
[1] 2.420391e-08
```

As $\xi = 31.12417$ with $PV = 2.420391 \times 10^{-8}$, it does not support H_0 because PV is less than the level of significance $\gamma = 5\%$ The appropriateness of both the distribution can also be compared on the basis of AIC (AIC $= -2\log(l) + 2p$), while the number of parameters involved are p.

The AIC values for the ILD and GILD can be obtained as 2*MLE0$minimu+2*1 $= 0{:}8291346$ and 2*MLE $minimu+2*2= -28.29503, respectively. The GILD shows minimum AIC value than the ILD. Hence, the GILD is a better model than ILD as it was expected. Similarly, the estimation for

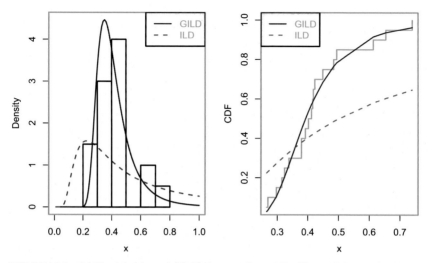

FIGURE 11.4 (A) Fitted density and (B) CDF curves of generalized inverse Lindley distribution (GILD) and ILD for flood-level data.

other competing models can be performed and compared with each other. For more details on comparing results, readers may be referred to Sharma et al. (2016).

Using the R codes given in Section 6, the fitted density and cumulative distribution curves can be easily plotted. Unlike Section 6, we define the pdf and cdf using function() command and then plot the curves in Fig. 11.4.

```
pdf=function(x,alpha,theta){((alpha*theta^2)/(1+theta))*((1+x^alpha)/(x^(2*
    alpha+1)))*exp(-theta/x^(alpha))}
cdf=function(x,alpha,theta){(1+(theta/((1+theta)*x^alpha)))*exp(-theta/x^(
    alpha))}
par(mfrow=c(1,2),mar=c(4,4,1,1),lwd=2,col="green")
curve(pdf(x,MLE$estimate[1],MLE$estimate[2]),0,1,lty=1,ylab="Density",col="
    blue")
curve(pdf(x,1,MLE0$estimate),0,1,lty=2,add=T,col="red")
hist(x,prob=TRUE,add=TRUE)
legend("topright", c("GILD","ILD"),lty=1:2,col=c("blue","red"))
Fn=c(1:n)/n                    # Empirical cdf estimate
plot(x,Fn,ylab="CDF",type="s")
lines(x,cdf(x,MLE$estimate[1],MLE$estimate[2]),lty=1,col="blue")
lines(x,cdf(x,1,MLE0$estimate),lty=2,col="red")
legend("topleft", c("GILD","ILD"),lty=1:2,col=c("blue","red"))
```

8.2 Censored case: head and neck cancer data

In this subsection, the modeling of a censored data is demonstrated along with R codes. The observations of the data denote the survival times of 45 patients suffering from HNC and treated using both radiotherapy and chemotherapy (RT + CT). Statistical analysis of the data was first performed by Efron (1988) using classical logistic regression model. It is also important to note that data are unimodal (upside-down bathtub-shaped) Sharma et al. (2014). Using lognormal distribution, Makkar et al. (2014) analyzed this data set under Bayesian paradigm. Sharma et al. (2015) and Sharma (2018) considered complete sample of observations only and fitted the data using the inverted Lindley and generalized inverted Lindley distributions, respectively.

The patients survival times (in days) were 37, 84, 92, 94, 110, 112, 119, 127, 130, 133, 140, 146, 155, 159, 169+, 173, 179, 194, 195, 209, 249, 281, 319, 339, 432, 469, 519, 528+, 547+, 613+, 633, 725, 759+, 817, 1092+, 1245+, 1331+, 1557, 1642+, 1771+, 1776, 1897+, 2023+, 2146+, 2297+, where + stands for censored observation.

First, we enter the data (x_i, δ_i), where x_i be the lifetime of ith item and δ_i takes 0 and 1 for censored and uncensored observations, respectively. The data can be easily entered using c() command.

```
Del=c(rep(1,14),0, rep(1,12), 0,0,0,1,1,0,1,0,0,1,0,0,1,0,0,0,0)
x=c(37, 84, 92, 94, 110, 112, 119, 127, 130, 133, 140, 146, 155, 159, 169,
173, 179, 194, 195, 209, 249, 281, 319, 339, 432, 469, 519, 528, 547, 613,
633, 725, 759, 817, 1092, 1245, 1331, 1557, 1642, 1771, 1776, 1897, 2023,
2146, 2297)
```

The maximum likelihood estimation can be performed using nlm() package as discussed in the last section. We note that out of 45 observations, 14 are censored. We define the negative log-likelihood as an R function of the GILD parameters as given below.

```
r=31;n=length(x)
LogL=function(th){
alp=th[1];theta=th[2]
Z1=r*log(alp)+2*r*log(theta)-r*log(1+theta)
Z2=sum(Del*log(1+x^alp))-theta*sum(Del*x^{-alp})-(2*alp+1)*sum(Del*log(x))
Z3=sum((1-Del)*log(1-((1+theta+theta*x^(-alp))/(1+theta))*exp(-theta*x^{-alp})
))
L=Z1+Z2+Z3
return(-L)
}
```

To set initial values of the parameters under optimization algorithm, we study the log-likelihood profile and trace the contour plots with varying values of the parameters. We take $\alpha = 0.10, 0.35, 0.60, 0.85, 1.10, 1.35, 1.60, 1.85$ and $\theta = 1, 11, 21, 31, \ldots 241$ arbitrarily and compute the negative log-likelihood values for the given set of parameters. For this purpose, we use nested for() loops as follows.

```
al=seq(0.1,2,by=.25)        # Values of the parameter alpha
th=seq(1,250,by=10)         # Values of the parameter theta
LogL_Mat=matrix(nrow = length(al), ncol = length(th))
for (i in 1:length(al)) {
    for (j in 1:length(th)) {
        LogL_Mat[i, j] = LogL(c(al[i], th[j]))
    }
}
```

The command `contour()` is useful for plotting contours of the given function. We need to supply the parameters values (`al`, `th` as above) and the corresponding functional values (`LogL_Mat`) and contour levels. As `contour()` creates a graphical environment, the other graphical parameters, such as line type, axis labels, colors of lines, etc., may be supplied as per need. For the GILD likelihood, the contour plot can be generated using the following R codes.

```
par(mar=c(4,4,1,1))
contour(al,th,LogL_Mat, levels =seq(100,1000,20),lty=1,xlab=expression(alpha),
col="purple", ylab=expression(theta))
```

Fig. 11.5 shows the contour plot of the log-likelihood function for the survival times. Note that the minimum level (height) of the log-likelihood is the maximum of log-likelihood. From Fig. 11.5, we can see that log-likelihood achieves its maximum at the level 240. Therefore, the possible values of the parameters α and θ may lie in (0.5,1) and (50,200), respectively. Although the optimization algorithm converses for any values of the parameters in these intervals, we supply $\alpha = 0.7$ and $\theta = 50$ in `nlm()` command as the initial guess. The `nlm(LogL,c(0.7,50))` is used as

```
MLE=nlm(LogL,c(0.7,50))
MLE$minimum
[1] 232.472
MLE$estimate
[1]  0.786959 80.613441
MLE$gradient
[1]  2.813749e-06 -8.109061e-09
```

The MLEs of α and θ are $\widehat{\alpha} = 0.786959$ and $\widehat{\theta} = 80.613441$, respectively. Now we plot the superimposed curve of the empirical and fitted survival functions. To obtain empirical estimate of the survival curve, one can use the

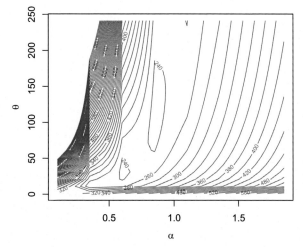

FIGURE 11.5 Contour plot of the generalized inverse Lindley distribution log-likelihood for head and neck cancer data.

routines (e.g., `survfit`) available in `survival` package (Therneau, 2015). For defining survival object, the `Surv` function, which combines survival times and censoring indicator, can be used. The `survfit` function returns a table having survival estimates, standard error, and confidence bounds. The codes are summarized below.

```
library(survival)
Model=survfit(Surv(time=x, event=Del)~1)
summary(Model)
Call: survfit(formula = Surv(time = x, event = Del) ~ 1)
```

time	n.risk	n.event	survival	std.err	lower 95% CI	upper 95% CI
37	45	1	0.978	0.0220	0.936	1.000
84	44	1	0.956	0.0307	0.897	1.000
92	43	1	0.933	0.0372	0.863	1.000
94	42	1	0.911	0.0424	0.832	0.998
110	41	1	0.889	0.0468	0.802	0.986
112	40	1	0.867	0.0507	0.773	0.972
119	39	1	0.844	0.0540	0.745	0.957
127	38	1	0.822	0.0570	0.718	0.942
130	37	1	0.800	0.0596	0.691	0.926
133	36	1	0.778	0.0620	0.665	0.909
140	35	1	0.756	0.0641	0.640	0.892
146	34	1	0.733	0.0659	0.615	0.875
155	33	1	0.711	0.0676	0.590	0.857
159	32	1	0.689	0.0690	0.566	0.838
173	30	1	0.666	0.0704	0.541	0.819
179	29	1	0.643	0.0716	0.517	0.800
194	28	1	0.620	0.0727	0.493	0.780
195	27	1	0.597	0.0735	0.469	0.760
209	26	1	0.574	0.0742	0.446	0.740
249	25	1	0.551	0.0747	0.423	0.719
281	24	1	0.528	0.0750	0.400	0.698
319	23	1	0.505	0.0752	0.377	0.676
339	22	1	0.482	0.0752	0.355	0.655
432	21	1	0.459	0.0750	0.333	0.633
469	20	1	0.436	0.0747	0.312	0.610
519	19	1	0.413	0.0742	0.291	0.588
633	15	1	0.386	0.0742	0.265	0.562
725	14	1	0.358	0.0739	0.239	0.537
817	12	1	0.328	0.0735	0.212	0.509
1557	8	1	0.287	0.0749	0.172	0.479
1776	5	1	0.230	0.0789	0.117	0.451

The following R codes can be used to draw survival curves, see Fig. 11.6.

```
par(mfrow=c(1,1),mar=c(4,4,1,1))
plot(Model,ylab="Survival",xlab="Time",col="red")
pdf=function(x,alpha,theta){((alpha*theta^2)/(1+theta))*((1+x^alpha)/(x^(2*
    alpha+1)))*exp(-theta/x^(alpha))}
cdf=function(x,alpha,theta){(1+(theta/((1+theta)*x^alpha)))*exp(-theta/x^(
    alpha))}
lines(x,1-cdf(x,MLE$estimate[1],MLE$estimate[2]),lty=1,lwd=2,col="blue")?
text(1300,.55,"95% upper bound",cex=.8)
text(1300,.15,"95% lower bound",cex=.8)
text(1300,.37,"K-M estimate",cex=.8)
arrows(350,.55,700,.8,length = 0.1,col="blue",code=2)
text(1000,.8,"GILD estimate",cex=.8)
```

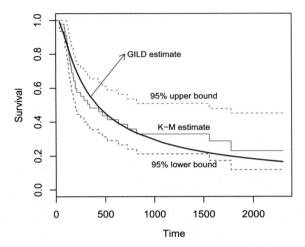

FIGURE 11.6 Empirical and estimated survival estimates for head and neck cancer data. *GILD*, generalized inverse Lindley distribution, *K−M*, Kaplan−Meier.

9. Conclusion

This chapter devoted to the implementation of R software for modeling practical lifetime data using probability distributions. The lifetimes refer to "times to failure" or "times to event" data that are continuous variables in range $(0, \infty)$. Therefore, the methods discussed here are also applicable to a wide variety of problems occurring in engineering, management, biomedical, and demography, etc. A brief introduction of R and statistical techniques are given with suitable illustrations so that scientists and engineers would not face any difficulty to use these tools for their study. R codes are suggested for data handling, curve plotting, loops, conditional statements, and data fitting. Step-by-step analyses along with R functions are performed for complete (maximum flood levels) and censored (survival of HNC patients) data sets. The GILD is used as the underlying probability model. For the sample generation and maximum likelihood estimation for the GILD, R codes are presented. Summing up, we can conclude that this chapter would serve as a guide to the scientists, engineers, and other practitioners to analyze real-life data sets using one of the important statistical software, R.

References

Balakrishnan, N., Cramer, E., 2014. The Art of Progressive Censoring: Applications to Reliability and Quality. Birkhauser, Boston.

Balakrishnan, N., Aggrawalla, R., 2000. Progressive Censoring: Theory, Methods and Applications. Birkhauser, Boston.

Becker, R.A., Chambers, J.M.S., 1984. An Interactive Environment for Data Analysis and Graphics. Wadsworth & Brooks/Cole, Pacific Grove, CA, USA.

Chambers, J.M., 1998. Programming with Data: A Guide to the S Language. Springer-Verlag, New York.

Crawley, M.J., 2013. The R Book. John Wiley & Sons, Ltd.

Dalgaard, P., 2008. Introductory Statistics with R, second ed. Springer, New York.

Dumonceaux, R., Antle, C., 1973. Discrimination between the lognormal and Weibull distribution. Technometrics 15, 923–926.

Efron, B., 1988. Logistic regression, survival analysis, and the Kaplan-Meier curve. Journal of the American Statistical Association 83 (402), 414–425.

Kaplan, E.L., Paul Meier, 1958. Nonparametric Estimation from Incomplete Observations. Journal of the American Statistical Association 53 (282), 457–481. http://www.jstor.org/stable/2281868.

Lawless, J., 1982. Statistical Models and Methods for Lifetime Data. John Wiley & Sons, New York.

Makkar, P., Srivastava, P.K., Singh, R.S., Upadhyay, S.K., 2014. Bayesian survival analysis of head and neck cancer data using lognormal model. Communications in Statistics – Theory and Methods 43, 392–407.

Maswadah, M., 2010. Conditional confidence interval estimation for the inverse weibull distribution based on censored generalized order statistics. Journal of Statistical Computation and Simulation 73, 887–898.

Matloff, N., 2011. The Art of R Programming: A Tour of Statistical Software Design. No Starch Press, Inc.

Mazucheli, J., Fernandes, L.B., Ricardo, P., de Oliveira, L.R., 2016. The Lindley Distribution and its Modifications. R Package Version 1.1.0. https://CRAN.Rproject.org/package=LindleyR.

Nelson, W., 1982. Applied Life Data Analysis. John Wiley & Sons, New York.

Ohri, A., 2012. R for Business Analytics. Springer, New York.

Peng, R.D., 2015. R Programming for Data Science. Leanpub.

R Core Team, R., 1996. A Language for Data Analysis and Graphics. R Foundation for Statistical Computing, Vienna, Austria. https://www.R-project.org/.

Rizzo, M.L., 2008. Statistical Computing with R. Chapman & Hall/CRC.

Ross, I., Robert G, R., 1996. A language for data analysis and graphics. Journal of Computational and Graphical Statistics 5 (3), 299–314.

Sharma, V.K., 2018. Bayesian analysis of head and neck cancer data using generalized inverse Lindley stress–strength reliability model. Communications in Statistics – Theory and Methods 47 (5), 1155–1180.

Sharma, V.K., Singh, S.K., Singh, U., 2014. A new upside-down bathtub shaped hazard rate model for survival data analysis. Applied Mathematics and Computation 239, 242–253.

Sharma, V.K., Singh, S.K., Singh, U., Agiwal, V., 2015. The inverse Lindley distribution: a stress-strength reliability model with application to head-and neck cancer data. Journal of Industrial and Production Engineering 32, 162–173.

Sharma, V.K., Singh, S.K., Singh, U., Merovci, F., 2016. The generalized inverse Lindley distribution: a new inverse statistical model for the study of upside-down bathtub data. Communications in Statistics – Theory and Methods 45 (19), 5709–5729.

Singh, S.K., Singh, U., Sharma, V.K., 2013. Bayesian prediction of future observations from inverse Weibull distribution based on Type-II hybrid censored sample. International Journal of Advanced Statistics and Probability 1, 32–43.

Therneau, T., 2015. A Package for Survival Analysis in S, Version 2.38. https://CRAN.Rproject.org/package=survival.

Upadhyay, S., Peshwani, M., 2003. Choice between weibull and log-normal models: a simulation based bayesian study. Communications in Statistics – Theory and Methods 32, 381–405.

Zuur, A., Ieno, E.N., Meesters, E., 2009. A Beginner's Guide to R, Use R! In: Gentleman, R., Hornik, K., Parmigiani, G. (Eds.).

Chapter 12

Probability-based approach for evaluating groundwater risk assessment in Sina basin, India

Thendiyath Roshni[1], Sourav Choudhary[1], Madan K. Jha[2], Nehar Mandal[1]

[1]*Department of Civil Engineering, National Institute of Technology Patna, Patna, Bihar, India;*
[2]*Agricultural and Food Engineering Department, Indian Institute of Technology Kharagpur, Kharagpur, West Bengal, India*

1. Introduction

Groundwater as a vital resource plays an important role for providing basic services to nature and mankind. As population is increasing day by day, this resource has to be properly managed so as to prevent it from getting extinct. This resource is under huge stress nowadays because of increasing dependency on this resource in the areas where surface water is scarce. The groundwater is being polluted more because of its excessive use and exposing risk on its quality and quantity factors. Hence, it is necessary to identify the risk factors of the groundwater resource and establish a reasonable risk analysis model to improve and control the present situation of groundwater resources. Groundwater risk assessment has gradually led to the groundwater pollution risk assessment and risk assessment of groundwater development and utilization. Thus, a methodology is adopted to calculate drought index and its risk factors.

This study mainly focuses on the methodology to create information for proactive identification of droughts. Any mathematical formulation of groundwater drought is associated with uncertainties and knowledge deficiency (e.g., overexploitation of groundwater). The precipitation is either high or low, droughts can occur. During high precipitation, drought can occur because of uneven distribution of temporal precipitation. Hence, to cope with this uncertainty, frequency analysis can be used to estimate the severity and duration of droughts for a return period of operational droughts (Mishra and Singh, 2010). The present study investigates the derivation of reliability analysis through a probability-based index to cope with uncertainties by using Monte Carlo sampling (Tanaka and Sugeno, 1992).

Handbook of Probabilistic Models. https://doi.org/10.1016/B978-0-12-816514-0.00012-6
289

Groundwater droughts refer to stored groundwater that is depleted without being sufficiently recharged; these are of the hydrological drought type but occur at a slower rate. The stages in groundwater droughts referred to as insufficient recharge of groundwater with time. This type of drought is called groundwater droughts, and its timescale is often years (Van Lanen and Peters, 2000; Sadeghfam et al., 2018). Groundwater drought is often due to lack of planning and overexploitation of aquifer resources, which accelerates groundwater droughts, and these together amplify their impacts.

For the present work, the study area is the Sina basin, where there is a serious need for water resource planning and management. The challenges for the study area include increasing water demand, overabstracting rivers, and wide spreading of aquifers. Participatory management plans are yet to be deployed in this particular basin. The present chapter takes up challenges on drought planning for increased resilience. Hence, the objective of the present study is to assess groundwater risk for mild, moderate, and severe droughts in the Sina basin using uncertainty analysis. Its data requirements are based on groundwater depth time series data (Kuo et al., 2007; Cancelliere et al., 2009; Low et al., 2011; Sadeghfam et al., 2018). Reliability analysis is framed in such a way that it accounts for safe operations of a system and also its failure as a function of load, which accounts for external actions and resistance, thereby accounting for system capacity (Tung et al., 2005). Load and resistance are of the same dimension, and a function of their differences provides an approach to measure safety or failure of the system, as elaborated later (Loucks, 1997; Bocchini et al., 2013).

2. Study area

The study area selected is the Sina basin (Fig. 12.1), a drought-prone area, located in Maharashtra, India, which is characterized under a semiarid region. The study area comprises latitude $17°28'N-19°16'N$ and longitude $74°28'E-76°7'E$. The total geographical area comprises 12,444 km^2 and average annual rainfall of 644 mm. The topography of the basin is 420–964 m above the mean sea level. The average maximum temperature is 40.5°C (May), and the average minimum temperature is 10.5°C (December). The Sina River, which is a large tributary of the Bhima River, flows through the basin. The basin consists of 133 observation wells which are uniformly distributed over the basin. The groundwater level data are collected from CGWB Pune for 133 wells for premonsoon and postmonsoon season for the period 1990–2009.

3. Methodology

Organizing the observed data or objects into meaningful groups with similar characteristics by maximizing the dissimilarity between the groups that were unknown initially is being carried out by the cluster analysis which is a self-explanatory data analysis tool. The application of clustering is seen in all fields such as segmentation of images to group the nodes in a network, to cluster the

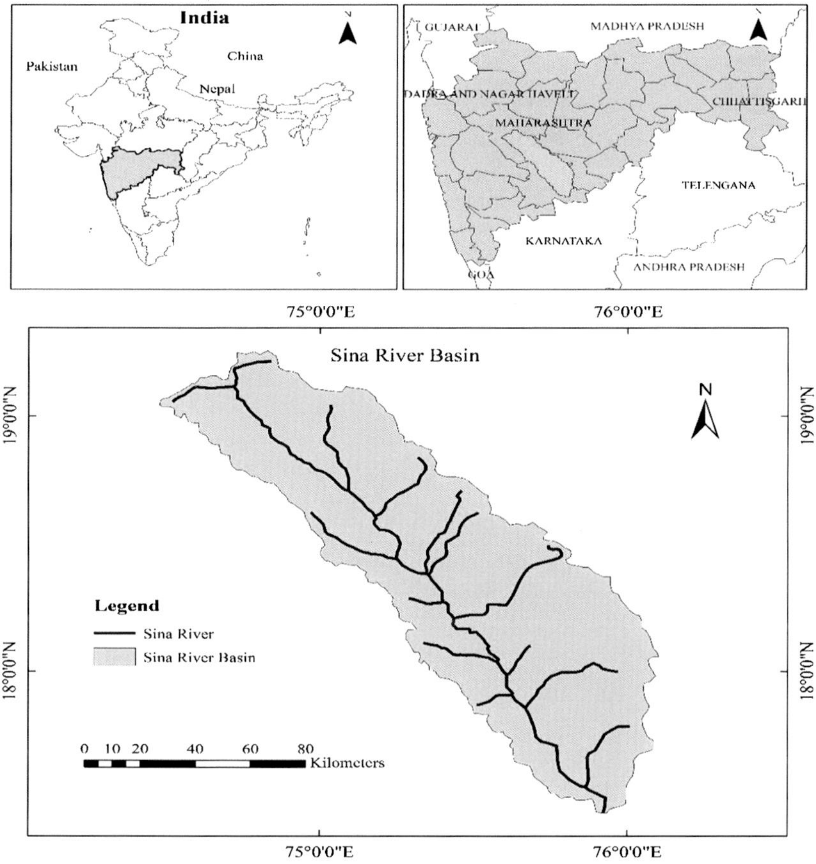

FIGURE 12.1 Study area map of the Sina basin.

documents in the web to enable efficient searching, etc. (Arthur and Vassil-vitskii, 2007). It is generally used as a preprocessing technique to improve the efficiency of computations. The clustering techniques are broadly categorized into partitioning methods, hierarchical methods, density based methods, model-based methods, grid-based methods, etc. To classify the well log site depending on its spatial and geological characteristics, a canopy technique is followed in the present study.

Canopy clustering (McCallum et al., 2000) acts as a preclustering technique to handle huge data sets. This simple and fast canopy clustering technique uses approximate distance metric and two threshold distance for processing, and its basic algorithm begins with the removal of points at random. The input data set and two thresholds T1 and T2 such that T1 > T2 is given as input to the clustering algorithm. Canopy is formed by a random point chosen from the given data set, and it is removed from the input data set.

The distance between each object with the original object in the canopy is determined, and the object is assigned to the canopy whenever the distance is less than T1 and removed from the input data set when the distance is less than T2. The process is repeated to form new canopies until the input data set is empty. This is a simple and efficient algorithm.

By canopy clustering technique, 133 wells of the Sina basin are grouped into 10 clusters. The groundwater levels in the imaginary wells of each 10 clusters are calculated by taking the average of the groundwater levels in the wells of the clusters. Hence, for the whole basin, 10 imaginary wells are selected and the averaged data are used for the drought analysis.

Groundwater drought is assessed by a drought index which is called the standardized water index (SWI). The SWI for different wells often shows scarce result because of limited data. Hence, random numbers were needed which were generated by random number simulation. The random number simulation was carried out by Monte Carlo simulation. For generating the random numbers, the ultimate value of generation was taken as the maximum obtained SWI values of different wells. The probability density function has been calculated by fitting normal or lognormal distribution curve. The area under the curve has been calculated for the given drought intensity ranges such as mild (0−1), moderate (1−1.5), severe (1.5−2), and extreme (>2).

The annual maxima of GWD for all the wells have been calculated, and their cumulative density function (CDF) has been plotted from year 1990−2009. The percentage of SWI values corresponding to different ranges such as mild, moderate, severe, and extreme is projected on the GWD annual maxima curve for obtaining the resistance. Resistance is the ability of the system to resist the stresses produced because of varying GWD values as loads.

The methodology consists of collecting the groundwater data from the past record. These data are required to note the groundwater fluctuation and have a standard reference level below which the groundwater data will impose drought conditions. The whole methodology has been grouped under the load module (Fig. 12.2), resistance module (Fig. 12.3), and the decision support system (Fig. 12.4).

The load system consists of getting the groundwater depth data for the past 20 years and calculating its SWI which is given by the following formula:

$$\text{SWI} = \frac{\text{GWD} - \text{GWD}*}{\text{S}*} \tag{12.1}$$

where GWD is the groundwater depth measured from the ground level, GWD* is the mean groundwater depth, and S* is the standard deviation of the groundwater depth data.

During calculation of SWI, the data obtained will be very scarce because negative SWI will not account for any drought. The negative SWI will indicate groundwater level above the base referenced level (mean groundwater depth level) which will not stress the groundwater aquifer resource. Having limited

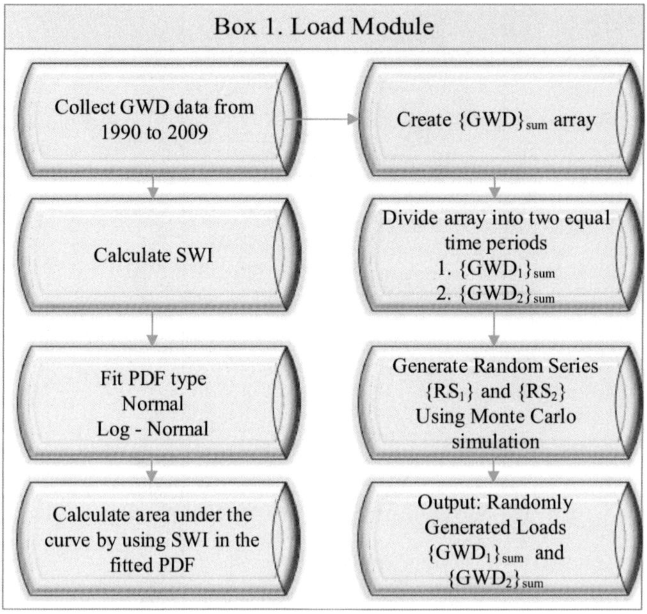

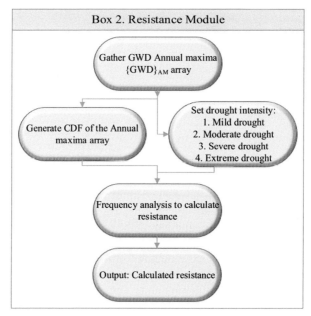

Box 1. Load Module

Collect GWD data from 1990 to 2009

Create $\{GWD\}_{sum}$ array

Calculate SWI

Divide array into two equal time periods
1. $\{GWD_1\}_{sum}$
2. $\{GWD_2\}_{sum}$

Fit PDF type
Normal
Log - Normal

Generate Random Series $\{RS_1\}$ and $\{RS_2\}$ Using Monte Carlo simulation

Calculate area under the curve by using SWI in the fitted PDF

Output: Randomly Generated Loads $\{GWD_1\}_{sum}$ and $\{GWD_2\}_{sum}$

FIGURE 12.2 Flowchart of the load module. *CDF*, cumulative density function; *PDF*, probability density function; *SWI*, standardized water index.

Box 2. Resistance Module

Gather GWD Annual maxima $\{GWD\}_{AM}$ array

Generate CDF of the Annual maxima array

Set drought intensity:
1. Mild drought
2. Moderate drought
3. Severe drought
4. Extreme drought

Frequency analysis to calculate resistance

Output: Calculated resistance

FIGURE 12.3 Flow chart of the resistance module.

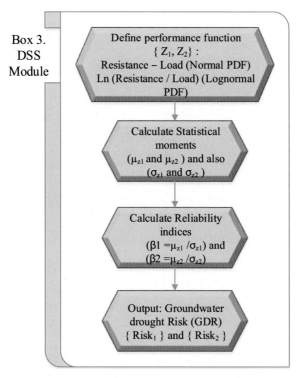

FIGURE 12.4 Flowchart of the decision support system.

the SWI, the positive value will hinder calculation for different drought bands such as mild (0−1), moderate (1−1.5), severe (1.5−2), and extreme (>2). Hence, a random number generator and a standard distribution of the generated number will be required.

The generation of random numbers by Monte Carlo simulation is given by the following equation:

$$X_{i+1} = \{aX_i + c\}(\bmod m) \quad i = 1, 2, ..., n \qquad (12.2)$$

where a is the multiplier, c is the increment, and m is an integer-valued modulus. The number of samples taken for simulation was 10,000. The value of a and c is taken according to the number of data for simulation. After the generation of random numbers by Monte Carlo simulation for 10,000 samples, a standard normal distribution of the generated data is plotted, and the area under each band mild (0−1), moderate (1−1.5), severe (1.5−2), and extreme (>2) is calculated. This calculated SWI will only account for the resistance which is the probability of exceedance of a given band.

Hence, for the calculation of randomly generated loads, an array of the sum of groundwater depth for the past years is formed and is denoted as $\{GWD\}_{sum}$ which will be accounted for defining the reliability throughout the study area. The random series array once generated is denoted as {RS}. For the calculation of groundwater drought risk, these random series are divided into two series as ${RS}_1$ and ${RS}_2$. The output of the randomly generated load is denoted as ${GWD_1}_{sum}$ and ${GWD_2}_{sum}$.

After the load module resistance module is considered which accounts for the system ability to resist the groundwater fluctuation, abstraction, etc., this module leads to the creation of annual maxima of groundwater depth array for all the representative wells. It is denoted as ${GWD}_{AM}$. A cumulative distribution of the annual maxima data is plotted, and the area under the curve is projected as a probability of exceedance to the annual maxima curve to get the resistance of the corresponding drought band (Sadeghfam et al., 2018).

The drought band intensity mild (0−1), moderate (1−1.5), severe (1.5−2), and extreme (>2) is set to predict its occurrence for the past selected years. Frequency analysis is processed to calculate the resistance based on the procedure of projection of SWI on the annual maxima curve. Hence, for a particular band, a resistance is obtained which accounts for the maximum depth to which a selected represented well will be stressed.

By the calculation of load and resistance, a decision support system module is followed which will calculate the groundwater drought risk (GDR) for the selected representative well.

The frequency analysis is performed for calculating resistance for different drought intensity ranges for different wells in the basin. Resistance is calculated as the probability of exceedance of the PDF curve. Thus, the exceedance value is projected on the GWD annual maxima curve to get the resistance value which will be the system extreme value up to failure. A performance function is defined which calculates the reliability indexes and GDR by considering load and resistance values and finally calculating the statistical moments.

The performance function is denoted by {Z} which is the difference between the resistance and the load for normal probability distribution and it is lognormal of resistance upon load for lognormal probability distribution. The statistical moments are calculated for the performance function ${Z_1}$ and ${Z_2}$. The reliability indices are given by

$$(\beta1 = \mu_{z1}/\sigma_{z1}) \text{ and } (\beta2 = \mu_{z2}/\sigma_{z2}) \tag{12.3}$$

where $\beta1$ and $\beta2$ are the reliability indices and μ_z and σ_z are the corresponding mean and standard deviation of the performance function, respectively. The GDR is denoted as {Risk} which can be calculated from ${Risk_1}$ and ${Risk_2}$ for two random series ${RS}_1$ and ${RS}_2$. The finally calculated risk is distributed by the inverse distance weighted (IDW) method throughout the basin.

4. Results and discussion

The groundwater level for premonsson and postmonsoon was obtained for 133 observation wells of the Sina basin. The distribution of 133 groundwater depth

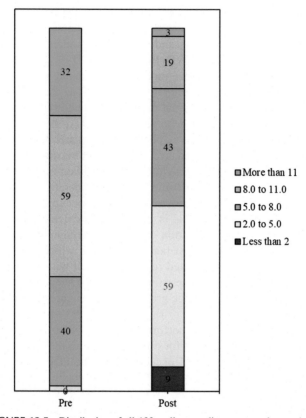

FIGURE 12.5 Distribution of all 133 wells according to groundwater depth.

data (in m) for premonsoon and postmonsoon season for the selected time period is shown in Fig. 12.5.

Clustering (Written and Frank, 2005; Han and Kamber, 2006) is the process of grouping the data based on similar features. In this principle, the intercluster similarity should be very low and the intracluster similarity should be very high. It is used to group document files, clustering nodes, and also segmentation images. To improve further computations, these clustering techniques can be used. In the present study, as the groundwater depth data are available for 133 wells, it is a tedious task to analyze it for each well separately. Hence, the whole data are grouped into 10 clusters, and the data are analyzed for representative wells of each cluster. All 133 wells were grouped using cluster analysis by canopy method, and imaginary wells were selected as a representative well which is tabulated (Table 12.1), and the number of wells in each cluster is also provided.

Fig. 12.6 represents the wells which are grouped into 10 clusters. All the 10 clusters are represented by different symbols. The properties of wells in the cluster will resemble the chosen representative imaginary well (IW).

TABLE 12.1 Imaginary well data for both premonsoon and postmonsoon.

Imaginary wells	Premonsoon				Postmonsoon				No. of wells in clusters
	1990 –1994	1995 –1999	2000 –2004	2005 –2009	1990 –1994	1995 –1999	2000 –2004	2005 –2009	
IW1	8.90	9.23	10.25	10.34	6.22	5.25	6.72	4.37	30
IW2	6.88	8.39	9.54	10.85	4.32	3.96	4.58	3.73	5
IW3	9.72	10.02	10.82	10.55	6.40	6.34	7.35	3.83	15
IW4	8.90	9.11	9.47	10.01	5.76	4.66	5.52	2.82	28
IW5	7.25	7.27	8.17	7.67	4.71	4.10	4.96	2.76	12
IW6	8.86	9.60	10.45	10.60	6.18	6.23	7.11	4.15	10
IW7	9.12	9.59	10.96	10.41	5.92	5.43	6.52	4.42	11
IW8	8.62	9.24	10.04	9.82	6.29	5.46	6.70	4.52	12
IW9	8.12	9.28	9.22	8.80	5.66	5.34	5.63	3.59	9
IW10	8.61	10.62	11.10	9.68	6.46	6.74	8.11	5.76	1
Total									133

IW, imaginary well.

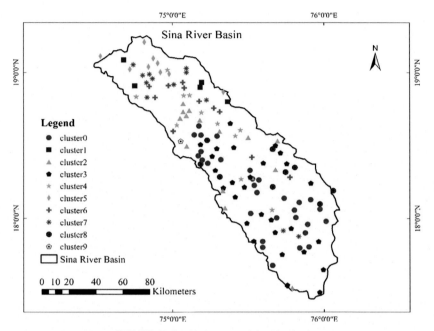

FIGURE 12.6 Cluster map of the Sina basin.

The uncertainty analysis by applying Monte Carlo sampling technique was carried out for all 10 IWs. The IW data are used for the calculation of annual maxima, which was useful for calculating resistance based on the frequency analysis. An SWI was calculated for each well to identify its drought band. A standard drought index was selected which was categorized as mild (0−1), moderate (1−1.5), severe (1.5−2), and extreme (>2). Based on the SWI values, the upper limit of the index was selected, and random numbers are generated and plotted by a standard normal distribution graph (Fig. 12.7). The simulation is carried out for 10,000 samples. The percentage of values under each band is calculated and is used to identify the resistance values for the cumulative distribution of the annual maxima curve.

The load is calculated from the best-fitted probability density function on the observed groundwater data. The range of the random load generator is given up to its worst possible case (resistance value) as the system will respond permanent failure beyond the systems resistance data. The load once calculated is put into performance function to get the probability of exceedance of the reliability index to get GDR.

The risk, reliability index, and resistance are calculated for all IWs using IDW interpolation technique which determines cell values using a linearly weighted combination of a set of sample points. The indices are calculated separately for mild, moderate, and severe drought. The spatial distribution of the risk, reliability, and resistance is plotted in Figs. 12.8−12.10 for mild, moderate, and severe drought bands.

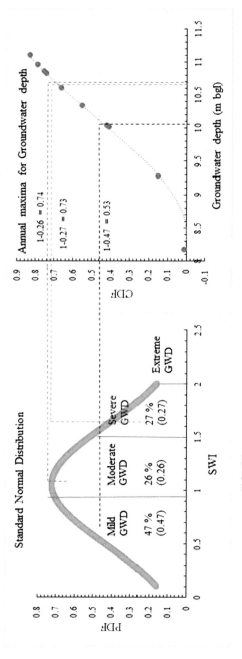

FIGURE 12.7 Standard normal distribution curve for obtaining resistance.

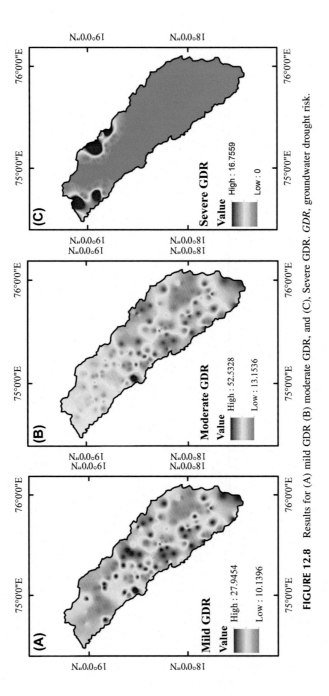

FIGURE 12.8 Results for (A) mild GDR (B) moderate GDR, and (C), Severe GDR. *GDR*, groundwater drought risk.

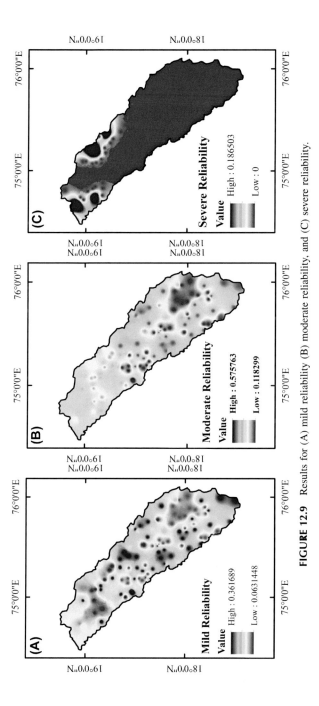

FIGURE 12.9 Results for (A) mild reliability (B) moderate reliability, and (C) severe reliability.

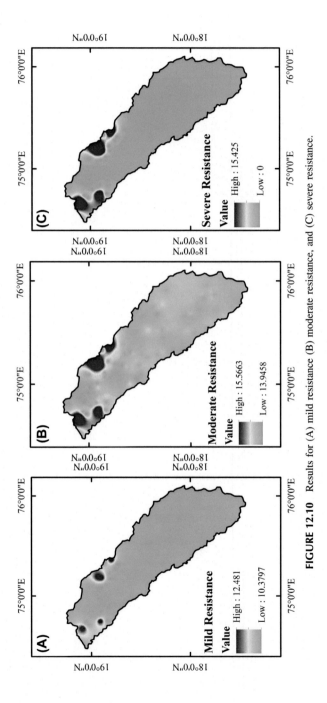

FIGURE 12.10 Results for (A) mild resistance (B) moderate resistance, and (C) severe resistance.

GDR is calculated as the probability of exceedance of reliability index which is calculated separately for different bands (mild, moderate, and severe), which is shown in Fig. 12.8. Hence, it is seen from the result that for the mild drought band, the GDR is distributed from 10.13% to 27.9%. The drought risk is observed more in the central region and at the extreme end of the basin (study area). But when the analysis is carried out for moderate drought band, the risk increases and is observed as 13.1%–52.5% for the same region. At severe drought condition, the risk observed is 16.755%. It can be clearly observed that extreme drought cases can occur after many years, which is not frequent, but if it happens, then the north part of the basin will be impacted more.

The reliability results (Fig. 12.9) showed the opposite trend to risk values and supported safe regions of groundwater extraction. The resistance values (Fig. 12.10) showed the system resistance to the load on the groundwater resource and went at a maximum depth of about 12.481 m for mild drought and a maximum of approximately 15 m for moderate and severe droughts.

5. Conclusions

The overexploitation of groundwater resource leads to more and more stress on groundwater. This raises serious groundwater issues in terms of groundwater quantity and quality. Hence, a framework for risk assessment of groundwater index is needed for sustainable management of groundwater resources.

In this chapter, a probability-based model is made to assess the groundwater risk to one of the most drought-prone areas, the Sina basin. The groundwater depth data of 133 wells were available for the present analysis, and hence, a clustering approach is selected for the representative (imaginary) wells. Canopy clustering technique was chosen because of its appropriateness for sampling large data. All the wells were grouped and represented within 10 clusters. Uncertainty analyses were carried out for the IWs only. The risk, reliability, and resistance indices were calculated for all IWs in the basin. The results showed that the GDR was observed more in the central part of the basin for mild and moderate droughts. The percentage increase in the GDR for moderate drought was a maximum of 48% compared with mild drought. Severe drought, although it not frequent, can impact the northern part of the basin. In addition, a 20% increase of resistance was observed from mild drought to moderate drought condition and recorded a maximum of 15 m of resistance for severe conditions whose return period is more than that of mild and moderate droughts.

References

Arthur, Vassilvitskii, S., 2007. k-means++: the advantages of careful seeding. In: Proceedings of the Eighteenth Annual ACM-SIAM Symposium on Discrete Algorithms, pp. 1027–1035.

Bocchini, P., Frangopol, D.M., Ummenhofer, T., Zinke, T., 2013. Resilience and sustainability of civil infrastructure: toward a unified approach. Journal of Infrastructure Systems 20 (2).

Cancelliere, A., Nicolosi, V., Rossi, G., 2009. Assessment of drought risk in water supply systems. Coping drought risk agric. Water Supply System 93−109.

Han, J., Kamber, M., 2006. Data Mining Concepts and Techniques. Morgan Kaufmann Publishers.

Kuo, J.T., Yen, B.C., Hsu, Y.C., Lin, H.F., 2007. Risk analysis for dam overtopping—feitsui reservoir as a case study. Journal of Hydraulic Engineering 133 (8), 955−963.

Loucks, D.P., 1997. Quantifying trends in system sustainability. Hydrological Sciences Journal 42 (4), 513−530.

Low, B.K., Zhang, J., Tang, W.H., 2011. Efficient system reliability analysis illustrated for a retaining wall and a soil slope. Computers and Geotechnics 38 (2), 196−204.

McCallum, Nigam, K., Ungar, L.H., 2000. Efficient clustering of high dimensional data sets with application to reference matching. In: Proceedings of the Sixth ACM SIGKDD International Conference on Knowledge Discovery and Data Mining ACM-SIAM Symposium on Discrete Algorithms, pp. 169−178.

Mishra, A.K., Singh, V.P., 2010. A review of drought concepts. Journal of Hydrology 391 (1), 202−216.

Sadeghfam, S., Hassanzadeh, Y., Khatibi, R., Moazamnia, M., Nadiri, A.A., 2018. Introducing a risk aggregation rationale for mapping risks to aquifers from point-and diffuse-sources—proof-of-concept using contamination data from industrial lagoons. Environmental Impact Assessment Review 72, 88−98.

Tanaka, K., Sugeno, M., 1992. Stability analysis and design of fuzzy control systems. Fuzzy Sets and Systems 45 (2), 135−156.

Tung, Y.K., Yen, B.C., Melching, C.S., 2005. Hydrosystems Engineering Reliability Assessment and Risk Analysis. McGraw−Hill Professional, New York.

Van Lanen, H.A.J., Peters, E., 2000. Definition, effects and assessment of groundwater droughts. In: Drought and Drought Mitigation in Europe. Springer, Netherlands, pp. 49−61.

Written, I.H., Frank, E., 2005. Data Mining: Practical Machine Learning Tools and Techniques. Morgan Kaufmann Publication.

Chapter 13

Novel concepts for reliability analysis of dynamic structural systems

J. Ramon Gaxiola-Camacho[1], Hamoon Azizsoltani[2], Achintya Haldar[3], S. Mohsen Vazirizade[3], Francisco Javier Villegas-Mercado[3]
[1]*Department of Civil Engineering, Autonomous University of Sinaloa, Culiacan, Sinaloa, Mexico;*
[2]*Department of Computer Science, North Carolina State University, Raleigh, NC, United States;*
[3]*Department of Civil and Architectural Engineering and Mechanics, University of Arizona, Tucson, AZ, United States*

1. Introduction

Estimation of dynamic response of structures under natural and man-made events is intrinsically very complicated and challenging. The task is more difficult for natural dynamic loading like earthquakes. Catastrophic damages caused by earthquakes all over the world will attest to the above statement. In spite of the considerable improvements in our understanding of the earthquake phenomenon and increasing computational power, it may not be practical to design completely seismic load-proof economical structures. The best alternative will be to design more seismic load-tolerant structures. In the opinion of the authors, the most important knowledge gap is the prediction of the design earthquake time history at a site for a particular structure. Other major challenges include the modeling of dynamic properties of structures at the time design earthquake occurs, nature and amount of energy dissipation during the excitation, dynamic amplification, and major sources of nonlinearity that develop as the excitation progresses caused by the time-domain application of the seismic loading.

Enormous amount of damages caused by recent earthquakes clearly indicate the weaknesses in our current design practices. The current design practice is essentially the protection of human life. The loss of economic activities during recent earthquakes was observed to be hundreds of billions of dollars. Following the Northridge earthquake of 1994 and Kobe earthquake of 1995, the Federal Emergency Management Agency (FEMA) initiated several studies to find an alternative to the current practices

Handbook of Probabilistic Models. https://doi.org/10.1016/B978-0-12-816514-0.00013-8
305

(FEMA-273, 1997; FEMA-350, 2000). One of the main outcomes of those studies was the introduction of a new design concept known as performance-based seismic design (PBSD). It is based on a performance level that a designer/owner is willing to take with an associated risk. It is a sophisticated risk-based design concept and reasonably developed for steel structures at present. However, acceptable risk estimation procures were not suggested or discussed. In several chapters of this handbook, a number of reliability evaluation procedures with various degrees of sophistication were presented. The authors believe that no such risk evaluation procedure is currently available for the dynamic application of seismic loading in time domain. It is a major knowledge gap. The authors and their team members attempted to fill this vacuum. Their overall efforts are summarized in this chapter.

In developing a new risk evaluation concept, the authors considered several factors. The most frequent complaint of the deterministic community is that currently available risk evaluation techniques are very crude; not equivalent to the sophisticated formulations used by the deterministic community. Related issues were addressed with limited success, particularly for the dynamic loading (Wen, 2001; Ghobarah, 2001). To represent structures under dynamic excitation as realistically as possible, the most rigorous analysis requires structural systems to be represented by finite elements (FEs). The excitation also needs to be applied in time domain. Furthermore, as the structure approaches the failure state, major sources of nonlinearity caused by the degradation of the material, geometric condition caused by large deformation, energy dissipation characteristics at the connections, etc., need to be considered appropriately. For steel structures, major sources of nonlinearity must be correctly considered. For example, beam to column connections are commonly represented as fully restrained (FR). However, it is documented in the literature that they are essentially partially restrained (PR) with different rigidities (Huh and Haldar, 2002; Gaxiola-Camacho et al., 2017a, 2017b, 2018). If rigidities of connections are not appropriately incorporated in the formulation, the dynamic properties such as damping, mode shape, frequency, etc., are also expected to be different resulting significant deviations in the estimation of the dynamic responses. If the major sources of nonlinearity are not incorporated following similar procedures as those used by the deterministic community, the estimated risk will also not be acceptable to them.

The above discussion indicates that the most commonly used first-order reliability method (FORM) may not be directly applicable to estimate risk for this class of problems. When structures are represented as accurately as practicable using FEs and are excited by dynamic loadings in time domain, the applicable limit state functions (LSFs) are expected to be implicit in nature. The implementation of FORM becomes relatively easier when LSFs are explicit in nature because their derivatives with respect to the basic random variables (RVs) present in the formulation will be readily available. For implicit LSFs, Haldar & Mahadevan (2000a) suggested several alternatives. As will be discussed in the following sections, those approaches also need further improvements. Some of these improvements are the major objectives of this chapter.

2. Challenges and trends in risk evaluation

Based on the discussions made in the previous section, it is necessary at this stage to summarize the challenges and trends in the risk estimation procedure that will be acceptable to both the deterministic and reliability communities. The available reliability estimation procedures for implicit LSFs, as suggested by Haldar and Mahadevan (2000a), including the sensitivity-based analysis, response surface (RS)—based approach, and the basic Monte Carlo simulation (MCS), may not satisfy the current needs of the profession. The stochastic finite element method (SFEM)—based approach proposed by them needs further improvements. The basic SFEM concept was developed using the assumed stress-based FE approach which is very rarely used in the profession. Using this FE approach, the stiffness matrix need not to be updated at every step required for nonlinear analyses and very large deformation configuration expected just before failure can be represented using very few FEs. With the help of numerous doctoral students, a complicated but sophisticated computer program was written. A comprehensive formulation of the assumed stress-based FE method can be found in the literature (Haldar and Nee, 1989; Gao and Haldar, 1995; Haldar and Mahadevan, 2000b; Reyes-Salazar and Haldar, 2001; Huh and Haldar, 2001, 2002; Lee and Haldar, 2003; Mehrabian et al., 2005, 2009; Reyes-Salazar et al., 2014; Gaxiola-Camacho et al., 2017a, 2017b, 2018; Azizsoltani and Haldar, 2017, 2018; Azizsoltani et al., 2018). At present, the computer program is not available for routine applications by users not in the research team of the authors. It may not also be practical to upgrade this program for various computer application platforms currently available. It will be extremely desirable if a user can use any available computer program capable of estimating nonlinear structural behavior for dynamic loadings applied in time domain and accurately extract the reliability information for specific LSFs within reasonable time period. It will also provide an option to compare the new concept presented in this chapter with the old SFEM concept using the assumed stress-based FE formulation.

The basic MCS method can also be used by users with limited expertise in the risk estimation procedure. The deterministic community will agree that one time-domain nonlinear dynamic analysis of a realistic structural system may require say about 1 h of computer time. For a low probability event like a major destructive earthquake, the underlying risk may need to be reasonably small.

For the sake of discussion, suppose one decides to conduct only 10,000 simulations to estimate the underlying risk, it will require about 10,000 h or 1.14 years of continuous running of a computer. Obviously, considering its inefficiency, the basic MCS procedure, even using several space reduction techniques, should be eliminated for further consideration. As will be discussed later, an alternative to MCS is necessary and will be presented in this chapter.

The response surface method (RSM)—based risk estimation concept appears to be very promising. It will be a major building block of the novel concept presented here. The authors believe that they proposed such a concept and verified it for different applications. The basic concept and its application potential are showcased in this chapter.

3. State-of-the-art in estimating risk of dynamic structural systems

A brief literature review on the stochastic behavior of complicated nonlinear dynamic engineering systems is necessary at this stage. To study stochastic behavior of dynamic structural systems, the classical random vibration approach (Lin, 1967; Lin and Cai, 2004) can be used. The basic random vibration concept, its derivatives, and perturbation methods to address uncertainty-related issues including the first- or second-order Taylor series, Neumann expansion, Karhunen—Loeve orthogonal expansion, polynomial chaos, etc., have numerous deficiencies that limit their application potential for the dynamic systems of interest. Most of them do not satisfy the physics-based representation of large nonlinear systems, are developed for relatively small numbers of dynamic degrees of freedom systems, ignore the distribution information of the system parameters, and are valid only for small amounts of randomness even when there is potential for significant amplification at the system level. Some of the recently proposed novel features to improve dynamic behavior, discussed earlier, also cannot be incorporated. The uncertainty in the dynamic loading is represented in the form of power spectral density functions; these are essentially applicable for linear systems and the dynamic loading cannot be applied in time domain. The development of the random vibration concept was an important research topic during the latter half of the 20th century and yet failed to make significant impact on stochastic response analysis of large dynamic engineering systems explicitly considering major sources of nonlinearity and uncertainty (Haldar and Kanegaonkar, 1986; Kanegaonkar and Haldar, 1987a, b, c; Peña-Ramos and Haldar, 2014a, b). One of the fundamental limitations is that the dynamic loading cannot be applied in time domain; an alternative is urgently needed.

To address this deficiency, several uncertainty quantification methods for large-scale computational models were proposed including reduced-order models, surrogate models, Bayesian methods, stochastic dimension reduction techniques, efficient Monte Carlo methods (e.g., importance sampling), etc. Some of them are problem-specific and will not satisfy the objectives of the current study. The state-of-the-art in the seismic risk analysis also needs a brief discussion at this stage. The group at the Pacific Earthquake Engineering Research Center developed a reliability approach known as the direct differentiation method (Der Kiureghian et al., 2006; Haukaas and Der Kiureghian, 2005, 2006; Koduru and Haukaas, 2006). They assumed the structural parameters at their mean values and the loading could not be applied in time domain. They used an out-crossing approach (Koo et al., 2005) with an importance sampling scheme (Au and Beck, 2001; Ching et al., 2005). Au and Beck (1999) used an adaptive importance sampling methodology to compute the multidimensional integrals encountered in reliability analysis. The multidimensional integrals are not available for the structural systems under consideration for implicit LSFs. Wen (2001) attempted to develop a method but did not use a physics-based analytical approach. These studies advanced the state-of-the-art but cannot be used to satisfy the objectives of the current study.

Although they may not be totally relevant to the discussion here, several new methods were proposed to address uncertainty-related issues in mechanical engineering applications, including high-dimensional model representation (HDMR) and explicit design space decomposition—support vector machines (EDSD-SVMs). In these studies, the general objective is to approximately develop multivariate expressions for the RS. One such method is HDMR (Alış and Rabitz, 2001; Li et al., 2001; Rabitz and Aliş, 1999; Sobol, 2003; Wei and Rahman, 2007; Xu and Rahman, 2005). It is also referred to as "decomposition methods," "univariate approximation," "bivariate approximation," "S−variate approximation," etc. HDMR is a general set of quantitative model assessment and analysis tools for capturing the high-dimensional relationships between sets of input and output model variables in such a way that the component functions of the approximation are ordered starting from a constant and adding terms such as first order, second order, and so on. The concept appears to be reasonable if higher-order variable correlations are weak, allowing the physical model to be captured by the first few lower-order terms. However, it cannot be applied for physics-based time-domain dynamic analysis and requires MCS for the reliability estimation. EDSD can be used when responses can be classified into two classes, e.g., safe and unsafe. A machine learning technique known as SVMs (Basudhar and Missoum, 2008; Basudhar et al., 2008; Missoum et al., 2007) is used to construct the boundaries separating distinct classes. The failure regions corresponding to different modes of failure are represented with a single SVM boundary, which is refined through adaptive sampling. It suffers similar deficiencies as that of HDMR.

The above brief discussions also clearly indicate that both HDMR and EDSD-SVM approaches have numerous assumptions and limitations and they use MCS to estimate the underlying reliability. They fail to incorporate explicitly the underlying physics, sources of nonlinearities, etc., and dynamic loadings cannot be applied in the time domain.

4. A novel structural risk estimation procedure for dynamic loadings applied in time domain

In developing the novel concept, the underlying challenge to the authors was to satisfy both the deterministic and reliability communities at the same time. To achieve this objective, several advanced mathematical concepts are necessary to be developed and integrated. In developing the concept, a physics-based algorithm satisfying all major dynamic characteristics must be formulated, parallel to the FE-based procedures used by the deterministic community. However, considering the size of the problem and the implementation potential, the number of deterministic evaluations must be kept to an absolute minimum requiring an alternative to MCS.

As mentioned earlier, risk is always estimated with respect to an LSF. However, for the class of problems discussed in this chapter, they are expected to be implicit in nature. It will be extremely desirable if an implicit LSF can be represented explicitly, even approximately, in terms of all the design RVs in the formulation. An RSM-based approach appears to be very attractive at this stage. The basic RSM concept was developed to study chemical reactions (Box, 1954). It was developed in the coded variable space, completely ignoring the distributional information of RVs. To satisfy the reliability community, the distributional information must be incorporated in the formulation of the RS and it must be generated in the failure region which may be unknown for most problems of interest. Because FORM incorporates distributional information and iteratively locates the most probable failure point (MPFP), the authors decided to integrate RSM with FORM. Once an approximate expression of a required RS is generated, it can be converted to an LSF, and then the underlying risk can be estimated using FORM. Because an RS is generated by an integrated approach eliminating the weaknesses of the original RS concept, it will be denoted hereafter as the integrated RS or IRS.

In generating an IRS, the basic objective will be to conduct nonlinear time-domain FE analyses to obtain the dynamic response information or points, as few as practicable without compromising the accuracy, and then fitting a polynomial through these points. This will follow the recent trend in basic research of using sophisticated intelligent multiple deterministic analyses to incorporate stochasticity in the formulation. Some of the major tasks in generating an IRS are to decide the degree of polynomial to represent it, efficient schemes to locate the center point around which samples will be generated, and schemes for the selection of sampling points.

The degree of the polynomial used to generate the IRS has significant effect on the efficiency of the algorithm, as will be discussed in more detail later. A linear IRS may not be appropriate for the dynamic problems under consideration. Higher-order polynomial may result in ill-condition of the system of equations and exhibit irregular behavior outside of the domain of samples (Gavin and Yau, 2008; Rajashekhar and Ellingwood, 1993). Considering success of their exploratory studies, the authors decided to use second-order polynomial, without and with cross terms. They can be represented as follows:

$$\widehat{g}(\mathbf{X}) = b_0 + \sum_{i=1}^{k} b_i X_i + \sum_{i=1}^{k} b_{ii} X_i^2 \tag{13.1}$$

and

$$\widehat{g}(\mathbf{X}) = b_0 + \sum_{i=1}^{k} b_i X_i + \sum_{i=1}^{k} b_{ii} X_i^2 + \sum_{i=1}^{k-1} \sum_{j>i}^{k} b_{ij} X_i X_j \tag{13.2}$$

where k is the number of RVs; X_i ($i = 1,2, \dots k$) is the ith RV; b_0, b_i, b_{ii}, and b_{ij} are the unknown coefficients; and $\widehat{g}(\mathbf{X})$ is the approximate LSF. Total number of coefficients that need to be evaluated to generate Eqs. (13.1) and (13.2) are $(2k + 1)$ and $(k + 1)(k + 2)/2$, respectively.

4.1 Generation of integrated response surface

The accuracy and efficiency of the proposed algorithm will depend on how the coefficients in Eqs. (13.1) and (13.2) are evaluated. Tasks in generating an IRS include the selection of the center point, the sampling points around it, and the intelligent sampling schemes for the selection of sampling points.

4.1.1 Selection of center point

To successfully generate an IRS, the authors decided to follow the same iterative procedure followed in FORM. The coordinates of the center point at the first iteration will be at the mean values of all RVs. At the same time, the information on the distribution needs to be incorporated to satisfy the requirements of FORM. In the basic RSM, the information on the uncertainty in RVs without considering their distributions is generally described as follows:

$$X_i = X_i^C \pm h_i x_i \sigma_{X_i} \text{ where } i = 1, 2, \dots, k \tag{13.3}$$

where X_i is the ith RV region or bound; X_i^C is the coordinate of the center point of RV X_i; σ_{X_i} is the standard deviation of RV X_i; h_i is an arbitrary factor controlling the experimental region; and x_i is the coded variable which assumes values of 0, ± 1, or $\pm \sqrt[4]{2^k}$, depending on the coordinates of the sampling point with respect to the center point and sampling schemes, and k is the number of RVs. As mentioned earlier, in the first iteration, the coordinates of the center point will be the mean values of all RVs. The experimental region must be kept to a minimum (Khuri and Cornell, 1996). This can be achieved by controlling h_i.

For the structural reliability evaluation, the distribution information of RVs also needs to be incorporated at this time. FORM is generally implemented in the normal variable space. All nonnormal variables are transformed to equivalent normal variables at the checking point. To generate an IRS, it will be the mean values of all RVs in the first iteration. For the X_i nonnormal RV, the corresponding equivalent normal mean $(\mu_{X_i}^N)$ and standard deviation $(\sigma_{X_i}^N)$ can be calculated at the checking point x_i^* as follows (Haldar and Mahadevan, 2000a, b):

$$\mu_{X_i}^N = x_i^* - \Phi^{-1}\left[F_{X_i}\left(x_i^*\right)\right]\sigma_{X_i}^N \tag{13.4}$$

and

$$\sigma_{X_i}^N = \frac{\phi\left\{\Phi^{-1}\left[F_{X_i}\left(x_i^*\right)\right]\right\}}{f_{X_i}\left(x_i^*\right)} \tag{13.5}$$

where $F_{X_i}\left(x_i^*\right)$ and $f_{X_i}\left(x_i^*\right)$ are the CDF and PDF of the original nonnormal X_i RV at the checking point x_i^*, respectively; $\Phi()$ and $\phi()$ are the CDF and PDF of the standard normal variable, respectively. Thus, the equivalent normal mean $(\mu_{X_i}^N)$ and standard deviation $(\sigma_{X_i}^N)$ values can be used in Eq. (13.3) to incorporate the distribution information of nonnormally distributed RVs.

4.1.2 Factorial design schemes

At this stage, it is necessary to select sampling points around the center point following intelligent schemes. They are generally denoted as the factorial design or experimental design schemes. For engineering applications, two commonly used factorial design schemes are (1) saturated design (SD) and (2) central composite design (CCD) (Khuri and Cornell, 1996). SD is less accurate but more efficient as it requires only as many sampling points as the total number of unknown coefficients to define an IRS. SD without cross terms consists of one center point and $2k$ axial points. Therefore, a second-order RS can be generated using $2k+1$ FE analyses using SD. SD with cross terms consists of one center point, $2k$ axial points, and $k(k-1)/2$ edge points (Lucas, 1974). Therefore, a second-order IRS with cross terms can be generated using $(k+1)(k+2)/2$ FE analyses.

CCD is a more efficient approach to the 3^k factorial design (Box et al., 1978). It consists of a center point, two axial points on the axis of each RV at distance $h = \sqrt[4]{2^k}$ from the center point, and 2^k factorial design points. It will require a total of $2^k + 2k + 1$ FE analyses to generate a second-order IRS. CCD is more accurate but less efficient for two major reasons: (1) it requires cross terms for the second-order polynomial and (2) regression analysis is necessary to generate an IRS requiring many sampling points. It can be observed that CCD will require significantly more FE deterministic analyses than SD and may not be practical to implement for large systems. Thus, an intelligent combination of SD and CCD must be used to generate the IRS.

4.1.3 Advanced factorial design schemes

The proposed algorithm is iterative in nature. The authors considered several advanced factorial design (AFD) schemes by exploiting the desirable features of SD and CCD. Among many alternatives, the following scheme appears to be very promising. Because a second-order RS without cross terms [Eq. (13.1)] can be generated very efficiently using SD, it can be used for the initial and intermediate iterations to generate an IRS. To improve the accuracy in an IRS, in the final iteration, CCD with an RS with cross terms (Eq. 13.2) can be used. It will be denoted hereafter as the AFD scheme.

In summary, for the AFD concept, during the first and intermediate iterations, RSs will be generated solving a set of linear equations using SD, and in the final iteration, the IRS will be generated using the regression analysis and CCD. The sampling points selection process using AFD is illustrated in Fig. 13.1 in terms of three RVs (X_1, X_2, and X_3).

4.2 Reliability estimation using IRS and AFD

As discussed earlier, during the FORM implementation process, the coordinates of the initial center point $\mathbf{x}_{C_1}$ will be at the mean or equivalent mean value μ_{X_i} of all X_i's. An IRS, denoted as $\hat{g}_1(\mathbf{X})$, can be explicitly generated by conducting deterministic FE analyses at all the experimental

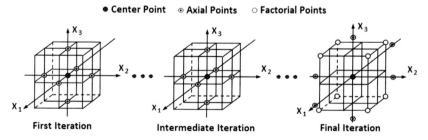

FIGURE 13.1 Sampling points selection process—advanced factorial design.

sampling points following the AFD concept. Once an explicit expression of the LSF is available, it may take few iterations before the reliability index converges satisfying a predetermined tolerance level to β_1 for the initial center point. The coordinates for the second center point can be generated using the information on β_1 and FORM. Following the same procedures as before, a new IRS, $\widehat{g}_2(\mathbf{X})$, can be generated and the corresponding β_2 can be obtained. If $|\beta_2 - \beta_1|$ does not satisfy the predetermined tolerance level, another new IRS will be generated and the corresponding β_n will be calculated. Once the successive reliability index values converge, the coordinates of the MPFP, $\mathbf{x}^*$, will be available. Using the information on $\mathbf{x}^*$, the final reliability index β which is the norm of the final checking or design point can be evaluated as follows (Haldar and Mahadevan, 2000b):

$$\beta = \sqrt{(\mathbf{x}^*)^t (\mathbf{x}^*)} \tag{13.6}$$

The corresponding probability of failure, P_f, can be calculated as follows:

$$P_f = \Phi(-\beta) = 1.0 - \Phi(\beta) \tag{13.7}$$

where Φ is the standard normal cumulative distribution function. In all subsequent discussions, the reliability index and the probability of failure are estimated using the above two equations. The reliability information extracted using FORM and IRS in the way discussed above will consider the distribution information of all RVs in the failure region and will also locate the MPFP (Gaxiola-Camacho et al., 2017a,b; Villegas-Mercado et al., 2017).

5. Proposed novel concept for reliability analysis of dynamic structural systems

The reliability evaluation procedure using IRS, AFD, and FORM discussed in the previous sections is expected to satisfy both the deterministic and reliability communities. Unfortunately, it cannot be used for the reliability estimation of realistic large structural systems. To elaborate on this, the following discussions are necessary.

5.1 Total Number of Deterministic Analyses

The implementation potential of the procedure discussed above needs further discussion. To generate an IRS using the AFD scheme, it will be required to conduct deterministic linear or nonlinear time-domain dynamic FE analyses to obtain information on the required responses. Thus, the feasibility of the procedure to generate the IRS will depend on the total number of deterministic analyses (TNDAs) required in the reliability evaluation. As previously discussed, if SD and Eq. (13.1) and CCD and Eq. (13.2) are used to generate an IRS, the process will require $2k + 1$ and $2^k + 2k + 1$ TNDA, respectively, where k is the total number of RVs. For the sake of discussion, if k is small, say $k = 6$, the two schemes will require 13 and 77 TNDA, respectively. They are reasonable and the reliability algorithm can be implemented without any problem. However, if $k = 40$, expected for a realistic large system, TNDA will be 81 and 1.0995×10^{12}, respectively. Obviously, the algorithm cannot be implemented in its current form of AFD. The above discussion indicates that if CCD is used in the final iteration to improve the accuracy, the algorithm cannot be used for the structural reliability estimation unless the k value is significantly reduced. The observation prompted the research team to explore ways to reduce it. Some of the options are discussed next.

5.2 Reduction of RVs

It is essential that the total number of RVs present in the formulation needs to be reduced as soon as practicable. Haldar and Mahadevan (2000a, b) suggested that the information on the sensitivity indexes of RVs can be used for this purpose. They showed that the sensitivity index of an RV is the directional cosine of the unit normal variable at the checking point if the RVs are uncorrelated. The information on directional cosines will be readily available from the FORM analyses. The efficiency of the algorithm can be significantly improved without sacrificing the accuracy by considering some of the RVs with smaller sensitivity indexes to be deterministic at their mean values. Hence, k can be reduced to k_r, representing the reduced number of RVs. The authors believe that if k_r is used, the efficiency in generating the IRS will be significantly improved. Reducing the total number of RVs in a problem using information on sensitivity indexes is a step in the right direction. However, because CCD will be used in the final iteration, TNDA could still be in thousands. Other steps are necessary to further reduce the required TNDA. It will be discussed further later in this chapter.

6. Accuracy in generating an IRS

At this stage, by conducting numerous linear or nonlinear time-domain dynamic FE analyses, a total of TNDA structural responses corresponding to the LSF of interest will be available. The task is how to fit a second-order polynomial through these points. The accuracy in generating an IRS will be very important in the reliability estimation. The commonly used concept to fit a polynomial through a number of points is the regression analysis (Haldar and Mahadevan, 2000a). It is essentially a global approximation technique in which all the sample points are assigned with the equal weight factor. In the context of generating an IRS using AFD, the equal weight factor concept used may not be appropriate; the weight factors should decay as the distances between the sampling points and the IRS increase. Instead of using the basic regression analysis, the authors explored the possibility of using the moving least squares method (MLSM) and a surrogate metamodel like Kriging method (KM) to improve accuracy in generating an IRS using the same response information. They are briefly discussed next.

6.1 Moving least squares method

The development of the IRS using the MLSM is related to the AFD where CCD and regression analysis are used to calculate the unknown coefficients in the final iteration. Hence, when MLSM is used, an IRS is generated using the LSM concept with different weights factors for the different experimental sampling points (Kim et al., 2005; Kang et al., 2010; Bhattacharjya and Chakraborty, 2011; Taflanidis and Cheung, 2012; Li et al., 2012; Chakraborty and Sen, 2014). It will assure the reliability index, β, estimated using MLSM will be more accurate.

Weight factors for experimental sampling points are reported in the literature (Bhattacharjya and Chakraborty, 2011; Chakraborty and Sen, 2014). Different types of weight factors are suggested including linear, quadratic, cubic, fourth-order polynomial, etc. The authors decided to use the following fourth-order polynomial to estimate the corresponding weight factor:

$$w(d_i) = 1 - 6r^2 + 8r^3 - 3r^4 \ \ if \ r \leq 1.0; \quad w(d_i) = 0 \ \ if \ r > 1.0 \qquad (13.8)$$

where d_i is the difference between the ith experimental sampling point and the point where IRS is required to be evaluated; r is equal to d_i/R_i representing the ratio of the difference (d_i) and region of influence (R_i) of the experimental sampling point under consideration. To avoid ill-conditioning, R_i needs to be selected to cover adequate experimental sampling points. Bhattacharjya and Chakraborty (2011) and Chakraborty and Sen (2014) suggested that R_i should be selected as twice the distance between the center point and the farthest experimental sampling point. For a total of N experimental sampling points, the regression model can be expressed as follows (Haldar and Mahadevan, 2000a):

$$\mathbf{Y} = \mathbf{Bb} + \boldsymbol{\varepsilon} \qquad (13.9)$$

where $\mathbf{Y}$ is a vector with N TNDA for constructing the IRS; $\mathbf{b}$ is the unknown coefficient vector to be calculated using MLSM; $\boldsymbol{\varepsilon}$ is a vector containing the residuals or errors; and $\mathbf{B}$ is a matrix containing the coefficients of the full second-order polynomial expressed as follows:

$$
\mathbf{B} =
\begin{bmatrix}
1 & x_{11} & x_{21} & \cdots & x_{k1} & x_{11}^2 & x_{21}^2 & \cdots & x_{k1}^2 & x_{11}x_{21} & x_{11}x_{31} & \cdots & x_{11}x_{k1} & \cdots & x_{k-1,1}x_{k1} \\
1 & x_{12} & x_{22} & \cdots & x_{k2} & x_{12}^2 & x_{22}^2 & \cdots & x_{k2}^2 & x_{12}x_{22} & x_{12}x_{32} & \cdots & x_{12}x_{k2} & \cdots & x_{k-1,2}x_{k2} \\
\cdot & \cdot & \cdot & \cdot & \cdot & \cdot & \cdot & \cdot & \cdot & \cdot & \cdot & \cdot & \cdot & \cdot & \cdot \\
\cdot & \cdot & \cdot & \cdot & \cdot & \cdot & \cdot & \cdot & \cdot & \cdot & \cdot & \cdot & \cdot & \cdot & \cdot \\
\cdot & \cdot & \cdot & \cdot & \cdot & \cdot & \cdot & \cdot & \cdot & \cdot & \cdot & \cdot & \cdot & \cdot & \cdot \\
1 & x_{1N} & x_{2N} & \cdots & x_{kN} & x_{1N}^2 & x_{2N}^2 & \cdots & x_{kN}^2 & x_{1N}x_{2N} & x_{1N}x_{3N} & \cdots & x_{1N}x_{kN} & \cdots & x_{k-1,N}x_{kN}
\end{bmatrix}
$$

(13.10)

where x_{ij} represents the observation on the ith RV in the jth data set. To estimate the vector **b** at the checking point, the weighted sum of the squared residuals at all the sampling points, S, must be minimized. It can be expressed as follows:

$$S = (\mathbf{Y} - \mathbf{Bb})^{\mathrm{T}} \mathbf{W} (\mathbf{Y} - \mathbf{Bb}) \tag{13.11}$$

where **W** is the weighting matrix which contains the weight factors along the diagonal. It can be expressed as follows:

$$\mathbf{W} = \begin{bmatrix} \mathrm{w}(d_1) & 0 & \cdots & 0 \\ 0 & \mathrm{w}(d_2) & \cdots & 0 \\ \cdot & \cdot & \cdot & \cdot \\ \cdot & \cdot & \cdot & \cdot \\ \cdot & \cdot & \cdot & \cdot \\ 0 & 0 & \cdots & \mathrm{w}(d_n) \end{bmatrix} \tag{13.12}$$

where $\mathrm{w}(d_i)$ represents the weight factor of the ith experimental sampling point.

Thus, if Eq. (13.11) is minimized, the unknown coefficient vector **b** can be expressed as follows (Haldar and Mahadevan, 2000a, b):

$$\mathbf{b} = [\mathbf{B}^{\mathrm{T}} \mathbf{W} \mathbf{B}]^{-1} \mathbf{B}^{\mathrm{T}} \mathbf{W} \mathbf{Y} \tag{13.13}$$

It is to be noted that Eq. (13.13) will provide improved regression coefficients. It is expected that the reliability index, β, estimated this way will be more accurate. However, TNDA used to generate the IRS will remain the same as that of using the ordinary regression analysis. This will be elaborated with the help of examples later. Further discussions on the topic can be found in Azizsoltani and Haldar (2017).

6.2 Kriging method

The KM was originally developed to improve the accuracy for tracing gold in ores. More discussions on the topic can be found in Azizsoltani and Haldar (2017) and Azizsoltani et al. (2018). It is a geostatistical method of interpolation with the prior assumption that the function to be estimated is a Gaussian process. It also assumes that the near sample points should get more weights (Lichtenstern, 2013). However, MLSM generates a polynomial in an average sense, but it will pass through all experimental sampling points. Two basic desirable characteristics of KM are that an IRS generated by it is uniformly unbiased and its prediction error is less than all the possible forms. These features make it the best linear unbiased surrogate for an IRS (Wackernagel, 2013). KM is basically the best linear unbiased predictor estimate of an IRS, and its gradients can be used to calculate information on unobserved data points. Several types of KM are reported in the literature. For the class of problems of interest, the mean values of the spatial data points are expected to depend on the location of the experimental sampling points. Hence, the authors decided to implement Universal KM in the present study because of its capabilities in incorporating external drift functions as additional variables (Hengl et al., 2007).

The Universal KM involves basically a linear weighted sum of the responses calculated by the FE analyses at the sampling points. It can predict the value of the response at any nonsampled point. The concept behind the generation of the IRS, denoted hereafter as $[\widehat{g}(\mathbf{X})]$ for this discussion, using KM can be expressed as follows:

$$\widehat{g}(\mathbf{X}) = \sum_{i=1}^{r} \omega_i Z(\mathbf{X}_i) \tag{13.14}$$

where ω_i is the ith ($i = 1, 2, ..., r$) unknown weight corresponding to the observation vector $Z(\mathbf{X}_i)$ which is estimated by performing r deterministic FE analyses.

Based on the above discussion, the observation vector, $Z(\mathbf{X})$, represents a Gaussian process and is assumed to be a linear combination in terms of a nonstationary deterministic drift function, $u(\mathbf{X})$, and a residual random function, $Y(\mathbf{X})$. Thus, $Z(\mathbf{X})$ can be expressed as follows (Cressie, 2015):

$$Z(\mathbf{X}) = u(\mathbf{X}) + Y(\mathbf{X}) \tag{13.15}$$

where $u(\mathbf{X})$ represents a second-order polynomial including cross terms, and the term $Y(\mathbf{X})$ is an intrinsically stationary function with underlying variogram function γ_Y and zero mean.

The corresponding relationship with respect to the available spatial data calculated with the help of the responses using the experimental sampling points can be extracted by the variogram function. The generation of the variogram cloud must be the first step in the calculation of the appropriate variogram function. Basically, the variogram cloud is the graphical representation of the dissimilarity function. Thus, the experimental variogram can be generated using the variations with the similar distance in the variogram cloud. The experimental variogram can be determined as the average of the dissimilarities with the similar distance l_i. Finally, the experimental variogram must be represented using a proper mathematical model to produce the variogram function. The dissimilarity function can be denoted as follows:

$$\gamma^*(l_i) = \frac{1}{2}[Z(x_i + l_i) - Z(x_i)]^2 \tag{13.16}$$

where $\gamma^*(l_i)$ represents the dissimilarity function for the corresponding ith RV which is separated by a distance l_i; and x_i is the coordinate of the experimental sampling point.

Because dissimilarity function is symmetric with respect to the distance l_i, only absolute values of l_i will be considered. The total number of responses used to generate the variogram function will determine its accuracy. Hence, the authors propose to integrate the responses calculated during the final and just before the final iteration to increase the total number of responses in generating the variogram function.

Several variogram functions including bounded linear models, exponential, nugget effect, and spherical, are reported in the literature. Least squares and weighted least squares regression methods are commonly utilized for fitting the above models to the experimental variogram. The group of stable anisotropic variogram models using weighted least squares regression method is selected in this research.

Based on this discussion, variograms are generated for every application using the above models; particularly the one with the highest coefficient of determination (Haldar and Mahadevan, 2000a) is carefully chosen to generate the IRS (Eq. (13.14)).

To guarantee uniform unbiasedness in Eq. (13.14), universality conditions must be satisfied by estimating the weight factors ω_i. It can be fulfilled as follows (Cressie, 2015):

$$\sum_{i=1}^{r} \omega_i f_p(\mathbf{X}_i) = f_p(\mathbf{X}_0) \quad \text{for } p = 0, 1, ..., P \tag{13.17}$$

where $f_p(\mathbf{X}_i)$ represents the regression function of a second-order polynomial including cross terms; $\mathbf{X}_i$ is the ith sampling point, and $\mathbf{X}_0$ represents the coordinates of the nonsampled point that will be used to predict the response of the structure.

For every regressor variable $\mathbf{X}_i$, a specific number of r sets of data are collected, and each one of these sets corresponds to P observations (Haldar and Mahadevan, 2000a). The weight factors are calculated by minimizing the variance of the prediction error with the help of the optimality criteria and the Lagrange multipliers as follows:

$$\mathbf{\Gamma}_Y \boldsymbol{\omega} + \mathbf{F}\boldsymbol{\lambda} = \boldsymbol{\gamma}_{Y,0} \tag{13.18}$$

where $\mathbf{\Gamma}_Y$ represents the symmetric residual variogram matrix, i.e., $(\mathbf{\Gamma}_Y)_{ij} \equiv \gamma_Y(\mathbf{X}_i - \mathbf{X}_j)$, $i, j = 1, ..., r$; $\boldsymbol{\omega}$ is the vector form of the unknown weight factors; $\mathbf{F}$ can be introduced in the form of $\mathbf{F} \equiv f_p(\mathbf{X}_i)$; $\boldsymbol{\lambda} \equiv (\lambda_0, \lambda_1, ..., \lambda_P)^T$ is the Lagrange multiplier; and $\boldsymbol{\gamma}_{Y,0}$ is the residual variogram vector.

Solving Eqs. (13.17) and (13.18) information on the unknown weights and Lagrange multipliers will be obtained as follows:

$$\begin{pmatrix} \boldsymbol{\omega} \\ \boldsymbol{\lambda} \end{pmatrix} = \begin{pmatrix} \boldsymbol{\Gamma}_Y & \mathbf{F} \\ \mathbf{F}^T & 0 \end{pmatrix}^{-1} \begin{pmatrix} \boldsymbol{\gamma}_{Y,0} \\ \mathbf{f}_0 \end{pmatrix} \tag{13.19}$$

If $\mathbf{F}$ is a full column rank matrix, all columns are linearly independent. Wackernagel (2013) demonstrated that Eq. (13.19) will provide a unique solution. Using Eq. (13.14) and solving Eq. (13.19), the required IRS $[\widehat{g}(\mathbf{X})]$ can be calculated as follows:

$$\widehat{g}(\mathbf{X}) = \left[\boldsymbol{\gamma}_{Y,0} - \mathbf{F}\left(\mathbf{F}^T\boldsymbol{\Gamma}_Y^{-1}\mathbf{F}\right)^{-1}\left(\mathbf{F}^T\boldsymbol{\Gamma}_Y^{-1}\boldsymbol{\gamma}_{Y,0} - \mathbf{f}_0\right) \right]^T \boldsymbol{\Gamma}_Y^{-1}\mathbf{Z} \tag{13.20}$$

Hence, the LSF will be explicitly available using Eq. (13.20), and the underlying reliability will be estimated using the generated LSF and FORM algorithm (Haldar and Mahadevan, 2000a). FORM is discussed in more detail in Chapter 1.

7. Reliability estimation

As mentioned earlier, with the availability of an IRS, it is now necessary to extract the reliability information using FORM.

7.1 Limit State Functions

The first step in the reliability analysis will be to define appropriate LSFs. For seismic design, serviceability-related limit states generally control the design. Considering the availability of space, only serviceability LSFs are considered in this chapter. With the availability of an appropriate IRS, a serviceability LSF can be defined as follows:

$$g(\mathbf{X}) = \delta_{allow} - y_{max}(\mathbf{X}) = \delta_{allow} - \widehat{g}(\mathbf{X}) \tag{13.21}$$

where δ_{allow} represents the allowable or permissible value to satisfy a specific performance requirement; $\widehat{g}(\mathbf{X})$ is the generated IRS; and $\mathbf{X}$ is a vector representing all the RVs. Different LSFs are discussed in more detail in Chapter 1.

7.2 Performance levels

In the context of seismic design of structures using PBSD criteria, the required LSFs need to be generated. As previously discussed, FEMA-350 (2000) defined three performance levels: immediate occupancy (IO), life safety (LS), and collapse prevention (CP). Because PBSD is implemented in terms of multiple target performance levels, FEMA-350 (2000) suggested allowable drift values (δ_{allow}) for CP, LS, and IO in terms of probability of exceedance

TABLE 13.1 Structural correlation of performance levels.

Performance level	Return period	Probability of exceedance	Allowable drift
Immediate occupancy	72 years	50% in 50 years	$0.007 \times H$
Life safety	475 years	10% in 50 years	$0.025 \times H$
Collapse prevention	2475 years	2% in 50 years	$0.050 \times H$

and return period. They are summarized in Table 13.1, where H represents the total or interstory height depending on the LSF under consideration. The information provided in Table 13.1 is essentially for LSFs in terms of overall lateral and interstory drift. Considering that generally dynamic structural systems may fail due to excessive lateral deflection, overall lateral and interstory drift LSFs are considered in verifying the proposed reliability-based approach.

7.3 Incorporation of uncertainties

Major sources of uncertainty in the resistance and load-related parameters need to be incorporated for the reliability evaluation of dynamic structural systems. The uncertainties related to resistance parameters are widely documented in the literature (Haldar and Mahadevan, 2000a, b). In this chapter, only steel frames made with W-sections available in the steel construction manual available in the United States (AISC, 2011) are considered. Young's modulus of steel (E), yield stress of columns (Fy_c) and girders (Fy_g), cross-sectional area (A), and moment of inertia (I_x) of W-sections are considered as RVs. They are considered to be lognormally distributed with coefficient of variation (COV) of 0.06, 0.10, 0.10, 0.05, and 0.05, respectively. Dead load (DL) and live load (LL) are considered as the gravity loads. The uncertainties associated with both are available in the literature (Haldar and Mahadevan, 2000a, b). DL and LL are represented by a normal and Type 1 distributions with COV of 0.10 and 0.25, respectively. As briefly discussed earlier, connections in steel frames are PR with different rigidities. To address this issue, initially the connections are considered to be FR. They are then considered to be PR type. To model PR connections, the rigidities of them are represented by the Richard 4-parameter model (Mehrabian et al., 2005). To define the Richard model, the four parameters are the initial stiffness (k), plastic stiffness (k_P), reference moment (M_0), and curve shape parameter (N). The information on them was generated using test results. They are considered as RVs. The parameters k, k_P, M_0, and N are normally distributed with COV of 0.15, 0.15, 0.15, and 0.05, respectively (Gaxiola-Camacho et al., 2018). Consideration of uncertainties in the seismic loading is very challenging. Uncertainty associated with the intensity and the frequency contents needs to be considered. To incorporate the uncertainty in the intensity, a factor g_e is used in this study. It is represented as an RV with a Type 1 distribution with a COV of 0.2. Uncertainty associated with the frequency contents is incorporated by considering several ground motion time histories; this will be discussed in detail later.

7.4 Reliability evaluation

The reliability using the proposed concept is evaluated in several ways. For the clarity of discussion, the following abbreviations will be used in the subsequent discussion.

- AFD—When the reliability is estimated using the AFD concept and in the final iteration the IRS is generated using the ordinary regression analysis.
- MLSM—When the reliability is estimated using the AFD concept and in the final iteration the IRS is generated using MLSM.
- KM—When the reliability is estimated using the AFD concept and in the final iteration the IRS is generated using KM.

The proposed basic concept of the reliability evaluation is summarized in Fig. 13.2.

8. Numerical examples—verifications and case studies

The proposed risk evaluation concepts for dynamic loading applied in time domain are relatively new. They need verifications and their capabilities need to be established at this stage. In the first example, a relatively small two-story steel frame is considered and is excited by several measured earthquake time histories. This example is taken to document the accuracy of AFD. The second example is a case study to demonstrate the capabilities of the concepts. The third example is considered to showcase how PBSD can be implemented.

8.1 Example 1—verification on AFD—two-Story steel frame

A two-story steel frame, as shown in Fig. 13.3, is considered. It was designed satisfying the US steel construction manual (AISC, 2011) (Gaxiola-Camacho et al., 2017a, 2017b, 2018). The geometry and sections of columns and girders are given in Fig. 13.3. The uncertainties associated with all RVs are summarized in Table 13.2. The uncertainties are expressed in terms of distribution, mean to nominal ratio $(\overline{X}/X_N)$, nominal value, mean value, and COV.

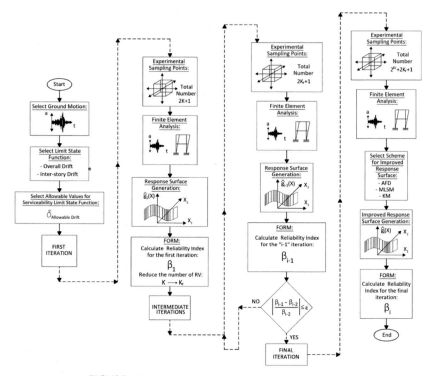

FIGURE 13.2 The proposed reliability evaluation method.

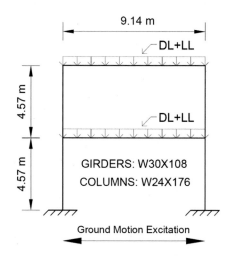

FIGURE 13.3 Two-Story steel frame subjected to seismic loading.

TABLE 13.2 Uncertainties related to random variables (RVs).

RV	Example	Distribution	$\overline{X}/X_N$	Nominal (X_N)	Mean ($\overline{X}$)	Coefficient of variation
E (kN/m^2)	1,2,3	Lognormal	1.00	1.9995E+08	1.9995E+08	0.06
F_{YG} (kN/m^2)*	3	Lognormal	1.35	2.4821E+05	3.3509E+05	0.10
F_{YC} (kN/m^2)*	3	Lognormal	1.15	3.4474E+05	3.9645E+05	0.10
F_Y (kN/m^2)	1,2	Lognormal	1.15	3.4474E+05	3.9645E+05	0.10
A (m^2)	1,2,3	Lognormal	1.00	**	***	0.05
I_x (m^4)	1,2,3	Lognormal	1.00	**	***	0.05
DL_1 (kN/m^2) ****	3	Normal	1.05	3.9740	4.1727	0.10
DL_2 (kN/m^2) ****	3	Normal	1.05	4.5965	4.8263	0.10
DL_3 (kN/m^2) ****	1	Normal	1.05	3.8304	4.0219	0.10
DL_4 (kN/m^2) ****	2	Normal	1.05	2.8977	3.0426	0.10
LL_1 (kN/m^2) ****	3	Type 1	0.40	2.3940	0.9576	0.25
LL_2 (kN/m^2) ****	3	Type 1	0.40	2.3940	0.9576	0.25
LL_3 (kN/m^2) ****	1	Type 1	0.50	2.3940	1.1970	0.25
LL_4 (kN/m^2) ****	2	Type 1	0.50	1.8176	0.9088	0.25
k (kN-m/rad)	3	Normal	1.00	**	*****	0.15

Continued

TABLE 13.2 Uncertainties related to random variables (RVs).—cont'd

RV	Example	Distribution	$\overline{X}/X_N$	Nominal (X_N)	Mean ($\overline{X}$)	Coefficient of variation
kp (kN-m/rad)	3	Normal	1.00	**	*****	0.15
Mo (kN-m)	3	Normal	1.00	**	*****	0.15
N	3	Normal	1.00	**	*****	0.05
g_e	1,2,3	Type 1	1.00	1.00	1.00	0.20

*Yield stress of girder (Fy_G) or column (Fy_C) cross section reported in FEMA-355C (2000). **Nominal value (X_N) is calculated using mean value ($\overline{X}$) and $\overline{X}/X_N$. ***Mean values of A and I_x can be found in steel construction manual (AISC, 2011). For every column and girder, A and I_x are considered as RVs. ****DL_1 and LL_1 are the dead and live load for the roof of the 3- and 9-story frames, respectively; DL_2 and LL_2 are the dead and live load for the intermediate levels of the 3- and 9-story frames, respectively; DL_3 and DL_4 are the dead load for the 2- and 13-story frames, respectively; LL_3 and LL_4 are the live load for the 2- and 13-story frames, respectively. *****Mean values of four Richard parameters are reported in Gaxiola-Camacho et al. (2017b).

The two-story frame is excited by the north—south (N—S) and east—west (E—W) components of four earthquake records measured during the Northridge earthquake of 1994. They were recorded at Santa Susana, Alhambra—Fremont time, Littlerock—Brainard, and Rancho Palos Verdes. To consider the strongest part of the ground motion records, the frame is subjected to the first 24 s of the excitation. During the first iteration of the reliability evaluation process, the total number of RVs is 14 or $k = 14$. Using the sensitivity analysis discussed in Section 5.2, only five RVs are found to be significant, i.e., $k_r = 5$. To define LSF represented by Eq. (13.21), the overall lateral and interstory allowable drifts are estimated to be 2.86 and 1.43 cm, respectively. The probability of failure of the frame for the two LSFs is estimated using AFD. The accuracy is then compared by conducting 100,000 MCS. The results are summarized in Table 13.3.

Based on the results summarized in Table 13.3, several important observations can be made. The probability of failure values obtained by AFD and MCS are very similar indicating that the AFD is a viable reliability estimation procedure. However, AFD required about 100 TNDA but MCS results are for 100,000 cycles, a significant improvement in the efficiency without sacrificing accuracy. The estimated probability of failure for the four earthquakes and N—S and E—W components is different indicating that the frequency contents of an earthquake record are very important. It clearly indicates that the risk assessment using only one earthquake history will be inadequate and unacceptable. From the results, it appears that AFD is an alternative to MCS.

8.2 A case study to document the capabilities of the proposed reliability evaluation concept—failure of 13-Story steel frame located in San Fernando Valley, CA

After the successful verification, it is now necessary to demonstrate the capabilities of AFD, MLSM, and KM. A 13-story steel building from the 1970s suffered significant damages during the 1994-Northridge earthquake. It is considered as a case study. The building was located in San Fernando Valley, CA. A typical frame of the building including its geometry and cross sections of columns and girders is shown in Fig. 13.4. Uncertainties related to RVs in terms of resistance-related and loading conditions were earlier summarized in Table 13.2. The frame was excited by a recorded earthquake time history very close to the building. It was recorded at the Canoga Park Station, as shown in Fig. 13.5. It is reported by Uang et al. (1995) that the floors between Plaza-first and sixth—seventh suffered considerable displacements. For this example, the interstory drift is considered to be the critical LSF. Following the recommendations suggested in ASCE 7-10 (2010), the allowable interstory drifts for the Plaza-first and sixth—seventh floors are considered to be 4.27 and 3.51 cm, respectively. The probability of

TABLE 13.3 Structural reliability of the two-story frame.

Ground motion	Limit state function	AFD			Overall lateral drift		
					MCS		
		β	p_f	TNDA	β	p_f	TNDA
Santa Susana N–S	Overall lateral drift	3.6137	1.5093E-04	94	3.5985	1.6000E-04	100,000
	Interstory drift	3.0545	1.1272E-03	94	3.0618	1.1000E-03	100,000
Santa Susana E–W	Overall lateral drift	3.8067	7.0417E-05	94	3.9444	4.0000E-05	100,000
	Interstory drift	3.4565	2.7362E-04	105	3.5149	2.2000E-04	100,000
Alhambra–Fremont N–S	Overall lateral drift	4.2430	1.1028E-05	105	4.1075	2.0000E-05	100,000
	Interstory drift	4.1201	1.8935E-05	105	4.0128	3.0000E-05	100,000
Alhambra–Fremont E–W	Overall lateral drift	5.2484	7.6713E-08	94	4.2649	1.0000E-05	100,000
	Interstory drift	5.4044	3.2513E-08	94	4.2649	1.0000E-05	100,000
Littlerock–Brainard N–S	Overall lateral drift	3.8286	6.4437E-05	105	3.7750	8.0000E-05	100,000
	Interstory drift	3.8287	6.4411E-05	105	3.7750	8.0000E-05	100,000
Littlerock–Brainard E–W	Overall lateral drift	4.2997	8.5515E-06	105	4.2649	1.0000E-05	100,000
	Interstory drift	3.8521	5.8555E-05	94	3.8906	5.0000E-05	100,000
Rancho Palos Verdes N–S	Overall lateral drift	4.8866	5.1296E-07	105	4.2649	1.0000E-05	100,000
	Interstory drift	3.8442	6.0473E-05	94	3.8906	5.0000E-05	100,000
Rancho Palos Verdes E–W	Overall lateral drift	3.5370	2.0235E-04	94	3.5401	2.0000E-04	100,000
	Interstory drift	3.8271	6.4831E-05	94	3.8082	7.0000E-05	100,000

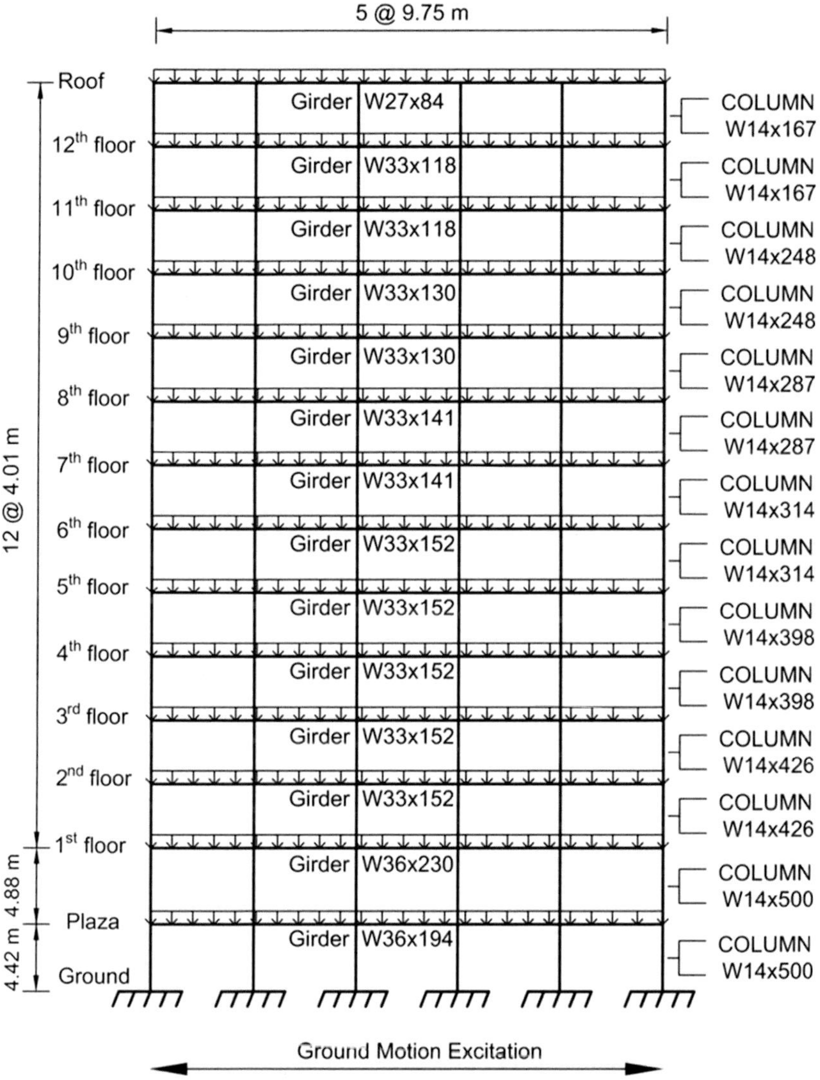

FIGURE 13.4   13-Story building located in San Fernando Valley, California.

failure for only the interstory drifts LSFs for the two floors is estimated using AFD, MLSM, KM, and 500,000 cycles of MCS.

In the first iteration, 72 RVs are used to generate the IRS, and for intermediate and final iterations 7 RVs were found to be the most important, i.e., $k = 72$ and $k_r = 7$. Results of the reliability analysis are summarized in Table 13.4. The results reported in Table 13.4 are very exciting and demonstrate the capabilities of the four alternatives. The probabilities of failure obtained by AFD, MLSM, and KM are very similar to that of the

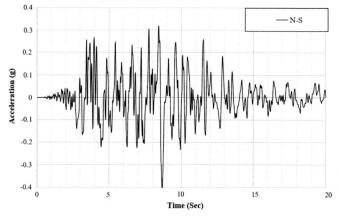

FIGURE 13.5 Canoga Park Station 1994-Northridge earthquake.

TABLE 13.4 Structural reliability of the 13-Story frame.

Interstory drift		AFD	MLSM	KM	MCS
Plaza	β	0.5280	0.5281	0.5327	0.5316
|	p_f	0.2987	0.2987	0.2971	0.2975
First floor	TNDA	179	179	179	500,000
Sixth floor	β	−0.0681	−0.0676	−0.0681	−0.0715
|	p_f	0.5272	0.5270	0.5272	0.5285
Seventh floor	TNDA	183	183	183	500,000

results obtained by 500,000 cycles of MCS. However, the results obtained by the proposed three methods are obtained by only 179 TNDA. The estimated probabilities of failure are extremely large clearly indicating the high damage potential of the floors, confirming the damages observed caused by the actual excitation. This example also confirms that the basic reliability estimation concept presented here is an alternative to the classical MCS method.

8.3 Example 3—implementation of the PBSD concept for 3- and 9-Story steel frames

This example showcases how the PBSD concept can be implemented and the proposed reliability evaluation concept will be a major building block. One multidisciplinary application potential of the proposed reliability

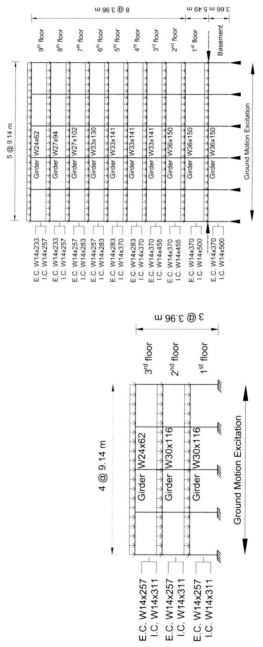

FIGURE 13.6 SAC steel frames (*EC*, exterior column; *IC*, interior column).

concept is demonstrated in this section with the help of two steel buildings designed using the PBSD philosophy. After the 1994-Northridge earthquake, to improve the technical knowledge about seismic-resistant design of structures, FEMA funded the SAC Joint Venture (FEMA-273, 1997; FEMA-350, 2000; FEMA-355C, 2000). During this project, several configurations of buildings were designed by practicing experts. They are expected to be used for future studies. For this study, the authors decided to estimate the reliability of two steel frames of 3- and 9-story high, as shown in Fig. 13.6. Both frames are described in detail in FEMA-355C (2000). The dimensions of the frames and the size of the members are shown in Fig. 13.6.

A large number of beam to column connections were significantly damaged during the Northridge earthquake of 1994. Richard et al. (1997) proposed a proprietary PR connection by cutting slots in the web of the beams to improve the connections. It will be denoted as the "post-Northridge" (Mehrabian et al., 2005) connection. The frames shown above in Fig. 13.6, as designed, did not consider the PR connections. In this study, the reliabilities of the frames are evaluated in the presence of both FR and "post-Northridge" PR connections. Reliabilities of the frames are estimated for three performance levels: CP (2475-year return period), LS (475-year return period), and IO (72-year return period). To represent these performance levels, three sets of 10 ground motions with both horizontal components, producing 60 time histories, suggested by Somerville et al. (1997) are considered. They are listed in Table 13.5. The uncertainties of every RV used in the process were described previously in Table 13.2. The overall lateral displacement and interstory drift of the two LSFs are considered. For the overall lateral displacement, the allowable displacement for CP, LS, and IO are considered to be 59.44, 29.72, and 8.32 cm for the 3-story building, and 185.85, 92.93, and 26.02 cm for the 9-story frames, respectively. For the interstory drift, the second and fourth floor levels for the 3- and 9-story frames, respectively, are considered. The corresponding permissible values for CP, LS, and IO levels are 19.81, 9.91, and 2.77 cm, respectively. The total number of RVs for the two frames, k values, is found to be 25 and 91, respectively. The significant number of RVs, k_r, for both frames is found to be 7. Each frame is then excited by the 60 earthquake time histories for the three performance levels and the underlying reliabilities are estimated for both LSFs in the presence of FR and PR connection. The results are summarized in Figs. 13.7–13.10.

In Figs. 13.7–13.10, the reliability index β values are plotted for each earthquake time histories. Several important observations can be made from these figures. The results clearly indicate that the frequency content of earthquake time histories is an important parameter. Designing a structure considering one time history is expected to be inadequate. Structures must be designed by considering multiple time histories, as suggested in ASCE 7-10

TABLE 13.5 — Ground motions.

Set 1: 2475-Year return period—CP performance level				Set 2: 475-Year return period—LS performance level				Set 3: 72-Year return period—IO performance level			
Eq	Name	SF	Time (sec)	Eq	Name	SF	Time (sec)	Eq	Name	SF	Time (sec)
1	1995 Kobe	1.2	25.0	21	Imperial Valley, 1940	2.0	25.0	41	Coyote Lake, 1979	2.3	12.0
2	1995 Kobe	1.2	25.0	22	Imperial Valley, 1940	2.0	25.0	42	Coyote Lake, 1979	2.3	12.0
3	1989 Loma Prieta	0.8	20.0	23	Imperial Valley, 1979	1.0	15.0	43	Imperial Valley, 1979	0.4	15.0
4	1989 Loma Prieta	0.8	20.0	24	Imperial Valley, 1979	1.0	15.0	44	Imperial Valley, 1979	0.4	15.0
5	1994 Northridge	1.3	14.0	25	Imperial Valley, 1979	0.8	15.0	45	Kern, 1952	2.9	30.0
6	1994 Northridge	1.3	14.0	26	Imperial Valley, 1979	0.8	15.0	46	Kern, 1952	2.9	30.0
7	1994 Northridge	1.6	15.0	27	Landers, 1992	3.2	30.0	47	Landers, 1992	2.6	25.0
8	1994 Northridge	1.6	15.0	28	Landers, 1992	3.2	30.0	48	Landers, 1992	2.6	25.0
9	1974 Tabas	1.1	25.0	29	Landers, 1992	2.2	30.0	49	Morgan Hill, 1984	2.4	20.0
10	1974 Tabas	1.1	25.0	30	Landers, 1992	2.2	30.0	50	Morgan Hill, 1984	2.4	20.0
11	Elysian Park (simulated)	1.4	18.0	31	Loma Prieta, 1989	1.8	16.0	51	Parkfield, 1966, Cholame	1.8	15.0
12	Elysian Park (simulated)	1.4	18.0	32	Loma Prieta, 1989	1.8	16.0	52	Parkfield, 1966, Cholame	1.8	15.0

Continued

TABLE 13.5 — Ground motions.—cont'd

Set 1: 2475-Year return period—CP performance level				Set 2: 475-Year return period—LS performance level				Set 3: 72-Year return period—IO performance level			
Eq	Name	SF	Time (sec)	Eq	Name	SF	Time (sec)	Eq	Name	SF	Time (sec)
13	Elysian Park (simulated)	1.0	18.0	33	Northridge, 1994, Newhall	1.0	15.0	53	Parkfield, 1966, Cholame	2.9	15.0
14	Elysian Park (simulated)	1.0	18.0	34	Northridge, 1994, Newhall	1.0	15.0	54	Parkfield, 1966, Cholame	2.9	15.0
15	Elysian Park (simulated)	1.1	18.0	35	Northridge, 1994, Rinaldi	0.8	14.0	55	North Palm Springs, 1986	2.8	20.0
16	Elysian Park (simulated)	1.1	18.0	36	Northridge, 1994, Rinaldi	0.8	14.0	56	North Palm Springs, 1986	2.8	20.0
17	Palos Verdes (simulated)	0.9	25.0	37	Northridge, 1994, Sylmar	1.0	15.0	57	San Fernando, 1971	1.3	20.0
18	Palos Verdes (simulated)	0.9	25.0	38	Northridge, 1994, Sylmar	1.0	15.0	58	San Fernando, 1971	1.3	20.0
19	Palos Verdes (simulated)	0.9	25.0	39	North Palm Springs, 1986	3.0	16.0	59	Whittier, 1987	1.3	15.0
20	Palos Verdes (simulated)	0.9	25.0	40	North Palm Springs, 1986	3.0	16.0	60	Whittier, 1987	1.3	15.0

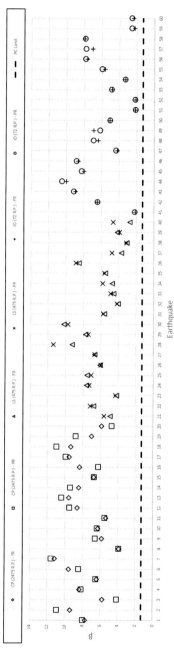

FIGURE 13.7 Overall lateral drift reliability for the three-story frame.

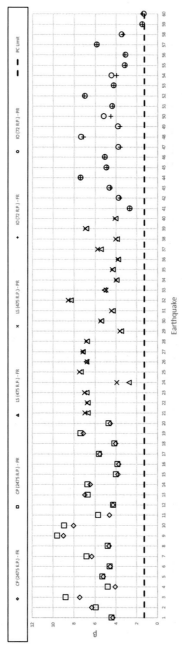

FIGURE 13.8 Interstory drift reliability for the three-story frame.

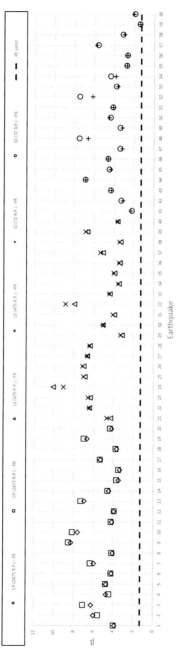

FIGURE 13.9 Overall lateral drift reliability for the nine-story frame.

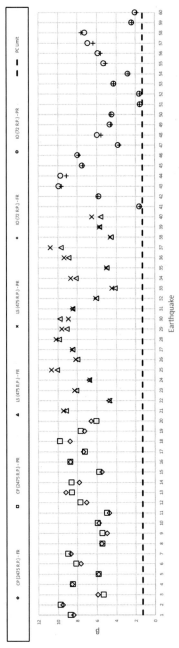

FIGURE 13.10 Interstory drift reliability for the nine-story frame.

(2010). For both frames, the interstory drift appears to be more critical than the overall displacement. The β values fall within a range, satisfying the intent of the code (ASCE 7-10, 2010) and reliability-based design. The allowable values suggested in FEMA-350 (2000) for both LSFs and performance levels are reasonable. In addition, β values for the frames with post-Northridge PR connections are very similar to that of FR connections indicating their beneficial effect. As observed by the authors, post-Northridge connections help improve the behavior of the steel frames and should be used in the future in the earthquake prone areas. It is also interesting to note that the probability of collapse for each case is less than 0.1, satisfying the intent of ASCE 7-10 (2010) (Azizsoltani and Haldar, 2017, 2018).

9. Multidisciplinary applications

The authors are very satisfied with the proposed methods for evaluating reliability of structures excited by the earthquake loading applied in time domain. The question remains if the proposed reliability evaluation method is problem-specific, i.e., they are suitable only for the seismic excitation, not applicable for other types of loadings applied in time domain. To address this issue, the authors considered the reliability of solder balls used in electronic packaging subjected to cyclic thermomechanical loading and offshore structures excited by the wave loading, as briefly discussed below.

9.1 Reliability of electronic packaging—thermomechanical loading

Reliability estimation of solder balls subjected to cyclic thermomechanical loading may not be of interest to most of the readers of this handbook. However, it is interesting to note that Azizsoltani and Haldar (2018) recently published a paper in the Journal of Electronic Packaging using the methods discussed in the previous sections. They represented solder ball using FEs. Major sources of nonlinearities are incorporated as realistically as practicable. Uncertainties in all design variables are quantified using available information. The thermomechanical loading is represented by five design parameters and uncertainties associated with them are incorporated. The implicit performance function is generated by universal KM. The accuracy, efficiency, and application potential of the procedure are established with the help of MCS and results from laboratory investigation reported in the literature. The study conclusively verified the proposed method. The study showcased how reliability information can be extracted with the help of multiple deterministic analyses in multidisciplinary applications.

9.2 Reliability evaluation of offshore structures subjected to wave loading

Reliability estimation of offshore structures represents a very complicated process mainly because of the presence of considerable amount of uncertainties in both the environmental conditions and the configuration of the structural systems used. The most important part in the environmental loading is the wave which is highly stochastic in nature. However, in general, offshore structures are designed by applying the wave loading statically neglecting the effect of the period (Dyanati and Huang, 2014). A brief literature review will indicate that the probability of failure of offshore structures was estimated by applying the wave loading statically (Singurdsson and Cramer, 1995; Singurdsson et al., 1996). Very recently, the authors represented offshore structures by FEs to incorporate major sources of nonlinearity and excited them in time domain with wave loading using constraint new wave technique. Using the wave scatter diagram for the region (Vazirizade et al., 2019), the maximum annual wave height distribution is estimated by a three-parameter Weibull distribution. The probability of failure of offshore structures is estimated by considering uncertainties in the wave height, hydrodynamic coefficients, modulus of elasticity, yield stress, strain-hardening ratio, marine growth, load and mass, current velocity, and section dimensions. Because of complexity of the problem, other available methods cannot consider all the sources of uncertainty. The method presented in the first part of this chapter consisting of AFD and KM was used to extract the reliability information. A typical jacket-type offshore structure considered in this example is shown in Fig. 13.11.

By conducting only 90 TNDAs, the authors estimated the probability of failure of the offshore structure. In this study, the deck drift is considered as the LSF for reliability estimation. The results are summarized in Table 13.6. The results match very well with MCS of 10,000 cycles. More complete discussion on the topic will be published in the near future. The authors are extremely pleased to inform the readers that the methods they proposed can be used to estimate reliability of various multidisciplinary applications.

10. Further improvements of Kriging method

If the number of RVs remains relatively large even after selecting the most sensitive variables, the use of CCD in the last iteration to generate an IRS may require a large number of deterministic evaluations making the proposed procedure very inefficient or impractical. The authors believe that the efficiency of KM can be significantly improved in the following way. In implementing CCD, the cross terms of the second-order polynomial and the necessary sampling points are needed to be considered only for the most significant RVs in sequence in order of their sensitivity indexes until the reliability index converges with a predetermined tolerance level. It will be a difficult topic and may be beyond the scope of this handbook.

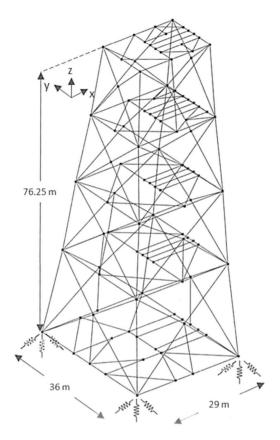

FIGURE 13.11 Offshore structure.

TABLE 13.6 Structural reliability of the offshore structure subjected to wave loading.

	AFD	KM	MCS
B	2.327231	2.331893	2.3301
p_f	9.9765E-03	9.8532E-03	9.9000E-03
TNDA	90	90	10,000

11. Conclusions

Based on the results presented in this research, several observations can be made. In terms of the novel concept discussed in this chapter, the basic reliability approach is found to be very efficient and accurate. Dynamic structural systems are represented using FEs and loading is applied in time domain. In addition, major sources of nonlinearity and uncertainty are incorporated in the formulation. The applicable potentials of the reliability technique are demonstrated for steel structures excited by the earthquake loading, offshore structures excited by the wave loading, and electronic packaging systems excited by the thermomechanical loading. The validations indicate that the reliability estimation concept can be used for routine applications and are expected to be acceptable to all concerned parties. Using the reliability approach, the structural risk can be extracted using hundreds instead of thousands of deterministic analyses. The methodology represents an alternative to MCS and the classical random vibration approaches. The results verify that designing structures using multiple ground motions, as suggested in current design codes, is a reasonable criterion that increases the possibility of success in producing more earthquake-resistant design. It is also necessary to design offshore structures by exciting them by wave loading in time domain.

Acknowledgments

This research was financially supported by several agencies of the government of Mexico: *Consejo Nacional de Ciencia y Tecnología* (CONACYT) under Grant No. A1-S-10088 and the *Universidad Autónoma de Sinaloa* (UAS). The study is also partially supported by the National Science Foundation under Grant No. CMMI-1403844. Any opinions, findings, or recommendations expressed in this paper are those of the authors and do not necessarily reflect the views of the sponsors.

References

AISC, 2011. Steel Construction Manual. American Institute of Steel Construction (AISC).

Alış, Ö.F., Rabitz, H., 2001. Efficient implementation of high dimensional model representations. Journal of Mathematical Chemistry 29 (2), 127–142.

ASCE 7-10, 2010. Minimum Design Loads for Buildings and Other Structures. American Society of Civil Engineers (ASCE), Reston, VA, USA.

Au, S.K., Beck, J.L., 1999. A new adaptive importance sampling scheme for reliability calculations. Structural Safety 21 (2), 135–158.

Au, S.K., Beck, J.L., 2001. First excursion probabilities for linear systems by very efficient importance sampling. Probabilistic Engineering Mechanics 16 (3), 193–207.

Azizsoltani, H., Haldar, A., Intelligent Computational Schemes for Designing more Seismic Damage-Tolerant Structures. Journal of Earthquake Engineering, https://doi.org/10.1080/13632469.2017.1401566, published online on 11/17/2017.

Azizsoltani, H., Haldar, A., 2018. Reliability Analysis of Lead-Free Solders In Electronic Packaging Using A Novel Surrogate Model And Kriging Concept. Journal of Electronic Packaging, ASME 140 (4), 041003-1–11. https://doi.org/10.1115/1.4040924.

Azizsoltani, H., Gaxiola-Camacho, J.R., Haldar, A., September 2018. Site-Specific Seismic Design of Damage Tolerant Structural Systems Using a Novel Concept. Bulletin of Earthquake Engineering 16 (9), 3819−3843. https://doi.org/10.1007/s10518-018-0329-5.

Basudhar, A., Missoum, S., 2008. Adaptive explicit decision functions for probabilistic design and optimization using support vector machines. Computers and Structures 86 (19−20), 1904−1917.

Basudhar, A., Missoum, S., Sanchez, A.H., 2008. Limit state function identification using support vector machines for discontinuous responses and disjoint failure domains. Probabilistic Engineering Mechanics 23 (1), 1−11.

Bhattacharjya, S., Chakraborty, S., 2011. Robust optimization of structures subjected to stochastic earthquake with limited information on system parameter uncertainty. Engineering Optimization 43 (12), 1311−1330.

Box, G.E., 1954. The exploration and exploitation of response surfaces: some general considerations and examples. Biometrics 10 (1), 16−60.

Box, G.E., Hunter, W.G., Hunter, J.S., 1978. Statistics for Experimenters.

Chakraborty, S., Sen, A., 2014. Adaptive response surface based efficient finite element model updating. Finite Elements in Analysis and Design 80, 33−40.

Ching, J., Au, S.K., Beck, J.L., 2005. Reliability estimation for dynamical systems subject to stochastic excitation using subset simulation with splitting. Computer Methods in Applied Mechanics and Engineering 194 (12−16), 1557−1579.

Cressie, N., 2015. Statistics for Spatial Data. John Wiley & Sons.

Der Kiureghian, A., Haukaas, T., Fujimura, K., 2006. Structural reliability software at the university of California, Berkeley. Structural Safety 28 (1−2), 44−67.

Dyanati, M., Huang, Q., 2014. Seismic Reliability of a fixed offshore platform against collapse. In: ASME 2014 33rd International Conference on Ocean, Offshore and Arctic Engineering. American Society of Mechanical Engineers (pp. V04BT02A020-V04BT02A020).

FEMA-273, 1997. NEHRP Guidelines for the Seismic Rehabilitation of Buildings. Federal Emergency Management Agency (FEMA).

FEMA-350, 2000. Recommended Seismic Design Criteria for New Steel Moment-Frame Buildings. Federal Emergency Management Agency (FEMA).

FEMA-355C, 2000. State of the Art Report on Systems Performance of Steel Moment Frames Subject to Earthquake Ground Shaking. Federal Emergency Management Agency (FEMA).

Gao, L., Haldar, A., 1995. Safety evaluation of frames with PR connections. Journal of Structural Engineering 121 (7), 1101−1109.

Gavin, H.P., Yau, S.C., 2008. High-order limit state functions in the response surface method for structural reliability analysis. Structural Safety 30 (2), 162−179.

Gaxiola-Camacho, J.R., Azizsoltani, H., Villegas-Mercado, F.J., Haldar, A., 2017. A Novel Reliability Technique for Implementation of Performance-Based Seismic Design of Structures. Engineering Structures 142, 137−147. http://doi.org/10.1016/j.engstruct.2017.03.076.

Gaxiola-Camacho, J.R., Haldar, A., Azizsoltani, H., Valenzuela-Beltran, F., Reyes-Salazar, A., 2018. Performance Based Seismic Design of Steel Structures Using Rigidities of Connections. ASCE-ASME Journal of Risk and Uncertainty in Engineering Systems, Part A: Civil Engineering 4 (1). https://doi.org/10.1061/AJRUA6.0000943.

Gaxiola-Camacho, J.R., Haldar, A., Reyes-Salazar, A., Valenzuela-Beltran, F., Vazquez-Becerra, G.E., Vazquez-Hernandez, A.O., April, 2018. Alternative Reliability-Based Methodology for Evaluation of Structures Excited by Earthquakes. Earthquakes and Structures, An International Journal 14 (4). https://doi.org/10.12989/eas.2018.14.4.361.

Ghobarah, A., 2001. Performance-based design in earthquake engineering: state of development. Engineering Structures 23 (8), 878–884.

Haldar, A., Kanegaonkar, H.B., 1986, January). Stochastic fatigue response of jackets under intermittent wave loading. In: Offshore Technology Conference. Offshore Technology Conference.

Haldar, A., Mahadevan, S., 2000a. Reliability Assessment Using Stochastic Finite Element Analysis. Wiley, New York, NY, USA.

Haldar, A., Mahadevan, S., 2000b. Probability, Reliability, and Statistical Methods in Engineering Design. Wiley, New York, NY, USA.

Haldar, A., Nee, K.M., 1989. Elasto-plastic large deformation analysis of PR steel frames for LRFD. Computers and Structures 31 (5), 811–823.

Haukaas, T., Der Kiureghian, A., 2005. Parameter sensitivity and importance measures in nonlinear finite element reliability analysis. Journal of Engineering Mechanics 131 (10), 1013–1026.

Haukaas, T., Der Kiureghian, A., 2006. Strategies for finding the design point in non-linear finite element reliability analysis. Probabilistic Engineering Mechanics 21 (2), 133–147.

Hengl, T., Heuvelink, G.B., Rossiter, D.G., 2007. About regression-Kriging: from equations to case studies. Computers and Geosciences 33 (10), 1301–1315.

Huh, J., Haldar, A., 2001. Stochastic finite-element-based seismic risk of nonlinear structures. Journal of Structural Engineering 127 (3), 323–329.

Huh, J., Haldar, A., 2002. Seismic reliability of non-linear frames with PR connections using systematic RSM. Probabilistic Engineering Mechanics 17 (2), 177–190.

Kanegaonkar, H.B., Haldar, A., 1987a. Non-Gaussian closure for stochastic response of geometrically nonlinear complaint platforms. In: Proceedings of the International Offshore Mechanics and Arctic Engineering Symposium. ASME.

Kanegaonkar, H.B., Haldar, A., 1987b. Nonlinear random vibrations of compliant offshore platforms. In: Nonlinear Stochastic Dynamic Engineering Systems. Springer, Berlin, Heidelberg, pp. 351–360.

Kanegaonkar, H.B., Haldar, A., 1987c. Non-Gaussian stochastic response of nonlinear compliant platforms. Probabilistic Engineering Mechanics 2 (1), 38–45.

Kang, S.C., Koh, H.M., Choo, J.F., 2010. An efficient response surface method using moving least squares approximation for structural reliability analysis. Probabilistic Engineering Mechanics 25 (4), 365–371.

Khuri, A.I., Cornell, J.A., 1996. Response Surfaces: Designs and Analyses. Marcel Dekker, New York, NY, USA.

Kim, C., Wang, S., Choi, K.K., 2005. Efficient response surface modeling by using moving least-squares method and sensitivity. AIAA Journal 43 (11), 2404–2411.

Koduru, S.D., Haukaas, T., 2006. Uncertain reliability index in finite element reliability analysis. International Journal of Reliability and Safety 1 (1–2), 77–101.

Koo, H., Der Kiureghian, A., Fujimura, K., 2005. Design-point excitation for non-linear random vibrations. Probabilistic Engineering Mechanics 20 (2), 136–147.

Lee, S.Y., Haldar, A., 2003. Reliability of frame and shear wall structural systems. II: dynamic loading. Journal of Structural Engineering 129 (2), 233–240.

Li, G., Wang, S.W., Rosenthal, C., Rabitz, H., 2001. High dimensional model representations generated from low dimensional data samples. I. mp-Cut-HDMR. Journal of Mathematical Chemistry 30 (1), 1–30.

Li, J., Wang, H., Kim, N.H., 2012. Doubly weighted moving least squares and its application to structural reliability analysis. Structural and Multidisciplinary Optimization 46 (1), 69–82.

Lichtenstern, A., 2013. Kriging Methods in Spatial Statistics. Ph.D. dissertation. Technische Universität München, München.

Lin, Y.K., 1967. Probabilistic Theory of Structural Dynamics. McGraw-Hill, New York, N.Y.

Lin, Y.K., Cai, G.Q., 2004. Probabilistic Structural Dynamics: Advanced Theory and Applications. Mcgraw-hill Professional Publishing.

Lucas, J.M., 1974. Optimum composite designs. Technometrics 16 (4), 561−567.

Mehrabian, A., Ali, T., Haldar, A., 2009. Nonlinear analysis of a steel frame. Nonlinear Analysis: Theory, Methods and Applications 71 (12), e616−e623.

Mehrabian, A., Haldar, A., Reyes-Salazar, A., 2005. Seismic response analysis of steel frames with post-north ridge connection. Steel and Composite Structures 5 (4), 271−287.

Missoum, S., Ramu, P., Haftka, R.T., 2007. A convex hull approach for the reliability-based design optimization of nonlinear transient dynamic problems. Computer Methods in Applied Mechanics and Engineering 196 (29−30), 2895−2906.

Peña-Ramos, C.E., Haldar, A., 2014a. Three dimensional response of RC bridges under spatially varying seismic excitation-methodology. Journal of Structural Engineering 41 (3), 251−264.

Peña-Ramos, C.E., Haldar, A., 2014b. Three dimensional response of RC bridges under spatially varying seismic excitation-numerical analysis and observations. Journal of Structural Engineering 41 (3), 265−278.

Rabitz, H., Aliş, Ö.F., 1999. General foundations of high-dimensional model representations. Journal of Mathematical Chemistry 25 (2−3), 197−233.

Rajashekhar, M.R., Ellingwood, B.R., 1993. A new look at the response surface approach for reliability analysis. Structural Safety 12 (3), 205−220.

Reyes-Salazar, A., Haldar, A., 2001. Energy dissipation at PR frames under seismic loading. Journal of Structural Engineering 127 (5), 588−592.

Reyes-Salazar, A., Soto-Lopez, M.E., Gaxiola-Camacho, J.R., Bojorquez, E., Lopez-Barraza, A., 2014. Seismic response estimation of steel buildings with deep columns and PMRF. Steel and Composite Structures 17 (4), 471−495. https://doi.org/10.12989/scs.2014.17.4.471.

Richard, R.M., Allen, C.J., Partridge, J.E., 1997. Proprietary slotted beam connection designs. Modern Steel Construction 28−33.

Singurdsson, G., Cramer, E., 1995. Guideline for offshore structural reliability analysis−examples for jacket platforms. Det Norske Veritas (report).

Singurdsson, G., Cramer, E., Lotsberg, I., Berge, B., 1996. Guideline for offshore structural reliability analysis: application to jacket platforms. Det Norske Veritas (report).

Sobol, I.M., 2003. Theorems and examples on high dimensional model representation. Reliability Engineering and System Safety 79 (2), 187−193.

Somerville, P., Smith, N., Punyamurthula, S., Sun, J., 1997. Development of Ground Motion Time Histories for Phase 2 of the FEMA/SAC Steel Project. Report SAC/BD-97/04 SAC Joint Venture.

Taflanidis, A.A., Cheung, S.H., 2012. Stochastic sampling using moving least squares response surface approximations. Probabilistic Engineering Mechanics 28, 216−224.

Uang, C.M., Yu, Q.S., Sadre, A., Bonowitz, D., Youssef, N., 1995. Performance of a 13-Story Steel Moment-Resisting Frame Damaged in the 1994 Northridge Earthquake. Report SSRP-95/04. University of California, San Diego.

Vazirizade, S.M., Haldar, A., Gaxiola-Camacho, J.R., 2019. Uncertainty Quantification of Sea Waves - an Improved Approach. Oceanography & Fisheries 9 (5). https://juniperpublishers.com/ofoaj/pdf/OFOAJ.MS.ID.555775.pdf.

Villegas-Mercado, F.J., Azizsoltani, H., Gaxiola-Camacho, J.R., Haldar, A., 2017. Seismic Reliability Evaluation of Structural Systems for Different Soil Conditions. International Journal of Geotechnical Earthquake Engineering (IJGEE) 8 (2), 23−38. https://doi.org/10.4018/IJGEE.2017070102 (Winner of IGI Global's Tenth Annual Excellence in Research Journal Award).

Wackernagel, H., 2013. Multivariate Geostatistics: An Introduction with Applications. Springer Science & Business Media.

Wei, D., Rahman, S., 2007. Structural reliability analysis by univariate decomposition and numerical integration. Probabilistic Engineering Mechanics 22 (1), 27−38.

Wen, Y.K., 2001. Reliability and performance-based design. Structural Safety 23 (4), 407−428.

Xu, H., Rahman, S., 2005. Decomposition methods for structural reliability analysis. Probabilistic Engineering Mechanics 20 (3), 239−250.

Chapter 14

Probabilistic neural networks: a brief overview of theory, implementation, and application

Behshad Mohebali[1], Amirhessam Tahmassebi[1], Anke Meyer-Baese[1], Amir H. Gandomi[2,3]
[1]*Department of Scientific Computing, Florida State University, Tallahassee, FL, United States;*
[2]*Faculty of Engineering and Information Technology, University of Technology Sydney, Ultimo, Australia;* [3]*School of Business, Stevens Institute of Technology, Hoboken, NJ, United States*

1. Introduction

Probabilistic neural networks (PNNs) are a group of artificial neural network built using Parzen's approach to devise a family of probability density function estimators (Parzen, 1962) that would asymptotically approach Bayes optimal by minimizing the "expected risk," known as "Bayes strategies" (Mood, 1950). In a PNN, there is no need for massive back-propagation training computations. Instead, each data pattern is represented with a unit that measures the similarity of the input patterns to the data pattern. PNNs have shown great potential for tackling complex scientific and engineering problems. Major categories of issues that researchers have attempted to address using PNN are as follows:

- Classification of labeled stationary data patterns
- Classification of data patterns where the data have a time-varying probabilistic density function
- Corresponding author email address: a.h.gandomi@stevens.edu (Amir H. Gandomi)
- Signal processing applications, dealing with waveforms as data patterns
- Unsupervised algorithms that work with unlabeled data sets

Handbook of Probabilistic Models. https://doi.org/10.1016/B978-0-12-816514-0.00014-X

The first category might be the simplest type among the mentioned four. Here, the assumption is that the probability density function of the data does not have significant and meaningful variations through the life span of the network. In Bankert (1994), the researchers used a PNN based algorithm to classify 16 × 16 pixel pictures of clouds into 10 categories. A set of 95 labeled pictures were used to form the PNN. In Wu et al. (2007), PNN is used to classify 1800 pictures of leaves into 32 categories of plants based on their geometrical properties. Although the feature extraction process in the latter is more sophisticated and application dependent, the underlying principles are the same as what was introduced in Specht (1990). In Nishanth and Ravi (2016), the authors used PNN to impute the missing categorical features in an incomplete data set. This application relies on PNNs' ability to produce satisfactory results even with small data sets.[1]

The assumption of time invariance may not hold in most of the practical use cases and the probability density function of the data shows nontrivial changes over time. In that regard, Rutkowski (2004) has introduced an adaptive PNN that can track the changes in the PDF of the data and adjust its inner parameters to take those changes into account. This approach has been used in Hazrati and Erfanian (2010) to recognize EEG signals coming from experiment subjects when they "think" about closing their hands to grab a virtual ball in virtual reality.

Another important category of applications for PNNs is signal processing. This can be recognizing the occurrence of an event (Tripathy et al., 2010), the prediction of severity of an event (Adeli and Panakkat, 2009; Asencio-Cortés et al., 2017), using the time-domain waveform of a parameter of interest, or classifying a set of events after preprocessing is done on the waveforms and the features are extracted as in Mishra et al. (2008) and Wang et al. (2013). In Tripathy et al. (2010), PNN with optimized smoothing factor is used to distinguish between two events: magnetizing inrush and internal fault, each of which would warrant a different course of action. In signal processing applications, raw data are usually a time-domain waveform or a wavelet. This means each data pattern might consist of hundreds of data points. In such scenarios, the role of pre-processing, feature extraction, and feature reduction becomes more prominent as these procedures can decrease the computational burden while improving the overall performance of the network at the same time. For example, in Wang et al. (2013), the data patterns are windows of 200 data points around the R peaks on an ECG waveforms obtained from multiple subjects with the sampling rate of 360 Hz. The window is chosen in a way so that the significant features of ECG

---

1. For example, if the data rows that actually have a value for a certain categorical feature is a small percentage of all the data rows.

beat that would come before and after the R peak are included. Then, the 200 elements of the data vector are normalized using z-score. Principal component analysis (PCA) and linear discriminant analysis are then applied to the normalized data to reduce the number of features that would be fed to the network while maximizing their discriminatory potential. The data are then used to categorize the waveforms into eight types of ECG beats.

Mishra et al. (2008) uses PNN to distinguish 11 types of disturbances in power quality[2] using waveforms of voltage magnitude, frequency, and the phase of the power supply. First, the S-transform of the waveforms is calculated. Then, the covariance matrix of the S-transform matrix (called S-matrix) is calculated. The extracted features are based on standard deviation and energy of the transformed signals as follows:

- Standard deviation of the vector that is formed by selecting the maximum value on each column of the S-matrix
- The energy of the vector that is formed by selecting the maximum value on each column of the S-matrix
- Standard deviation of the vector that is formed by selecting the maximum value on each row of the S-matrix
- Standard deviation of the phase contour

Mishra et al. (2008) shows the importance of feature selection by demonstrating that a PNN can work properly even with small number of features as long as the features contain enough discriminatory information about the model that the data represent.

Although PNNs need labeled data to operate, unlabeled data can be used in special cases after being labeled by unsupervised techniques. In Song et al. (2007), the unlabeled, unstructured set of MRI images are labeled by a self-organizing map (SOM). The SOM algorithm "softly" labels the images in the sense that each image can contribute to multiple classes. However, the degree of contribution to each class might be different.

The rest of this chapter serves as an application-oriented introduction to PNNs and is structured as follows. First, the fundamental statistical concepts that were introduced by Parzen (1962) and used by Specht (1990) to develop PNNs are briefly reviewed in Section 2. Then, Section 3 shows how PNNs use those mentioned concepts for classification. Section 4 deals with some of the practical challenges of implementing PNNs. A simple example of a PNN classifier written in Python is included in Section 5. This chapter is concluded in Section 6.

---

2. Power quality is a measure of how steady the power supply is. It shows how closely the nominal values of the voltage magnitude and frequency and the sinusoidal waveform is followed by the output of the power supply.

2. Preliminary concepts: nonparametric estimation methods[3]

For classification tasks, we need to estimate class-related PDFs as they determine the classifier's structure. Thus, each PDF is characterized by a certain parameter set. For Gaussian distributions, the covariance and the mean value are needed, and they are estimated from the sample data. Let us also assume that we have a set of training sample representative of the type of features and underlying classes, with each labeled as to its correct class. This yields a learning problem. When the form of the densities is known, we are faced with a parameter estimation problem.

In nonparametric estimation, there is no information available about class-related PDFs, and so they have to be estimated directly from the data set. There are many types of nonparametric techniques for pattern recognition. One procedure is based on estimating the density functions $p(\mathrm{x}|\omega_i)$ from sample patterns. If the achieved results are good, they can be included in the optimal classifier. Another approach estimates directly the posteriori probabilities $P(\omega_i|\mathrm{x})$, and is closely related to nonparametric decision procedures. They bypass probability estimation and go directly to decision functions.

The following nonparametric estimation techniques related in its concepts to PNN will be reviewed:

- Parzen windows
- k nearest neighbor
- Potential function

2.1 Parzen windows

One of the most important nonparametric methods for PDF estimation is "Parzen windows" (Meisel, 1972; Poggio and Girosi, 1990). For a better understanding, we will take the simple one-dimensional case. The goal is to estimate the PDF $p(x)$ at the point x. This requires to determine the number of the samples N_h within the interval $[x - h, x + h]$ and then to divide by the total number of all feature vectors M and by the interval length $2h$. Based on the described procedure, we will obtain an estimate for the PDF at x

$$\widehat{p}(x) = \frac{N_h(x)}{2hM} \qquad (14.1)$$

As a support function k_h, we will choose

$$K_h = \begin{cases} 0,5 & : \quad |m| \leq |1| \\ 0 & : \quad |m| > |1| \end{cases} \qquad (14.2)$$

---

3. This section is taken from Meyer-Baese and Meyer-Baese (2004).

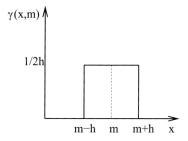

FIGURE 14.1   Clustering process of a two-dimensional vector table. *From Anke Meyer-Bäse, Statistical and syntactic pattern recognition. In Pattern recognition for medical imaging, 147–255, 2004.*

From Eq. (14.1) we get

$$\widehat{p}(x) = \frac{1}{hM} \sum_{i=1}^{M} K\left(\frac{x - m_i}{h}\right) \tag{14.3}$$

with the *i*th component of the sum being equal to zero if m_i falls outside the interval $[x - h, x + h]$. This leads to

$$\gamma(x, m) = \frac{1}{h} K\left(\frac{x - m}{h}\right) \tag{14.4}$$

as it can be seen from Fig. 14.1.

If $\widehat{p}(x)$ is considered to be a function corresponding to the number of samples, we obtain thus

$$\widehat{p}(x) = \widehat{p}(x, M) \tag{14.5}$$

Parzen showed that the estimate $\widehat{p}$ with $M \to \infty$ is bias free, if $h = h(M)$ and

$$\lim_{x \to \infty} h(M) = 0 \tag{14.6}$$

In practice, where only a finite number of samples are possible, a right compromise between M and h has to be made. The choice of h is crucial, and it is recommended to start with an initial estimate of h and then modify it iteratively to minimize the misclassification error. Theoretically, a large M is necessary for acceptable performance. But in practice, a large number of data points increase the computational complexity unnecessarily.

Typical choices for the function $K(m)$ are

$$K(m) = (2\pi)^{\frac{-1}{2}} e^{-\frac{m^2}{2}} \tag{14.7}$$

$$K(m) = \frac{1}{\pi(1 + m^2)} \tag{14.8}$$

or

$$K(m) = \begin{cases} 1 - |m| & : & |m| \leq |1| \\ 0 & : & |m| > |1| \end{cases} \tag{14.9}$$

2.2 k nearest neighbor density estimation

In the Parzen windows estimation, the length of the interval is fixed, while the number of samples falling inside an interval varies from point to point. For the k nearest neighbor density estimation, exactly the reverse holds: the number of samples k falling inside an interval is fixed, while the interval length around x will be varied each time, to include the same number of samples k. We can generalize for the n-dimensional case: in low density areas the hypervolume $V(x)$ is large, while in high density areas it is small.

The estimation rule can be given now as

$$\widehat{p}(x) = \frac{k}{NV(x)} \tag{14.10}$$

and reflects the dependence of the volume $V(x)$. N represents the total number of samples, while k describes the number of points falling inside the volume $V(x)$.

This procedure can be very easily elucidated based on a two-class classification task: an unknown feature vector x should be assigned to one of the two classes ω_1 or ω_2. The decision is made by computing its Euclidean distance d from all the trainings vectors belonging to various classes. With r_1, we denote the radius of the hypersphere centered at x that contains k points from class ω_1, while r_2 is the corresponding radius of the hypersphere belonging to class ω_2. V_1 and V_2 are the two hypersphere volumes.

The k nearest neighbor classification rule in case of two classes ω_1 and, respectively, ω_2 can now be stated

$$\text{Assign x to class } \omega_1(\omega_2) \text{ if } \frac{V_2}{V_1} > (<) \frac{N_1}{N_2} \frac{P(\omega_2)}{P(\omega_1)} \tag{14.11}$$

If we adopt the Mahalanobis distance instead of the Euclidean distance, then we will have hyperellipsoids instead of hyperspheres.

2.3 Potential functions

Potential functions represent a useful method for estimating an unknown PDF $\widehat{p}(x)$ from the available feature vectors (Andrews, 1972 and Meisel, 1972). The estimated PDF is given by a superposition of potential functions $\gamma(\mathbf{x,m})$

$$\widehat{p}(x) = \frac{1}{M} \sum_{j=1}^{M} \gamma(\mathbf{x}, \mathbf{m}_j) \tag{14.12}$$

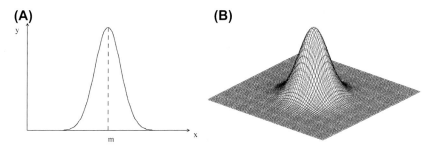

FIGURE 14.2 Potential functions for the (A) one-dimensional and (B) two-dimensional case. *From Anke Meyer-Bäse, Statistical and syntactic pattern recognition. In Pattern recognition for medical imaging, 147–255, 2004.*

Fig. 14.2A and B illustrate an example of a potential function for the one-dimensional and two-dimensional cases.

A possible potential function is

$$\gamma(\mathbf{x}, \mathbf{m}) = \frac{1}{(2\pi)^{\frac{n}{2}}\sigma^n} e^{\left(\frac{-\|\mathbf{x}-\mathbf{m}\|^2}{2\sigma^2}\right)} \tag{14.13}$$

where $\|\mathbf{x}\|$ is a norm in the n-dimensional space. The potential function defines a distance measure (Mahalanobis distance) between two feature vectors $\mathbf{x}$ and $\mathbf{m}$.

Eq. (14.12) describes the complete algorithm for a prespecified potential function. $\widehat{p}(x)$ can be estimated for every x directly from Eq. (14.12).

The choice of the potential function is not so trivial because the width of the potential function plays herein an important role. The smaller its width, the more it peaks and the higher it is. This means it considers only feature vectors in its immediate neighborhood. Larger widths produce a potential function of a smoother shape, as it can be seen from Fig. 14.3.

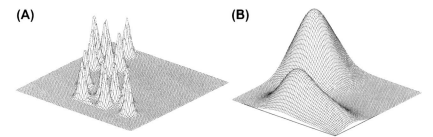

FIGURE 14.3 Importance of the width of the potential function (A) sharper surfaces and (B) smoother surfaces. *From Anke Meyer-Bäse, Statistical and syntactic pattern recognition. In Pattern recognition for medical imaging, 147–255, 2004.*

The choice of a certain variance has a considerable importance on the overlapping degree of neighboring potential functions. This is especially critical when potential functions describing feature vectors belonging to different classes overlap.

There are several possible ways of determining σ Batchelor (1974):

1. Let us assume that within the distance σ from a specific feature vector, there are L other feature vectors. The average value describes the distance $D_L(\mathbf{m})$ from the feature vector to the Lth feature vector:

$$\sigma = \frac{1}{M} \sum_{i=1}^{M} D_L(\mathbf{m_i}) \tag{14.14}$$

To determine L, we have to take into account the distribution of the feature vectors. In many practical problems, $L = 10$ is a good choice.

2. σ can also be chosen as a multiple of the minimal distance between two feature vectors. This is necessary to achieve a certain overlapping degree. In Batchelor (1974), it is recommended to set the multiple equal to 4.

The potential function has the following general properties (Meisel, 1972):

1. $\gamma(\mathbf{x,m})$ has its maximum at $\mathbf{x} = \mathbf{m}$.
2. $\gamma(\mathbf{x,m})$ goes asymptotically toward zero if the distance between the two feature vectors is very large. This is of special importance for multimodal distributions.[4]
3. $\gamma(\mathbf{x,m})$ is a continuous function decreasing monotonically on both sides from the maximum.
4. If $\gamma(\mathbf{x_1,m_1}) = \gamma(\mathbf{x_2,m_1})$, then the feature vectors $\mathbf{x_1}$ and $\mathbf{x_2}$ have the same similarity degree with respect to $\mathbf{m_1}$.

There are several types of known potential functions: besides the above-mentioned unimodal or multimodal normal distributions, potential functions built from orthonormal functions are also of interest (Meisel, 1972). They have the following form

$$\gamma(x, m) = \sum_{i=1}^{R} \lambda_i^2 \Phi_i(x)\Phi_i(\mathbf{m}) \tag{14.15}$$

where $\Phi_i(x)$ is an orthonormal function and λ is a constant. Orthonormal functions fulfill

$$\int \Phi_i(x)\Phi_j(x)dx = \begin{cases} 1 & i=j \\ 0 & \text{else} \end{cases} \tag{14.16}$$

with $\lambda_i = 1$.

---

4. There are more than one clusters for each class.

It is easy to see that they are potential functions because

$$\sum_{i=1}^{\infty} \Phi_i(\mathbf{x})\Phi_i(\mathbf{m}) = \delta(\mathbf{x} - \mathbf{m}) \tag{14.17}$$

where $\delta(\mathbf{x})$ is a Dirac function and $\{\Phi_i\}$ describes a set of orthonormal functions.

The resulting discriminant function can be determined from Eq. (14.16) and is given by

$$\widehat{p}(\mathbf{x}) = \frac{1}{M}\sum_{i=1}^{M}\gamma(\mathbf{x},\mathbf{m}_i) = \frac{1}{M}\sum_{i=1}^{M}\sum_{k=1}^{R}\Phi_k(\mathbf{x})\Phi_k(\mathbf{m}_i) \tag{14.18}$$

where $\widehat{p}(\mathbf{x})$ is the estimated PDF, that is,

$$\widehat{p}(\mathbf{x}) = \sum_{k=1}^{R} c_k \Phi_k(\mathbf{x}) \tag{14.19}$$

The estimated coefficients c_k are

$$c_k = \frac{1}{M}\sum_{i=1}^{M}\Phi_k(\mathbf{m}_i) \tag{14.20}$$

It's important to note that the potential function estimator is both unbiased and asymptotically consistent. We also can easily see that the potential function method is related to Parzen windows. In fact, the smooth function used for estimation is known as either *kernels* or potential functions or Parzen windows.

3. Structure of probabilistic neural networks

To better understand the inner mechanism of the PNNs, one has to look back to Parzen (1962), in which Parzen showed that the probability density function of a set of random variables $X_1, X_2, \ldots, X_n$ with unknown PDF $f(X)$ is estimated with a family of estimators in the form of

$$f_n(x) = \frac{1}{nh(n)}\sum_{j=1}^{n}K\left(\frac{x - X_j}{h(n)}\right) \tag{14.21}$$

where $h(n)$ is a sequence of numbers that satisfies:

$$\lim_{n \to \infty} h(n) = 0 \tag{14.22}$$

Fig. 14.1 shows a simple example of a function $K(y)$ that can be used in Eq. (14.21). However, $K(y)$ is generally a Borel function that satisfies these conditions:

$$1. \quad \sup_{-\infty < y < \infty} K(y) < \infty \qquad (14.23)$$

$$2. \quad \int_{-\infty}^{\infty} |K(y)| dy < \infty \qquad (14.24)$$

$$3. \quad \lim_{y \to \infty} |yK(y)| dy = 0 \qquad (14.25)$$

$$4. \quad \int_{-\infty}^{\infty} K(y) dy = 1 \qquad (14.26)$$

then

$$\lim_{n \to \infty} E[f_n(x) - f(x)]^2 = 0 \qquad (14.27)$$

Eq. (14.27) shows that the Parzen's family of PDFs can estimate the unknown PDF of variable X as $n \to \infty$. The reader is encouraged to study Parzen (1962) for its fundamental importance for PNNs. Specht used this concept in Specht (1990) to formulate an approach to classify patterns of data with unknown class based on initial set of patterns, the real class of which is known.

To explain how a PNN will address that task, consider a sequence of independent identically distributed pairs of random variables as $\{X_i, Y_i\}, i = 1, 2, \dots, n$, known as the training sequence, where $Y_i \in \{1, 2, \dots, M\}$ is the class associated with the pattern $X_i \in A \subset R^p$. The random variable X_i has the unknown probability density function $f_m(x)$ based on its class (fm(x) and is called "class conditional density" in Rutkowski (2004)).

Define the discriminant function of class j as follows:

$$d_j(x) = p_j f_j(x) \qquad (14.28)$$

where p_j is the prior probability of class j and is given by:

$$p_j = \frac{n_j}{n} \qquad (14.29)$$

where n is the number of patterns in the training sequence and n_j is the number of patterns belonging to class j. $f_j(x)$ is the probability density function of variable x. The class of a given pattern x with unknown class is determined as m if:

$$d_m(x) > d_j(x)$$
$$\forall j \in \{1, 2, \dots, M\}, j \neq m \qquad (14.30)$$

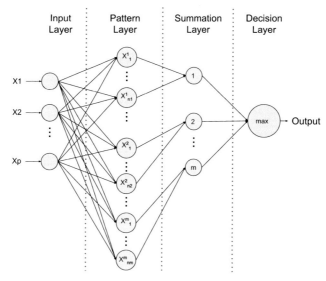

FIGURE 14.4   The basic structure of a probabilistic neural network according to Specht (1990). The network has four basic layers: the input layer that grabs and distributes the input vector, the pattern layer that applies the kernel to the input, the summation layer that gets the average of the output of the pattern units for each class, and the decision layer that declares the class assigned to input vector based on the unit with the maximum output from the summation layer.

Since $f_j(x)$ is usually unknown, its estimator is used in its stead. Using Eq. (14.21), we can estimate $f_j(x)$ as follows:

$$\widehat{f}_{j,n}(x) = \frac{1}{h_{nj}^p} \sum_{i=1}^{n_j} K\left(x, X_i^{(j)}\right) \tag{14.31}$$

where $X^{(j)}{}_i$ is the ith pattern in the training sequence that belongs to class j. p is the dimension of the input vector and the patterns. Fig. 14.4 shows a basic illustration of what has been formulated so far. The network that Specht (1990) introduced has four layers:

1. Input layer
2. Pattern layer
3. Summation layer
4. Decision layer

The input layer is just a set of p junctions for getting the input vector and distributing the input to the next layer. The pattern layer is where the bulk of the calculations takes place. Each pattern in the training sequence has a dedicated pattern unit that applies a two-step process to the input vector before passing its result to the next layer. In the first step, the pattern unit compares its dedicated pattern to the input vector. This is the distance measure that is explained for the potential functions. Define $D\left(x, X_i^{(j)}\right)$ as the distance measure of input vector x and the ith pattern in the training sequence belonging to class j:

- Dot product, defined as

$$D\left(x, X_i^{(j)}\right) = \sum_{k=1}^{p} x_k \cdot X_{i,k}^{(j)} \tag{14.32}$$

where x_k is the kth element of input vector, and $X_{i,k}^{(j)}$ is the kth element of the ith pattern in the training sequence that belongs to class j. Both vectors are normalized to unit length before the dot product (Specht, 1990).

- Euclidean distance, defined as

$$D\left(x, X_i^{(j)}\right) = \sqrt{\sum_{i=1}^{p} \left(x_i - X_i^{(j)}\right)^2} \tag{14.33}$$

- Manhattan (or city block) distance:

$$D\left(x, X_i^{(j)}\right) = \sum_{i=1}^{p} \left|x_i - X_i^{(j)}\right| \tag{14.34}$$

The second step is applying a nonlinear function (called kernel) to the distance between input and the training pattern. Define the kernel as $K(x,u) = K(D(x,u))$. The kernel function needs to satisfy the four conditions that were put for potential function:

1. The kernel has to have its maximum at $x = u$
2.

$$\lim_{D(x,u) \to \infty} K(x, u) = 0 \tag{14.35}$$

3. $K(x,u)$ is continuous on $-\infty < x < \infty$
4. If $K(x\_1,u) = K(x\_2,u)$, meaning $D(x\_1,u) = D(x\_2,u)$, then x_1 and x_2 vectors have the same degree of similarity with respect to u.

The implication from the first and second conditions is that the appropriate kernel has to be chosen with respect to the choice of the distance measure. The implication of the forth condition is that the vectors with the same distance from the data pattern u will have the same result after the kernel is applied to them. These are some examples of acceptable kernels for the mentioned distance measures:

- For dot product:

$$K(x, u) = \exp\left[\frac{x.u - 1}{2\sigma^2}\right] \tag{14.36}$$

Note that both x and u are normalized before the dot product. Therefore, the maximum value of the dot product can be 1 that occurs at $x = u$.

- For Euclidean distance:

$$K(x, u) = \exp\left[\frac{-D(x, u)}{2\sigma^2}\right] \qquad (14.37)$$

where $D(x,u)$ is the Euclidean distance between x and u. The Euclidean distance is always positive which means the maximum value for the kernel happens when $xx = u$ and $D(x,u) = 0$. This is also true for the Manhattan distance measure.

- For the Manhattan distance:

$$K(x, u) = \exp\left[\frac{-D(x, u)^2}{2\sigma^2}\right] \qquad (14.38)$$

The pattern units will pass the result of their calculations to the summation units. Each class $j \in \{1, 2, \ldots, M\}$ has a dedicated unit in the summation layer. The summing unit of class j calculates the average of the values coming from the pattern units which had a pattern that was associated with class j, as Fig. 14.4 shows.

Another feature that can be incorporated in the summation layer is determining the significance of a false decision for any given class j. Define l_j as the loss coefficient associated with the decision that a given input x belongs to class j while its actual class is different. This coefficient can be added to the discriminant function as follows:

$$\widehat{d}_j(x) = p_j l_j \widehat{f}_j(x) \qquad (14.39)$$

The value of l_j cannot be deducted from the data and is subjectively set depending on the application and significance of the false positive decision for class j. In an application where there is no difference between the classes in that regard, l_j can be set to 1 for all $j \in \{1, 2, \ldots, M\}$.

The role of the decision layer is to pick the largest $\widehat{d}_j(x)$ and declare j as the class of input vector x.

4. Improving memory performance

The PNNs have the advantage of not needing extensive training computation time that is associated with the networks that work with back-propagation training method. However, this advantage comes at the cost of requiring massive memory for operation. Each row of the data set needs an independent unit on the pattern layer to compare the similarity between the input vector and its corresponding data set row. When the PNNs were formulated by Specht (1990), the size of the data sets was significantly smaller than today. Still, the memory requirement was taxing for the technology of the day. This fact still holds today despite the advancement of the hardware technology since early 1990s. We explore two families of approaches to reduce the size of the PNN problem without significant loss of performance.

4.1 Feature reduction using principal component analysis

The conventional PCA can be used to reduce the size of the data patterns and the inputs without losing much of the information embedded in the data set. In that regard, PCA projects a high dimensional data vector $X \in R^n$ into a lower dimension subspace $R^q \subset R^n$, knowing $n > q$. Here, a brief explanation about applying PCA to data patterns is included. The details of the method and why it works can be found in Jolliffe (2011).

Suppose $X \in R^{d \times n}$ is a matrix of data patterns where each row represents a pattern and each column represents an individual feature. The goal is to find a mapping $M \in R^{n \times q}: R^n \rightarrow R^q$ that can map the rows of matrix X to new rows of matrix $\widehat{X}$ where $\widehat{X} \in R^{d \times q}$ is the lower dimension approximation of matrix X. The matrix W is obtained in this fashion:

1. Center the data matrix by subtracting each element by the mean value of its corresponding column:

$$X_{ij\_new} = X_{ij\_old} - \mu_j \tag{14.40}$$

μ_j is given by:

$$\mu_j = \frac{1}{d} \sum_{i=1}^{d} X_{ij} \tag{14.41}$$

where d is the number of rows in matrix X, which is the number of data patterns.

2. Form the covariance matrix as $V = X^T X$
3. Find the eigenvalues of matrix V and their corresponding eigenvectors. Suppose $\lambda_1, \lambda_2, ... \lambda_n$ are eigenvalues of the covariance matrix V in descending order and $v_1, v_2, ... v_n$ are their corresponding eigenvectors (v_i corresponding with λ_i and $\lambda_1 > \lambda_2 >, ..., > \lambda_n$).
4. Form the mapping matrix M, the columns of which are the first q eigenvectors of matrix V:

$$M = [v_1 | v_2 | ... | v_q]$$

Choosing q is a tradeoff between the level of dimension reduction and the preservation of the information. As q increases toward n, the information loss will reduce. However, smaller values of q will reduce the size of the data more.

5. Calculate the mapped data using matrix M:

$$\widehat{X} = XM \qquad (14.42)$$

This approach can reduce the computation time by reducing the number of input units and the weight matrix. This reduction can reduce the calculation time that is spent in the input layer.

PCA has been used in Wu et al. (2007) to reduce the dimension of the input vector from 12 to 5 before using the input to classify leaf images into 32 categories of plants. Also Othman and Basri (2011) used PCA to extract features from a data set of medical images. Because the original inputs are vectors made from individual images of $M \times N$ pixels, each input vector has $M \times N$ elements. This is dramatically more than the 12 input features in Wu et al. (2007). Then PCA is used to reduce the dimension of the input vectors from $M \times N$ to $d \ll M \times N$. In Wang et al. (2013), the input vector is a waveform obtained by applying a window of 200 data points to ECG signals. The mentioned windows are centered on R peaks (100 points on each side of the R peak). This means the original input vector has 200 elements. The vector elements are then normalized using Z-score method to decrease superficial differences between the sample waveforms using this formula:

$$x_{z\_score} = \frac{x_{original} - \mu}{\sigma} \qquad (14.43)$$

where μ is the mean value of $x_{original}$ and σ is the standard deviation. Then, PCA is used to reduce the features before the data were used to form the pattern layer.

4.2 Pattern layer size reduction using clustering

In a normal PNN, each data pattern is assigned to a pattern unit in the pattern layer that individually measures the similarity between the input vector and its assigned data pattern. One can imagine that in this architecture the pattern layer can get extremely (and possibly impractically) large as the size of the data gets bigger. We know from Parzen (1962) that larger data sets can increase the accuracy of the estimation of the probability density function of the data. So a larger data set should be welcome and not a concern for practicality. One prominent approach to reduce the number of pattern units without losing too much information is creating clusters from data patterns using k-means algorithm. This comes from the idea that not all the data patterns contain original, independent, and discriminating information.[5]

---

5. This is independent from PCA, which can be used in conjunction with k-means to reduce the computation burden of the PNN even more.

With this approach, each pattern unit has a group of one or more data patterns assigned to it. Each pattern group (known as cluster) is represented by its centroid. Here is a brief explanation of the k-means clustering algorithm for PNN:

1. Determine the number of the clusters, which will be the same as the number of the pattern units.
2. Assign a random centroid to each cluster. The centroid of the cluster is the mean value of all the data patterns in the cluster. Initially, it has a random value that will be adjusted by the data patterns that will be added.
3. Add a new data pattern to the cluster that can minimize this formula:

$$\frac{(A^j)^2}{(A^j + 1)^2}\|x^k - x^j\|^2 \tag{14.44}$$

where A^j is the number of patterns already in the cluster, x^j is the coordinates of the cluster centroid, and x^k is the new data pattern vector that is being added to the cluster.

4. Update the cluster centroid using the coordinates of the newly added pattern that minimized Eq. (14.44):

$$x^j_{new} = \frac{A^j_{old}x^j_{old} + x^k}{A^j_{old} + 1} \tag{14.45}$$

$$A^j_{new} = A^j_{old} + 1 \tag{14.46}$$

5. After all the data patterns are added to their clusters, use the cluster centroids instead of the data patterns to form the input weights in the pattern layer.

Note that the output of each pattern unit needs to be multiplied by the number of data patterns in its associated cluster (A^j) before it is added to the summation unit.

5. Simple probabilistic neural network example in Python

In this example, we have included three clusters (in red, yellow, and green) in two-dimensional coordinates (feature 1 and feature 2). Table 14.1 presents the data points along with the class label for each point. In addition to this, Fig. 14.5 presents the two-dimensional feature space for the simple PNN example. The data points are clustered into red, yellow, and green. The desired point to be clustered is shown as black star. In this example, we tried to reflect

TABLE 14.1 Simple probabilistic neural network example: Data.

Feature 1	Feature 2	Class label
0.1	0.9	1 (red)
0.5	0.9	1 (red)
0.2	0.7	1 (red)
0.6	0.6	2 (yellow)
0.8	0.8	2 (yellow)
0.4	0.5	2 (yellow)
0.8	0.5	3 (green)
0.6	0.3	3 (green)
0.3	0.2	3 (green)

the architecture of the PNN as simple as we can. The code is written in Python (Tahmassebi, 2018). It includes three different functions: (1) a function to create a simple dummy data in two-dimensional, (2) a function to calculate all the PNN stages including input layer, pattern layer, kernel, and summation, and (3) a function to visualize the results and data points. All these functions are called in the main function. In this example, we simply employed a two-dimensional Gaussian distribution as the kernel. However, any other kernels can be used. As shown, the desired point (shown in black star) is obviously in the green cluster and the PNN model as seen did correctly cluster this point as green.

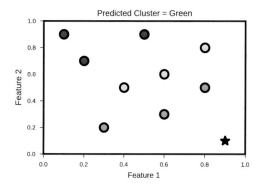

FIGURE 14.5   The two-dimensional feature space for probabilistic neural network. The data points are clustered into red, yellow, and green. The desired point to be clustered is shown as black star.

```python
# Loading Libraries
import numpy as np
import pandas as pd
import matplotlib.pyplot as plt
import matplotlib as mpl
import seaborn as sns
sns.set_style("ticks")
mpl.rcParams['axes.linewidth'] = 3
mpl.rcParams['lines.linewidth'] =7

# Function to create the data
def Create_DataFrame():
    # defining the features and class labels as a dictionary
    data = {
            "Feature_1" : [0.1,0.5,0.2,0.6,0.8,0.4,0.8,0.6,0.3],
            "Feature_2" : [0.9,0.9,0.7,0.6,0.8,0.5,0.5,0.3,0.2],
            "Class_Label" : [1,1,1,2,2,2,3,3,3]
            }

    # converting the dictionary into a dataframe
    df = pd.DataFrame(data = data)

    return df

# Function to calculate PNN
def PNN(df, DesiredPoint):
    # defining a group for each class labels
    Clusters = df.groupby("Class_Label")
    # defining the number of classes as clusters
    NumClusters = len(Clusters)
    # an empty dictionary for calculating the sum of Gaussian for
    each class
    GaussianSums = dict()
    # defining the number of features
    NumFeatures = df.shape[1] - 1
    # defining the standard deviation for Gaussian distribution
    Sigma = 1.0
    # creating features array
    Features = df.drop(["Class_Label"], axis = 1).values

    # INPUT LAYER OF PNN
    # defining a row variable for moving over the data row by row
    _row = 0
    # loop over the number of clusters
    for i in range(1, NumClusters + 1):
        # initialize the GaussianSum for each class
        GaussianSums[i] = 0.0
        # defining the number of points per cluster
        PointsPerCluster = len(Clusters.get_group(i))

        # PATTERN LAYER OF PNN
        # definining temporary sum for holding the sum of X and Y
        elements
        TempSum = 0.0
        # loop over points of each cluster and GaussianSum
        calculation
```

```
        for j in range(1, PointsPerCluster + 1):
            # calculating the X element of Gauassian
            TempX = ( DesiredPoint[0] - Features[_row][0]  )**2
            # calculating the y element of Gauassian
            TempY = ( DesiredPoint[1] - Features[_row][1]  )**2
            # calculating the Gaussian
            TempCoeff = -(TempX + TempY)/(2.0 * Sigma**2)
            # adding the calculated Gaussian for all the points per
    cluster
            TempSum += TempCoeff
            # incrementing the row to cover all points per cluster
            _row += 1
        # storing the GaussianSum per cluster in a dictionary
        GaussianSums[i] = TempSum

    # returning the key of the maximum GaussianSum per cluster
    CalculatedClass = max(GaussianSums, key = GaussianSums.get)

    # Visualization
    Visualization(df, Features, DesiredPoint, CalculatedClass)
    print("Calculated Class = " + str(CalculatedClass))

# Function to visualize the data
def Visualization(df, Features, DesiredPoint, CalculatedClass):

    color_dict = {1 : "Red", 2 : "Yellow", 3 : "Green"}
    plt.figure(figsize=(6,4))
    plt.scatter(Features[:,0],
                Features[:,1],
                s = 200.,
                c = df["Class_Label"],
                cmap=plt.cm.prism,
                marker = "o",
                lw = 3,
                edgecolor='k')
    plt.scatter(DesiredPoint[0],
                DesiredPoint[1],
                s = 200.,
                c = "k",
                marker = "*",
                lw = 3,
                edgecolor='k')
    plt.xlabel("Feature 1", fontsize = 20)
    plt.ylabel("Feature 2", fontsize = 20)
    plt.title("Predicted Cluster = " + color_dict[CalculatedClass],
        fontsize = 20)
    plt.xlim([0,1])
    plt.ylim([0,1])
    plt.show()

# Main Function
def main():
    # desired point for clustering
    DesiredPoint = [0.4, 0.7]
    df = Create_DataFrame()
    PNN(df, DesiredPoint)
```

6. Conclusions

This chapter has been an introduction to PNNs and their numerous applications in science and engineering from a practical point of view. These applications range from simple pattern recognition to complex waveform classification. The PNNs are most effective when used alongside methods of feature extraction and feature reduction, the latter of which can reduce the volume of required calculation and memory as well. Because the pattern units of the same class operate independently from the ones of another class in the pattern layer, the PNN can be considered a great use case for parallel computing. By parallel computing gaining traction during the last decade, the PNNs can emerge again as an attractive alternative to the feedforward back-propagation networks in applications that have massive amounts of data available for training.

References

Adeli, H., Panakkat, A., 2009. A probabilistic neural network for earthquake magnitude prediction. Neural Networks 22, 1018−1024.

Andrews, H., 1972. Mathematical Techniques in Pattern Recognition. J. Wiley Verlag.

Asencio-Cortés, G., Martínez-Álvarez, F., Troncoso, A., Morales-Esteban, A., 2017. Medium-large earthquake magnitude prediction in Tokyo with artificial neural networks. Neural Computing and Applications 28, 1043−1055.

Bankert, R.L., 1994. Cloud classification of avhrr imagery in maritime regions using a probabilistic neural network. Journal of Applied Meteorology 33, 909−918.

Batchelor, B., 1974. Practical Approach to Pattern Classification. Plenum Press Verlag.

Hazrati, M.K., Erfanian, A., 2010. An online eeg-based brain-computer interface for controlling hand grasp using an adaptive probabilistic neural network. Medical Engineering and Physics 32, 730−739.

Jolliffe, I., 2011. Principal component analysis. In: International Encyclopedia, of Statistical Science. Springer, pp. 1094−1096.

Meisel, W., 1972. Computer-Oriented Approaches to Pattern Recognition. Academic Press.

Meyer-Baese, A., Meyer-Baese, A., 2004. Pattern Recognition for Medical Imaging. Academic Press.

Mishra, S., Bhende, C., Panigrahi, B., 2008. Detection and classification of power quality disturbances using s-transform and probabilistic neural network. IEEE Transactions on Power Delivery 23, 280−287.

Mood, A.M., 1950. Introduction to the Theory of Statistics.

Nishanth, K.J., Ravi, V., 2016. Probabilistic neural network based categorical data imputation. Neurocomputing 218, 17−25.

Othman, M.F., Basri, M.A.M., 2011. Probabilistic neural network for brain tumor classification. In: Intelligent Systems, Modelling and Simulation (ISMS), 2011 Second International Conference on. IEEE, pp. 136−138.

Parzen, E., 1962. On estimation of a probability density function and mode. The Annals of Mathematical Statistics 33, 1065−1076.

Poggio, T., Girosi, F., 1990. Networks and the best approximation property. Biological Cybernetics 63, 169−176.

Rutkowski, L., 2004. Adaptive probabilistic neural networks for pattern classification in time-varying environment. IEEE Transactions on Neural Networks 15, 811−827.

Song, T., Jamshidi, M.M., Lee, R.R., Huang, M., 2007. A modified probabilistic neural network for partial volume segmentation in brain mr image. IEEE Transactions on Neural Networks 18, 1424−1432.

Specht, D.F., 1990. Probabilistic neural networks. Neural Networks 3, 109−118.

Tahmassebi, A., 2018. ideeple: deep learning in a flash. In: Disruptive Technologies in Information Sciences, 10652. International Society for Optics and Photonics, p. 106520S.

Tripathy, M., Maheshwari, R.P., Verma, H., 2010. Power transformer differential protection based on optimal probabilistic neural network. IEEE Transactions on Power Delivery 25, 102−112.

Wang, J.-S., Chiang, W.-C., Hsu, Y.-L., Yang, Y.-T.C., 2013. Ecg arrhythmia classification using a probabilistic neural network with a feature reduction method. Neurocomputing 116, 38−45.

Wu, S.G., Bao, F.S., Xu, E.Y., Wang, Y.-X., Chang, Y.-F., Xiang, Q.-L., 2007. A leaf recognition algorithm for plant classification using probabilistic neural network. In: Signal Processing and Information Technology, 2007 IEEE International Symposium on. IEEE, pp. 11−16.

Chapter 15

Design of experiments for uncertainty quantification based on polynomial chaos expansion metamodels

Subhrajit Dutta[1,2], Amir H. Gandomi[3,4]
[1]*Assistant Professor, Department of Civil Engineering, National Institute of Technology Silchar, Silchar, Assam, India;* [2]*National Institute of Technology Silchar, Department of Civil Engineering, Silchar, Assam, India;* [3]*Professor, University of Technology Sydney, Faculty of Engineering and IT, Ultimo, Australia;* [4]*Stevens Institute of Technology, School of Business, Hoboken, NJ, United States*

Chapter points

- This chapter gives an overview of the experimental designs used to construct metamodels for uncertainty quantification of complex systems.
- Here polynomial chaos expansion is adopted as the metamodeling technique.
- The accuracy of PCE metamodel compared with the true computational model is estimated using several error estimates.
- Numerical examples based on uncertainty quantification of structural systems are implemented with increasing level of complexity.

1. Introduction

Uncertainties in real-life engineering problems are unavoidable. As an engineer or a scientist, it is important to identify the various sources of uncertainty, characterize them, and finally propagate them through the system or model to get meaningful statistical estimates of the responses under consideration. To this end, uncertainty quantification (UQ) frameworks are developed to probabilistically tackle uncertainties in complex systems. In most cases, a computational/numerical (say, finite element) model is used to evaluate the system responses. However, in case of large-scale systems, UQ may become prohibitive in the

Handbook of Probabilistic Models. https://doi.org/10.1016/B978-0-12-816514-0.00015-1
369

presence of multiple runs of computationally expensive-to-evaluate models (Dutta et al., 2017). In this context, *surrogate models or metamodels* have gained popularity in the recent past. A computationally inexpensive surrogate model mathematically approximates the original numerical model, thereby reducing the overall computational effort in UQ problems. Examples of some popular metamodels include polynomial chaos expansion (PCE), Kriging or Gaussian process modeling, support vector regression (SVR), and radial basis function (RBF). Here, PCEs are used as the metamodeling tool because it was found to be one of the most efficient methods in computing the stochastic responses of complex engineering systems (Dutta et al., 2018; Blatman and Sudret, 2011a; Dutta and Gandomi, 2019).

In general, the construction of a PCE metamodel consists of the following steps: choosing the design of experiments (DoEs) to generate the trained data set, building the metamodel, i.e., computation of metamodel parameters based on the trained samples, assessing the quality of built metamodel, and finally evaluating statistical estimates for the system responses under consideration using the metamodel. PCE metamodels can be built up using intrusive or nonintrusive approaches (Dutta et al., 2018). Here, a nonintrusive approach has been adopted based on least-square minimization in which the computational/ numerical model is taken as a black box function. To compute the metamodel parameters (PCE coefficients in this case) by least-square technique, a DoE of the input random variables is performed. In computer experiments, the input (random) variables are sampled using a particular probability distribution to investigate the system responses. In case of correlated random variables, joint probability distributions are required to characterize the input uncertainty. The accuracy of any metamodel is greatly influenced by the sampling technique used to generate the experimental designs (EDs) of input random variables. Some of the studies on the influence of EDs on metamodel predictions are reported in the studies by Goel et al. (2008) and Simpson et al. (2001). In this line, researchers have suggested the use of space-filling sampling schemes, wherein the design space of input random variables needs to be filled based on some criteria. Two broad categories of space-filling criteria exists—uniform filling criteria and distance-based criteria. A uniformity-based design is the one in which the number of samples in a subspace of a domain is proportional to the volume of that subspace. On the other hand, the distance-based criteria sample by maximizing the minimum distance between two sample points and vice versa. In this chapter, some of the well-known space-filling DoEs are reviewed in the context of metamodel (PCE) construction. The performance comparison of these DoEs in PCE-based UQ problems are illustrated using analytical and numerical benchmark problems.

This chapter is organized as follows. In Section 3, the basic concept of PCE is introduced. In Section 4, some of the well-known DoEs are reviewed. In Section 5, a comparative study of the different DoE is performed on some benchmark problems with increasing computational complexity.

2. Polynomial chaos expansions

2.1 Basic formulation

Let us consider a physical system which can be idealized by a computational model $\mathcal{M}$. Consider a vector of random variables ξ with support $\mathcal{D}_\xi$ and described by the marginal/joint probability density function (PDF) f_ξ. Considering a finite variance of the physical model $Y = \mathcal{M}(\xi)$ such that

$$\mathbb{E}[Y^2] = \int_{D_\xi} \mathcal{M}^2(\xi) f_\xi d\xi \tag{15.1}$$

the PCE of $\mathcal{M}(\xi)$ is defined as (Ghanem, 1991; Soize and Ghanem, 2011)

$$Y = \mathcal{M}(\xi) = \sum_\alpha y_\alpha \psi_\alpha(\xi) \tag{15.2}$$

where ψ_α is the multivariate orthonormal polynomial, α are is multiindex that maps the multivariate ψ_α to its corresponding bases coordinate, denoted as the deterministic PCE coefficient y_α. In PCE, the model uncertainty is characterized in a decoupled form by the random basis function, and the deterministic coefficients are given by the expansion in Eq. (15.2). A square-integrable random variable, random vector, or random process can be written in a mean-square convergent series using random orthonormal polynomial bases known as PCE for Hermite bases and generalized PCE for other bases. For Hermite bases, the random variables should be normally distributed, whereas nonnormal random variables must be transformed to standard normal random space using transformation schemes. For practical purpose, the expansion of Y must be truncated with a limited number of polynomial terms. Extending to M random variables, the complete N-order PCE is defined as the set of all multidimensional Hermite polynomials ψ whose degree does not exceed N (Ghanem, 1991):

$$Y(\xi_1, \xi_2 \ldots \xi_M) = \sum_{\alpha=0}^{P-1} y_\alpha \psi_\alpha(\xi) \tag{15.3}$$

where P is the number of terms whose degree $\leq N$ such that

$$P = \sum_{k=0}^{N} C_k^{M+k+1} = \frac{(M+N)!}{M!N!} \tag{15.4}$$

The key point in characterizing uncertainty by PCE lies in the determination of its coefficients, which can be achieved by techniques such as Galerkin method by spectral projection (e.g., Monte Carlo sampling, importance sampling, Gauss quadrature, Smolyak's coarse tensorization) or by collocation (Blatman and Sudret, 2011a; Soize and Ghanem, 2011). Recent studies on PCE illustrated that nonintrusive methods are computationally efficient for coefficient determination (Blatman and Sudret, 2011a). Therefore, the nonintrusive method is discussed here considering its robustness and efficiency.

2.2 Polynomial chaos expansion coefficient computation

In general, the PCE coefficient y_α is computed using two approaches: intrusive approaches (e.g., Galerkin scheme) and nonintrusive approaches (e.g., projection, least-square regression). Here, the focus is laid on least-square methods, under the statistical name regression. The term nonintrusive term indicates that the chaos coefficients are evaluated over a set of input realizations $\xi = \{\xi_1, \xi_2, ..., \xi_M\}$, referred to as the experimental design (ED). Considering $Y = \{\mathcal{M}(\xi_1), \mathcal{M}(\xi_2), ..., \mathcal{M}(\xi_M)\}$ as the set of outputs of the computational model for each sample point, the set of coefficients y_α is then calculated by minimizing the least-square residual of the polynomial approximation of "true" computational model:

$$\widehat{y} = \underset{y_\alpha}{\text{argmin}} \frac{1}{M} \sum_{i=1}^{M} \left(\left(\mathcal{M}(\xi_i) - \sum_{\alpha=0}^{P-1} y_\alpha \psi_\alpha(\xi) \right) \right)^2 \tag{15.5}$$

2.3 Polynomial chaos expansion model accuracy

Once the PCE metamodel is constructed, it is required to assess its accuracy in comparison to the "true" computational model. Several statistical error estimators exist, out of which an effective measure is given by the mean square error (MSE)

$$\text{MSE} = \mathbb{E}[\mathcal{M}(\mathbf{X}) - \mathcal{M}^{\text{PCE}}(\mathbf{X})]^2 \tag{15.6}$$

where the validation set $\mathbf{X} = \{x_1, x_2, ..., x_{\text{val}}\}$ consists of samples of the input variables that do not belong to the ED used to construct the PCE metamodel $\mathcal{M}^{\text{PCE}}$. A lower value of MSE corresponds to a more accurate metamodel. A variant estimator of MSE is the relative mean square error (RMSE)

$$\text{RMSE} = \frac{\text{MSE}}{\sigma_Y^2} \tag{15.7}$$

with σ_y^2 being the variance of the computational model response.

In practice, a sufficiently large validation set (typically in the range of $10^4 - 10^5$) is needed to achieve a stable estimate of the MSE. This error estimate is computationally demanding for analytical functions with closed-form expressions as it requires computing model responses for a large sample set. To overcome this issue, an error estimate is developed based on the ED used for PCE construction. In statistical learning, the leave-one-out error (Err_{LOO}) or the relative leave-one-out error (ε_{LOO}) estimate is computed, which performs very well in terms of the estimation bias and the MSE (Blatman and Sudret, 2011a, 2011b). Statistically, ε_{LOO} gives a measure of the coefficient of determination ($R^2 \approx Q^2 \approx 1 - \varepsilon_{LOO}$), while Err_{LOO} can be related to the well-known error estimator, predicted residual sum of squares (PRESS). The prediction accuracy is considered to be higher with a lower value of ε_{LOO} and vice versa (Box 15.1).

3. Design of experiments

3.1 Monte Carlo sampling

Monte Carlo sampling (MCS) is a popular method developed by Metropolis and Ulum for generating random samples from a vector. In this case, following the probability distributions of the input variables X_i, each sample value x_i is generated to fill the design space. MCS used "pseudo" random number generator (PRNG) with the objective of space filling for the input variables under consideration. The term "pseudo" is used here because computer systems use a formula to generate a sequence of random numbers. The following steps are used in MCS to generate N random samples from an input vector $\mathbf{X} = \{x_1, x_2, \ldots, x_N\}$

MCS is a robust DoE approach; however, for practical problems, the use of MCS is inefficient. Some of the issues with MCS are (1) PRNG is reproducible and repeats after a long interval; (2) clustering problem occurs for small sample sizes; (3) for uniform space filling, large sample size is often required, making the simulation computation intensive. To circumvent these issues, researchers proposed other variants of MCS such as quasi-Monte Carlo sampling (QMCS).

BOX 15.1 MCS scheme

Step 1. Define the input vector, X_i, with its probability density function f_{X_i}
Step 2. Draw random samples (using PRNG): $\{x_{11}, x_{12}, \ldots, x_{n1}\}$
Step 3. Feed these samples into the deterministic computational model, $\mathcal{M}$
Step 4. Calculate the output vector, $\mathbf{Y}$ (set of first random sample of the output variables)

$$\{y_{11}, y_{12}, \ldots, y_{n1}\} = \mathcal{M}\{x_1, x_{12}, \ldots, x_{n1}\}$$

Step 5. Repeat from Step 2 for a large number (N) of samples
Step 6. The propagated random variable $Y_1 = \{y_{11}, y_{12}, \ldots, y_{n1}\}$ as its random samples is then characterized for uncertainty quantification
MATLAB built-in functions for generating pseudo random numbers:
- rand(m, n) gives $m \times n$ uniform random numbers
- normrnd(μ, σ) gives normal random numbers
- [exp/bino/nbin/gam/geo/logn/wbl]rnd (par1[a], par2[b], ...) for different types of random numbers
  μ represents mean with σ being standard deviation and [a,b] distribution parameters.

3.2 Quasi-Monte Carlo sampling

QMCS, also known as quasi-random low discrepancy sequence (QRLDS), uses a deterministic sampling scheme to fill the space uniformly. QMCS quantifies the uniformity in terms of discrepancy, i.e., closeness to the uniform distribution. Hence, it avoids the localization/clustering of sampling points. The discrepancy or error bound is valid independent of the input dimensionality. The advantage of deterministic sequence with low discrepancy includes a faster convergence rate of random estimators with respect to the standard MCS method. Several QRLD sequences, Halton, Hammersley, and Sobol and their variants, exist in the literature (Sobol, 1967; Halton, 1960). It is noteworthy that even if these methods inherently aims at space filling, they do not quantify of space-filling measures during the process. Hence, other deterministic methods based on stratified space-filling criteria have been proposed.

3.3 Latin hypercube sampling

The sampling technique using MCS and QMCS discussed in aforementioned sections is powerful and robust in quantifying uncertainty, albeit at the cost of computation. Latin hypercube sampling (LHS) is a stratified sampling scheme used to reduce the number of simulations in quantifying response uncertainty. In this ED method, the input space is partitioned in different "strata," and a representative value is selected from each stratum. The representative values for the domain are then combined to check for any repetition in the complete simulation. Following steps are used in LHS to generate random samples from an input vector $\mathbf{X}$ (Box 15.2).

3.4 Importance sampling

Importance sampling (IS) is one of the popular variance reduction techniques that use additional apriori information about the problem at hand. The basic idea of IS is sampling only in the region of interest. For example, in case of low probability of failure (reliability) estimates, sampling region of interest is close to the failure/safe boundary. In IS technique, an expectation with respect to target density function $f_{\mathbf{X}}(\mathbf{x})$ is approximated by a weighted average of random samples drawn from another distribution $h_{\mathbf{V}}(\mathbf{x})$, which is termed as "importance sampling" density function. The selection of importance sampling function is crucial to produce a good estimate of quantifying uncertainty for system responses. An efficient importance sampling function $h_{\mathbf{V}}()$ should have the following properties: (1) $h_{\mathbf{V}}()$ should be positive for nonzero target distribution; (2) $h_{\mathbf{V}}() \approx |f_{\mathbf{X}}()|$; (3) Computation of $h_{\mathbf{V}}()$ must be simple for any random sample. Some of the features of IS scheme of experimental design include

BOX 15.2 LHS scheme

Step 1. Partition the input sample space of each random variable (RV) into L ranges of equal probability $= 1/L$. It is not necessary to divide the domain with equal probability

Step 2. Generate one representative random sample from each range. Sometimes, the midvalue is used instead of a random sample from range

Step 3. Randomly select one value from L values of each RV to get the first sample s_1

Step 4. Randomly select one value from the remaining L—one value of each RV to get the second sample s_2, and so on upto L sample s_L

Step 5. Repeat Steps 1 to 4 for all the RVs

Step 6. The rest sampling technique is the same as in MCS

MATLAB built-in functions for LHS design:

- $X = $ lhsdesign(L,K) gives L random samples of each $X_1, X_2, ..., X_K$ in an $L \times K$ matrix
- $X = $ lhsdesign(L,K,'smooth','off') gives the median value for each stratum
- $X = $ lhsdesign(L,K,'iterations',J) runs the simulation for J iterations
- lhsnorm generates multivariate Gaussian distributions

- The sampling scheme has a far lower variance than in MCS
- A number of IS random samples are in the order of 10^2 as compared with 10^5–10^6 MCS samples
- The estimate is not affected by
  - Distribution of the input RVs
  - Correlation among the RVs
- Unlike MCS, sampling with correlation does not enter the scheme (as long as $f_X()$ is available)

4. Examples

In this section, PCE-based metamodels are built-up for the uncertainty quantification (propagation) for response(s) of one analytical problem and two numerical examples from the literature. A nonintrusive approach is adopted to find out the PCE coefficients using design of experiments.

4.1 Analytical problem: Ishigami function

The Ishigami function is a well-known benchmark problem for uncertainty quantification. It is a three-dimensional nonmonotonic and highly nonlinear analytical function, given by the following equation:

$$f(x) = \sin(x_1)a\sin^2(x_2) + bx_3^4\sin(x_1). \tag{15.8}$$

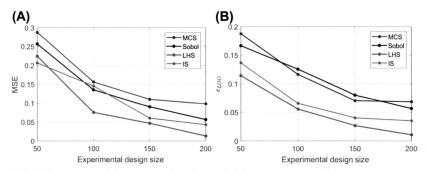

FIGURE 15.1 Error estimates for various DoE: (A) Mean square error; (B) Leave-one-out error. *DoE*, design of experiment; *IS*, importance sampling; *LHS*, Latin hypercube sampling; *MCS*, Monte Carlo sampling; *MSE*, mean square error.

The input vector consists of three independent and identically distributed (i.i.d) uniform random variables $X_i \sim \mathcal{U}(-\pi, \pi)$. The coefficient values (a = 7, b = 0.1) are chosen for this example.

The performances of the various sampling strategies with respect to the error estimates introduced in the previous section are compared. For validation purpose, an MCS-based approach with a validation set of size $N_v = 1000$ is considered. Fig. 15.1A and B provide the MSE and ε_{LOO} calculated considering $N = 50$ to 200 sizes of the ED.

As expected, the LHS ED shows a decrease in both MSE and ε_{LOO} across all ED sizes. All the other design schemes show a consistent decrease in the error estimates, except MCS which shows steady values. In addition, LHS generally shows a more stable behavior with smaller variability between repetitions, especially as the size of the ED increases. In general, the metamodel accuracy seems to be more for larger EDs. Also, comparable results are obtained with IS and Sobol sequence.

4.2 Numerical problems: finite element models

4.2.1 Truss structure

A simply supported truss structure shown in Fig. 15.2 is considered next. This planar truss has 23 members and 23 degrees of freedom with horizontal and

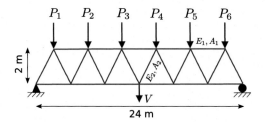

FIGURE 15.2 A planar truss structure with 23 members.

TABLE 15.1 Characterization of input random variables.

RV	Distribution	Mean	CoV
A_1 (m^2)	Lognormal	0.002	0.10
A_2 (m^2)	Lognormal	0.001	0.10
E_1, E_2 (MPa)	Lognormal	2.1×10^5	0.10
$P_1 \cdots P_6$(kN)	Gumbel	50	0.15

vertical displacements along the global directions. The input random variables considered in this cases are the vertical loads, P; the cross-sectional area, A; and Young's modulus E. Ten independent and identically distributed random variables ($M = 10$) are considered whose distributions and parameters are given in Table 15.1 (Fajraoui et al., 2017).

In the underlying deterministic problem, the truss system response considered is the deflection V (Fig. 15.2). This deflection is computed using a finite element MATLAB code considering linear elastic and isotropic material properties.

A PCE approach with degree in the range ($N = 3-10$) is chosen. The accuracy of the PCE is compared with an MCS-based validation set of size $N_v = 250$. Fig. 15.3A and B depict the performance of different sampling methods in terms of the MSE and ε_{LOO} for varying ED sizes from $N = 50$ to 200. This comparison is made with respect to the MCS-based validation set. It is observed from the error estimates that Latin hypercube design outperforms other sampling schemes. Again, the PCE metamodel accuracy improves for larger size of EDs.

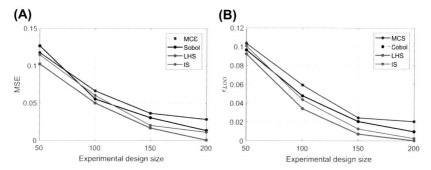

FIGURE 15.3   Error estimates for various DoE: (A) Mean square error; (B) Leave-one-out error. *DoE*, design of experiment; *IS*, importance sampling; *LHS*, Latin hypercube sampling; *MCS*, Monte Carlo sampling; *MSE*, mean square error.

4.2.2 Tensile membrane structure

The next example considered here is real conic tensile membrane structure, which is adopted from a recent studies on tensile membrane structure (TMS) design optimization under uncertainty (Dutta et al., 2017, 2018). The details of this structure is given in Fig. 15.4. Additional details for analysis are

- The membrane yarn directions are warp along radial and fill along circumferential directions.
- Thickness of the membrane is 1.0 mm.
- Material properties of the membrane: Modulus of elasticity, $E = 600.0$ kN/m; Poisson's ratio, $v = 0.4$; Design/nominal yield stress, $f_y = 40.0$ kN/m.
- Design wind load intensity, $W_n = 1.0$ kN/m^2.

The membrane is discretized with constant-strain triangular (CST) elements that have three translational degrees of freedom per node $\{U_X, U_Y, U_Z\}$ along the global $\{X, Y, Z\}$ directions. Proper symmetric boundary conditions are applied for the quarter part of the membrane in addition to the support boundary conditions. The finite element analysis of TMS is implemented in MATLAB to obtain the nodal deformations in global directions. The initial prestress is applied along the element warp (radial) and fill (circumferential) directions as shown in Fig. 15.4A. The quarter part of TMS meshed with triangular CST elements is shown in Fig. 15.4D.

The flexible membrane remains in stable condition because of the existing initial prestress applied along yarn (warp and fill) directions. Membrane structures are primarily designed to resist the action of gusty winds. A TMS

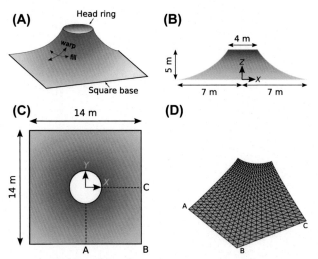

FIGURE 15.4 (A) Frame-supported tensile membrane structure; (B) side elevation; (C) plan; (D) finite element model.

subjected to a design wind load needs to remain in a stable shape and avoid wrinkling and tearing failures. Such failures are governed by the membrane principal stress values. For an optimal performance of these structures, the membrane deformation must be minimized (Dutta et al., 2018). Hence, the structural response parameters of interest for the optimum design of the TMS are the absolute maximum principal stress ($p_{1_{max}}$) to ensure no tearing, the absolute minimum principal stress ($p_{2_{min}}$) to ensure no wrinkling/slackness, and the total nodal deformation (f_δ). These are defined as

$$f_\delta = \sum_j \delta_j = \sum_j \sqrt{U_{X_j}^2 + U_{Y_j}^2 + U_{Z_j}^2} \qquad (15.9)$$

$$\text{for } j = 1, 2, \ldots, J_{node}$$

$$p_{1_{max}} = \max_l p_{1_l} \quad \text{for} \quad l = 1, 2, \ldots, L_{elem} \qquad (15.10)$$

$$p_{1_{min}} = \min_l p_{2_l} \quad \text{for} \quad l = 1, 2, \ldots, L_{elem} \qquad (15.11)$$

where L_{elem} and J_{node} are the total number of elements and nodes, respectively, used in the finite element model of the TMS. U_{X_j} is the jth nodal displacement along X direction.

Owing to the inherent uncertainty in wind load, there is a definite need to quantify the uncertainty that propagates to the aforementioned structural responses. For this uncertainty quantification, the wind load intensity (W)—which needs to be characterized probabilistically—is the only input random variable considered here. The cumulative distribution function (CDF) for W is obtained based on past statistical studies. W follows an extreme type I/Gumbel distribution with a coefficient of variation, $V_W = 0.37$ and a bias factor, $\lambda_W = 0.78$ (Ellingwood and Tekie, 1999).

The PCE-based metamodel (as in Eq. 15.3) is built up to propagate the uncertainty. For a selected set of initial prestress values, the PCE metamodel is constructed using the experimental designs of the wind load intensity (i.e., random variable, $M = 1$). Here, a third-order PCE ($N = 3$) is constructed for the response parameter using ordinary least-square technique, and PCE coefficients are obtained. The PCE metamodels created this way has to be verified against results from more robust but computation-intensive MCS. In this case, MCS-based validation set of size $N_v = 200$ is chosen. The PCE metamodel performance of the different experimental designs are compared in Fig. 15.5A and B. A comparable result is observed for both LHS and Sobol sequence as opposed to MCS and IS.

Next, the density functions are plotted just to compare the system responses obtained from PCE-based metamodel with the "true" computational model counterpart. Fig. 15.6A and B show comparisons of the PCE-based PDF with the "true" PDF (obtained from validation set size of 200 MCS) for f_δ and

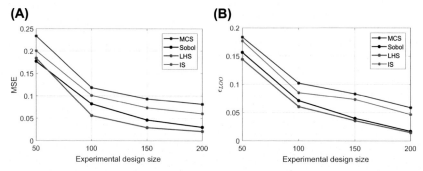

FIGURE 15.5 Error estimates for various DoE: (A) Mean square error; (B) Leave-one-out error. *DoE*, design of experiment; *IS*, importance sampling; *LHS*, Latin hypercube sampling; *MCS*, Monte Carlo sampling; *MSE*, mean square error.

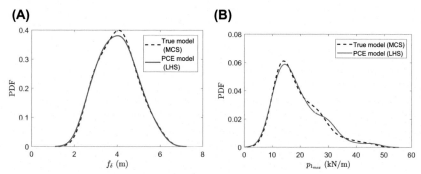

FIGURE 15.6 Comparison of PCE metamodel with the "true" model PDF. *MCS*, Monte Carlo sampling; *PCE*, polynomial chaos expansion; *PDF*, probability density function.

$p1_{\max}$, respectively, for a particular combination of initial prestress. For both cases, the PCE-based metamodel provides a satisfactory accuracy with a leave-one-out error (ε_{LOO}) of 0.01,461 for f_δ and 0.03,166 for $p1_{\max}$.

5. Summary

This chapter introduces the DoE schemes used for quantifying uncertainties in physical system responses. For uncertainty quantification, a metamodel-based approach using PCE is selected. Nonintrusive approach based on least-square technique is used for the determination of PCE parameters/coefficients. For verification of the PCE metamodels, few of the universally used error estimate measures are explained. State-of-the-art experimental design schemes were introduced, and their computer implementations along with features were discussed. Comparative studies of the well-known sampling strategies are performed on numerical/analytical models with varying input dimensionality and computational complexity.

References

Blatman, G., Sudret, B., 2011. Adaptive sparse polynomial chaos expansion based on least angle regression. Journal of Computational Physics 230 (6), 2345−2367.

Blatman, G., Sudret, B., 2011. An adaptive algorithm to build up sparse polynomial chaos expansions for stochastic finite element analysis. Probabilistic Engineering Mechanics 25 (2), 183−197.

Dutta, S., Ghosh, S., Inamdar, M.M., 2017. Reliability-based design optimization of frame-supported tensile membrane structures. ASCE-ASME Journal of Risk and Uncertainty in Engineering Systems, Part A: Civil Engineering 3 (2), G4016001.

Dutta, S., Ghosh, S., Inamdar, M.M., 2018. Optimisation of tensile membrane structures under uncertain wind loads using PCE and kriging based metamodels. Structural and Multidisciplinary Optimization 57 (3), 1149−1161.

Dutta, S., Gandomi, A.H., 2019. Surrogate model-driven evolutionary algorithms: theory and applications. In: Banzhaf, W., et al. (Eds.), Evolution in Action − Past, Present, and Future: A Festschrift in Honor of Erik Goodman's 75th Birthday. Springer-Verlag.

Ellingwood, B.R., Tekie, P.B., 1999. Wind load statistics for probability-based structural design. Journal of Structural Engineering, ASCE 125 (4), 453−463.

Fajraoui, N., Marelli, S., Sudret, B., 2017. Sequential design of experiment for sparse polynomial chaos expansions. SIAM/ASA Journal on Uncertainty Quantification 5 (1), 1061−1085.

Ghanem, R., 1991. Spanos PD Stochastic Finite Elements: A Spectral Approach. Springer- Verlag, Berlin, Germany.

Goel, T., Haftka, R.T., Shyy, W., 2008. Watson LT Pitfalls of using a single criterion for selecting experimental designs. International Journal for Numerical Methods in Engineering 75, 125−155.

Halton, J.H., 1960. On the efficiency of certain quasi-random sequences of points in evaluating multidimensional integrals. Numerische Mathematik 2 (1), 84−90.

Simpson, T.W., Poplinski, J., Koch, P.N., 2001. Allen JK metamodels for computer-based engineering design: survey and recommendations. Engineering Computations 17, 129−150.

Sobol, I.M., 1967. On the distribution of points in a cube and the approximate evaluation of integrals. Zhurnal Vychislitel noi Matematiki i Matematicheskoi Fiziki 7 (4), 784−802.

Soize, C., Ghanem, R., 2011. Physical systems with random uncertainties: chaos representations with arbitrary probability measure. SIAM Journal on Scientific Computing 26 (2), 395−410.

Chapter 16

Stochastic response of primary—secondary coupled systems under uncertain ground excitation using generalized polynomial chaos method

Anoop Kodakkal[1], Pravin Jagtap[2], Vasant Matsagar[2]
[1]*Chair of Structural Analysis, Department of Civil, Geo and Environmental Engineering, Technische Universität München (TUM), Munich, Germany;* [2]*Department of Civil Engineering, Indian Institute of Technology (IIT) Delhi, New Delhi, India*

1. Introduction

1.1 Introduction to stochastic analysis of structural systems

Civil engineering structures are planned, analyzed, designed, and built in modern times using the predictions from computational models. Similar to the most of the other engineering disciplines, civil engineering practice also follows deterministic analyses for structural analysis and design. Real-world structural engineering problems have, however, number of uncertainties associated with them and deterministic models alone are not sufficient for the prediction of these uncertainties. A computational model that quantifies the uncertainties from various sources can provide clear insight into the behavior of the structural system and make reliable predictions of the quantities of interest (QoI). Based on the set objectives, there are two types of uncertainty quantification problems, namely (1) forward propagation of uncertainty problem and (2) inverse problem. In forward propagation of uncertainty (also known as push-forward problem), the uncertainties in the QoI are evaluated from the uncertainties in the inputs. The known uncertainties in the inputs are propagated through the model to obtain the uncertainties in the output QoI (Smith, 2014). However, in case of the inverse problem, from some

Handbook of Probabilistic Models. https://doi.org/10.1016/B978-0-12-816514-0.00016-3
383

observations of the output QoI or other observable quantities, the corresponding uncertainties in the input parameters are decided, in such a way that, when these input parameters are fed through the system, those will produce the given output QoI (Vogel, 2002). In many cases of the structural engineering practice, the structure to be designed is mostly unique and no data are available about the QoI during the analysis and design; in many cases, the situation continues even after construction of the structure. A brief introduction to the uncertainties in structural engineering design for practicing engineers can be found in the work reported by Bulleit (2008).

A full-scale test is infeasible to conduct before construction, which compels engineers and researchers to rely more on computational modeling to acquire details on the structural behavior. Hence, a forward propagation of uncertainty is of prime importance in structural engineering problems. The stochastic model provides better understanding of the phenomena involved and is helpful in providing a reliable prediction with quantified uncertainties in such cases. Each variable is called as a random variable in the stochastic analysis and the uncertainty associated with each random variable is represented by a probabilistic measure such as a probability density function (PDF).

Fig. 16.1 depicts a schematic representation for the forward propagation of uncertainty in the form of a flowchart. Three steps of a typical forward uncertainty quantification are described as (1) identification and characterization of the input parameters, (2) development of a suitable deterministic/stochastic model, and (3) forward propagation of the uncertainty. In the first step, input uncertain parameters are identified and the sources of uncertainty for each input are properly characterized by probability measures such as probability density functions (PDFs). Uncertainties in structural engineering problems can arise from various sources. The geometry itself (Gumbert et al., 2002) or the geometric parameters of the system, such as the area of cross section, moment of inertia, and length of the members, can be uncertain.

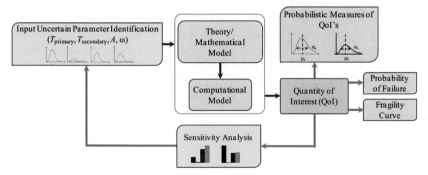

FIGURE 16.1 Forward propagation of uncertainty (Sudret, 2007).

The material properties such as the modulus of elasticity and Poisson's ratio of different materials used in the structure possess inherent uncertainties. The material properties assumed to be having a unique value in the computational model is impossible to achieve in execution. Liel et al. (2009) considered uncertainties in the strength of the material while conducting a seismic analysis. The loading parameters of the system such as the velocity of the wind acting on the structure, the earthquake load that may be induced during the life span of the structure, etc., are also associated with large variability. Cheng et al. (2005) presented an analysis of the suspension bridge considering the uncertainties in wind velocity, geometry, and materials. Jacob et al. (2013b) presented a probabilistic ground motion model developed for the purpose of computing stochastic response of the base-isolated buildings. The boundary conditions and initial values imposed in the computational model also contribute to the uncertainties. The theoretical support conditions idealized in a computational model are impossible to achieve in site conditions. Chakraborty and Dey (1996) have presented examples considering uncertainties in the foundation reaction. The uncertainties listed above contribute to the input uncertainties in structural engineering. The sources of these input uncertainties are identified and the uncertainty in the inputs is characterized from the information on the inputs obtained from experimental observations, expert knowledge, etc. The uncertainty in the inputs may be aleatoric uncertainty that is inherent in the system or epistemic uncertainty that arises from the lack of knowledge of the system.

In the second step, theoretical/mathematical models are developed that represent the real-world problem in hand. The computational model is derived from these mathematical models and is used to predict QoI. There can be a number of outputs derived from a computational model. However, for an engineer/researcher, certain quantities are of more interest as compared with the others, as they govern the decision-making. These quantities are called QoI in this chapter. Developing a stochastic model for a new problem may be time consuming and would require experts' knowledge. In structural engineering applications, the computational model may be proprietary or the user has limited knowledge or access to inside of the code. In these cases, nonintrusive methods for uncertainty quantification that uses the well-tested deterministic models or legacy codes prove handy.

In the third step, the probability measures of the QoI are computed by propagating the well-characterized uncertainties in the inputs through the developed model. There are different methods to propagate the uncertainties through the developed models. Brief discussion on these methods is provided in the upcoming section. The statistical measures of the QoI are obtained from the forward propagation of uncertainty and these quantities are used in analysis and design subsequently.

The sensitivity analysis aims to compute how the uncertainties of the QoI are derived from the uncertainties in input parameters. Generally, in case of the complex problems, only some of the input parameters are observed to have major contribution toward uncertainties in the QoI. The basic aim of the sensitivity analysis is to identify the input parameters, which have greater influence on the QoI. The sensitivity information helps in identifying the important design parameters and is useful in design decisions and optimizations. In addition, by controlling the uncertainty in the important input parameters, by better quality control, for example, the uncertainties in the output QoI can be improved upon. Sensitivity information can be used to reduce the number of uncertain inputs in the subsequent analysis. The probability density function of the QoI can be used for obtaining the reliability of the structure subjected to defined limit state function. Structural reliability analysis computes the probability of limit state violations/exceedance of a structure during a reference period (Choi et al. 2007). Reliability-based approaches have been used in structural analysis and design quite commonly (Gupta and Manohar, 2006; Chaudhuri and Chakraborty, 2006). Reliability-based design approaches are useful in case of some unique structures and situations, for which the design codes are not available.

Fragility curves are important in dynamic analysis of the base-excited structures, which are developed from the stochastic analysis and are measure of the seismic performance of a structure. Numerous studies have been conducted on the seismic performance of bridges, buildings, liquid storage tanks, base-isolated structures, etc., in stochastic domain (Choi et al., 2004; Padgett and DesRoches, 2007; Saha et al., 2016).

1.2 Forward propagation of uncertainty—an introduction to different approaches

The different methods available for forward propagation of uncertainty are briefly described here. In case of the linear parameterized models, the output response uncertainties can be computed explicitly. The mean and standard deviation of the QoI can be obtained explicitly from the mean and standard deviation of the inputs for the linear problems. Mueller and Siltanen (2012) have applied this method for X-ray tomography problems.

Sampling methods are easy to implement and are often used in uncertainty quantification. The classical Monte Carlo (MC) method is a widely used sampling-based method. The MC method was first introduced by Metropolis and Ulam (1949). The MC simulation is based on random sampling from the input uncertain parameters and getting the response for each sampling point through the model. Once the ensemble of the QoI is obtained, the statistical quantities and probability measures of the output QoI can be easily obtained using the theory of statistics. Sampling method is suitable in case of the problems having large number of input uncertainties. Schuëller and Pradlwarter (2009) presented a review of various sampling and nonsampling approaches of stochastic analysis used for structural systems having uncertain input parameters. Alhan and Gavin (2005) presented a study on the reliability analysis of a four-story structure under the uncertainties in base-isolation system characteristic, eccentricity of the isolator, and the ground motion characteristics. The MC simulation technique is used to evaluate the PDF of the responses and probability of failure. Li and Chen (2004) considered uncertainties in both the system and earthquake excitation function parameters. The responses are then calculated using the MC simulations. This technique has wide application in structural engineering problems. The accuracy of the sampling method depends on the number of samples; hence, large number of model evaluations are required in this case which leads to huge computational cost and becomes extremely expensive computationally. However, advancements in scientific computing, availability of robust algorithms, and suitability of parallel processing with ideal efficiency alleviate this limitation to a larger extent, and thus direct MC simulation is powerful (and perhaps the only universal) tool for treating complex stochastic finite element method (FEM) applications.

In perturbation method, a truncated multidimensional Taylor's series expansion for the output QoI yields approximate uncertainty criteria. The accuracy of this method depends on the order of the Taylor expansion (Cacuci, 2003). The Taylor series expansion is developed around the mean value of the input parameters. The early application of this method in the field of structural mechanics can be found in the works reported by Handa and Andersson (1981), Hisada and Nakagiri (1981), and others. This method can be extended to FEM models, and the formulations for higher moments are available in the literature (Cacuci, 2003; Stefanou, 2009). Chaudhuri and Chakraborty (2006) presented a general procedure for uncertainty quantification of multi-degree-of-freedom (MDOF) systems using a perturbation approach.

Reliability-based methods investigate the tails of the response PDF by computing the probability of exceedance of a prescribed threshold probability of failure. First-order reliability methods (FORMs) and second-order reliability methods (SORMs) are used to compute the probability of failure. Li and Der Kiureghian (1995) have used the FORM for nonlinear structure subjected to random excitations. However, these methods do not provide the full stochastic characteristics of the QoI.

Response surface method is used to represent the relationship between several inputs and one output in the form of a polynomial function and is used in stochastic response of structures (Gupta and Manohar, 2004).

Let $\xi = \{\xi_1, \xi_2, \ldots, \xi_n\}$ be a vector of random variables, then the general equation of any response quantity (y) can be written as

$$y = a_0 + \sum_{i=1}^{N} a_i \xi_i + \sum_{i=1}^{N} a_{ii} \xi_i^2 + \sum_{i=1}^{N} \sum_{j=1, j \neq i}^{N} a_{ij} \xi_i \xi_j \qquad (16.1)$$

where a_0, a_i, a_{ii}, and a_{ij} are deterministic coefficients.

Another class of methods, called spectral methods, uses the uncertain parameter to be represented as an expansion called spectral expansion (Ghanem and Spanos, 1991). The input and output uncertain parameters thus represent series expansion and the properties of the orthogonal basis, which are used during the uncertainty propagation process. Wiener (1938) introduced polynomial chaos (PC) expansion to represent functions of Gaussian random processes for constructing a physical theory of chaos. The nomenclature of chaos is a misnomer in the context of uncertainty quantification as the system under consideration is not really chaotic. However, the nomenclature of generalized polynomial chaos (gPC) will still be followed in this chapter.

1.3 Introduction to multiply supported secondary systems

Stochastic analysis details of the two sets of structures are presented in this chapter: (1) fixed-base primary structure (FBPS) with fixed-base secondary system (FBSS) and (2) base-isolated primary structure (BIPS) with FBSS. In both the cases, the secondary systems (SSs) are multiply connected (supported) spatially and the connections are fixed. A brief introduction to the multiply supported SS is presented here.

In case of the important structures, such as nuclear power plants (NPPs), high-rise residential towers, and industrial facilities, relatively lighter structures are attached to heavier supporting structures. The lighter structures are traditionally called as SS as compared with the supporting structures, which are called as primary structures (PSs) or main supporting structures. An equipment housed on the floor of a multistory building is an example of the SS. In addition, a piping system circulating spatially in a building or liquid storage tank supported by a multistory building are examples of other SSs.

In a global perspective, the SS in case of the floor-mounted equipment is connected with the PS at one point only, making it a singly supported SS. However, the another example of SS, wherein a piping system is attached to the generating system in NPPs or a pipeline carrying fluid/liquid along the height of the structure, they are supported or connected to the PS at different floors and referred as multiply supported SSs.

Safety of the SSs remains vital for proper functioning of the power plants, industrial facilities, hospitals, and other important structures especially during and after the unfortunate catastrophic event of an earthquake. The seismic response of the multiply supported MDOF SSs is required to be computed to ensure not only its safety and survivability but also its continual functioning. When the material carried by these systems is sensitive in nature, the limiting failure conditions are quite stringent, to prevent any cascading (secondary) hazard, such as fire following an earthquake. There are many parameters that are uncertain in these types of strategically important structures. The uncertainties in the various system parameters, both PS and SS, contribute to the uncertainty of the global behavior of the coupled system. Chen and Ahmadi (1992) presented a stochastic study of the base-isolated structures with SS using equivalent linearization technique to obtain the mean square statistics of the response quantities. A major contribution toward uncertainties arises from the earthquake excitation functions. For unconventional structures such as this, a stochastic analysis can provide better insight into the behavior of the complex system and is helpful in providing a reliable dynamic response prediction, based on which the design decisions can suitably be made. Challenges from uncertainty quantification, modeling challenges of highly nonlinear behavior of the isolation system in case of the base-isolated buildings, and the challenges from the coupling of PS and multiply supported SS are dealt in the numerical example presented herein.

Details of the computational framework for modeling and estimating the stochastic response of a coupled primary–secondary system are described in the following sections. The details of the gPC method and nonintrusive collocation method based on regression are dealt with in Section 2. The details of the modeling aspects of the SSs and base-isolated buildings are presented in Section 3. The stochastic modeling of the FBPS with FBSS and BIPS with FBSS is presented in details along with a procedure for nonintrusive collocation method in Section 4. Details of the considered numerical example and results are presented in Section 5. The conclusions and outlook are presented in Section 6 of this chapter.

2. Details of the generalized polynomial chaos method

In case of the gPC method, the input/output uncertain parameters are represented by a series expansion, called as the PC expansion series. The PC expansion is a truncated series of polynomials consisting of deterministic coefficients and orthogonal basis (Ghanem and Spanos, 1991). The basic idea of the gPC method is to project the random variable of interest into a stochastic space defined by the orthogonal base functions (Ψ) as the functions of random vector (ξ). The random space is in general multidimensional, and the random vector consists of more than one random variable. Therefore,

$$\xi = \{\xi_1, \xi_2, ..., \xi_n\}. \tag{16.2}$$

Each of these random variables has a random space associated with it, i.e., $\xi_i \in \Omega_i$, $i = 1, 2, ..., n$.

Xiu and Karniadakis (2002) presented a general expression for PC in a probability space (Ω, F, P), where, Ω is the sample space, F is the σ algebra on Ω, and P is a probability measure on Ω. Then,

$$X = x_0\psi_0 + \sum_{i_1=1}^{\infty} x_{i_1}\psi_1\left(\xi_{i_1}\right) + \sum_{i_1=1}^{\infty}\sum_{i_2=1}^{\infty} x_{i_1 i_1}\psi_2\left(\xi_{i_1}, \xi_{i_2}\right) + \qquad (16.3)$$
$$\sum_{i_1=1}^{\infty}\sum_{i_2=1}^{\infty}\sum_{i_3=1}^{\infty} x_{i_1 i_2 i_3}\psi_3\left(\xi_{i_1}, \xi_{i_2}, \xi_{i_3}\right) + \cdots$$

or,

$$X = \sum_{i=0}^{\infty} x_i\psi_i(\xi) \qquad (16.4)$$

where $\Psi_i(\xi)$ are the orthogonal basis which are a set of complete multidimensional polynomials in terms of multidimensional random vector, ξ. For practical purposes, Eq. (16.4) is truncated to a finite number of $N+1$ terms.

$$X \cong \sum_{i=0}^{N} x_i\psi_i(\xi) \qquad (16.5)$$

where N depends on the dimensionality (n) and order of expansion (p). Here, N has a large variation with these quantities, such that

$$N = \frac{(n+p)!}{n!p!} - 1. \qquad (16.6)$$

Table 16.1 shows the total number of terms in the PC expansion with different dimensionality (d) and order of expansion (p). It can be seen that as the number of dimensions or order increases, the number of terms in the PC

TABLE 16.1 Total number of terms in PC expansion with dimensionality d and order of PC expansion p.

		Order				
		$p=1$	$p=2$	$p=3$	$p=4$	$p=5$
Dimensionality	$d=1$	2	3	4	5	6
	$d=2$	3	6	10	15	21
	$d=3$	4	10	20	35	56
	$d=4$	5	15	35	70	126
	$d=5$	6	21	56	126	252

expansion increases rapidly. The number of dimensions (n) is equal to the number of uncertain inputs.

2.1 Details of orthogonal basis

There exists orthogonality property for the base function Ψ such that

$$E\big[\psi_i, \psi_j\big] = E\big[\psi_i^2\big]\delta_{ij} \tag{16.7}$$

where δ_{ij} is the Kronecker delta function, and E is the expectation defined as

$$E\big[\psi_i, \psi_j\big] = \int_\Omega \psi_i\psi_j\rho(\xi)d\xi \tag{16.8}$$

where ρ is the probability density function of ξ. The PC expansion involves orthogonal basis of standard distribution and deterministic coefficients. An assumption made during the solution procedure is that, the weight function corresponding to the orthogonal polynomials and the PDF belonging to the random variable are from the same random space. This implies that there exists an orthogonal polynomial basis corresponding to each random variable of standard distribution (Sepahvand et al., 2010). Table 16.2 shows the orthogonal polynomial and corresponding standard distributions. In cases, where the weight function and the PDF of the random variable belong to the same random space, the PC expansion of the uncertain parameter is an optimal representation. The expansion includes a minimum number of coefficients or terms. In general, an optimal representation has two terms. The higher coefficients are all found to be zeros.

If the weight function and the PDF of the random variable do not belong to the same random space, a space transformation is required and the PC representation thus derived is a nonoptimal PC expansion. In this case, calculation of more coefficients is required for a desirable accurate representation of the

TABLE 16.2 Some common uncertain parameter distributions and the corresponding optimal polynomials (Xiu and Karniadakis, 2002).

Distribution	Density function	Polynomial	Weight function	Support range
Normal	$\frac{1}{\sqrt{2\pi}}e^{\frac{-x^2}{2}}$	Hermite	$e^{\frac{-x^2}{2}}$	$(-\infty, \infty)$
Uniform	$\frac{1}{2}$	Legendre	1	$(-1, 1)$
Beta	$\frac{x^{\alpha-1}(1-x)^{\beta-1}}{B(\alpha,\beta)}$	Jacobi	$x^{\alpha-1}(1-x)^{\beta-1}$	$(0, 1)$
Gamma	$\frac{x^\alpha e^{-x}}{\Gamma(\alpha+1)}$	Laguerre	$x^\alpha e^{-x}$	$(0, \infty)$

random variable. This transformation may be achieved by transforming one variable to the other space. A commonly adopted strategy is to transform both the variables to the same space, for example, to a uniform space.

2.2 Convergence of PC expansion

The selection of order of the representation depends on the accuracy demanded by the user. Fig. 16.2 represents a block diagram of the gPC method. The solution generally starts with a first-order representation of the random variable. The PC coefficients of the output random variable are determined by intrusive or nonintrusive techniques, which will be explained later subsequently. As the PC expansion of the output random variable is determined, convergence criteria are checked to determine how exact the representation is. If the convergence criteria are not fulfilled, a coefficient is added to the PC expansion and the process is continued until the convergence criteria are fulfilled. Ghanem and Spanos (1991) presented that the mean and variance, the first and second moments, are the important parameters to estimate the accuracy of the PC expansion. Because the PC approximation converges in the mean square sense, the convergence of higher order moments is not guaranteed.

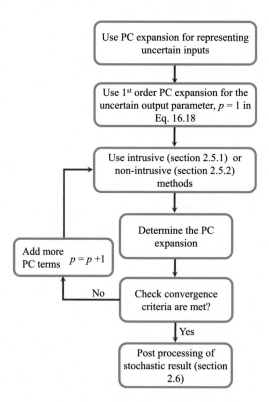

FIGURE 16.2 The generalized polynomial chaos method workflow (Sepahvand et al., 2010).

Different convergence requirements suggest varying convergence criteria for the random variable. Increasing the number of terms in the PC expansion improves the convergence of the random variable. When the PC expansion of a random variable converges to the uncertain parameter and the convergence in the PDF and all statistical moments are guaranteed, in such case the convergence is called a full convergence. If the convergence of other statistical properties is not guaranteed, then it is partially converged. The following convergence definitions can be used for the convergence requirement of the PC expansion (Sepahvand et al., 2010). The PC expansion of an uncertain parameter X is said to converge if

$$\lim_{N \to \infty} \left(\sum_{i=0}^{N} x_i \psi_i(\xi) \right) = X. \tag{16.9}$$

This is the simplest and most powerful definition for convergence of a random variable.

The PC expansion of an uncertain parameter X is also said to converge if

$$\lim_{N \to \infty} \left(\left| \sum_{i=0}^{N} x_i \psi_i(\xi) - X \right| \right) = 0. \tag{16.10}$$

This definition is based on the convergence of the PDF of the random variable, where $P(\ldots)$ represents the probability distribution. An estimate of the central statistical moments can be used for the estimation of error and the convergence of the PC expansion. The estimate of error for the moment of order m is given as

$$\varepsilon_m = E \left[\left| \sum_{i=0}^{N} x_i \psi_i(\xi) - X \right| \right]^m. \tag{16.11}$$

2.3 Normal distribution and Hermite polynomial expansion

A Gaussian distribution is most commonly used in case of the problems that involve natural processes. The optimal polynomial corresponding to a normally distributed parameter is the Hermite polynomial. The following section briefly deals with the Hermite polynomial expansion of a random variable. Hermite polynomials are the classical orthogonal polynomial sequence, named after Charles Hermite and defined as

$$H_n(x) = (-1)^n e^{\frac{x^2}{2}} \frac{d^n}{dx^n} e^{-\frac{x^2}{2}}. \tag{16.12}$$

In Eqs. 16.12, n is assigned the values from 0 to 5 and the resulting first six Hermite polynomials are as listed below.

$$H_0(\xi) = 1$$
$$H_1(\xi) = \xi$$
$$H_2(\xi) = \xi^2 - 1$$
$$H_3(\xi) = \xi^3 - 3\xi \qquad (16.13)$$
$$H_4(\xi) = \xi^4 - 6\xi^2 + 3$$
$$H_5(\xi) = \xi^5 - 6\xi^3 + 15\xi.$$

In practice and computer implementations, the Hermite polynomial can be generated from the recursive relation as

$$H_0(\xi) = 1$$
$$H_{k+1}(\xi) = \xi \cdot H_k(\xi) - k \cdot H_{k-1}(\xi). \qquad (16.14)$$

A normalizing constant, $\sqrt{E\left[H_K^2\right]} = \sqrt{k!}$ is used for obtaining the normalized Hermite polynomials. Fig. 16.3 shows the first six normalized Hermite polynomials.

2.4 PC expansion for multidimensional random variable

If the output response depends on more than one input random variable, the PC expansion includes a complete basis of polynomials. The response may be represented by the truncated PC expansion as

$$Y \cong \sum_{i=0}^{P} y_i \psi_i(\xi). \qquad (16.15)$$

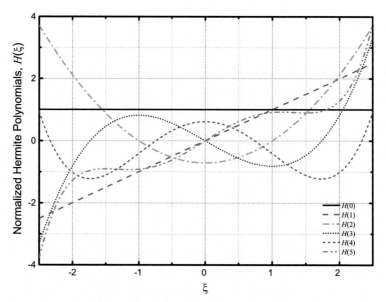

FIGURE 16.3 Normalized Hermite polynomials.

For multivariate cases, the orthogonal basis used to represent the response will be the tensor product of the orthogonal basis used for univariate case. The orthogonal basis of the multivariate case for an order p has the tensor product of univariate basis with the sum of their order less than or equal to p. The number of terms increases with increasing the dimensionality. For a two-dimensional case, the number will be increasing according to the terms of Pascal's triangle. The number of terms required for the expansion depends on the order and dimensionality of the problem in hand. This number of terms for a given order and given dimensionality can be determined from Eq. (16.6).

The orthogonal basis for a second-order expansion over two random dimensions is listed below.

$$\begin{aligned}
\psi_0(\xi) &= H_0(\xi_1)H_0(\xi_2) = 1 \\
\psi_1(\xi) &= H_1(\xi_1)H_0(\xi_2) = \xi_1 \\
\psi_2(\xi) &= H_0(\xi_1)H_1(\xi_2) = \xi_2 \\
\psi_3(\xi) &= H_2(\xi_1)H_0(\xi_2) = \xi_1^2 - 1 \\
\psi_4(\xi) &= H_1(\xi_1)H_1(\xi_2) = \xi_1\xi_2 \\
\psi_5(\xi) &= H_0(\xi_1)H_2(\xi_2) = \xi_2^2 - 1
\end{aligned}$$ (16.16)

For high dimensional problems, the evaluation of the orthogonal basis can be troublesome. Sudret and Der-Kiureghian (2000) presented a ball algorithm for implementing the multivariate PC. The problem of determining the PC of order p and dimension n is to find all possible sequences of nonnegative integers whose sum is equal to or less than p. This problem is equivalent to that of filling $(n + p-1)$ boxes with $(n-1)$ balls. The equivalence of the ball sample, integer sequence, and polynomial basis can be found in Table 16.3. The integer sequence can be obtained from the ball sample as the number of empty boxes between the two consecutive balls. The Algorithm-1 presented here depicts the procedure in details. An example of the algorithm can be referred from the work reported by Sudret and Der-Kiureghian (2000).

Algorithm-1

1. Make $n+p$-1 boxes.
2. Fill the first $(n-1)$ boxes, which is equivalent to the sequence $(0, 0, ..., p)$.

TABLE 16.3 Ball Algorithm—relation between ball samples, integer sequence, and Hermite polynomial basis for $d = 4$, $p = 2$.

Ball sample	Integer sequence	Polynomial basis
	0 0 2 0	$H_2(\xi_3)$
	1 0 1 0	$H_1(\xi_1)H_1(\xi_3)$
	1 0 0 1	$H_1(\xi_1)H_1(\xi_4)$

3. Shift the rightmost ball by one box to the right.
4. If the rightmost ball is already on the last box, then
    4.1 find the rightmost ball that can be shifted to the right. Then shift this ball to the right and,
    4.2 shift all the balls lying to its right to the immediate right.
5. For each ball position, the corresponding integer sequence is stored. Repeat Step 3 and Step 4 to get the next ball position.
6. When all the balls are on the rightmost position, equivalent to the sequence $(p, 0, ..., 0)$ - End.

2.5 Uncertainty propagation in gPC method

The uncertainty propagation refers to the process of determining the output uncertain random variable from the uncertain input random variable. Consider the model that produces output $Y = \eta(X)$. The input is uncertain and X is the random variable, so is the output Y. As we expressed the input random variable X in Eq. (16.4), the output parameter Y can also be represented by a PC expansion. It is usual to employ the same orthogonal basis $\Psi_i(\xi)$.

$$Y = \sum_{i=0}^{\infty} y_i \psi_i(\xi) \tag{16.17}$$

$$Y \cong \sum_{i=0}^{P} y_i \psi_i(\xi). \tag{16.18}$$

Because the PC expansion is truncated to order n, this may be called as PC-n representation. The output parameters are represented as a PC expansion. The coefficients in this expansion can be obtained through two classes of methods namely intrusive and nonintrusive methods.

2.5.1 Intrusive methods—stochastic Galerkin

In intrusive methods, the coefficients in Eq. (16.18) are determined by Galerkin projection method. From the original equation, $Y = \eta(X)$, a system of stochastic equations by method of Galerkin projection is represented as

$$\langle Y, \psi_k \rangle = \langle \eta(X), \psi_k \rangle. \tag{16.19}$$

Substituting both sides for X,Y, with their corresponding PC expansion,

$$\left\langle \sum_{i=0}^{P} y_i \psi_i(\xi), \psi_k \right\rangle = \left\langle \eta \left(\sum_{i=0}^{N} x_i \psi_i(\xi), \right) \psi_k \right\rangle \tag{16.20}$$

for $k = 1, 2, ..., P$.

Now, using orthogonality property of the polynomials, the left-hand side is simplified as $y_i \langle \psi_i(\xi), \psi_i(\xi) \rangle$. The right-hand side should be evaluated; the terms therein can be cumbersome to evaluate if the function η is complicated. New code needs to be developed to solve these equations. The intrusive methods cannot be used in cases where the model equations are not explicitly known. The method becomes more complicated as the model equation is complex.

2.5.2 Nonintrusive methods

Unlike intrusive method, the nonintrusive method does not require any modification to the existing codes. The method aims at obtaining the polynomial representation of the QoI through a number of evaluations of the deterministic model. This method is more appealing to engineers and researchers as it allows to use legacy codes, validated, and tested models without requiring any modifications. Because the method is based on multiple independent evaluations of the deterministic model, it is an embarrassingly parallelizable computation as compared with the Galerkin projection.

2.5.2.1 Discrete projection

The output is represented as a PC expansion in Eq. (16.18). The coefficients of the expansion can be evaluated through application of the projection theorem, which gives

$$y_i = E[\eta(\xi)\psi_i(\xi)] = \int_{\Gamma^m} \eta(\xi), \psi_i(\xi)\rho_m(\xi)d\xi. \tag{16.21}$$

In quadrature methods, the coefficients are evaluated by approximating the integral in Eq. (16.21) with numerical quadrature. This method is also called pseudospectral approach and nonintrusive spectral projection. Ghanem and Spanos (1991) explained the approach for determining the coefficients via this method in details. Gauss quadrature may be used for evaluating the integral in Eq. (16.21). The integral of $\eta(\xi)\rho(\xi)$ [a, b] can be represented by a quadrature rule as

$$\int_a^b \eta(\xi)\rho(\xi)d\xi \approx \sum_{k=1}^q w_k\eta(\zeta_k) \tag{16.22}$$

where q is the order of the quadrature, ζ_k are the quadrature points, and w_k are the quadrature weights. For multidimensional integrals, the product rule requires quadrature points in each dimension. Hence, the number increases exponentially with the dimension, known as the curse of dimensionality.

2.5.2.2 Collocation method—least square regression

This approach is based on the least square minimization of the discrepancy between the uncertain parameter and its truncated PC expansion. In collocation methods, the error is forced to make zero at certain points and these points are called collocation points. Let $\xi^1, \xi^2, ..., \xi^n$ be the n number of collocation points and $Y^1, Y^2, ..., Y^n$ be the responses at these collocation points. Then,

$$
\begin{bmatrix}
\sum_{i=0}^{N} \psi_0(\xi^1)\psi_i(\xi^1) & \cdots & \sum_{i=0}^{N} \psi_0(\xi^n)\psi_i(\xi^n) \\
\vdots & & \vdots \\
\sum_{i=0}^{N} \psi_n(\xi^1)\psi_i(\xi^1) & \cdots & \sum_{i=0}^{N} \psi_n(\xi^n)\psi_i(\xi^n)
\end{bmatrix}
\begin{Bmatrix} y_0 \\ \vdots \\ y_N \end{Bmatrix}
=
\begin{Bmatrix}
\sum_{i=0}^{N} Y^0 \psi_i(\xi^0) \\
\vdots \\
\sum_{i=0}^{N} Y^n \psi_i(\xi^n)
\end{Bmatrix}.
$$

$$
\begin{bmatrix}
\psi_0(\xi^1) & \cdots & \psi_P(\xi^1) \\
\vdots & & \vdots \\
\psi_0(\xi^1) & \cdots & \psi_P(\xi^1)
\end{bmatrix}
\begin{Bmatrix} y_0 \\ \vdots \\ y_P \end{Bmatrix}
=
\begin{Bmatrix} Y^0 \\ \vdots \\ Y^n \end{Bmatrix}
\rightarrow Ay = Y
$$

(16.23)

This results in a system of linear equations which on solving yields the polynomial coefficients, y_i. The number of sampling points must be at least equal to $P + 1$. If more number of collocation points are selected, it will result into more number of equations. Hence, the solution is required to be determined by regression analysis. The left-hand side matrix is called an information matrix, $[A]$. The solution by the least square minimization problem reduces to

$$
y = (A^T A)^{-1} A^T Y. \tag{16.24}
$$

For multivariate problems, i.e., problems with more than one random variable being uncertain, a common methodology adopted is to find the collocation points from the full tensor product space. Because the number of points in the full tensor product grid increases exponentially, these grids suffer from curse of dimensionality.

2.6 Postprocessing of polynomial chaos expansion

The PC expansion of the QoI is obtained as a series at the end of the analysis in Section 2.5. This series expansion can be used to evaluate the moments of the distribution, PDF, etc., of the QoI. This section explains how the PC expansion obtained for the QoI can be used in the postprocessing of the stochastic problem. The PC expansion can also be used to sample the QoI. Sampling from the PC expansion is inexpensive as compared with the direct MC evaluation, as one only needs to evaluate a polynomial.

2.6.1 Statistical moments

The statistical moments such as the mean and variance can be obtained directly from the PC expansion of the QoI using the orthogonal property of the polynomial with normalized basis,

$$\mu_Y = E\left[\sum_{i=0}^{p} y_i \psi_i(\xi)\right] = y_0 \tag{16.25}$$

$$E[Y^2] = \sum_{i=0}^{p} y_i^2. \tag{16.26}$$

Hence, the variance can be obtained as

$$Var(Y) = E[Y^2] - E[Y^2] = \sum_{i=0}^{p} y_i^2 - y_0^2. \tag{16.27}$$

The mean and standard deviations are obtained only using the coefficients of the PC expansion. The higher-order moments can also be obtained from the PC expansion coefficients. Sudret (2007) presented the closed-form solutions of the moments from the PC expansion coefficients.

2.6.2 Estimation of the densities—PDFs

The probability density function of the QoI can be obtained from the PC expansion inexpensively. The PC expansion may be used to sample large number of outputs, and these samples can be used to evaluate the PDF numerically using kernel density estimates. The kernel density estimator of the PDF is $fY_k(y)$, where y is obtained using the following expression

$$fY_k(y) = \frac{1}{n_s h} \sum_{i=1}^{n_s} K\left(\frac{y - y_k^{(i)}}{h}\right). \tag{16.28}$$

In this expression, the function K is called as kernel and $h > 0$ is the bandwidth. More details on this can be referred from the work reported by Wand and Johnes (1995). Inbuilt kernel density functions (*ksdensity*) are available in popular programming languages such as MATLAB, Python, R, etc.

2.6.3 Sensitivity analysis—Sobol' indices

Sensitivity analysis explores the sensitivity of the QoI to the inputs. The variability of the QoI with respect to the inputs is quantified. Sobol' is a variance-based sensitivity analysis. Sudret (2008) presented a method to calculate the Sobol' indices from PC expansion. The Sobol' indices can be evaluated from the PC expansion easily with negligible additional computational cost. The variance from the PC expansion can be determined as

$$D_{\text{PC}} = \sum_{j=1}^{P} y_j^2 E\left[\psi_j^2\right].$$ (16.29)

The PC-based Sobol' indices are determined from the PC expansion with an easy computation. The response expansion coefficients are gathered according to the dependency (α) of each basis polynomial on j^{th} input, which are square-summed and normalized to obtain the sensitivity indices. The total PC-based Sobol' indices can be determined as

$$S_j^{\text{T}} = \sum y_\alpha^2 E\left[\psi_\alpha^2\right]/D_{\text{PC}}.$$ (16.30)

2.6.4 Reliability analysis

Some fundamentals of reliability analysis are presented in this section along with computing the reliability of a structure from the PC expansion. Reliability analysis predicts the probability of failure of a structure under the defined limit state function. The limit state function can be ultimate strength, stability of structures, fatigue strength, serviceability such as allowable deformation, maximum acceleration for human comfort, etc.

Let $g(\xi)$ represent a limit state function where $\xi = [\xi_1, \xi_2..., \xi_n]$ is the random vector. The probability of failure, P_f, can be computed by integrating the joint probability density function of ξ, $f(\xi)$, over the failure domain $\Omega_F = g(\xi) \leq 0$ such that

$$P_f = P[g(\xi) \leq 0] = \int_{\Omega_F} f(\xi)d\xi_1 d\xi_2...d\xi_n.$$ (16.31)

A typical structural limit state function having capacity R and demand S is given as

$$g(R, S) = R - S.$$ (16.32)

FORM and SORM are based on the linear and quadratic approximation of the limit state function $g(\xi) = 0$ at the most probable point. It is seen that, for engineering problems, FORM and SORM are accurate enough to compute the failure probabilities. A good introduction to structural reliability can be referred from the works reported by Thoft-Christenses and Baker (1982), Ditlevsen and Henrik (1996), and Rackwitz (2001).

Sampling methods based on the MC sampling are also used for computing the failure probabilities. The methods are independent of the dimension of the problem and are easy to implement. The failure probability may be computed as

$$P_f = \frac{1}{N_{\text{mc}}} \sum_{i=1}^{N_{\text{mc}}} I[g \cdot (\xi) \leq 0]$$ (16.33)

where N_{mc} is the total number of MC samples, I is the indicator function whose value is one, if the event is true, or zero otherwise. There are many improvements to the standard MC method and interested readers may refer to such advanced simulation-based techniques such as importance sampling (Mori and Ellingwood, 1993) and subset simulation (Au and Beck, 2001).

Failure probability can be computed from the PC expansion of the QoI. The simplest way for it is to use the MC approach for estimating the probability of failure from the PC expansion. In Eq. (16.33), the actual limit state function may be approximated by a PC expansion computed earlier as

$$P_f = \frac{1}{N_{mc}} \sum_{i=1}^{N_{mc}} I\left\{ \left[Y_{\lim} - \sum_{i=0}^{P} y_i \psi_i(\xi) \right] \leq 0 \right\} \tag{16.34}$$

where $Y_{\lim}$ is the limiting value of Y defining the limit state function. Corresponding reliability index can be computed as (Jacob et al., 2013a)

$$\beta = -\Phi^{-1}(P_f) \tag{16.35}$$

where Φ^{-1} is the cumulative distribution function (CDF) of the standard normal random variable. Sampling is done from the PC surrogate, which is inexpensive to compute. The accuracy of the computed reliability index depends on the accuracy of the PC expansion. A hybrid approach is proposed by Li and Xiu (2010) to enhance the accuracy of reliability computed from the PC expansion while sampling more from the PC surrogate and sampling less from the original system.

2.6.5 Fragility analysis

Sudret and Mai (2013) and Saha et al., (2016) have used PC-based fragility analysis for dynamical systems under base excitation from earthquake-induced loading. The PC expansion of the QoI can be developed for different amplitude levels and the probability of failure is computed as explained in the previous section. The fragility curves are developed from the probabilities of failures of each of the peak ground accelerations (PGAs).

3. Deterministic model of base-isolated SDOF and base-isolated MDOF structure with secondary system

Two examples are presented here in the numerical study. A base-isolated building is modeled as a single-degree-of-freedom (SDOF) system. In addition, the BIPS and FBPS with FBSS under uncertain base excitations are studied. The details of the modeling and governing equations of motion for these systems are described in the following sections.

3.1 Details of modeling base-isolated SDOF system

The base-isolated SDOF system is shown in Fig. 16.4A, in its simplified version, wherein assumption of the rigid superstructure behavior is employed (especially in the first-isolation mode). The mass of the superstructure and that of the base-isolation system is lumped together and is assumed to act together. The governing equation of motion for this system is written as

$$m_b\ddot{x}_b(t) + F_b(t) = -m_b\ddot{x}_g(t) \qquad (16.36)$$

where $\ddot{x}_b(t)$ is the absolute acceleration for the displacement of the mass with respect to the ground at the isolator level $(x_b(t))$; m_b is the total mass of the base-isolated structure; $F_b(t)$ is the restoring force in the isolation system; and $\ddot{x}_g(t)$ is the ground acceleration. The restoring force is calculated based on the equivalent linear elastic-viscous damping model of the isolator (ASCE-7: 2010 and IBC, 2012). The linear equivalent restoring force developed in the isolator can be expresses as (Matsagar and Jangid, 2004)

$$F_b(t) = c_b\dot{x}_b(t) + k_b x_b(t) \qquad (16.37)$$

where c_b and k_b are the damping and stiffness of the base-isolation system, respectively.

3.2 Details of fixed-base primary structure with fixed-base secondary system and base-isolated primary structure with fixed-base secondary system

In general, seismic analysis of the multiply supported MDOF SSs is conducted using two approaches, and those are classified as coupled and decoupled (cascading) approaches. In coupled analysis, the combined primary–secondary system is considered as a single integrated dynamic unit. The input base excitation is considered in the form of a ground acceleration time history or ground response spectrum. Because of complexity in modeling, the combined structural system and practical difficulties in carrying out the coupled dynamic analysis, decoupled analysis of the primary–secondary system is traditionally adopted. In case of the decoupled analysis, complex structures are commonly subdivided into two subsystems: primary and secondary. However, the decoupled analysis has certain shortcoming, such as it ignores the interaction effect between primary–secondary systems, which is an important investigation parameter for design of the SSs connected in case of the NPP-type important structures. In the present study, response of the coupled primary–secondary systems is determined while duly considering the interaction effect between the two. Because actual piping systems are generally supported at several locations on the supporting PSs, in this scenario it is important to evaluate the actual governing response quantities for which the realistic piping design can be exposed to. This study is conducted while modeling the MDOF secondary (piping) system connected to the MDOF

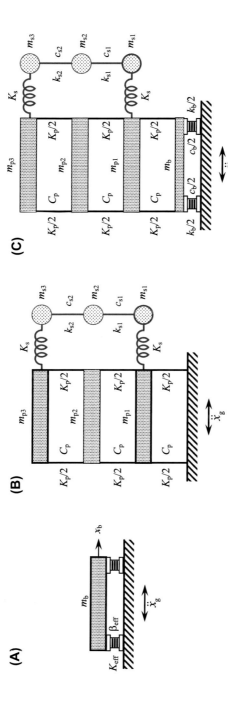

FIGURE 16.4 Details of (A) base-isolated single-degree-of-freedom system, (B) fixed-base primary structure with fixed-base secondary system (FBSS), and (C) base-isolated primary structure with FBSS.

fixed-base and base-isolated primary (supporting) structure. The exact details of the number of primary and secondary piping stories investigated in the numerical study of this chapter are provided in the upcoming discussion. The comparison between the response quantities obtained from the fixed-base and base-isolated conditions of the PS evaluates the effectiveness of the base isolation for protecting the multiply supported secondary (piping) system when subjected to the earthquake ground motion.

A lumped mass modeling approach with one lateral degree-of-freedom at every floor level is adopted to carry out the numerical investigations of the FBPS and BIPS with multiply connected secondary (piping) systems. A three-storied building frame in the form of a three-degree-of-freedom structure with 5 m bay width and story height is considered as a PS in the present investigations. It is assumed that the floors of this structure are rigid in their own plane owing mainly to the diaphragm action of the slab, and the inextensible weightless columns provide only lateral stiffness to the structure. Fig. 16.4B and C, respectively, show the mathematical models for the FBPS with FBSS and BIPS with FBSS.

In case of the BIPS, the superstructure is mounted on the corresponding isolation system, where an additional lateral degree-of-freedom is introduced into the structure at isolation level, at base mass of raft. The SS in the form of a piping system is running parallel along the height of the PS and it is considered connected at the first and top floor locations of the PS. The piping system is idealized into a three-degree-of-freedom system while equally dividing the total mass of the piping system into three lumped masses, which are assumed to be lumped at the equal floor location to that of the PS (though this is not always necessary). The basic properties of the lumped mass model in the form of mass, stiffness, and damping for the considered structural system of the PS and SS are reported in Tables 16.4 and 16.5, respectively (Biondi et al., 1996).

TABLE 16.4 Lumped mass modeling properties of primary structure.

Sr. No.	Story mass, kg $(M_p = m_{p1} = m_{p2} = m_{p3})$	Story stiffness, N/m (K_p)	Damping ratio (ξ_p)
1.	50×10^3	80×10^5	0.05

TABLE 16.5 Lumped mass modeling properties of secondary (piping) system.

Sr. No.	Story mass, kg $(M_s = m_{s1} = m_{s2} = m_{s3})$	Story stiffness, N/m $(K_s = k_{s1} = k_{s2})$	Damping ratio (ξ_s)
1.	4×10^2	10×10^3	0.05

TABLE 16.6 Properties of laminated rubber bearing.

Sr. No.	Isolator mass, kg (m_b)	Isolator stiffness, N/m (k_b)	Isolator damping ratio (ξ_b)
1.	2.5×10^2	80×10^4	0.1

TABLE 16.7 Modal frequencies of FBPS and BIPS with FBSS.

	Modal frequency (Hz)	
Mode no.	FBPS with FBSS	BIPS with FBSS
1	0.60730	0.33956
2	0.89824	0.61239
3	1.12540	1.12485
4	1.47061	1.47039
5	2.51139	2.05865
6	3.62794	3.49496
7	–	29.92205

In this study, linear modeling of the isolation system is carried out while calculating the effective stiffness ($K_{eff} = k_b$) and effective damping ratio ($\beta_{eff} = \xi_b$) of the isolator. The isolation damping coefficient is $c_b = 2\xi_b M \omega_b$, where $\omega_b = 2\pi/T_b$ is the isolation angular frequency, T_b is the isolation time period, and M denotes the total mass supported by the isolators. The basic properties of the isolation system considered in the present study are reported in Table 16.6 (Biondi et al., 1996). The modal frequencies of the respective models developed using these tabulated properties of the FBPS and BIPSs attached with a FBSS are summarized at Table 16.7.

3.3 Governing equations of motion

3.3.1 Fixed-base primary structure with secondary system

The governing equations of motion for the FBPS with FBSS modeled in the form of an MDOF system subjected to ground excitation are written in the matrix form as

$$[M_T]\{\ddot{x}_T(t)\} + [C_T]\{\dot{x}_T(t)\} + [K_T]\{x_T(t)\} = -[M_T]\{r\}\ddot{x}_g(t) \qquad (16.38)$$

where $[M_T]$ is the mass matrix, $[K_T]$ is the stiffness matrix, and $[C_T]$ is the damping matrix of the total system comprising of the PS and the SS; $\{r\}$ is the vector of influence coefficients; and $\ddot{x}_g(t)$ is the input ground motion acceleration, respectively. Here, $\{x_T(t)\}$ is the total displacement vector comprising of lateral translational degrees of freedom of the PS and degrees of freedom of the SS.

The total displacement vector is of the form $\{x_T(t)\} = \{x_P(t), x_L(t), x_S(t)\}^T$. The subscript P is considered for the primary DOF (except the linked DOF); the subscript L is considered for the primary DOF which are linked to the SS; and the subscript S is considered for the DOF of the SS. In addition, $x(t)$ represents the relative displacement of the DOF considered with respect to the ground. The detailed form of the mass matrix $[M_T]$, stiffness matrix $[K_T]$, and damping matrix $[C_T]$ is written as

$$[M_T] = \begin{bmatrix} M_P & 0 & 0 \\ 0 & M_L & 0 \\ 0 & 0 & M_S \end{bmatrix}; [K_T] = \begin{bmatrix} K_{PP} & K_{PL} & 0 \\ K_{LP} & K_{LL} & K_{LS} \\ 0 & K_{SL} & K_{SS} \end{bmatrix}; [C_T] = \begin{bmatrix} C_{PP} & C_{PL} & 0 \\ C_{LP} & C_{LL} & C_{LS} \\ 0 & C_{SL} & C_{SS} \end{bmatrix}$$

(16.39)

It may be noted here that the damping matrix is not known explicitly; therefore, it is evaluated by assuming the modal damping ratio, which is kept constant in each mode of vibration.

3.3.2 Base-isolated primary structure with secondary system

On similar lines to that of the abovementioned equations, the governing equations of motion of the superstructure for the BIPS with FBSS modeled in the form of an MDOF system subjected to ground excitation are written in the matrix form as

$$[M_T]\{\ddot{x}_T(t)\} + [C_T]\{\dot{x}_T(t)\} + [K_T]\{x_T(t)\} = -[M_T]\{r\}\left(\ddot{x}_g(t) + \ddot{x}_b(t)\right)$$

(16.40)

where $\ddot{x}_b(t)$ is the acceleration of the base mass for the isolation level displacement, $x_b(t)$ with respect to the ground; and $x_T(t)$ is the relative displacement of the DOF considered with respect to the base-isolation level (i.e., m_b). Correspondingly, the governing equation of motion for the base-isolator mass is expressed as

$$m_b\ddot{x}_b(t) + F_b(t) - C_p\dot{x}_1(t) - K_p x_1(t) = -m_b\ddot{x}_g(t)$$

(16.41)

where F_b is the restoring force developed in the isolator which is dependent on the type of isolation system and its modeling characteristics. The stiffness and damping of the first story are represented by C_p and K_p, respectively, which are kept the same for rest of the stories in this study.

3.3.3 Details of laminated rubber bearing base-isolation system

The most commonly used elastomeric isolation system is represented by the laminated rubber bearings (LRBs). The LRBs consist of steel and rubber plates

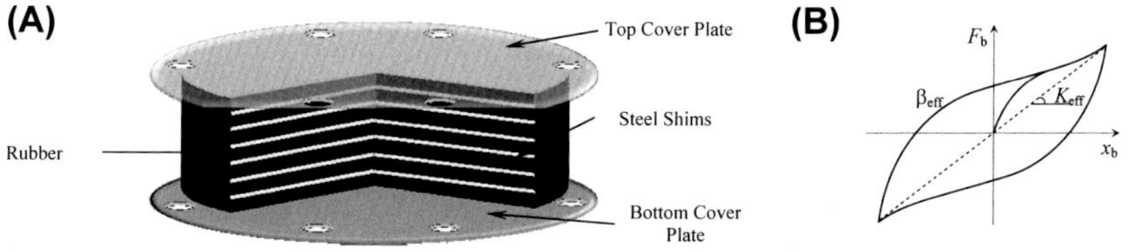

(A)

Top Cover Plate

Steel Shims

Rubber

Bottom Cover
Plate

(B)

F_b

β_{eff} K_{eff}

x_b

FIGURE 16.5 (A) Schematic of laminated rubber bearing (LRB) isolation system used in present study for seismic isolation of PS, (B) force deformation behavior of LRB.

bonded together in alternate layers to form a unit through vulcanization in a single operation under heat and pressure in a mold (Datta, 2010; Matsagar and Jangid, 2011). The LRBs shown in Fig. 16.5A also consist of thick steel plates (shims) bonded to the top and bottom surfaces of the bearing to facilitate the connection of the bearing to the foundation below and the superstructure above. Manufacturing of these bearings uses either natural rubber or synthetic rubber (such as neoprene), which has inherent damping, usually 2%−10% of the critical viscous damping. The horizontally aligned steel shims prevent the rubber layers from bulging out, so that the unit can support high vertical loads with relatively small vertical deflections (1−3 mm under full gravity load). The internal steel shims do not restrict horizontal deformations of the rubber layers in shear and so the bearing exhibits much more flexibility under lateral loads than vertical loads (Kelly, 2001). In general, the mechanical characteristic of the LRB consists of high damping capacity, large horizontal flexibility due to low shear modulus of elastomer, and high vertical stiffness. Fig. 16.5B represents the equivalent linear behavior of the LRB with viscous damping. At large deformations, the LRB exhibits hysteretic and stiffening behavior. The restoring force (F_b) developed in the bearing is given by Eq. (16.37). In general, the stiffness and damping of the LRB are selected to provide some specific values of the isolation time period (T_b) and damping ratio (ξ_b) which are defined as

$$T_b = 2\pi \sqrt{\frac{M}{k_b}} \tag{16.42}$$

$$\xi_b = \frac{c_b}{2M\omega_b} \tag{16.43}$$

where $M = \left(m_b + \sum_{i=1}^{N_{PS}} m_i + \sum_{j=1}^{N_{SS}} m_j \right)$ is the total mass of the base-isolated structure above base-isolation device; m_i is the mass of ith floor of the PS; and m_j is the mass of the jth mass of the SS, whereas

$$\omega_b = \frac{2\pi}{T_b} \tag{16.44}$$

is the angular isolation frequency.

3.4 Solution of equations of motion

In case of the base-isolated structures, there is a large difference in the damping of the superstructure and the isolation device, which makes the system nonclassically damped and it averts the use of classical modal superposition method for solving the equations of motion. Here, in this chapter, Newmark's method of step-by-step integration is adopted as the numerical solution procedure for solving the equations of motion. In this procedure, a linear variation of acceleration over a small time interval of Δt is considered.

4. Stochastic structural dynamics using gPC method

The equation of motion of the base-isolated SDOF under sinusoidal base excitation of amplitude A and angular frequency ω is

$$m_b \ddot{x}_b + c_b + \dot{x}_b + k_b x_b = -m_b[A \sin(\omega t)]. \qquad (16.45)$$

The stochastic modeling using the gPC method and the details of the solution procedure is explained in details in this section. In this chapter, the isolator parameters and the excitation parameters are considered uncertain. Therefore, Eq. (16.45) is rewritten with the stochastic variables as

$$\ddot{x}_b(t, \xi) + 4\pi/T_b(\xi_1)\beta_{\text{eff}}(\xi_1)\dot{x}_b(t, \xi) + \{2\pi/T_b(\xi_1)\}^2 x_b(t, \xi)$$
$$= -A(\xi_2)\sin\{\omega(\xi_3)t\} \qquad (16.46)$$

where $\dot{x}_b(t, \xi)$ and $x_b(t, \xi)$ are the unknown uncertain velocity and displacement at the isolation level relative to the ground, respectively; and T_b is the effective time period of the base-isolation system.

The displacement and acceleration of the SDOF are considered as QoI. They are the function of time and the random vector, $\xi = \{\xi_1, \xi_2, \xi_3\}$. The uncertainties in the time period (T_b) and damping ratio (β_{eff}) are represented by a random variable, ξ_1, whereas ξ_2 and ξ_3 are used to represent the uncertainties in amplitude (A) and angular frequency (ω) of the earthquake base-excitation function, respectively. The random variables are assumed to be independent and identically distributed (iid). A detailed procedure used in the collocation method for solution using the least square regression is presented here, and the same will be used for the numerical study in the present work.

4.1 PC expansion of time-independent input uncertainties

The input parameters are time independent and are represented by a PC expansion. The jth input uncertain parameter (X^j) is represented by the gPC expansion as (Ghanem and Spanos, 1991)

$$X^j(\xi_j) \approx \sum_{i=0}^{n} x_i^j \psi_i(\xi_j) \qquad (16.47)$$

where x_i^j represents the gPC expansion coefficient and $\psi_i(\xi_j)$ is the orthogonal basis used to model the input uncertainty. The orthogonal basis may be selected for the corresponding distribution of the input parameters. It is preferred to choose the optimal polynomials from Table 16.2. A Hermite polynomial is used as the orthogonal basis for normal random variable.

4.1.1 Determination of PC expansion coefficients of input parameters

The gPC expansion coefficients in Eq. (16.47) can be obtained using the orthogonality property of the Hermite polynomials (H_i) as

$$x_i^j = \frac{1}{\langle H_i^2 \rangle} \int_{-\infty}^{\infty} X_j H_i(\xi_j) \rho(\xi_j) d\xi_j \qquad (16.48)$$

where $\langle H_i^2 \rangle$ denotes the inner product of the Hermite polynomial and $\rho(\xi_j)$ is the probability density function of the jth variable. For normally distributed input parameters, Hermite polynomials are the optimal polynomials, and the input representation has only two terms because all the other terms vanish. The PC representation of a normal random variable with Hermite random variable is

$$X^j(\xi_j) = \mu_j + \sigma_j \xi_j \qquad (16.49)$$

where μ_j is the mean of the distribution, and σ_j is the standard deviation. The uncertainties in the loading parameters are having a lognormal distribution and when a lognormal distribution is represented by a Hermite polynomial, more number of terms are required in the PC expansion. The PC expansion coefficients of a lognormal random variable with the Hermite polynomial are obtained from Eq. (16.47) as

$$X^j(\xi_j) = \sum_{i=0}^{N} \mu_1 \frac{\sigma_{nor}^i}{i!} H_i(\xi_j) \qquad (16.50)$$

where μ_1 is the mean of the lognormal distribution, and σ_{nor} is the standard deviation of the underlying normal distribution. Fig. 16.6 shows the variance evaluated from the PC expansion of the lognormal random variable that represents the amplitude and frequency of the sinusoidal base excitation. It can be seen that as the number of PC terms are increased the variance can be better approximated. The PC-6 representation is sufficient to predict the variance accurately.

4.2 PC expansions of time-dependant QoI

The output QoI is also represented by a PC expansion as Eq. (16.18) at each time instant. The PC coefficients have to be evaluated at each instant of time for a dynamics problem. The orthogonal basis for the output parameter is

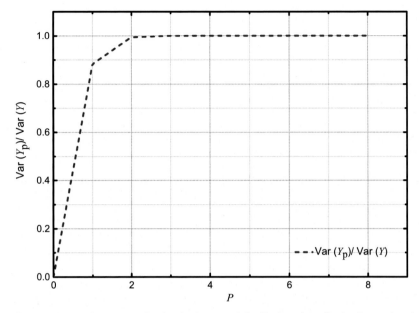

FIGURE 16.6 Variance approximation for lognormal distribution of amplitude of ground excitation with normalized Hermite polynomials.

chosen same as that of the input parameters in the cases where there is no information available about the distribution of the QoI. The order of the PC expansion is decided according to the convergence criteria as shown in Fig. 16.2.

The PC coefficients in Eq. (16.18) are determined by nonintrusive methods due to its simplicity and lower computational time requirement. The collocation points are selected and the model is evaluated at every collocation point at each time instant. A linear regression approach is adopted for determining the PC coefficients. At each collocation point, Eq. (16.18) may be written and hence a set of equations equal to the number of collocation points are obtained as in Eq. (16.23). The solution of Eq. (16.23) is obtained by a least square minimization as presented in Eq. (16.24). The QoI may also be time-averaged quantities or peak values. In these cases, PC expansion of the peak quantities is represented by Eq. (16.18) and is time independent.

4.2.1 Selection of collocation points

At least P collocation points need to be chosen for evaluating the PC expansion coefficients. For one-dimensional cases, the collocation points are chosen as the roots of the orthogonal basis. For multidimensional cases, one strategy is to use points from the full grid space. The computational cost of the full tensor grid grows exponentially with increase in dimensions. The collocation points can be chosen from the Smolyak's sparse grid (Smolyak, 1963).

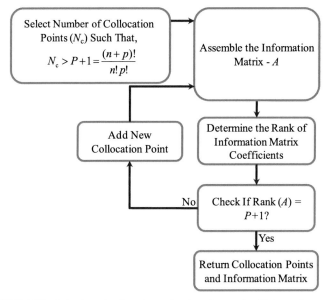

FIGURE 16.7   Selection of collocation points based on rank of information matrix.

Berveiller et al. (2006) have shown that the choice from the full tensor grid space, which is closest to the origin, is better than other selection schemes. These points are exactly the integration points used in Gauss−Hermite quadrature in Eq. (16.21). This approach uses the most important points for regression. The number of points chosen is $(n-1)$ times the number of unknowns.

Sudret (2008) proposed a methodology to select less number of points based on the invertability of the information matrix. This method is adopted in the current study as the computational model to be evaluated is expensive and time consuming. The adopted strategy is shown in Fig. 16.7. Firstly, the points from the full grid space are ordered according to their increasing norm. The information matrix is then iteratively assembled until it becomes invertible. The rank of the information matrix is made equal to the total number of unknowns. Eq. (16.24) then yields the PC expansion coefficients of the output QoI. Convergence criteria are checked and additional PC terms are added if the convergence criteria are not met, and the abovementioned procedure is continued for increased order as shown in Fig. 16.2.

4.3 Postprocessing of results

The postprocessing can be done as explained in Section 2.5 to obtain the moments of the QoI, PDF of QoI, sensitivity, reliability index, etc.

5. Numerical study

Numerical investigations conducted herein include (1) a base-isolated SDOF system with uncertain system parameters subjected to uncertain sinusoidal base excitation and (2) a MDOF FBPS with an MDOF FBSS and an MDOF BIPS with an MDOF FBSS subjected to uncertain base excitation. The details of these studies are described in the following sections.

5.1 Base-isolated SDOF system with random inputs

A base-isolated SDOF with uncertain time period and damping, subjected to uncertain sinusoidal base excitation, is modeled using the gPC method. The effective damping ratio (ξ_b) and effective time period (T_b) of the isolator are considered uncertain. The loading parameters of the sinusoidal base excitation in the form of amplitude (A) and angular frequency (ω) are also considered uncertain in this study. The isolation system parameters are assumed to have a truncated normal distribution. These uncertain parameters are truncated at mean and three times the standard deviation corresponding to 99.7% confidence interval. The selected effective time period and effective damping ratio of the SDOF are same as that of the BIPS as detailed in Table 16.8. The loading parameters are adopted from Saha et al. (2016). It is observed from the analysis of the 100 randomly selected earthquakes that the PGA and frequency are best represented by a lognormal distribution. The fitted lognormal distributions for amplitude and frequency of base-excitation function are adopted in this study. The details of the loading parameters and their distribution are tabulated in Table 16.8. It can be seen that the uncertainty in loading parameters are significantly large as compared with the uncertainties in the isolator parameters. This is close to reality as one has more control over variability of the isolator parameters (being factory made) as compared with the variability in the expected loading that may occur naturally during the life span of the structure.

A sinusoidal loading with the amplitude as a PGA of the earthquake and the frequency as a predominant frequency content of the earthquake is used in this study. Even though the sinusoidal loading does not fully represent the earthquake scenario, this assumption is made to simplify the problem. In the stochastic modeling of the SDOF, the time period and damping are represented by the same random variable (ξ_1). The random variables are selected for each of the loading parameter's amplitude (ξ_2) and frequency (ξ_3). The Hermite polynomials are selected as the orthogonal basis. For effective time period and effective damping ratio of the isolator, the chosen polynomial is the optimal one and the PC expansion of these inputs has only two terms. However, for the amplitude and frequency of the loading, a lognormal random variable is used to represent the uncertainty. Higher-order PC expansion is selected for the best representation of the characteristics. A sixth-order PC expansion (PC-6) is

TABLE 16.8 Details of uncertainties in the input parameters for fixed-base and base-isolated structure.

Uncertain input	Fixed-base primary structure		Base-isolated primary structure		Distribution type
	Mean	Standard deviation	Mean	Standard deviation	
Time period of the primary structure, T_p (sec)	1.12	0.0746	2.93	0.09766	Normal
Damping ratio of the base-isolation system, ξ_b	–	–	0.1	0.00083	Normal
Time period of the secondary system, T_s (sec)	1.64	0.10933	1.64	0.10933	Normal
Damping ratio of the superstructure (PS: ξ_p and SS: ξ_s)	0.05	0.00083	0.05	0.00083	Normal
Amplitude of the excitation (g)—of underlying normal	−1.57	0.71	−1.57	0.71	Lognormal
Frequency of the excitation (Hz)—of underlying normal	0.82	0.8	0.82	0.8	Lognormal

selected for the amplitude and frequency of the loading. The acceleration and displacement of the SDOF system are the QoI and the stochastic response is obtained using gPC method.

5.1.1 Stochastic time history response of base-isolated SDOF

Only the amplitude of the excitation function is considered uncertain in this section. The PC expansion coefficients are obtained at each time instant using the stochastic collocation approach as explained earlier. A fifth-order PC expansion (PC-5) is required for both the QoI. The mean and variance of the displacement are plotted as function of time in Fig. 16.8, and the mean and variance of acceleration is shown in Fig. 16.9. The mean response is following the deterministic response of an SDOF system. The damping in the responses is notable with time. The variation also reduces as time passes proportional to the mean displacement due to the damping noticeable in the responses. The mean acceleration is proportional to the ground acceleration. The variance is

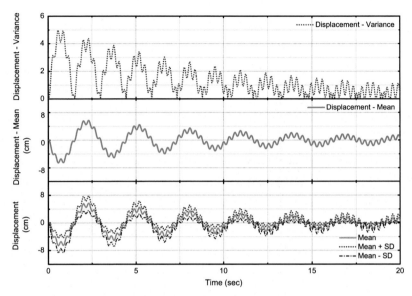

FIGURE 16.8 Displacement time history of single-degree-of-freedom system.

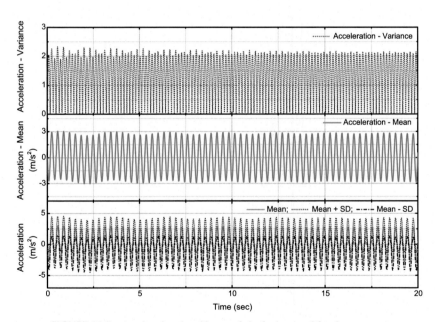

FIGURE 16.9 Acceleration time history of single-degree-of-freedom system.

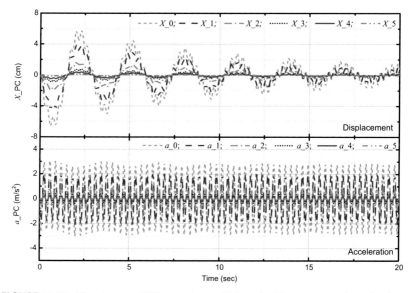

FIGURE 16.10 Time history of PC expansion coefficients for displacement and acceleration of single-degree-of-freedom system.

more around the peaks and reduces to zero at zero displacement and acceleration. The mean with one standard deviation bounds are also plotted in the same figures. The analysis is carried out up to a total of 30 s on time scale. However, only a part of the time history is shown in the figure for clarity. The time history of the PC expansion coefficients is shown in Fig. 16.10. It can be seen that the first, second, and third coefficients are also predominant as the zero coefficient (mean). This indicates that the variance is also higher for the problem at hand. The last two PC coefficients are not dominant as the others indicating convergence.

5.1.2 Probability measures of QoI

The PC expansion of the peak displacement and peak acceleration of the SDOF is obtained considering the uncertainties in the isolator parameters and loading parameters. A fifth-order PC expansion (PC-5) is used for both the QoI. The probability density functions of the QoI are obtained from the PC expansion. The probability density function of these peak responses is shown in Fig. 16.11. The PDF of the peak response QoI obtained from standard MC estimation with 5000 samples are also plotted for reference. The deterministic response of the SDOF under mean isolator parameters and mean loading parameters is presented as a vertical line corresponding to the deterministic value. The mean value is also indicated in the figure. The major contribution of the response appears from the fact that the amplitude and frequency have a

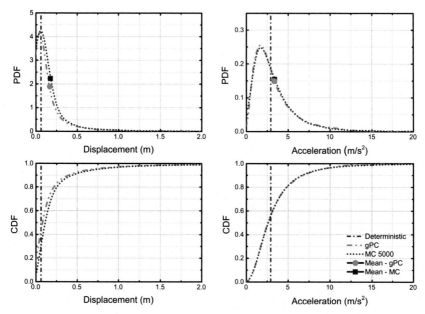

FIGURE 16.11 PDF of peak displacement and peak acceleration for SDOF system.

large variation. The variation in frequency may cause the system to substantially vary the response and change the behavior significantly in cases where the loading frequency is less than the natural frequency and when the loading frequency is more than the natural frequency. In addition, the natural frequency of the SDOF falls in the variation range of the loading frequency. This is the cause of the increased mean value as compared with the deterministic response. The PDF of the peak displacement and peak acceleration obtained from the gPC method is in close agreement with that obtained from 5000 MC samples. The mean and standard deviation obtained from the PC expansion and MC 5000 are tabulated in Table 16.9 for comparison; the deterministic value is also presented. The mean and standard deviation from the gPC method and MC

TABLE 16.9 Stochastic response of QoI for base-isolated SDOF.

		gPC		MC 5000	
QoI	Deterministic	Mean	Standard deviation	Mean	Standard deviation
Displacement (cm)	6.26	17.09	40.59	17.45	43.14
Acceleration (m/s²)	2.91	3.37	3.23	3.39	3.25

5000 are well in agreement indicating the suitability of the gPC method for stochastic dynamic response analysis of the base-isolated systems. The mean value of both peak displacement and peak acceleration of the SDOF is large as compared with the deterministic value. There is a considerably large variation of the QoI from the mean values, which arises from the large variation in the loading parameters.

The accuracy of the MC simulation method largely depends on the number of samples considered. To achieve a reasonable convergence, the deterministic model is evaluated for large set of input parameter values. The MC simulation with increasing number of modal evaluations and the PDFs are plotted in Fig. 16.12. The reference PDFs are the PDFs obtained at the MC 5000. The PDF converges as the number of samples are increased. Fig. 16.13 shows the normalized mean, variance, and kurtosis for increasing number of model evaluations. The higher moments of the QoI also achieve convergence with 5000 model evaluations. The large number of model evaluations required for the MC simulation is impractical in complex and highly nonlinear structural engineering problems. The adequately good accuracy of the stochastic response predicted by the gPC method similar to the MC simulation at a fraction of the cost indicates the suitability of the method for practical and complex structural engineering problems.

5.2 Deterministic response of fixed-base primary structure with fixed-base secondary system and base-isolated primary structure with fixed-base secondary system

Response of the SS connected to the FBPS and BIPS is also investigated under the recorded earthquake acceleration. The earthquake ground motion selected for the present study is, 180° component of the Imperial Valley, 1940 earthquake recorded at El- Centro, Array #9 station. PGA of this earthquake ground motion is recorded as 0.313g, where g denotes gravitational acceleration. The important earthquake response quantities such as base shear in the PS, top floor acceleration in the PS, shear force in the SS, and bending moment in the SS. Notably, the shear force and bending moment are the most important parameters while designing the secondary (piping) systems, in addition to the floor acceleration from supporting PS connected with. Moreover, the shear force induced in the PS is its seismic design parameter. While comparing the time history plots of the shear force and bending moment demand in case of the SS attached with the FBPS and BIPS, it is observed that the peak response values are reduced in case of the base-isolated structure which shows the remarkable effectiveness of the base-isolation technology in safeguarding the spatially distributed piping-type SSs attached to the supporting structure at different locations. The deterministic response at the mean system parameter values of the PS outputs are shown in Fig. 16.14 and the same for SS outputs are shown in Fig. 16.15.

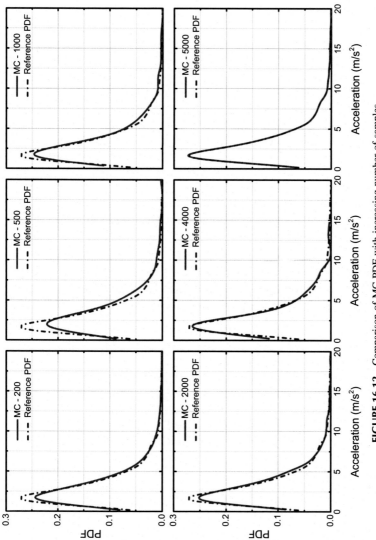

FIGURE 16.12   Comparison of MC PDF with increasing number of samples.

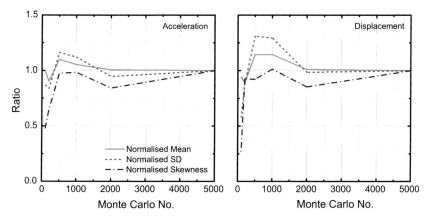

FIGURE 16.13 Convergence of MC with increasing number of samples.

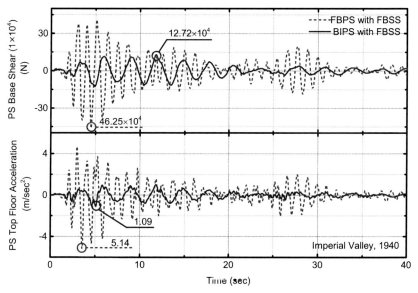

FIGURE 16.14 Primary structure (PS) deterministic response under Imperial Valley, 1940 earthquake excitation.

5.3 Stochastic response of fixed-base primary structure with fixed-base secondary system and base-isolated primary structure with fixed-base secondary system

The stochastic response of the FBPS with SS and the BIPS with SS are obtained using the gPC method. The time period and damping ratio of the PS and SSs are considered uncertain, as well as the damping ratio of the isolator

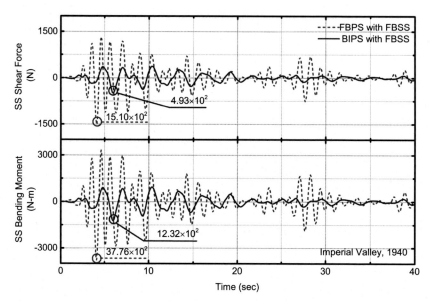

FIGURE 16.15 Secondary system (SS) deterministic response under Imperial Valley, 1940 earthquake excitation.

is uncertain. The details of the uncertain input distribution are shown in Table 16.8. The uncertain structural parameters have a truncated normal distribution with mean and three times standard deviation corresponding to 99.7% confidence interval. The standard deviation of the time period and damping of the base isolator is assumed half that of the fixed-base case. This is prudent, considering the fact that the base-isolation systems are generally factory made and have better quality control during manufacturing as compared to the fixed-base structure, which is constructed in situ.

The time period and damping of the SS are also considered uncertain and the details of the mean and standard deviation values are given in Table 16.8. For the FBPS with FBSS, the time period of the fixed-base structure is represented by one random variable, ξ_1. The time period of the SS and the damping ratio of the superstructure (of both the PS and the SS) are represented by another random variable, ξ_2 . For the BIPS with FBSS, the time period of the base-isolated structure and the damping ratio of the base-isolation system are represented by one random variable, ξ_1. The time period of the SS and the damping ratio of the superstructure (of both the PS and SS) are represented by another random variable, ξ_2.

The loading parameters are having a lognormal distribution as obtained from the 100 historical earthquake data (Saha et al., 2016; http://web.iitd.ac.in/~matsagar/EqDataBase.htm). One random variable each (ξ_3, ξ_4) is respectively used for representing the uncertainties in the amplitude (A) and frequency (ω) of the base-excitation function.

Thus, there are six uncertain parameters, both in the system parameters (time period of the PS, damping ratio of the isolator, time period of the SS, and damping ratio of the superstructure) and base-excitation parameters (amplitude and frequency). The random vector, $\xi = \{\xi_1, \xi_2, \xi_3, \xi_4\}$ have four random variables to represent these uncertainties. The uncertain inputs are represented by a PC expansion and the coefficients are evaluated as explained in Section 4. The amplitude and frequency of the base-excitation function are evaluated to have a lognormal distribution, and the parameters of this distribution are obtained from the statistical analysis of an ensemble of observed 100 earthquake data. Hence, these two parameters are found to have very large variation as compared with the system parameters. A sixth-order PC expansion (PC-6) is required to represent both these random variables of the excitation. The prediction of the exact variance with increasing order of the PC expansion for the considered lognormal distribution is shown in Fig. 16.6. It is shown that the sixth-order PC expansion is able to represent the mean and standard deviation of the assumed lognormal base-excitation function accurately. The PC expansion coefficients of the uncertain inputs are obtained by using Galerkin projection. The PC expansion of the normally distributed random variable is straightforward as shown in Eq. (16.49) and has only two terms, as the Hermite polynomials are the optimum polynomials for normal distribution. For lognormally distributed parameters represented by Hermite polynomials, the PC expansion will have more number of terms and the PC expansion coefficients can be obtained by utilizing the orthogonality property of the orthogonal basis and performing a Galerkin projection. Lognormally distributed random variable, represented by the Hermite polynomial, is represented by Eq. (16.50) and can be used directly to find the PC expansion coefficients.

Once the PC expansion of the input parameters is obtained, the procedure explained in Fig. 16.7 is used to identify the collocation points in such a way that the information matrix is well-conditioned. The deterministic model is evaluated at each of the collocation points and the least squares regression is followed to obtain the PC expansion of the output QoI.

5.3.1 Stochastic response of the primary structure and secondary system

There are two QoI for the PS, the peak base shear and the peak top floor acceleration. Base shear is used for the design of the base-isolation system and the FBPS. The top floor acceleration is a measure of the human comfort in the structure during an earthquake; additionally, top floor acceleration is also an important parameter for the sensitive (especially high-frequency) equipments housed inside the PS. The QoI chosen hence represent both strength and serviceability criteria of the structure. The QoI for the SS are the shear force and bending moment, as they are important for the design of the SSs, especially for piping-type SSs. The stochastic response of the SS in the form of shear force and bending moment are calculated at two locations, i.e., near bottom and top mass of the SS. The maximum of these two locations will be used for the design, and hence at the end of the stochastic analysis, the maximum location is identified among these two locations by comparing both the mean and standard deviation of the QoI at these locations. Hence, there are in total six output QoI and are obtained by stochastic collocation as explained in Section 4. A second-order PC expansion (PC-2) is used for the QoI in the beginning and the order is increased in the subsequent steps. The collocation points are selected based on the algorithm presented in Fig. 16.7, and the deterministic model is evaluated at these collocation points. The PC expansion of the output is obtained for both the PS and SSs for both BIPS and FBPSs and is presented in Figs. 16.16 and 16.17. For the FBPS with FBSS, the mean and standard deviation of shear force and bending moment in the SS are found to be critical near the top SS mass. However, for the BIPS with FBSS, the mean and standard deviation of the shear force and bending moment of the SS are found to be critical near the bottom SS mass, and hence these are presented in Fig. 16.16. The higher-order combinatorial terms are omitted from the PC-5 expansion and a reduced base PC-5 is also used (PC-5 RB). The mean and standard deviation values of all the six QoI are presented in Tables 16.10 and 16.11. It can be seen that the mean and standard deviation of the PC-4 and PC-5 are in close agreement in all the QoI for both FBPS and BIPS with SS.

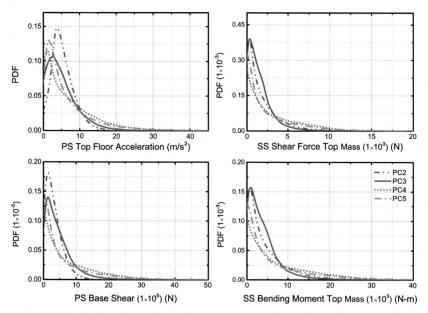

FIGURE 16.16 Stochastic response of QoI for different order PC expansion (fixed-base primary structure with fixed-base secondary system).

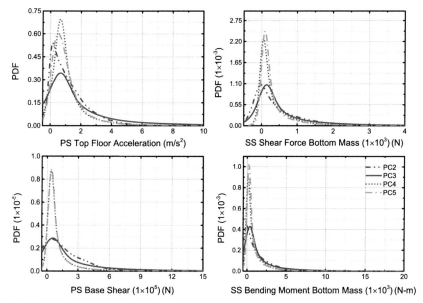

FIGURE 16.17 Stochastic response of QoI for different order PC expansion (base-isolated primary structure with fixed-base secondary system).

Hence, a PC-4 expansion is used for conducting the further analysis. However, a convergence of mean and standard deviation does not guarantee full convergence of the probability density function. This observation can also be made from Figs. 16.16 and 16.17.

5.3.2 Probability measures of QoI

The probability density function of all the QoI of the PS and SS is plotted and shown in Figs. 16.18 and 16.19 for both FBPS with SS and BIPS with SS. A fourth-order PC expansion (PC-4) is used for all of the QoI. The QoI for the PS are the peak base shear and the peak top floor acceleration. The PDF of these QoI for both base-isolated and fixed-base cases are plotted and shown in Fig. 16.18. For the BIPS with SS case, the mean value of base shear and the top floor acceleration is found to be significantly less as compared with that of the fixed-base case proving the effectiveness of the base-isolation system for earthquake response mitigation. The mean and standard deviation of these QoI are tabulated in Table 16.12. The response in both the base-isolated and fixed-base cases has large standard deviation from the mean. This is attributed to the fact that the input loading parameters, namely the amplitude and frequency of the base-excitation function, have a large variation. However, the response of the fixed-base structure has large standard deviation as compared with that of the base-isolated structure. The mean and standard deviation of the base shear and top floor acceleration for the FBPS with FBSS is found to be

TABLE 16.10 Stochastic response of QoI for different order PC expansion (fixed-base primary structure with secondary system).

QoI	PC-2		PC-3		PC-4		PC-5	
	Mean	SD	Mean	SD	Mean	SD	Mean	SD
Base shear (kN)	289	278	420	538	532	822	547	889
Top floor acceleration (m/s^2)	4.82	3.71	2.93	7.13	6.42	7.69	6.47	8.28
Shear force top mass—secondary system (N)	1205	2149	1446	2044	1786	3126	1920	3508
Shear force bottom mass—secondary system (N)	1133	2251	1175	1828	1210	1956	1309	2285
Bending moment top mass—secondary system (N-m)	3012	5371	3649	5102	4466	7819	4802	8768
Bending moment bottom mass—secondary system (N-m)	2832	5626	2936	4570	3023	4901	3272	5714

TABLE 16.11 Stochastic response of QoI for different order PC expansion (base-isolated primary structure with secondary system).

QoI	PC-2		PC-3		PC-4		PC-5	
	Mean	SD	Mean	SD	Mean	SD	Mean	SD
Base shear (kN)	106	144	179	580	111	187	112	187
Top floor acceleration (m/s²)	1.20	1.23	1.27	4.50	1.14	1.46	1.12	1.40
Shear force top mass—secondary system (N)	384	724	426	1005	317	590	337	621
Shear force bottom mass—secondary system (N)	434	789	565	1405	341	591	357	620
Bending moment top mass—secondary system (N-m)	962	1807	1070	2515	798	1474	842	1553
Bending moment bottom mass—secondary system (N-m)	1085	1967	1407	3505	853	1476	892	1549

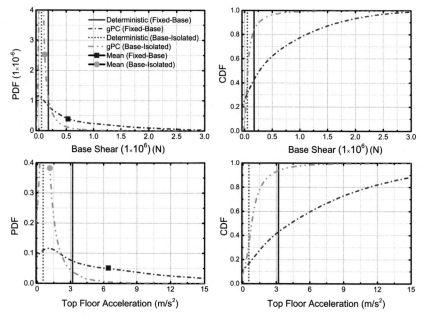

FIGURE 16.18 Stochastic response of QoI for primary structure.

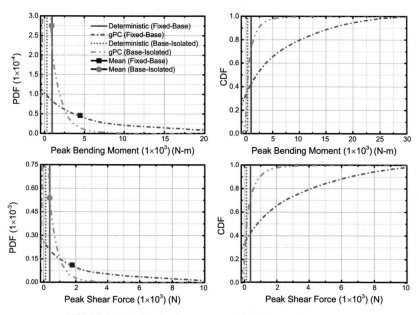

FIGURE 16.19 Stochastic response of QoI for secondary system.

TABLE 16.12 Stochastic response of QoI for fixed-base primary structure (FBPS) with fixed-base secondary system (FBSS) and base-isolated primary structure (BIPS) with FBSS.

QoI	FBPS with FBSS			BIPS with FBSS		
	Deterministic	gPC		Deterministic	gPC	
		Mean	Standard deviation		Mean	Standard deviation
Base shear of PS (kN)	172.2	531.6	822	48.6	111.3	186.6
Top floor acceleration of PS (m/s²)	3.19	6.42	7.69	0.53	1.14	1.46
SF# near top mass—SS (N)	373.7	1785.5	3126.4	93.16	316.9	589.6
SF# near bottom mass—SS (N)	271.5	1209.8	1956.3	117.6	341.2	590.7
BM* near top mass—SS (N-m)	934.4	4466.1	7818.5	232.9	792.6	1498.5
BM* near bottom mass—SS (N-m)	678.7	3023.2	4901.2	294.01	852.8	1476.2

BM* = bending moment; SF# = shear force.

significantly more than the corresponding quantities of the BIPS with FBSS. For both the fixed-base and base-isolated cases, the deterministic value of the dynamic response underpredicts the mean values, which is of immense concern. The shape of the PDFs for all the QoI for the base-isolated structure is skewed and with a large tail extending till the fixed-base case, having similar shape as that of the loading parameters.

For the SS, the QoI chosen for the analysis are the shear force and bending moment at two locations. The PC expansion coefficients for all the QoI at all these locations are obtained using the nonintrusive collocation approach. By observing the mean and standard deviation at these locations (Table 16.13), it can be observed that the mean and standard deviation of both shear force and bending moment are more near the top SS mass location for the fixed-base

TABLE 16.13 Sensitivity analysis Sobol' indices for QoI for fixed-base primary structure (FBPS) with fixed-base secondary system (FBSS) and base-isolated primary structure (BIPS) with FBSS.

QoI	FBPS with FBSS				BIPS with FBSS			
	ξ_1	ξ_2	ξ_3	ξ_4	ξ_1	ξ_2	ξ_3	ξ_4
Base shear of PS (kN)	0.113	0.128	0.303	0.744	0.003	0.002	0.538	0.768
Top floor acceleration of PS (m/s²)	0.113	0.082	0.551	0.547	0.013	0.013	0.662	0.589
Shear force near top mass—SS (N)	0.039	0.101	0.254	0.850	0.017	0.059	0.425	0.782
Shear force near bottom mass—SS (N)	0.036	0.079	0.249	0.853	0.017	0.059	0.467	0.766
Bending moment near top mass—SS (N-m)	0.039	0.101	0.253	0.850	0.028	0.098	0.425	0.782
Bending moment near bottom mass—SS (N-m)	0.036	0.080	0.250	0.852	0.017	0.059	0.467	0.766

case and near the bottom SS mass location for the base-isolated case. Only the PDF of shear force and bending moment of the SS at the critical location is reported in Fig. 16.18, as the maximum shear force and bending moment will be crucial in the design of the SS. It is worthwhile to note that, unlike deterministic approach, in a stochastic design, along with the mean value the variation in the QoI is also considered for decision-making. Similar to the QoI of the PS, the shear force and bending moment values are found to be less for the BIPS with FBSS as compared to the FBPS with FBSS. The response distribution of the base-isolated structure is found to be skewed and the mean of these QoI are found to be more than the deterministic values.

The mean and standard deviation of all the QoI (base shear in the PS, top floor acceleration of the PS, maximum bending moment, and shear force in the SS) are significantly less for the BIPS with FBSS case as compared with the FBPS with FBSS. This clearly indicates that the base-isolation system is effective in mitigating the adverse effects of an unfortunate earthquake event during the life time of a coupled primary–secondary system.

5.3.3 Sensitivity analysis

A variance-based sensitivity analysis is carried out by evaluating the Sobol' indices for all the QoI using the procedure mentioned in Section 2.6.3. The Sobol' indices are tabulated in Table 16.13 for the QoI of both the PS and SS in both the fixed-base and base-isolated cases. The sensitivity indices are a measure of how much each of the QoI is sensitive to the input uncertain parameters. All the QoI are found to be more sensitive to the loading parameters. This, combined with the large variation in the loading parameters, results in the large variation of all the QoI. For the FBPS, the base shear and top floor acceleration are found to be more sensitive to the random variable representing the time period and damping of the PS. The shear force and bending moment in the SS are found to be more sensitive to the random variable representing the time period and damping of the SS. The base shear and top floor acceleration are found to have almost equal contribution from the uncertainty of the amplitude and frequency of the base-excitation function. The shear force and bending moment of the SS are found to be more sensitive to the frequency content of the base-excitation function as compared with that of the amplitude of the excitation. For the base-isolated structure, it is observed that all the system parameters have considerably less contribution toward the uncertainty in the QoI. The uncertainties in the loading parameters are found to contribute more toward the uncertainty in the outputs. It is worth to note that the sensitivity of the frequency is more as compared to that of the amplitude. The loading amplitude and frequency have considerable contribution on all of the QoI. This combined with the fact that large variation is present in the loading parameters explains the skewed distribution of the parameters and their large variation.

5.3.4 Computing probability of failure from PC expansion

The probability of failure and reliability index of the FBPS with FBSS and the BIPS with FBSS are computed as explained in Section 2.6.4. The strength design criteria are adopted for the PSs in both of these cases with respect to an admissible maximal base shear value. The limit state function is given by

$$\text{Base Shear in PS(kN):} \quad g_1 = BS_{max} - BS(\xi) \quad (16.51)$$

where BS_{max} is the threshold considered at 4000 kN, and $BS(\xi)$ is the base shear capacity of the fixed-base or base-isolated structure. The reliability of both the FBPS with FBSS and the BIPS with FBSS with respect to this limit state function is evaluated by using a MC-based strategy as explained in Section 2.6.4. Sampling is done from the PC expansion of the response and the probability of failure of the structure is computed as in Eq. (16.34). The results of both the FBPS with FBSS and the BIPS with the FBSS are tabulated in Table 16.14. From the PC expansion of the base shear 5×10^4 samples are drawn. It is computationally very cheap to sample from the PC expansion as it involves only evaluating a polynomial expansion. The probability of failure of the FBPS with SS is found be two order of magnitude more than that of the BIPS with SS structure. The reliability index and probability of failure for both the base-isolated and fixed-base structure are tabulated in Table 16.14.

Similarly, the probability of failure and reliability index (limit state functions for a defined threshold) are calculated for the other response quantities (QoI) as follows.

$$\text{Top Floor Acceleration of PS}\left(\text{m}/\text{s}^2\right): \quad g_2 = A_{max} - A(\xi) \quad (16.52)$$

where A_{max} is the threshold considered at 10 m/s^2, and $A(\xi)$ is the top floor acceleration in the fixed-base or base-isolated structure.

$$\text{SF}^\#\text{Near Top Mass} - \text{SS(N):} \quad g_3 = SFT_{max} - SFT(\xi) \quad (16.53)$$

where SFT_{max} is the threshold considered at 10,000 N, and $SFT(\xi)$ is the shear force near the top mass of the SS.

$$\text{SF}^\#\text{Near Bottom Mass} - \text{SS(N):} \quad g_4 = SFB_{max} - SFB(\xi) \quad (16.54)$$

where SFB_{max} is the threshold considered at 10,000 N, and $SFB(\xi)$ is the shear force near the bottom mass of the SS.

$$\text{BM}^*\text{Near Top Mass} - \text{SS(N}-\text{m):} \quad g_5 = BMT_{max} - BMT(\xi) \quad (16.55)$$

where BMT_{max} is the threshold considered at 35,000 N-m, and $BMT(\xi)$ is the bending moment near the top mass of the SS.

$$\text{BM}^*\text{Near Bottom Mass} - \text{SS(N}-\text{m):} \quad g_6 = BMB_{max} - BMB(\xi) \quad (16.56)$$

where BMB_{max} is the threshold considered at 35,000 N-m, and $BMB(\xi)$ is the bending moment near the bottom mass of the SS.

TABLE 16.14 Reliability analysis results for fixed-base primary structure (FBPS) with fixed-base secondary system (FBSS) and base-isolated primary structure (BIPS) with FBSS.

Limit state function	FBPS with FBSS		BIPS with FBSS		$\beta_{BI}-\beta_{FB}$
	Probability of failure	Reliability index (β_{FB})	Probability of failure	Reliability index (β_{BI})	
Base shear in PS (kN) $g_1 = 4000-BS(\xi)$	1.13×10^{-3}	3.053	4.8×10^{-5}	3.900	0.847
Top floor acceleration of PS (m/s²) $g_2 = 10-A(\xi)$	0.234	0.7249	3.6×10^{-3}	2.685	1.9601
SF# near top mass—SS (N) $g_3 = 10,000-SFT(\xi)$	1.89×10^{-2}	2.076	4.2×10^{-5}	3.933	1.857
SF# near bottom mass—SS (N) $g_4 = 10,000-SFB(\xi)$	1.08×10^{-4}	3.699	6.4×10^{-5}	3.830	0.131
BM* near top mass—SS (N-m) $g_5 = 35,000-BMT(\xi)$	1.02×10^{-3}	3.714	8×10^{-6}	4.314	0.6
BM* near bottom mass—SS (N-m) $g_6 = 35,000-BMB(\xi)$	2.4×10^{-5}	4.065	1×10^{-5}	4.264	0.199

BM* = bending moment; SF# = shear force.

The limit state function and the obtained probability of failure and reliability index are tabulated in Table 16.14. For all the six cases (QoI of both PS and SS), the probability of failure of the FBPS with SS is found be two order of magnitude more than that of the BIPS with SS. The reliability index for bending moment and shear force of the SS at critical locations are compared between the FBPS with FBSS as well as BIPS with FBSS in Table 16.14. The reliability index in case of the BIPS is found to be more ($\beta_{BI} > \beta_{FB}$) as compared with the respective FBPS case for all the seismic response quantities. This clearly demonstrates the effectiveness of the base-isolation system in mitigating the earthquake effects and improving reliability in both strength and serviceability limit states for the QoI not only in the PS but also in the SS.

6. Conclusions

The basic formulation and implementation details of the nonintrusive PC using stochastic collocation are explained in details. The stochastic framework based on the PC expansion can be used for computing the mean, standard deviation, higher-order moments, probability density function (PDF), and cumulative density function (CDF) of the QoI. Moreover, the computation for sensitivity with respect to the random input parameters, reliability with respect to a specified limit state function, and fragility curves of the QoI are also possible using the stochastic framework based on the PC expansion. The presented stochastic framework for dynamic analysis can be used for a different deterministic problem easily, as the framework is nonintrusive. In addition, for computationally expensive simulations, the evaluation of deterministic results at different collocation points is an embarrassingly parallel.

Stochastic response of the base-isolated building modeled simplistically as an SDOF and the FBPS with SS and BIPS with SS is computed using gPC expansion technique. The PC expansion results are compared for the base-isolated SDOF with the standard MC simulation with 5000 samples and the applicability of the method for the structural dynamics problem involving base-isolated structure under random ground/base excitations is demonstrated. From this study, the following salient conclusions are drawn.

1. The mean, standard deviation, and probability of failure (for same limit state function) of the base shear, top floor acceleration, maximum bending moment in the SS, and maximum shear force in the SS are found to be significantly less for the BIPS with FBSS than that of the FBPS with FBSS, which demonstrates the effectiveness of the base-isolation system in mitigating the earthquake-induced effects in both the PS and the SS.

2. The deterministic responses are underpredicted in all the cases as compared with the stochastic mean for the FBPS with multiply supported spatial SS and for BIPS with multiply supported spatial SS, emphasizing the significant importance of the stochastic analysis over the deterministic counterpart.

3. The stochastic dynamic analysis is carried out with actual earthquake parameters obtained from the 100 real earthquake data, which have revealed large variations. This has resulted in large variation in the dynamic responses of the PS and SS. The dynamic responses of the structures in fixed-base and base-isolated cases are found to be more sensitive toward the uncertainty in the loading parameters than the system parameters. This together with the large variation in the loading parameters results in large variation in the QoI, which must be accounted for in the design of the PS and SS.

References

Alhan, C., Gavin, H.P., 2005. Reliability of base isolation for the protection of critical equipment from earthquake hazards. Engineering Structures 27 (9), 1435–1449.

ASCE/SEI 7-10, 2010. Minimum design loads for buildings and other structures. In: ASCE Standard, American Society of Civil Engineers (ASCE), Structural Engineering Institute (SEI), Reston, Virginia (VA), USA.

Au, S.K., Beck, J.L., 2001. Estimation of small failure probabilities in high dimensions by subset simulation. Probabilistic Engineering Mechanics 16 (4), 263–277.

Berveiller, M., Sudret, B., Lemaire, M., 2006. Stochastic finite element: a non intrusive approach by regression. European Journal of Computational Mechanics/Revue Européenne de Mécanique Numérique 15, 81–92.

Biondi, B., Aveni, A.D., Muscolino, G., 1996. Seismic response of combined primary- secondary systems via component-mode synthesis. In: 11th World Conference on Earthquake Engineering (11WCEE), Acapulco, Mexico, p. 38.

Bulleit, W.M., 2008. Uncertainty in structural engineering. Practice Periodical on Structural Design and Construction, American Society of Civil Engineers (ASCE) 13 (1), 24–30.

Cacuci, D.G., 2003. Sensitivity and Uncertainty Analysis: Theory. Chapman and Hall, CRC Press, Boca Raton, Florida (FL), USA.

Chakraborty, S., Dey, S.S., 1996. Stochastic finite element simulation of random structure on uncertain foundation under random loading. International Journal of Mechanical Sciences 38 (11), 1209–1218.

Chaudhuri, A., Chakraborty, S., 2006. Reliability of linear structures with parameter uncertainty under non-stationary earthquake. Structural Safety 28 (3), 231–246.

Chen, Y., Ahmadi, G., 1992. Stochastic earthquake response of secondary systems in base-isolated structures. Earthquake Engineering and Structural Dynamics 21 (12), 1039–1057.

Cheng, J., Cai, C.S., Xiao, R.C., Chen, S.R., 2005. Flutter reliability analysis of suspension bridges. Journal of Wind Engineering and Industrial Aerodynamics 93 (10), 757–775.

Choi, E., Desroches, R., Nielson, B., 2004. Seismic fragility of typical bridges in moderate seismic zones. Engineering Structures 26 (2), 187–199.

Choi, S.K., Ramana, G.V., Canfield, R.A., 2007. Reliability Based Structural Design. Springer Science and Business Media, Springer-Verlag, London, United Kingdom (UK).

Datta, T.K., 2010. Seismic Analysis of Structures. John Wiley and Sons (Asia) Private Limited, Singapore.

Ditlevsen, O., Henrik, O.M., 1996. Structural Reliability Methods. Wiley, New York (NY), USA.

Ghanem, R., Spanos, P., 1991. Stochastic Finite Elements: A Spectral Approach. Springer- Verlag, New York (NY), USA.

Gumbert, C.R., Newman, P.A., Hou, G.J.-W., 2002. Effect of random geometric uncertainty on the computational design of a 3-D flexible wing. In: 20th AIAA Applied Aerodynamics Conference, St. Louis, Missouri (MO), USA. https://doi.org/10.2514/6.2002-2806.

Gupta, S., Manohar, C.S., 2004. An improved response surface method for the determination of failure probability and importance measures. Structural Safety 26 (2), 123–139.

Gupta, S., Manohar, C.S., 2006. Reliability analysis of randomly vibrating structures with parameter uncertainties. Journal of Sound and Vibration 297 (3–5), 1000–1024.

Handa, K., Andersson, K., 1981. Application of finite element methods in the statistical analysis of structures. In: 3rd International Conference on Structural Safety and Reliability (ICOSSAR81), Trondheim, Norway, pp. 409–420.

Hisada, T., Nakagiri, S., 1981. Stochastic finite element method developed for structural safety and reliability. In: 3rd International Conference on Structural Safety and Reliability (ICOSSAR81), Trondheim, Norway, pp. 395–408.

International Building Code, (IBC), 2012. International Code Council Incorporation. Illinois (IL), USA.

Jacob, C., Dodagoudar, G.R., Matsagar, V.A., 2013a. Seismic reliability analysis of base- isolated buildings. In: Proceedings of the International Symposium on Engineering under Uncertainty: Safety Assessment and Management (ISEUSAM – 2012), Springer, Kolkata, India, pp. 1251–1265.

Jacob, C., Sepahvand, K., Matsagar, V.A., Marburg, S., 2013b. Stochastic seismic response of base-isolated buildings. International Journal of Applied Mechanics 5 (1), 1–21.

Kelly, T.E., 2001. Base Isolation of Structures: Design Guidelines. Holmes Consulting Group Limited, Wellington, New Zealand.

Li, C., Der-Kiureghian, A., 1995. Mean out-crossing rate of nonlinear response to stochastic input. In: 7th International Conference on Applications of Statistics and Probability in Civil Engineering. ICASP7, Paris, France, pp. 1135–1141.

Li, J., Chen, J.B., 2004. Probability density evolution method for dynamic response analysis of structures with uncertain parameters. Computational Mechanics 34 (5), 400–409.

Li, J., Xiu, D., 2010. Evaluation of failure probability via surrogate models. Journal of Computational Physics 229, 8966–8980.

Liel, A.B., Haselton, C.B., Deierlein, G.G., Baker, J.W., 2009. Incorporating modelling uncertainties in the assessment of seismic collapse risk of buildings. Structural Safety 31 (2), 197–211.

Matsagar, V.A., Jangid, R.S., 2004. Influence of isolator characteristics on the response of base-isolated structures. Engineering Structures 26 (12), 1735–1749.

Matsagar, V.A., Jangid, R.S., 2011. Earthquake Base-Isolated Buildings. LAP Lambert Academic Publishing, Germany.

Metropolis, N., Ulam, S., 1949. The Monte Carlo method. Journal of the American Statistical Association 44 (247), 335–341.

Mori, Y., Ellingwood, B., 1993. Methodology for Reliability Based Condition Assessment: Application to Concrete Structures in Nuclear Plants. Report NUREG/CR-6052: ORNL/Sub/ 93-SD684, Washington (DC), USA.

Mueller, J.L., Siltanen, S., 2012. Linear and nonlinear inverse problem with practical application. In: Computational Science and Engineering Series, Society for Industrial and Applied Mathematics, Philadelphia, Pennsylvania (PA), USA.

Padgett, J.E., DesRoches, R., 2007. Sensitivity of response and fragility to parameter uncertainty. Journal of Structural Engineering, *American Society of Civil Engineers (ASCE)* 133 (12), 1710−1718.

Rackwitz, R., 2001. Reliability analysis: a review and some perspectives. Structural Safety 23 (4), 365−395.

Saha, S.K., Sepahvand, K., Matsagar, V.A., Jain, A.K., Marburg, S., 2016. Fragility analysis of base-isolated liquid storage tanks under random sinusoidal base excitation using generalised polynomial chaos expansion based simulation. Journal of Structural Engineering, *American Society of Civil Engineers (ASCE)* 04016059. https://doi.org/10.1061/(ASCE)st.1943-541x.0001518.

Schuëller, G.I., Pradlwarter, H.J., 2009. Uncertainty analysis of complex structural systems. International Journal for Numerical Methods in Engineering 80, 881−913.

Sepahvand, K., Marburg, S., Hardtke, H.J., 2010. Uncertainty quantification in stochastic systems using polynomial chaos expansion. International Journal of Applied Mechanics 2 (2), 305−353.

Smith, R.C., 2014. Uncertainty quantification: theory, implementation, and applications. In: Computational Science and Engineering Series, Society for Industrial and Applied Mathematics, Philadelphia, Pennsylvania (PA), USA.

Smolyak, S., 1963. Quadrature and interpolation formulas for tensor products of certain classes of functions. Doklady Akademii Nauk SSSR 4, 240−243.

Stefanou, G., 2009. The stochastic finite element method: past, present and future. Computer Methods in Applied Mechanics and Engineering 198 (9−12), 1031−1051.

Sudret, B., 2007. Uncertainty Propagation and Sensitivity Analysis in Mechanical Models - Contributions to Structural Reliability and Stochastic Spectral Methods. PhD Thesis. Ecole Nationale Des Ponts et Chaussées (ENPC), Paris, France.

Sudret, B., 2008. Global sensitivity analysis using polynomial chaos expansions. Reliability Engineering and System Safety 93 (7), 964−979.

Sudret, B., Der-Kiureghian, A., 2000. Stochastic finite element methods and reliability: a state-of-the-art report. In: Structural Engineering Mechanics and Materials. Department of Civil and Environmental Engineering, University of California, Berkeley. Report No. UCB/SEMM-2000/08, California (CA), USA.

Sudret, B., Mai, C.V., 2013. Computing seismic fragility curves using polynomial chaos expansions. In: 11th International Conference on Structural Safety and Reliability (ICOSSAR'2013), New York (NY), USA.

Thoft-Christensen, P., Baker, M., 1982. Structural Reliability Theory and its Applications. Springer-Verlag, Berlin Heidelberg, Germany.

Vogel, C.R., 2002. Computational Methods for Inverse Problems. Society for Industrial and Applied Mathematics, Philadelphia, Pennsylvania (PA), USA.

Wand, M.P., Johnes, M.C., 1995. Kernel Smoothing, Monographs on Statistics and Applied Probability 60. Chapman and Hall/CRC, New York (NY), USA.

Wiener, N., 1938. The homogeneous chaos. American Journal of Mathematics 60 (4), 897−936.

Xiu, D., Karniadakis, G., 2002. The Wiener-Askey polynomial chaos for stochastic differential equations. Journal of Scientific Computing 24 (2), 619−644.

Chapter 17

Stochastic optimization: stochastic diffusion search algorithm

Saman Maroufpoor[1], Rahim Azadnia[2], Omid Bozorg-Haddad[3]
[1]*Department of Irrigation and Reclamation Engineering, University of Tehran, Tehran, Iran;*
[2]*Department of Agricultural Machinery Engineering, University of Tehran, Tehran, Iran;* [3]*Dept. of Irrigation & Reclamation Engineering, Faculty of Agricultural Engineering & Technology, College of Agriculture & Natural Resources, University of Tehran, Karaj, Iran*

1. Introduction

Stochastic optimization methods are procedures for maximizing or minimizing objective functions when the stochastic problems are considered. Over the past few decades, these methods have been proposed for engineering, business, computer science, and statistics as essential tools. In particular, these methods have various applications in different fields. Stochastic optimization plays an important role in the analysis, design, and performance of modern systems. Stochastic optimization usually looks at problems from two perspectives: through the objective functions (cost functions) or through limitations. Stochastic optimization algorithms have a wide range of applications in statistical problems. In this chapter, we discuss the stochastic diffusion search (SDS) algorithm and its fundamental principles. The SDS algorithm with practical significance was presented first in 1989 as a kind of intelligent algorithms for solving a transform invariant pattern recognition problem within a search space. The search was carried out in parallel by a series of elements (agents). These agents are independent and collaborate with others in parallel to achieve the position of the target in search space. Agents are divided into two groups based on their efficiency. The first group is an active agent, and the second one is a passive (inactive) agent. The active agent points potentially correct position within a search space and tries to communicate with other agents, but the inactive agent is equal to 0. In the following section, both agents and the algorithm of SDS will be described in detail.

Handbook of Probabilistic Models. https://doi.org/10.1016/B978-0-12-816514-0.00017-5
437

2. Stochastic diffusion search

Invariant pattern recognition can be solved by various methods. In this case, SDS is an intermittent method to solve invariant pattern recognition. In fact, the SDS is extracted from stochastic diffusion network (SDN) (which is used in the visual area for specific properties of a position) that has been defined as a connectionist pattern recognition method (Nasuto and Bishop, 1999). The SDN involves real-time positioning, and it also explores the lip images in video sequences (Grech-Cini, 1995). As shown in Fig. 17.1, the SDN contains a set of n-tuple weightless neurons.

3. Search and optimization

Search and optimization are two factors that interchangeably can be used in swarm intelligence science. Furthermore, three different types of search are available (Mitchell, 1998).

In the first type, the main aim of the search is to explore a model in a search space. But the goal of an algorithm in search space is to try to find a closest match to the target; in the world of computer science technology, the search is defined as data search (Knuth, 1973). In the second type, the main aim is to find a path (path search) and determine the steps that the algorithm tries to achieve. In this search type, the mentioned path is developed during the search. Solution search is the third type of the search. The goal of this search is to explore a solution among a large problem space. This type of the search acts the same as the second type. The search can be considered as a competitive cooperation process in which all of agents explore for the best solution. A group of agents are attracted to seek the appropriate potential for data model. This procedure transforms competition smoothly into cooperation. Generally, the progress of the search is specified by the number of active agents pointing to the position of target in search space.

In addition, it should be mentioned that the search in SDS is conducted by competitive collaborative procedures. One of the mentioned processes is be simple agents which transmit potential location in the search space, and the

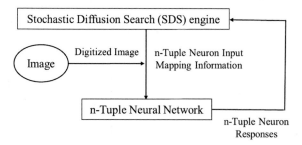

FIGURE 17.1 Block diagram of SDN. *SDN*, stochastic diffusion network.

other one is information diffusion that provides appropriate solutions. In SDS, agents are banned to connect to weighted links which play an important role in artificial neural networks. The aim of SDS is to determine a predefined data pattern without any search space (Nasuto et al., 1998). In other words, the SDS is an appropriate optimization technique that has been used for various problems such as site selection for wireless network (Hurley and Whitaker, 2002) and object recognition (Bishop and Torr, 1992). Previous evaluation of the SDS has shown its isotropy (Nasuto and Bishop, 1999), linear time complicacy (Nasuto et al., 1998), and diversity of search condition. The SDS strongly defines the application of algorithms by examining its resource allocation. Generally, the algorithm of SDS contains several steps that are summarized as follows (Al-Rifaie and Bishop, 2013):

01: Initialising agents ()
02: While not terminate (stopping condition is not met)
03: Testing agents ()
04: Determining agents' activities (active/inactive)
05: Diffusing agents ()
06: Exchanging of information
07: End while

From the exact definition of the SDS, it is concluded that we can describe two stages for operation. The first stage is a pure random search, and the second one is the activity of an agent, mentioning to the solution. In the SDS, the search engine is purely random, while distribution generates an equilibrium state of the SDS, which is important for formulating the criterion. This property illustrates that the SDS can be so efficient in reaching equilibrium in case of large spaces. It has also been proven that, in the case of search spaces that are surrounded by noise, the average number of inactive agents that participate in random search will decrease because of disturbances of diffusion activity. An agent that is specified by a pointer to a place in search space is called an activity. The value of an activity for agents that pointed the correct position within a search is 1 (agent is active), otherwise it is equal to 0 (agent is passive).

An example of the SDS is placing a familiar person in a 2D image of a population. Thus, obstruction of the instantiation of the template organizes only a match. The main goal is to place this person in spite of other matches on the position (e.g., other people with the same or similar appearance). Generally, this kind of problem is unusual in an artificial visual system.

The position mentioned previously can be described as other potential settings. The search can be conducted in an n-dimensional space, in which objects were defined by a set of points with microproperties of basic building blocks in the space. Also, we can imagine that the search space is represented by a graph in which one is interested in playing a pattern described as another subgraph. As well as a data model is that described in concept of the fundamental microproperties, the search space in SDS is defined in terms of these features too. In other words, the macroscopic properties of objects in a search space can be efficiently described in terms of appropriate microproperties and their connection. If we consider the search space as a bit map pictures or some features such as angles, then the microproperties can be thought of pixel intensities. It should be noticed that the positions of the mentioned microproperties common to the object in search space are solutions to the search.

As it mentioned previously, the search in SDS is carried out in parallel by a set of elements call agents. According to their performance in the search, the elements (agents) can become active if they find the correct position within the search space or they are inactive.

The main aim of these agents is to reach the search space and interpret the target. The agents independently search for sectional solutions. Therefore, the search can be thought of as a compactible process in which all of the elements (agents) independently search for reaching appropriate solution. The mentioned agents try for placing other agents in their position. Thus, this competition among the agents becomes a corporation when the agents are used to seek the potential for the data model. This cooperation illustrates that the best promising agent that takes apart in this competition will examine all of the potential positions related to the object. In the next step, the best position for the data model will be extracted from independent positions in the search space. Thus, it is concluded that the efficiency of the agents will appear when they cluster appropriate position in the search space as the first agent starts contributing to position other agents.

The mining game is simple and appropriative metaphor that illustrates the application of the agents in the SDS. A group of miners search for gold in a hill of a mountain range, but they have no information about the propagation of gold on the hill. The mountain contains a set of discrete hills, and every hill consists of a set of paths to the mine. The possibility of reaching the goal on the mentioned paths is proportionate to its net wealth. For increasing its wealth, the miner should identify the hill with valuable paths that consist of gold, so a maximum number of miners can work there.

To solve this problem, they decide to utilize an appropriate SDS. In the beginning of the mining operation, every miner is randomly allocated a hill to mine (the miners' hill hypothesis, h). The next day, each miner is allocated a randomly selected path on his hill to mine. At the end of each day, the amount of gold that the miner has found is proportional to the probability of his happiness. At the end of the day, the miners gather together, and each miner who was sad chooses another miner at random to talk to. If the selected miner is happy, then he happily explains his coworker about the identification of the hill he is mining (he communicates his hill hypothesis, as a result, that hill will be selected). If the selected miner is unhappy, he will not be able to explain his coworker about the hill, so he once more randomly chooses another hill to mine the next day.

On the SDS perspective, agents acts as miners; happy miners are considered to be active agents, and unhappy miners are inactive agents. Furthermore, the agent's hypothesis takes the role of a miner's hill hypothesis. This can be proved that the aforementioned procedure is isomorphic to SDS, and thus, the miners can organize themselves and gather together at the top of the hill with high concentration of gold. The algorithm of the mining game contains several steps that are summarized as follows (Al-Rifaie and Bishop, 2013):

01: Initialisation phase
02: Allocate each miner (agent) to a random
03: Hill (hypothesis) to pick a region randomly
04: Until (all miners congregate over the highest concentration of gold)
05: Test phase
06: Each miner evaluates the amount of gold they have mined
    (hypotheses evaluation)
07: Miners are classified into happy (active) and unhappy (inactive) groups
08: Diffusion phase
09: Unhappy miners consider a new hill by either communicating with another
    miner;
    Or, if the selected miner is also unhappy, there will be no information flow
    between the
    Miners; instead the selecting miner must consider another hill (new hypothesis)
    at random
10: END

Thus, we have concluded from the aforementioned process that if there is gold, the agent will be active, otherwise it will be inactive.

4. Markov chain model

Consider a noiseless search space where a unique object with a nonzero overlap with template-appropriate solution exists. Assume that upon a random choice of an appropriate feature for testing, a hypothesis about the solution locating an agent may fail to recognize the sufficient solution with a probability $p^->0$. Assume further that in the nth iteration, m out of a total of N agents are active. The following matrix determines one-step evolution of a agents (Nasuto et al., 1998):

$$p_n = \begin{array}{c} a_a \\ n_a \end{array}\begin{bmatrix} 1-p^- & p^- \\ p_1^n & 1-p_1^n \end{bmatrix} \tag{17.1}$$

$$p_1^n = \frac{m}{N}(1 - p^-) + \left(1 - \frac{m}{N}\right)p_m(1 - p^-) \tag{17.2}$$

where the active agent pointing to the best solution is illustrated by a_a and the passive agent that failed the test pointing to the correct solution is denoted with n_a.

Generally, the SDS has been suggested to place the target or the template. From the aforementioned matrix, the active agent pointing to the best solution is illustrated by a_a and the passive agent that failed the test pointing to the correct solution is denoted by n_a. Let the size of search space be donated by M (the number of possible location of objects). Also, in the previous iteration, p_m be the probability of placing target in a uniformly random and the suboptimal object be p_n. Let the probability of a wrong positive be p^+ and the probability of wrong negative be p^-. Imagine that N is the number of exist agents. The number of active agents mentioning toward the position of the target and also the active agents pointing toward the wrong positive (the number of passive agents will be equal to the differences among the total number of agents and the sum of these two numbers) specified the state of the search in the nth step. Thus, if an agent be active in the previous iteration, it will start to test the pattern at the same position with using various agents, randomly chosen subcomponent. Also, with probability $(1-p^-)$, it may stay active (test positive), otherwise it remains inactive. By selecting an active agent randomly and testing positively at its position, the first term in p_1^n passive agents may become active. Obviously, the mentioned stochastic process models a nonhomogenous Markov chain. The entire second row of the probability transition matrix P is not dependent on time, but by the change in the search process, the matrix P_n is dependent on time, which calls this nonhomogeneity expression. It should be noticed that only active agents contain useful potentially information, and they affect the search directions of other agents. The strong halting criterion obtains their important information from active agents. The evolution state and testing state are two factors of the one-step evolution of passive agents. In many different ways, passive agents can change its state. For example, through the diffusion state it can choose, with probability a/M, an active agent pointing toward the solution, and after that it can remain passive with the probability of p^-. Other possibilities act and follow in a similar way. The evolution of the SDS is generally specified by the evolution of particular agents. There are various possible methods in which SDS can transfer from one state to another. By summing up the probabilities of all possible methods, the probability of one-step transition for Markov chain model of SDS will be obtained. The state space of a Markov chain demonstrating stochastic diffusion search can be determined by a pair of integers (a,w), in which "a" denotes the number of active agents pointing the target and "w" explained the number of agents pointing the false positives. From the aforementioned, it is concluded that for $p^- \neq 0$ and $p^+ \neq 0$ the probability of transition of SDS from the state (v,b) to (r,a) in one iteration is given by the formula (Nasuto et al., 1998):

$$P\{X_{n+1} = (r,a) \| X_n = (v,b)\} = \sum_{k_2}^{min(v,r)} \sum_{k_1}^{min(b,a)} Bin(k_2, p^-) \tag{17.3}$$

$$Bin(k_1, p^+) Mult(k_1, k_2, r, a, v, b)$$

$$p_1^n = \frac{m}{N}(1 - p^-) + \left(1 - \frac{m}{N}\right) p_m (1 - p^-) \tag{17.4}$$

where

$$Bin(k_2, p^-) = \binom{v}{k_2}(1 - p^-)^{k_2}(p^-)^{v-k_2} \tag{17.5}$$

$$Bin(k_1, p^+) = \binom{v}{k_2}(p^+)^{k_1}(1 - p^+)^{b-k_1} \tag{17.6}$$

$$Mult(k_1, k_2, r, a, v, b) = \binom{M-v-b}{r-k_2} p_{ab}^{r-k_2} \binom{M-v-b-r+k_2}{a-k_1} \tag{17.7}$$

$$p_{af}^{a-k_1}(1 - p_{ab} - p_{af})^g$$

and

$$p_{ab} = \frac{v}{M}(1 - p^-) + \left(1 - \frac{v}{M} - \frac{b}{M}\right) p_m (1 - p^-) \tag{17.8}$$

$$p_{af} = \frac{b}{M} p^+ + \left(1 - \frac{v}{M} - \frac{b}{M}\right) p_d p^+ \tag{17.9}$$

$$g = M - r - a - v - b + k_1 + k_2 \tag{17.10}$$

and double summation in the aforementioned formula is over k_1, $k_2 \geq 0$ such that $g \geq 0$. The term Bin (k_2, p^-) denotes the probability in which k_2 out of v agents of the type (a,m) will preserve their state during testing phase and $v - k_2$ agents will evolve to the state $(n, *)$. Same of the mentioned state the term Bin (k_1, p^+) denotes the probability, in which k_1 out of b agents of the state (a,b) will preserve the states of agents during the testing phase and $b - k_1$ agents will become passive and derive to the state $(n, *)$. The term $Mult$ (k_1, k_2, r, a, v, b) explains the probability of $r - k_2$ out of $M - v - k$ passive agents deriving to the state (a, m) and also $a - k_1$ out of $M - v - b - r + k_2$ evolving to the state (a, d) while those remaining ones stay passive.

From the aforementioned formula, it can be achieved that for the case in which p^+ and p^- are equal to zero by computing the limit of the transition probability with p^-, p^+ tending to zero. Let S be a search space. Let f_n^s be the number of active agents pointing to the same target in the search space S in the

nth iteration. It is obvious to see that the mentioned condition is to fulfill $\sum_{s \in S} f_n^{rs}$
$\leq M$ in which M is the total number of the agents. Let the maximal number of active agents in the nth duplication mentioning to the same target be Z_n. Thus, $S_n^z \in s$ in the search space, i.e., (Nasuto et al., 1998):

$$Z_n = max_{s \in s}\left(f_n^s\right) \qquad (17.11)$$

From the aforementioned expression, it can be concluded that the definition of the convergence of the stochastic diffusion search is formulated as follows:

$$\exists_{a,b>0}(2b > M \cap b + a \leq M \cap a - b \geq 0) \; \exists_{n_0}$$
$$\forall_{n>n_0}\left((Z_n - a) < b\right) \qquad (17.12)$$

and the solution is the position pointed at by Z_n. Thus, the SDS will have gained an equilibrium if there exists a time instant n_0 and an interval (specified by a and b) such that after n_0, the maximal number of agents pointing to the same target will enter and will remain within the specified interval. It will also be noted that it is not essential to use the convergence of process for the aforementioned definition to a fixed state. Furthermore, the interval (specified by a and b) explains a tolerance region. If all of the oscillation of the maximal number of the agents pointing to the same target in the search space occurs in the mentioned interval, then the agents will be useless and not essential. In the study by Nasuto and Bishop (1999), it has been proved that two mentioned parameters (a, b) do not play an important role in the search space.

5. Convergence of the stochastic diffusion search

Nasuto et al. (1998) have analyzed the convergence of SDS in two separate cases. In the first case in which there exists an ideal, the instantiation of the target in the search space was inspected. In the presence of the target in the search space, the testing phase for the agents pointing to the target becomes deterministic (there is a perfect competition, so the agents that pointing to this position can fail the test). Let the position of the model be denoted as S^m in the search space. According to Markov chain of the SDS, the presence of the object in the search space is equivalent p^- to zero, so from the aforementioned explanation for $p^- = 0$,

$$p = \left\{\lim_{n \to \infty} Z_n = N\right\} = 1 \text{ Moreover } p\{S_n^z = S_m\} = 1 \qquad (17.13)$$

in which

$$Z_n = max_{s \in s}\left(f_n^s\right) \qquad (17.14)$$

The mentioned proposition proves the intuition that all of the agents will eventually converge on its own position in the existence of the target in the search space. Thus, in this case, two parameters (a and b) do not affect the convergence of the SDS. The following result can be proven when the target does not exist in the search space. By considering $p^- \neq 0$, the strong convergence criterion does not hold in the SDS. To prove of the aforementioned statement, refer to the study by Nasuto and Bishop (1999), it follows that in the case of $p^- \neq 0$, the model of SDS is a Markov chain model (Kemeny and Snell, 1983). Thus, it is easy to see that SDS accomplishes another model. The weaker convergence property is as follows:

By giving $p^- \neq 0$, the SDS converges in an appropriate state, i.e.,

$$(\exists a > 0)\left(\lim_{n \to \infty} Ez_n = a \right) \tag{17.15}$$

To prove the aforementioned formula, refer to the study by Nasuto and Bishop (1999). The aforementioned statement shows that in the most general case, the stochastic diffusion convergence has been considered as approaching an equilibrium in a statistical sense. It means that after approaching a constant state, all of the possible configurations of the agents pointing to the appropriate instantiation of the target as well as disturbances happen infinitely often considering to limiting probability distribution (it should be noted that some of them happen scarcely). In practice, terms with sufficient estimates for halting parameters a and b algorithm will be stable for a long term, thus authorizing restriction. Also, SDS causes to find appropriate solutions using a certain amount of computational resources. These remaining resources probe the search space, and also, they attempt to explore other sufficient potential solutions. The aforementioned solution can be investigated in two separate ways—in the first way, the solution fit to the target may decrease because the similarity criteria change through the time or sufficient solution can be found during the case of the time in the search space. It is also concluded that in both of the cases, the SDS will be able to create a better optimal solution due to the agents probing the search space. If once created, it will reallocate more of these resources toward these new optimal solutions rapidly.

6. Time complexity

From the study by Nasuto et al. (1998), it has been proven that the convergence rate of the SDS was specified in the case in which there exists a target in the search space and there was not any noise. However, the results that was achieved should expand over a wide range of case that both of these hypotheses were proved. The aforementioned conclusion follows from attributes of transition Markov chain model probability matrix. The time complexity estimates were obtained for a synchronous state of SDS. From this definition, the dynamical state of operation of the SDS has been proven, in which all of the agents carry out their tasks and simultaneously update their mode. It was found out using the Markov chain model described previously that the convergence rate of SDS is as follows (Nasuto et al., 1998):

$$
\begin{cases}
o\left(N - log\left(\left(1 - \dfrac{1}{M}\right)\right)\right) & if\ M > M(N) \\[4mm]
o\left(-log\left(\dfrac{1}{N}\left(1 - \dfrac{1}{M}\right)\right)\right) & if\ M < M(N)
\end{cases}
\tag{17.16}
$$

in which M (N) is given by

$$
M(N) = \left[\frac{N^{\left(\frac{1}{N-1}\right)}}{N^{\left(\frac{1}{N-1}\right)} - 1}\right]
\tag{17.17}
$$

As an important consequence, it follows that the convergence time of the SDS is $o\left(\frac{M}{N}\right)$ for M > M (N), but for M < M (N), with increase of the size of the search space, the search time grows nonlinearly and it is $O\left(\frac{1}{logN}\right)$ (Nasuto et al., 1998).

Also, it can be achieved from the theory to reach a clear saturation of the average search time for sufficiently small size of the search space. It may also be noted that the standard deviation of the convergence time illustrates a similar linear trend, and the ratio of the standard deviation to the mean being close, it follows that, in a wide range of cases, the SDS convergence times are distributed close to the value in which we are expected; however, there are some cases that in these modes, the SDS convergence time is very slow or maybe very fast, so it causes this pattern of the behavior occur. Also, from these experiments, it can be concluded that the existence of noise protected the overall functional dependence of the convergence time on the search space size.

7. Conclusions

In this chapter, the SDS is described as a connectionist method in which it can suggest an appropriate searching for an object in the search space. In contrast to the most connectionist search method, the SDS is able to explore either the object or if the object is not in the search space, and the SDS starts finding the objects with best instantiation without creating any problem. In fact, the SDS is a nature-inspired approach that solves the problem by creating a communication among the agents. The algorithm that the SDS uses can easily be expanded to a wide range of applications. As mentioned previously, agents are categorized into two different groups in which they act independently and collaborate with others in parallel to achieve the best position of the target in search space. The agents can be categorized into two separate groups: active and inactive. The experiments conducted in this case illustrate that the SDS is not only used for visual domain but also can be successfully used as an appropriate solution for the text string matching problems. The SDS is called an adaptive algorithm because of its weak convergence property and also it is capable of probing the best solution in a dynamic condition. The SDS belongs to a class of algorithms in which agents can be classified into two separate groups. The first class (the active agent) supports the solution found so far, and the second class (the passive agent) probes the rest of the search space. Therefore, it is concluded that the SDS suggested the best solution found so far, but it is still capable of finding the best solution if the target exists in the search space due to its change. From the theoretical assumption, it follows that the SDS can easily solve a problem that changes dynamically and that is not static. The weak convergence of the SDS is capable of solving the problems successfully (Grech-Cini, 1995). The problem of large state space is quite in opposite with Markov chain theory. In this case, the researchers address this problem by studying the various conditions, in which it will be possible to exchange the Markov chain model with a wide range of state space by another one with significantly reduced space that was achieved from the original one by partitioning it and aggregating all states belonging to same partition together. Different forms of aggregation are available: lump ability, weak lump ability (Kemeny and Snell, 1983), exact lump ability (Kim and Smith, 1995; Schweitzer, 1983), and approximate lump ability (Buchholz, 1994). Unfortunately, none of the aforementioned techniques are applicable to the Markov chain describing the SDS. In future research, we will concentrate on evaluation of the agent's behavior, so we hope to present reliable algorithm in this case.

References

Al-Rifaie, M.M., Bishop, J.M., 2013. Stochastic diffusion search review. Paladyn. Journal of Behavioral Robotics 4, 155−173.

Bishop, J., Torr, P., 1992. The Stochastic Search Network, Neural Networks for Vision, Speech and Natural Language. Springer, pp. 370−387.

Buchholz, P., 1994. Exact and ordinary lumpability in finite Markov chains. Journal of Applied Probability 31, 59−75.

Grech-Cini, E., 1995. Motion Tracking in Video Films. PhD Thesis. University of Reading.

Hurley, S., Whitaker, R.M., 2002. An agent based approach to site selection for wireless networks. In: Proceedings of the 2002 ACM Symposium on Applied Computing. ACM, pp. 574−577.

Kemeny, J.G., Snell, J.L., 1983. Finite Markov Chains: With a New Appendix "Generalization of a Fundamental Matrix". Springer.

Kim, D.S., Smith, R.L., 1995. An exact aggregation/disaggregation algorithm for large scale Markov chains. Naval Research Logistics 42, 1115−1128.

Knuth, D.E., 1973. Sorting and Searching. The Art of Computer Programming, vol. 3. Ch. 6.

Mitchell, M., 1998. An Introduction to Genetic Algorithms. MIT press.

Nasuto, S., Bishop, M., 1999. Convergence analysis of stochastic diffusion search. Parallel Algorithms and Application 14, 89–107.

Nasuto, S.J., Bishop, J.M., Lauria, S., 1998. Time Complexity Analysis of the Stochastic Diffusion Search, pp. 260–266. NC.

Schweitzer, P.J., 1983. Aggregation methods for large Markov chains. In: Proceedings of the International Workshop on Computer Performance and Reliability. North-Holland Publishing Co., pp. 275–286

Chapter 18

Resampling methods combined with Rao-Blackwellized Monte Carlo Data Association algorithm

Soheil Sadat Hosseini[1,2], Mohsin M. Jamali[3]

[1]*Department of Electrical Engineering, Capitol Technology University, Laurel, MD, United States;*
[2]*Department of Electrical Engineering and Computer Science, The University of Toledo, Toledo,
OH, United States;* [3]*College of Engineering, The University of Texas of the Permian Basin,
Odessa, TX, United States*

1. Introduction

Multiple target tracking (MTT) (Bar-Shalom et al., 2001; Blackman and
Popoli, 1999) first originated in the 1970s, and since then it has been one of the
most challenging research areas with applications in many crucial real-world
tasks. Various applications of MTT involved image processing (Lane et al.,
1998; Cox et al., 1993), video surveillance (Yilmaz et al., 2006), mobile
robotics (Durrant-Whyte and Bailey, 2006; Bailey and Durrant-Whyte, 2006),
finance (Hans and Dijk, 2000), and biomedicine (Hammarberg et al., 2002).

The MTT problem involves when more than one target is subject to
tracking at the same time. Such MTT scenarios are of great importance, and
they happen in many applications. It turns out that the MTT problem is
challenging in comparison with a single target case. It consists of filtering and
data association. MTT is reduced to a group of estimation problems if the
sequence of measurements associated with each target is known. The association between targets and measurements is, unfortunately, unknown. Finding
out which measurements were produced by which targets is the data association problem.

When tracking a single target, then it has a certain amount of computational costs. If there are n targets, then their tracking will be more than n times
of the cost of tracking one target. In fact, the complexity of the simplest MTT
methods is, at least, proportional to the square of the number of targets tracked,
when assuming that in each scan n new observations are received. The second

Handbook of Probabilistic Models. https://doi.org/10.1016/B978-0-12-816514-0.00018-7
449

serious difficulty of MTT is not of algorithmic nature. It concerns with the uncertainty created by sensor inabilities such as sporadic miss detections or the delivery of false alarms. The series of measurements comprises of a set of reports, each of which relates to a scan of the surveillance area. These reports provide some or all information about position, velocity, or range information about possible targets.

Each measurement is related to a number of possible sets of data, each representing a hypothesis to describe the measurement source. Measurement may be

- The false alarm set, meaning that the measurement is unreal
- The new target set, meaning that the measurement is real and relates to a target for which there are no previous measurements
- An existing set of previous measurements related to a single target.

To update the predicted target states, we need to utilize information from the measurements provided at the current instance of time. To do so, it is necessary to determine which measurements are produced by targets and which are spurious. We then need to assign the former measurements to the corresponding targets. These two steps form the data association problem (Bar-Shalom et al., 2001; Blackman and Popoli, 1999; Blackman, 1986). There are number of data assignment techniques such as the nearest neighbor (NN) (Bar-Shalom et al., 2011), global nearest neighbor (GNN) (Konstantinova et al., 2003), probabilistic data association (Bar-Shalom and Tse, 1975), multiple hypothesis tracking (Reid, 1979), etc.

Another solution for MTT problems is RBMCDA (Särkkä et al., 2004, 2007). In this method, particle filter (PF) is only used to evaluate the data association indicators instead of computing everything by pure Monte Carlo sampling. Therefore, the required number of particles can be significantly reduced. This can be a method with good performance for tracking a single target or multiple separated targets in sparsely cluttered environment. However, we found out that by using a validation gate we can reduce the number of calculations in RBMCDA even more.

Resampling step is an important step in particle filtering algorithms, since after a few iterations, propagating trivial weights becomes computationally expensive. Therefore, using an appropriate resampling technique is vital. As mentioned in the study by Särkkä et al. (2007), the resampling technique utilized is not necessarily the most efficient in all conditions, but it just worked well in their applications. One problem in the study by Särkkä et al. (2007), is that the authors did not use any practical or real problems to test their technique. Therefore, it might be possible the specified resampling technique used in the study by Särkkä et al. (2007), may not be useful for other problems. To improve this method and use the best of its features, several different resampling techniques will be added to RBMCDA. Using the modified RBMCDA shows the effect of each resampling technique on three-dimensional case study.

The rest of the article is organized as follows. The RBMCDA for MTT is presented in Section 2. Section 3 describes several resampling methods combined with RBMCDA. The simulation results for the proposed MTT algorithm are presented in Section 4. Finally, conclusions are drawn in Section 5.

2. Rao-Blackwellized Monte Carlo Data Association

A famous technique for solving MTT is the Rao-Blackwellized Monte Carlo Data Association (RBMCDA) (Särkkä et al., 2004, 2007). The RBMCDA uses Kalman filter for prediction and PF for data association. This algorithm allows use of any probability distribution function for modeling of clutter and any false alarm measurement. This work assumes that clutter is uniformly distributed and target measurements are also assumed to have Gaussian distribution (Särkkä et al., 2004, 2007). The RBMCDA algorithm can be described as follows (Särkkä et al., 2004, 2007):

(1) Target state priors can be calculated as a weighted importance sample set

$$p\left(x_{0,j}\right) = \sum_i w^{(i)} N\left(x_{0,j}\middle|\widehat{x}_{0,j}^{(i)}, P_{0,j}^{(i)}\right) \tag{18.1}$$

where $\widehat{x}_{0,j}^{(i)}$ and $P_{0,j}^{(i)}$ are mean and covariance of the target state.

The RBMCDA approach uses exactly Rao-Blackwellized particle filtering framework (Doucet et al., 2000) where the sampled latent variable, λ_k, is designed to include the data association indicators, da_k at time step k.

$$\lambda_k = da_k \tag{18.2}$$

(2) The state model of targets is linear Gaussian and can be shown as follows:

$$p(x_k|x_{k-1}) = N(x_k|A_{k-1}x_{k-1}, Q_k) \tag{18.3}$$

where A_{k-1} is a transitional matrix defined as state models of targets and Q_k is the process noise covariance matrix.

(3) The observation model of the targets can be computed as follows:

$$p(z_k|x_k, da_k) = N(z_k|H_k(da_k)x_k, v_k(da_k)) \tag{18.4}$$

where $H_k(da_k)$ is the observation model matrix and it is conditional on the data association da_k. $v_k(da_k)$ is the observation noise covariance of the target da_k.

(4) The clutter measurements are independent of states where a measurement model of the form is defined as follows:

$$p(z_k | da_k = 0) \tag{18.5}$$

(5) The data association step of RBMCDA is computed as a Markov chain of latent variables.

The optimal importance distribution for each particle i is computed as follows:

$$p(da_k | z_{1:k}, da_{1:k-1}^{(i)}) \tag{18.6}$$

(6) The marginal observation likelihood is provided by Särkkä et al. (2004):

$$p\left(z_k \middle| da_k, z_{1:k-1}, da_{1:k-1}^{(i)}\right) \tag{18.7}$$

$$= \begin{cases} \dfrac{1}{V}, & \text{if } da_k = 0, \\[2ex] KF_{lh}\left(z_k, \widehat{x}_{j,k}^{-(i)}, P_{j,k}^{-(i)}, H_{j,k}, v_{j,k}\right) & \text{if } da_k = j \end{cases}$$

where V is the area of the region covered by a sensor (e.g., radar), $j = 1, ..., N_T$ (number of targets) and $KF_{lh}()$ represents the observation likelihood estimation using Kalman filter. For target j, $H_{j,k}$ and $v_{j,k}$ denote the observation model matrix and the observation covariance matrix, respectively. For $j = 1, ..., N_T$, we have

$$\left[\widehat{x}_{j,k}^{-(i)}, P_{j,k}^{-(i)}\right] = KF_p\left(\widehat{x}_{j,k-1}^{-(i)}, P_{j,k-1}^{-(i)}, A_{j,k-1}, Q_{j,k-1}\right) \tag{18.8}$$

where $KF_p()$ represents the Kalman filter prediction section and for target j in particle i, $\widehat{x}_{j,k-1}^{-(i)}, P_{j,k-1}^{-(i)}$ denote the means and the covariance. For the target j, $da_{1:k-1}^{(i)}$, $A_{j,k-1}$ and $Q_{j,k-1}$ denote the transition matrix of state model and the noise covariance matrix, respectively. The Bayes' rule is used to calculate the posterior distribution of da_k as follows:

$$p(da_k | z_{1:k}, da_{1:k-1}^{(i)}) \propto p(z_k | da_k, z_{1:k-1}, da_{1:k-1}^{(i)}) p\left(da_k \middle| da_{k-mk-1}^{(i)}\right) \tag{18.9}$$

Särkkä et al. (2007) argue that there is no relation between an association da_k and the previous observations $z_{1:k-1}$. da_k only depends on the m previous associations $da_{k-m:k-1}$ if the order of the Markov model is m. The optimal importance distribution is sampled as follows:

$$\pi_j^{(i)} = p(z_k | da_k^{(i)} = j, z_{1:k-1}, da_{1:k-1}^{(i)}) p(da_k^{(i)} = j | da_{k-mk-1}^{(i)}) \qquad (18.10)$$

To use RBMCDA algorithm for an unknown number of targets, it is assumed that there is always a (very large) constant number of targets N_∞. In fact, the algorithm is tracking targets called "alive targets." When an observation is given, target births only may occur, and a birth occurs with probability p_b.

It is proposed that RBMCDA algorithm should be combined with resampling methods for tracking three-dimensional case study with unknown number of targets. Resampling is an important step in particle filtering algorithms, since after a few iterations, propagating trivial weights becomes computationally expensive. Therefore, using an appropriate resampling technique is vital. As mentioned in the study by Särkkä et al. (2007), the resampling technique utilized is not necessarily the most efficient in all conditions, but it just worked well in their applications. One problem in the study by Särkkä et al. (2007), is that the authors did not use any practical or real problems to test their technique. Therefore, it might be possible the specified resampling technique used in the study by Särkkä et al. (2007), may not useful for other problems. To improve this method and use the best of its features, several different resampling techniques will be added to RBMCDA. Using the modified RBMCDA shows the effect of each resampling technique on three-dimensional case study.

3. Resampling techniques

The resampling step is essential for PF; PF will quickly produce a degenerate set of particles without this step, i.e., a set in which a few particles dominate the rest of the particles with their weights. This means that the obtained estimates will be inaccurate and will have unacceptably large variances. With resampling, such drawbacks are eliminated, which is why it is highly critical to PF. Consequently, this step has been extensively studied. A pseudocode for RBMCDA combined with resampling methods is shown in Fig. 18.1.

Initialize the RBMCDA
Compute predicted next states
FOR k = 1 : Y (Number of measurements)
FOR i = 1 : NP (Number of particles)
Compute the unnormalized clutter association probability
Compute the unnormalized target association probabilities for each target j = 1,. . . ,T
Normalize the importance distribution
Sample a new association $c_k^{(i)}$
END
Perform Resampling methods
END

FIGURE 18.1 A pseudocode for Rao-Blackwellized Monte Carlo Data Association combined with resampling methods.

We are going to use different resampling techniques to see its effect on our MTT problem. A comparative study is performed between following seven frequently encountered resampling techniques for the RBMCDA algorithm.

1. Cumulated sum weight (CSW) resampling
2. Stratified resampling
3. Systematic resampling
4. Residual resampling
5. Residual systematic resampling (RSR)
6. Optimal resampling
7. Reallocation resampling

3.1 Cumulated sum weight resampling

The particles are resampled using the computed weights. A new set of particles $x_k^{(i*)}$ are generated based on the relative weight ω_k^s. Arulampalam et al. (2002) recommend the following two steps for the CSW resampling:

1. A random number r is generated uniformly distributed on [0,1].
2. The weight ω_k^s is accumulated, one at a time, and stops when the accumulated sum is more than r. Which is $q_k^{(j-1)} = \sum_{s=1}^{j-1} \omega_k^s < r$ but $q_k^{(j)} = \sum_{s=1}^{j} \omega_k^s \geq r$. Then, the new particle $x_k^{(i*)}$ has the value of the old particle $x_k^{(j)}$.

This resampling idea can be summarized as follows:

$$x_k^{(i*)} = x_k^{(j)} \text{ with probability } \omega_k^j (i*, j = 1, ..., B) \qquad (18.11)$$

3.2 Stratified/systematic resampling method

The stratified resampling algorithm presented by Kitagawa (1996), splits the whole population of particles into B subpopulations. The random numbers $\{o_k^s\}_{s=1}^{B}$ are independently sampled from uniform distribution on each of B disjoint subintervals $\left(0, \frac{1}{B}\right] \cup ... \cup$ of (0, 1], i.e.,

$$o_k^{(s)} \sim O\left(\frac{s-1}{B}, \frac{s}{B}\right), \ s = 1, 2, ..., B \qquad (18.12)$$

Moreover, the bounding procedure is defined as follows:

$$q_k^{(m-1)} < o_k^{(s)} \le q_k^{(m)} \tag{18.13}$$

where

$$q_k^{(m)} = \sum_{s=1}^{m} \omega_k^{(s)} \tag{18.14}$$

The systematic resampling technique was first developed by Kitagawa (1996) (Kitagawa, 1996) and further studied by Carpenter et al. (1999). It is the basic and simplest resampling technique. First, $o_k^{(1)}$ is chosen randomly from the uniform distribution on $(0, 1/B]$, and the rest of the o numbers are deterministically computed, i.e.,

$$o_k^{(1)} \sim O\left(\frac{0, 1}{B}\right], \tag{18.15}$$

$$o_k^{(s)} = o_k^{(1)} + \frac{s-1}{B}, \; s = 2, 3, ..., B.$$

Because of the smaller number of random numbers sampled, the systematic resampling algorithm is computationally more effective than the stratified technique.

3.3 Residual resampling (remainder resampling)

The residual resampling algorithm developed by Beadle and Djuric (1997) is an alternative technique to systematic resampling method. It includes two steps of resampling. The first step is to locate which particle's weight is above $\frac{1}{B}$, and the second step is to draw randomly utilizing the remaining particles whose weights are below $\frac{1}{B}$ (referred to as residuals). For the first step, set $B_k^{(s)} = B\omega_k^{(s)}$, and this step is named the deterministic replication section. The second stage is residual resampling method, assigning the residual of the weight as follows:

$$\widehat{\omega}_k^{(m)} = \omega_k^{(m)} - \frac{B_k^{(m)}}{B} \tag{18.16}$$

Then, the new particles are sampled by calling the CSW technique with the parameters. The total number of replicated particles in the first step is $B_k = \sum_{s=1}^{M} B_k^{(s)}$ and in the second step $R_k = B - B_k$.

In the second stage, CSW resampling is applied to choose the particles based on the residual weights. The probability for choosing $x_k^{(s)}$ is proportional to the residual weight of that particle $\left(\text{Probability}(x_k^n = x_k^s) = \widehat{\omega}_k^{(s)}\right)$.

Step 1) Specify parameters

FOR s=1 to M

$$B_k^{(s)} = \lfloor B\omega_k^{(s)} \rfloor$$

$$\hat{\omega}_k^{(s)} = \omega_k^{(s)} - \frac{B_k^{(s)}}{B}$$

END

Step 2) Deterministic replication of partic

b=0

FOR s=1 to M

FOR h=1:$B_k^{(s)}$

b = b + 1

$$\hat{x}_k^{(b)} = x_k^{(s)}$$

END

END

$B_k = b$

Step 3) Update weights

FOR s=1:M

$$\hat{\omega}_k^{(s)} = \omega_k^{(s)} - \frac{B}{(B - B_k)}$$

END

Step 4) Multinomial resampling method

$$[Q_t^{(m)}] = CSW \left[\{w_t^{(m)}\}_{m=1}^{M} \right]$$

n = 0

WHILE (n ≤B)

$\delta \sim \underline{U}(0,1]$; s =1

WHILE ($Q_k^{(s)} < \delta$)

s = s + 1

END

n = n + 1

$$\tilde{x}_k^n = x_k^{(s)}$$

FIGURE 18.2   Pseudocode for residual resampling method.

The first stage is a deterministic replication, and thus, the variation of the number of times a particle resampled only happens on the second step. Thus, if the second stage is employed utilizing CSW resampling, the lower and upper limits of the number of times that the sth particle is resampled are $B\omega_k^{(s)}$ and $B\omega_k^{(s)} + R_k$, respectively (Beadle and Djuric, 1997) (Fig. 18.2).

3.4 Residual systematic resampling technique

This technique has been developed and studied by many researchers (Beadle and Djuric, 1997; Bolic et al., 2005). For instance, Bolic, Djuric et al. (Bolic

STEP 1) Sampling

A random number o is generated using uniform distribution $O(0, \frac{1}{B}]$

STEP 2) Updating

For s = 1 to B

$\hat{B}_k^s = \lfloor B \times (w_k^s - o) \rfloor + 1$

$o = o + \dfrac{\hat{B}_k^s}{B} - w_k^s$

End

Step 3) Replication

$b = 0$
FOR s=1 to B
FOR h=1:$B_k^{(s)}$
$b = b + 1$
$\hat{x}_k^{(b)} = x_k^{(s)}$
END
END
$B_k = b$

FIGURE 18.3 Pseudocode for residual systematic resampling.

et al., 2005) proposed a new algorithm RSR for the cases, where the systematic resampling technique is used in the second stage. Different from residual resampling algorithm that proceeds in two separate steps, RSR proceeds in only one step with the integer replication. Moreover, there is no additional procedure needed for the residuals. Therefore, the computational complexity of RSR is of order O(N). The concept of RSR is like the idea, which is utilized, in systematic resampling algorithm. That is to collect the fractional parts of each particle in the search list until it is large enough to create a sample. The algorithm of the resampling step of RSR is shown in Fig. 18.3

3.5 Threshold-/grouping-based resampling

In this technique, particles are put into different groups using weight thresholds, and one utilizes different sampling techniques for each group to give more flexibility for resampling. One or more thresholds can be chosen dynamically/adaptively or fixed.

3.5.1 Dynamic threshold

The optimal resampling specifies a threshold value dyn_t, where dyn_t is the unique root of following equation (Fearnhead and Clifford, 2003).

$$B = \sum_{s=1}^{M} min\left(\frac{\omega_k^{(s)}}{dyn_k}, 1\right) \tag{18.17}$$

where B is provided, and $B < M$. A feature of optimal resampling is that there

Step 1) Determine the value of dyn_k using the equation as follows: $B = \sum_{s=1}^{M} min\left(\frac{\omega_k^{(s)}}{dyn_k}, 1\right)$

Step 2) Weights and states Update
$s = 1; n = 0; h = 0$
WHILE $s < M + 1$
IF $w_k^{(s)} \geq dyn_k$
$n = n + 1$
$\tilde{x}_k^{(n)} = x_k^{(s)}; \tilde{w}_k^{(n)} = w_k^{(s)}$
ELSE
$h = h + 1$
$E^{(h)} = x_k^{(s)}; F^{(h)} = w_k^{(s)}$
END
$s = s + 1$
END
Step 3) Apply Stratified resampling method
$B_1 = n$
$\left[\{\tilde{x}_k^{(n)}\}_{n=B_1+1}^{B}\right] = $ Stratified resampling method $\left[\{E^{(r)}, F^{(r)}\}_{r=1}^{h}, B - B_1\right]$
Step 4) Update the remaining weights
FOR $n = B_1 + 1 : B$
$\tilde{w}_k^{(n)} = dyn_k$
END

FIGURE 18.4 Pseudocode of the resampling step of optimal resampling.

are no multiple duplicates of particles in the final set of B particles, because all particles whose weights are larger than this threshold are entirely maintained. The most important advantage of the algorithm is that it is optimal in terms of minimizing the squared error-loss function

$$E\left(\sum_{s=1}^{M} \left(\tilde{\omega}_k^{(s)} - \omega_k^{(s)}\right)^2\right) \tag{18.18}$$

where $\tilde{\omega}_k^{(s)}$ is used as the new weight of $x_k^{(s)}$ when it is resampled, otherwise $\tilde{\omega}_k^{(s)}$ is equal to zero. An advantage is that optimal resampling decreases the number of particles to $B < M$ for PF, when it utilizes increased number of propagated particles. However, one disadvantage of this technique is the necessity for calculating the value of dyn_t at each iteration. The pseudocode of the resampling step of optimal resampling is shown in Fig. 18.4.

3.5.2 Fixed threshold

This section describes a fixed threshold-resampling algorithm. Reallocation resampling algorithm was presented in Liu et al. (2001). In their technique, one selects $\frac{1}{B}$ as the fixed threshold. Then, if the weight of the jth particle is above $\frac{1}{B}$, the jth particle will be sampled repeatedly $B\omega_k^{(s)}$ or $(B\omega_k^{(s)} + 1)$ times. After resampling, the new weights are $\frac{\omega_k^{(s)}}{B\omega_k^{(s)}}$ or $\omega_k^{(s)} \Big/ B\omega_k^{(s)} + 1$. If the weight of the jth particle is below 1/B, the jth particle will be generated with probability $B\omega_k^{(s)}$ repeatedly. In addition, it is assigned $\frac{1}{B}$ as new weight. The algorithm is shown in Fig. 18.5.

Step 1) Parameter initialization
$n = 0$
Step 2) Weights and states Update
FOR s=1 to M
IF $w_k^{(s)} \geq 1/B$ (Fixed Threshold)$N_k^{(s)} = \lfloor B \times w_k^{(s)} \rfloor$ or $N_k^{(s)} = \lfloor B \times w_k^{(s)} \rfloor + 1$
FOR $h = 1 : B_k^{(s)}$
$n = n + 1 \tilde{x}_k^{(n)} = x_k^{(s)}; \tilde{w}_k^{(n)} = w_k^{(s)}/B_k^{(s)}$
END
ELSE (Weight is less than Fixed Threshold)
$o \sim O(0, 1/B]$
IF $w_k^{(s)} \geq o$
$n = n + 1$
$\tilde{x}_k^{(n)} = x_k^{(s)}; \tilde{w}_k^{(n)} = 1/B$
END
END
END
Step 3) Update the particle number
$B^* = n$

FIGURE 18.5   Pseudocode for reallocation resampling.

From above methods, it is seen that the resampled particles are not equally weighted. In addition, the sum of all weights is not equal to one; therefore, in the whole resampling portion, the normalization step is performed as an additional requirement.

4. Experiments and simulation results

4.1 Tracking unknown number of targets in 3D

In this thesis, a MTT problem in 3D is presented. Then, different forms of RBMCDA algorithms are used to estimate number of targets from noisy measurements. There are two objects that appear and disappear in different times in this example.

The state model of each target is modeled based on the discretized Wiener process velocity model (Bar-Shalom, 1987), where the state can be written as

$$\boldsymbol{x}_k = \begin{pmatrix} x_k & y_k & z_k & \dot{x}_k & \dot{y}_k & \dot{z}_k \end{pmatrix}^T \qquad (18.19)$$

where (x_k, y_k, z_k) is the targets's position and $(\dot{x}_k, \dot{y}_k, \dot{z}_k)$ is the velocity in three-dimensional Cartesian coordinates. The discretized dynamics can be written with a linear, time-invariant equation as follows (Bar-Shalom, 1987):

$$\boldsymbol{x}_k = \begin{pmatrix} 1 & 0 & 0 & \Delta t & 0 & 0 \\ 0 & 1 & 0 & 0 & \Delta t & 0 \\ 0 & 0 & 1 & 0 & 0 & \Delta t \\ 0 & 0 & 0 & 1 & 0 & 0 \\ 0 & 0 & 0 & 0 & 1 & 0 \\ 0 & 0 & 0 & 0 & 0 & 1 \end{pmatrix} \boldsymbol{x}_{k-1} + \boldsymbol{q}_{k-1} \qquad (18.20)$$

where $\boldsymbol{q}_{k-1}$ is discrete Gaussian white process noise having moments

$$E[\boldsymbol{q}_{k-1}] = 0 \tag{18.21}$$

$$\left[\boldsymbol{q}_{k-1}\boldsymbol{q}_{k-1}^T\right] = \begin{pmatrix} \frac{1}{3}\Delta t^3 & 0 & 0 & \frac{1}{2}\Delta t^2 & 0 & 0 \\ 0 & \frac{1}{3}\Delta t^3 & 0 & 0 & \frac{1}{2}\Delta t^2 & 0 \\ 0 & 0 & \frac{1}{3}\Delta t^3 & 0 & 0 & \frac{1}{2}\Delta t^2 \\ \frac{1}{2}\Delta t^2 & 0 & 0 & \Delta t & 0 & 0 \\ 0 & \frac{1}{2}\Delta t^2 & 0 & 0 & \Delta t & 0 \\ 0 & 0 & \frac{1}{2}\Delta t^2 & 0 & 0 & \Delta t \end{pmatrix}$$

The size of time step is set to $\Delta t = 1$ and the power spectral density of process noise to $q = 0.01$. The measurements are simulated as measurements of the ground truth plus a white Gaussian noise component.

To model the clutter measurements, the data association indicator da_k is now defined to have value 0 if the measurement is clutter, value 1 if the measurement is from the actual target. The measurement model for the actual target is linear with additive Gaussian noise. Thus, we can express the joint measurement likelihood as

$$p(z_k|\boldsymbol{x}_k, da_k) = N(z_k|\boldsymbol{H}_k(da_k)\boldsymbol{x}_k, \boldsymbol{v}_k(da_k)) \tag{18.22}$$

where the measurement model matrix is

$$\boldsymbol{H} = \begin{pmatrix} 1 & 0 & 0 & 0 & 0 & 0 \\ 0 & 1 & 0 & 0 & 0 & 0 \\ 0 & 0 & 1 & 0 & 0 & 0 \end{pmatrix} \tag{18.23}$$

and the noise covariance matrix

$$\boldsymbol{R} = \begin{pmatrix} 0.05 & 0 & 0 \\ 0 & 0.05 & 0 \\ 0 & 0 & 0.05 \end{pmatrix} \tag{18.24}$$

4.2 Simulation results

In this study, we investigate the effect of resampling techniques on RBMCDA. Figs. 18.6 and 18.7 show results for NN and GNN. These techniques are not able to locate any targets and separate true targets from noisy measurements.

In addition, based on Table 18.1, it shows that combing reallocation with RBMCDA provides the best answer with the least error and faster process. These techniques were applied on unknown MTT problem in 3D.

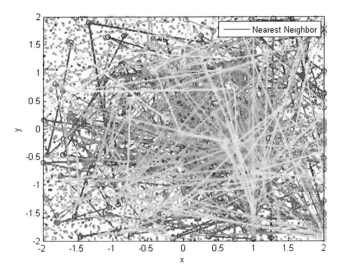

FIGURE 18.6   Nearest neighbor applied to 2D case study.

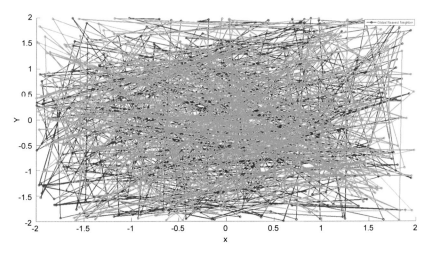

FIGURE 18.7   Global nearest neighbor applied to 2D case study.

TABLE 18.1 MSE and Time values for resampling methods combined with Rao-Blackwellized Monte Carlo Data Association compared with nearest neighbor (NN) and global nearest neighbor (GNN).

Technology	Cumulated sum weight	Residual	Residual systematic resampling	Systematic	Stratified	Optimal	Reallocation	NN	GNN
MSE	0.0565	0.0565	0.0565	0.0565	0.0565	1.7238	0.0524	Fail	Fail
Time (sec)	31.08	32.88	32.72	32.57	32.28	19.94	30.82	Fail	Fail

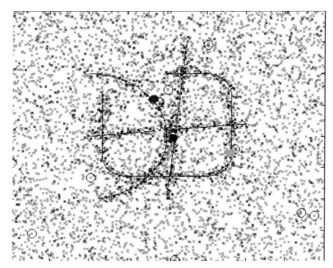

FIGURE 18.8   Reallocation resampling combined with Rao-Blackwellized Monte Carlo Data Association.

Figs. 18.6–18.8 show the pictures of each algorithm applied on this case study. Because of limited space, only a few figures of results have been shown in this article.

5. Conclusions

RBMCDA is developed for solving MTT case study. The RBMCDA uses Kalman filter for prediction and PF for data association. It is also proposed to experiment with various resampling techniques. In PF, the resampling step is important because without this step, PF will quickly result in a degenerate set of particles. That means PF only uses a few particles, which dominate the rest of the particles with their weights. Degeneration of particles contributes toward inaccurate results and unacceptably large variances. To eliminate such deteriorations, it is highly important to use proper resampling method in PFs. These techniques may provide better approximation and benefits in practice. The resampling techniques can hardly provide much different results if they meet the unbiasedness condition, maintain a constant number of particles, and equally weight the resampled particles. Some benefits may be acquired, if these restrictions are eliminated, e.g., preservation of particle diversity, and alleviation of impoverishment. For instance, based on weight thresholds, particles are categorizes in different groups, and a sampling strategy is used for each group to perform more flexibility for resampling. The threshold can be dynamic/adaptive or fixed. The model for grouping is usually according to weight thresholds in order that particles with similar weights are placed in the same group.

Several resampling algorithms are added to currently available RBMCDA method. These combinations are applied to three-dimensional case study. Results show that reallocation resampling was able to track targets with less MSE than CSW and provided faster results in compared with other techniques.

References

Arulampalam, M.S., et al., 2002. A tutorial on particle filters for online nonlinear/non-Gaussian Bayesian tracking. IEEE Transactions on Signal Processing 50 (2), 174−188.

Bailey, T., Durrant-Whyte, H., 2006. Simultaneous localization and mapping (SLAM): part II. Robotics and Automation Magazine, IEEE 13 (3), 108−117.

Bar-Shalom, Y., 1987. Tracking and Data Association. Academic Press Professional, Inc.

Bar-Shalom, Y., Tse, E., 1975. Tracking in a cluttered environment with probabilistic data association. Automatica 11 (5), 451−460.

Bar-Shalom, Y., Rong Li, X., Kirubarajan, T., 2001. Estimation with Applications to Tracking and Navigation. Wiley, New York.

Bar-Shalom, Y., Willett, P.K., Tian, X., 2011. Tracking and Data Fusion. YBS publishing.

Beadle, E.R., Djuric, P.M., 1997. A fast-weighted Bayesian bootstrap filter for nonlinear model state estimation. IEEE Transactions on Aerospace and Electronic Systems 33 (1), 338−343.

Blackman, S., 1986. Multiple-Target Tracking with Radar Applications. Artech House, USA.

Blackman, S., Popoli, R., 1999. Design and Analysis of Modern Tracking Systems, vol. 685. Artech House Noorwood, MA.

Bolic, M., Djuric, P.M., Sangjin, H., 2005. Resampling algorithms and architectures for distributed particle filters. IEEE Transactions on Signal Processing 53 (7), 2442−2450.

Carpenter, et al., 1999. Improved Particle Filter for Nonlinear Problems, vol. 146. ROYAUME-UNI: Institution of Electrical Engineers, Stevenage.

Cox, I., Rehg, J., Hingorani, S., 1993. A Bayesian multiple-hypothesis approach to edge grouping and contour segmentation. International Journal of Computer Vision 11 (1), 5−24.

Doucet, A., et al., 2000. Rao-Blackwellised particle filtering for dynamic Bayesian networks. In: Proceedings of the Sixteenth Conference on Uncertainty in Artificial Intelligence. Morgan Kaufmann Publishers Inc.

Durrant-Whyte, H., Bailey, T., 2006. Simultaneous localization and mapping: part I. Robotics and Automation Magazine, IEEE 13 (2), 99−110.

Fearnhead, P., Clifford, P., 2003. On-line inference for hidden Markov models via particle filters. Journal of the Royal Statistical Society: Series B 65 (4), 887−899.

Hammarberg, B., Forster, C., Torebjork, E., 2002. Parameter estimation of human nerve C-fibers using matched filtering and multiple hypothesis tracking. IEEE Transactions on Biomedical Engineering 49 (4), 329−336.

Hans, F.P., Dijk, D.V., 2000. Non-linear Time Series Models in Empirical Finance. Cambridge University Press.

Kitagawa, G., 1996. Monte Carlo filter and smoother for non-Gaussian nonlinear state space models. Journal of Computational and Graphical Statistics 5 (1), 1−25.

Konstantinova, P., Udvarev, A., Semerdjiev, T., 2003. A study of a target tracking algorithm using global nearest neighbor approach. In: Proceedings of the International Conference on Computer Systems and Technologies (CompSysTech'03).

Lane, D.M., Chantler, M.J., Dongyong, D., 1998. Robust tracking of multiple objects in sector-scan sonar image sequences using optical flow motion estimation. IEEE Journal of Oceanic Engineering 23 (1), 31−46.

Liu, J.S., Chen, R., Logvinenko, T., 2001. A theoretical framework for sequential importance sampling with resampling. In: Doucet, A., de Freitas, N., Gordon, N. (Eds.), Sequential Monte Carlo Methods in Practice. Springer New York, New York, NY, pp. 225−246.

Reid, D., 1979. An algorithm for tracking multiple targets. IEEE Transactions on Automatic Control 24 (6), 843−854.

Särkkä, S., Vehtari, A., Lampinen, J., 2004. Rao-Blackwellized Monte Carlo data association for multiple target tracking. In: Proceedings of the Seventh International Conference on Information Fusion, vol. I.

Särkkä, S., Vehtari, A., Lampinen, J., 2007. Rao-Blackwellized particle filter for multiple target tracking. Information Fusion 8 (1), 2−15.

Yilmaz, A., Javed, O., Shah, M., 2006. Object tracking: a survey. ACM Computing Surveys 38 (4), 13.

Chapter 19

Back-propagation neural network modeling on the load–settlement response of single piles

Zhang Wengang[1,2,3], Anthony Teck Chee Goh[4], Zhang Runhong[2], Li Yongqin[2], Wei Ning[2]
[1]*Key Laboratory of New Technology for Construction of Cities in Mountain Area, Chongqing University, Chongqing, China;* [2]*School of Civil Engineering, Chongqing University, Chongqing, China;* [3]*National Joint Engineering Research Center of Geohazards Prevention in the Reservoir Areas, Chongqing University, Chongqing, China;* [4]*School of Civil and Environmental Engineering, Nanyang Technological University, Singapore*

1. Introduction

As an important type of deep foundation, piles are long, slender structural elements used to transfer the loads from the superstructure aboveground through weak strata onto more suitable bearing strata including the stiffer soils or rocks. Therefore, the safety and stability of pile-supported structures depend largely on the behavior of the piles. The evaluation of the load–settlement performance of a single pile is one of the main aspects in the design of piled foundations. Consequently, an important design consideration is to check the load–settlement characteristics of piles, under influences of several factors, such as the mechanical nonlinear behavior of the surrounding soil, the characteristics of the pile itself, and the installation methods (Berardi and Bovolenta, 2005).

In relation to the settlement analysis of piles, Poulos and Davis (1980) have demonstrated that the immediate settlements contribute the major part of the final settlement, and this also takes into account the consolidation settlement for saturated clay soils and even for piles in clay as presented by Murthy (2002). As for piles in sandy soils, immediate settlement accounts for almost the entire final settlement. In addition, Vesic (1977) suggested a semiempirical method to compute the immediate settlement of piles.

Handbook of Probabilistic Models. https://doi.org/10.1016/B978-0-12-816514-0.00019-9

There are also theoretical and experimental methods for predicting the settlement of piles. Recently, soft computing methods, including the commonly used Artificial neural networks (ANNs), have been adopted with varying degrees of success to predict the axial and lateral bearing capacities of pile foundations in compression and uplift, including the driven piles (Chan et al., 1995; Goh 1996; Lee and Lee 1996; Teh et al., 1997; Abu-Kiefa 1998; Goh et al., 2005; Das and Basudhar 2006; Shahin and Jaksa 2006; Ahmad et al., 2007; Ardalan et al., 2009; Shahin 2010; Alkroosh and Nikraz 2011(a) (b); Tarawneh and Imam 2014; Shahin 2014; Zhang and Goh 2016).

Shahin (2014) developed an ANN model for load—settlement modeling of axially driven steel piles using recurrent neural networks (RNNs). These models were then calibrated and validated using 23 in situ, full-scale pile load tests, as well as cone penetration test (CPT) data. Nevertheless, Shahin's model focused solely on driven steel piles and include only a single input of the average of CPT's cone tip resistance q_c to account for the variability of soil strength along the pile shaft.

Nejad and Jaksa (2017) developed an ANN model to predict the pile behavior based on the results of CPT data, based on approximately 500 data sets from the published articles and compared the results with those values from a number of traditional methods. They claimed that the developed ANN model with the full 21 input variables are the optimal, based on which the complete load—settlement behavior of concrete, steel, and composite piles, either bored or driven, is examined. However, they neglected the input parameter combinations and the descriptive uncertainty due to redundancy of parameter information. Consequently, they fail to take into account the sub-models, i.e., the models developed through less input variables.

The aims of the book chapter are to (1) develop a BPNN model for accurately estimating the load—settlement behavior of single, axially loaded piles over a wide range of applied loads, pile characteristics, and installation methods, as well as soil and ground conditions; (2) examine the influence of selection of descriptive factors, categorical or numerical, on modeling accuracy; and (3) explore the relative importance of the factors affecting pile behavior by carrying out sensitivity analyses.

2. Back-propagation neural network methodologies

A three-layer, feed-forward neural network topology shown in Fig. 19.1 is adopted in this study. As shown in Fig. 19.1, the back-propagation algorithm involves two phases of data flow. In the first phase, the input data are presented forward from the input to output layer and produces an actual output. In the second phase, the error between the target values and actual values are propagated backward from the output layer to the previous layers and the connection weights are updated to reduce the errors between the actual output values and the target output values. No effort is made to keep track of the

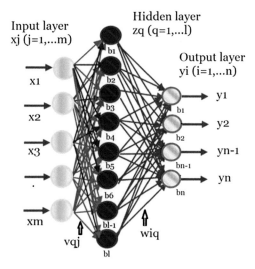

FIGURE 19.1   Back-propagation neural network architecture used in this study.

characteristics of the input and output variables. The network is first trained using the training data set. The objective of the network training is to map the inputs to the output by determining the optimal connection weights and biases through the back-propagation procedure. The number of hidden neurons is typically determined through a trial-and-error process; normally, the smallest number of neurons that yields satisfactory results (judged by the network performance in terms of the coefficient of determination R^2 of the testing data set) is selected. In the present study, a MATLAB-based back-propagation algorithm BPNN with the Levenberg–Marquardt (LM) algorithm (Demuth and Beale, 2003) was adopted for neural network modeling.

3. Development of the back-propagation neural network model

The development of BPNN models requires the determination of the model inputs and outputs, division and preprocessing of the available data, determination of the appropriate network architecture, stopping criteria, and model verification. Nejad and Jaksa (2017) used the NEUFRAME, version 4.0, to simulate the ANN operation, and the database used to calibrate and validate the neural network model was compiled from pile load tests from the published literature. To obtain the accurate prediction of the pile responses including the settlement and the capacity, an understanding of the factors affecting pile behavior is essential. Conventional methods include the following fundamental parameters: pile geometry, pile material properties, soil properties, and applied load for estimation of settlement, as well as additional factors including the pile installation methods, the type of load test, and whether the pile tip is closed or open. In view that pile behavior depends on soil strength and compressibility, and CPT is one of the most commonly used tests in practice for quantifying these soil characteristics, the CPT results in terms of q_c and f_s along the embedded length of the pile are used.

3.1 The database

Suitable case studies were those involving pile load tests that include field measurements of full-scale pile settlements, as well as the corresponding information relating to the piles and soil characteristics. The database compiled by Nejad and Jaksa (2017) contains a total of 499 cases from 56 individual pile load tests. The descriptive 21 variables including 5 parameters on pile information, 11 parameters on soil information, the applied load, type of test, type of pile, type of installation, and type of pile end are regarded as inputs to estimate the target pile settlement. The references used to compile the database are given in Table 19.1. The details of the database, for each pile load test, are given in Table 19.2, while a summary of the input variables and output as well as the descriptions are listed in Table 19.3. The numerical values for the corresponding categorical variables are listed in Table 19.4. The full database can be referred to the supplementary material of Nejad and Jaksa (2017). In addition, the applied load, P, and the corresponding pile settlement, δ_m, values were obtained by selecting a series of points from the load—settlement curve associated with each pile load test.

3.2 Data division

The cross-validation method suggested by Stone is implemented by Nejad and Jaksa (2017) to divide the data into three sets: training, testing, and validation. The training set is used to adjust the connection weights, whereas the testing set is used to check the performance of the model at various stages of training and to determine when to stop training to avoid overfitting. The validation set is used to estimate the performance of the trained network in the deployed environment. This study used the three data sets chosen by Nejad and Jaksa (2017). To eliminate the data bias and any possibility of extrapolation beyond the range of the training data sets, several random combinations of the training, testing, and validation sets are assessed until statistical consistency in terms of the mean, the standard deviation, minimum, maximum, and range, as suggested by Shahin et al. are obtained.

3.3 Back-propagation neural network model architecture

Determining the network architecture is one of the most important and difficult tasks in BPNN model development. In this study, even one hidden layer is adopted for simplicity, apart from the selection of the optimal number of nodes (neurons) in this hidden layer, and determination of the proper number of

TABLE 19.1 Database references.

References	Location of test(s)	No. of pile load tests
Nottingham (1975)	USA	2
Tumay and Fakhroo (1981)	Louisiana, USA	5
Viergever 1982	Almere, the Netherlands	1
Campnella et al.1981	Vancouver, Canada	1
Gambini 1985	Milan, Italy	1
Horvitz et al. (1986)	Seattle, USA	1
CH2M Hill (1987)	Los Angeles, USA	1
Briaud and Tucker (1988)	USA	12
Haustoefer and Plesiotis 1988	Victoria, Australia	1
Tucker and Briaud (1988)	USA	1
Reese et al. (1988)	Texas, USA	1
O'Neil (1988)	California, USA	1
Ballouz et al. (1991)	Texas A and M University, USA	2
Avasarala et al. (1994)	Florida, USA	1
Harris and Mayne 1994	Georgia, USA	1
Matsumoto et al. (1995)	Noto Island, Japan	1
Florida Department of Transportation (FDOT) 2003	USA	8
Paik and Salgado (2003)	Indiana, USA	2
Fellenius et al. (2004)	Idaho, USA	1
Poulos and Davis (2005)	UAE	1
Brown et al. (2006)	Grimsby, UK	2
McCabe and Lehane (2006)	Ireland	1
Omer et al., 2006	Belgium	4
U.S. Department of Transportation, 2006	Virginia, USA	3
South Aust. Dept. of Transport, Energy and Infrastructure[a]	Adelaide, Australia	1

[a]*Unpublished, based on Nejad and Jaksa (2017)*
Adapted from Nejad, F.P., Jaksa, M.B., 2017. "Load-settlement behavior modeling of single piles using artificial neural networks and CPT data." Computers and Geotechnics 89 9–21.

TABLE 19.2 Details of pile load test database.

References	Test type	Pile type	Method	Pile end	EA (MN)	A_{tip} (10^3 mm^2)	O (mm)	L (mm)	L_{embed} (m)	Max.Load (kN)	S_m (mm)
Nottingham (1975)	ML	Conc.	Driven	C	4263	203	1800	8	8	1140	24.5
Nottingham (1975)	ML	Steel	Driven	C	797	59	858	22.5	22.5	1620	37
Tumay and Fakhroo (1981)	ML	Conc.	Driven	C	13,356	636	2830	37.8	37.8	3960	12
Tumay and Fakhroo (1981)	ML	Conc.	Driven	C	4263	203	1800	36.5	36.5	2950	18.5
Tumay and Fakhroo (1981)	ML	Steel	Driven	C	1302	126	1257	37.5	37.5	2800	20
Tumay and Fakhroo (1981)	ML	Steel	Driven	C	1138	96	1010	31.1	31.1	1710	11.5
Tumay and Fakhroo (1981)	ML	Conc.	Driven	C	11,823	563	3000	19.8	19.8	2610	7.5
Campnella et al. 1981	ML	Steel	Driven	C	1000	82	1018	13.7	13.7	290	18.8
Viergever 1982	ML	Conc.	Driven	C	1323	63	1000	9.25	9.25	700	100
Gambini 1985	ML	Steel	Driven	C	1072	86	1037	10	10	625	19
Horvitz et al. (1986)	ML	Conc.	Bored	C	2016	96	1100	15.8	15.8	900	37
CH2M Hill 1987	ML	Conc.	Driven	C	6468	308	2020	25.8	25.8	5785	66
Briaud and Tucker (1988)	ML	Conc.	Driven	C	2583	123	1400	5.5	5.5	1050	58

Briaud and Tucker (1988)	ML	Conc.	Driven	C	3360	160	1600	8.4	8.4	1240	52.5
Briaud and Tucker (1988)	ML	Conc.	Driven	C	3360	160	1600	21	21	1330	28
Briaud and Tucker (1988)	ML	Conc.	Driven	C	4263	203	1800	10.3	10.3	1250	27
Briaud and Tucker (1988)	ML	Conc.	Driven	C	4263	203	1800	15	15	1420	28
Briaud and Tucker (1988)	ML	Conc.	Driven	C	4263	203	1800	10.4	10.4	1070	34
Briaud and Tucker (1988)	ML	Conc.	Driven	C	3360	160	1600	11.3	11.3	870	31
Briaud and Tucker (1988)	ML	Steel	Driven	O	2100	10	1210	19	19	1370	46.5
Briaud and Tucker (1988)	ML	Conc.	Driven	C	2583	123	1400	25	25	1560	18
Briaud and Tucker (1988)	ML	Conc.	Driven	C	3360	160	1600	19.2	19.2	1780	18
Briaud and Tucker (1988)	ML	Steel	Driven	O	2100	10	1210	9	9	2100	12
Briaud and Tucker (1988)	ML	Conc.	Bored	C	2016	96	1100	12.5	12.5	1100	27
Haustoefer and Plesiotis 1988	ML	Conc.	Driven	C	2646	126	1420	10.2	10.2	1300	60
O'Neil (1988)	ML	Steel	Driven	C	805	59	585	9.2	9.2	490	84
Reese et al., 1988	ML	Conc.	Bored	C	10,563	503	2510	24.1	24.1	5850	50

Continued

TABLE 19.2 Details of pile load test database.—cont'd

					1081	96	1100	14.4	14.4	1300	75
Tucker and Briaud (1988)	ML	Steel	Driven	C	1081	96	1100	14.4	14.4	1300	75
Ballouz et al., 1991	ML	Conc.	Bored	C	16,493	785.4	3142	10.7	10	4130	137.9
Ballouz et al., 1991	ML	Conc.	Bored	C	13,809	657.56	2875	10.7	10	3000	68.43
Avasarala et al. (1994)	ML	Conc.	Driven	C	2016	96	1100	16	16	1350	33
Harris and Mayne (1994)	ML	Conc.	Bored	C	9073	453.65	2388	16.8	16.8	2795	20.94
Matsumoto et al. 1995	ML	Steel	Driven	O	3167	41	2510	11	8.2	4700	40
FDOT (2003)	CRP	Conc.	Driven	O	32,387	729.66	9576	39.78	33.5	9810	15.98
FDOT (2003)	CRP	Conc.	Driven	O	25,909	583.73	7661	56.08	42.58	4551	7.87
FDOT (2003)	CRP	Conc.	Driven	O	25,909	583.73	7661	44.78	31.97	6000	4.8
FDOT (2003)	CRP	Conc.	Driven	O	33,106	745.87	7341	24.38	15.39	11,000	66.04
FDOT (2003)	CRP	Conc.	Driven	O	33,106	745.87	7341	24.38	14.02	16,000	9.4
FDOT (2003)	CRP	Conc.	Driven	O	32,387	729.66	9576	56.39	28.65	10,000	10.47
FDOT (2003)	CRP	Conc.	Driven	O	25,909	583.73	7661	44.87	23.52	7500	10.31
FDOT (2003)	CRP	Conc.	Driven	O	32,387	729.66	9576	53.34	32	10,000	13.49
Paik and Salgado (2003)	ML	Steel	Driven	C	6840	32.572	2036	8.24	7.04	1140	57.5
Paik and Salgado (2003)	ML	Steel	Driven	C	2876	99.538	1118	8.24	6.87	1620	62.5

Reference											
Fellenius et al. (2004)	ML	Comp.	Driven	C	7005	129.46	1276	45.86	45	1915	13
Poulos and Davis (2005)	ML	Conc.	Bored	C	19,085	636.17	2827	40	40	30,000	32.52
McCabe and Lehane (2006)	ML	Conc.	Driven	C	2110	62.5	1000	6	6	60	8.21
Omer et al. (2006)	ML	Conc.	Driven	C	2773	132	1288	10.66	8.45	2670	35.94
Omer et al. (2006)	ML	Conc.	Driven	C	2773	132	1288	10.63	8.45	2796	41.74
Omer et al. (2006)	ML	Conc.	Driven	C	2773	132	1288	10.74	8.52	2257	40.68
Omer et al. (2006)	ML	Conc.	Driven	C	2773	132	1288	10.64	8.53	2475	47.65
Brown et al. (2006)	ML	Conc.	Bored	C	8101	282.74	1885	12.76	9.96	1800	23.05
Brown et al. (2006)	CRP	Conc.	Bored	C	8101	282.74	1885	12.76	9.96	2205	26.78
U.S. D T et al., 2006	ML	Conc.	Driven	C	8200	372.1	2440	18	16.76	3100	15.52
U.S. D T et al., 2006	ML	Comp.	Driven	C	7360	303.86	1954	18.3	17.22	2572	35.84
U.S. D T et al., 2006	CRP	Comp.	Driven	C	3200	275.25	1860	18.3	17.27	2500	80.6
SA DPTI (unpublished)	ML	Conc.	Bored	C	11,874	282.74	1885	16.8	7.2	518	2.31

C, closed; Comp., composite; Conc., corcrete; CRP, constant rate of penetration; ML, maintained load; O, open.

TABLE 19.3 Summary of input variables and outputs.

Inputs and output	Parameters and parameter descriptions		
Input variables	Pile information	Axial rigidity of pile, EA (MN)	Variable 1 (x1)
		Cross-sectional area of pile tip, A_{tip} (m^2)	Variable 2 (x2)
		Perimeter of pile, O (mm)	Variable 3 (x3)
		Length of pile, L(m)	Variable 4 (x4)
		Embedded length of pile, L_{embed} (m)	Variable 5 (x5)
	Soil information from CPT	f_{s1} (kPa)	Variable 6 (x6)
		q_{c1} (MPa)	Variable 7 (x7)
		f_{s2} (kPa)	Variable 8 (x8)
		q_{c2} (MPa)	Variable 9 (x9)
		f_{s3} (kPa)	Variable 10 (x10)
		q_{c3} (MPa)	Variable 11 (x11)
		f_{s4} (kPa)	Variable 12 (x12)
		q_{c4} (MPa)	Variable 13 (x13)
		f_{s5} (kPa)	Variable 14 (x14)
		q_{c5} (MPa)	Variable 15 (x15)
		q_{ctip} (MPa)	Variable 16 (x16)
	Applied load, P (kN)		Variable 17 (x17)
	Categorical information for piles and the testing methods	Type of test (TT)	Variable 18 (x18)
		Type of pile (TP)	Variable 19 (x19)
		Type of installation (TI)	Variable 20 (x20)
		Type of pile end (PE)	Variable 21 (x21)
Output	Measured piles settlement, δ_m (mm)		

inputs out of the full 21 variables is an essential task because there is no unified rule for determination of an optimal BPNN architecture. In view of these two issues, a "trial-and-error" procedure is carried out to determine the optimal BPNN model architecture from aspects of the proper number of inputs and the number of nodes in the hidden layer, as listed in Table 19.5.

TABLE 19.4 Numerical values for the corresponding categorical variables.

Categorical factors	Description	Input value
Type of test (x18, TT)	Maintained load	0
	Constant rate of penetration	1
Type of pile (x19, TP)	Steel	0
	Concrete	1
	Composite	2
Type of installation (x20, TI)	Driven	0
	Bored	1
Type of pile end (x21, PE)	Open	0
	Closed	1

TABLE 19.5 Selection of the optimal BPNN model architecture.

Case no.	Combination of numerical and categorical variables	Number of hidden nodes
1	17 (x1,...x17)	1, 2, ..., 33, 34
2	17 + TT	1, 2, ..., 33, 34, 35, 36
3	17 + TP	1, 2, ..., 33, 34, 35, 36
4	17 + TI	1, 2, ..., 33, 34, 35, 36
5	17 + PE	1, 2, ..., 33, 34, 35, 36
6	17 + TT + TP (optimal)	1, 2, ..., 35, 36, 37, 38
7	17 + TP + TI	1, 2, ..., 35, 36, 37, 38
8	17 + TI + PE	1, 2, ..., 35, 36, 37, 38
9	17 + TT + PE	1, 2, ..., 35, 36, 37, 38
10	17 + TT + TI	1, 2, ..., 35, 36, 37, 38
11	17 + TP + PE	1, 2, ..., 35, 36, 37, 38
12	17 + TT + TP + TI	1, 2, ..., 37, 38, 39, 40
13	17 + TP + TI + PE	1, 2, ..., 37, 38, 39, 40
14	17 + TT + TI + PE	1, 2, ..., 37, 38, 39, 40
15	17 + TT + TP + PE	1, 2, ..., 37, 38, 39, 40
16	17 + TT + TP + TI + PE	1, 2, ..., 39, 40, 41, 42

BPNN, back-propagation neural network.

3.4 Training and stopping criteria for back-propagation neural network models

Training, or learning, is the process of optimizing the connection weights, based on first-order gradient descent. Its aim is to identify a global solution to what is typically a highly nonlinear optimization problem. The BPNN model has the ability to escape local minima in the error surface and, thus, produces optimal or near-optimal solutions. Stopping criteria determine whether the model has been optimally or suboptimally trained. Various methods can be used to determine when to stop training. The training set is used to adjust the connection weight, whereas the resting set measures the ability of the model to generalize, and using this set, the performance of the model is checked at many stages during the training process and training is stopped when the testing set error begins to increase. The preset rules as for the transfer function, the maximum epoch, and the stopping criteria are as follows, in MATLAB language:

logsig transfer function from the input layer to the hidden layer;
tansig transfer function from the hidden layer to the output layer;
maxepoch =500;
learning rate=0.01;
min_grad=1e-15;
mu_dec=0.7;
mu_inc=1.03.

3.5 Validations

Once model training has been successfully accomplished, the performance of the trained model should be validated against data that have not been used in the learning process, known as the validation set, to ensure that the model has the ability to generalize within the limits set by the training data in a robust fashion, i.e., to the new situations, rather than simply having memorized the input—output relationships that are contained in the training sets.

4. The optimal back-propagation neural network model

Based on the "trial-and-error" procedure in section 3.3, different BPNN model architectures have been tried and it is assumed that the BPNN model with the highest coefficient of determination R^2 value for the testing data sets is considered to be the optimal model. Table 19.6 lists the R^2 values of the testing data sets for the BPNN models with different number of inputs and nodes. It can be observed that the BPNN model with 17 numerical variables andand categorical TT + TP variables as the inputs with two hidden nodes is the optimal one.

5. Modeling results

Fig. 19.2A and B show the BPNN predictions for the training and testing data patterns, respectively. For pile settlement prediction, considerably high R^2 (approximately 0.9) is obtained for both the training (R^2=0.856) and testing

TABLE 19.6 The optimal BPNN model selection.

Case no.	Combination of numerical and categorical variables	R^2 for the testing sets
1	17 (x1,...x17)	0.773
2	17 + TT	0.888
3	17 + TP	0.761
4	17 + TI	0.833
5	17 + PE	0.824
6	17 + TT + TP (optimal)	0.908
7	17 + TP + TI	0.816
8	17 + TI + PE	0.738
9	17 + TT + PE	0.827
10	17 + TT + TI	0.788
11	17 + TP + PE	0.789
12	17 + TT + TP + TI	0.785
13	17 + TP + TI + PE	0.743
14	17 + TT + TI + PE	0.829
15	17 + TT + TP + PE	0.801
16	17 + TT + TP + TI + PE	0.763

BPNN, back-propagation neural network.

(R^2=0.908) patterns. Based on the plot, it is obvious that the developed BPNN model is less accurate in predicting small pile settlement mainly as a result of the bias (errors).

6. Parametric relative importance

The parametric relative importance determined by the BPNN is based on the method by Garson (1991) and discussed by Das and Basudhar (2006). Fig. 19.3 gives the plot of the relative importance of the input variables for the BPNN models. It can be observed that pile settlement is mostly influenced by the input variable x17 (applied load, P), followed by x8 (f_{s2}) and ×4 (length of pile). It is marginally influenced by x15 (q_{c5}) and x19 (type of pile), which also explains that input variable x19 slightly enhances the predictive capacity of the BPNN model from 0.888 for case no. Two to 0.908 for case no. 6.

7. Model interpretabilities

For brevity, only the developed optimal 17 + TT + TP model is interpreted. The BPNN model is expressed through the trained connections weights, the

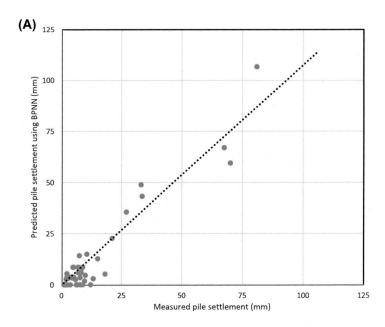

Training data sets

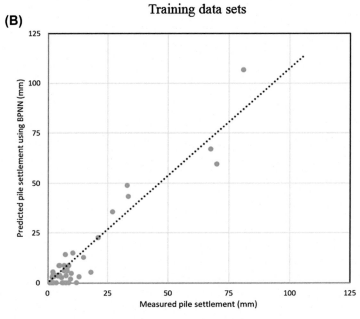

Testing data sets

FIGURE 19.2 Prediction of pile settlements using BPNN. *BPNN*, back-propagation neural network.

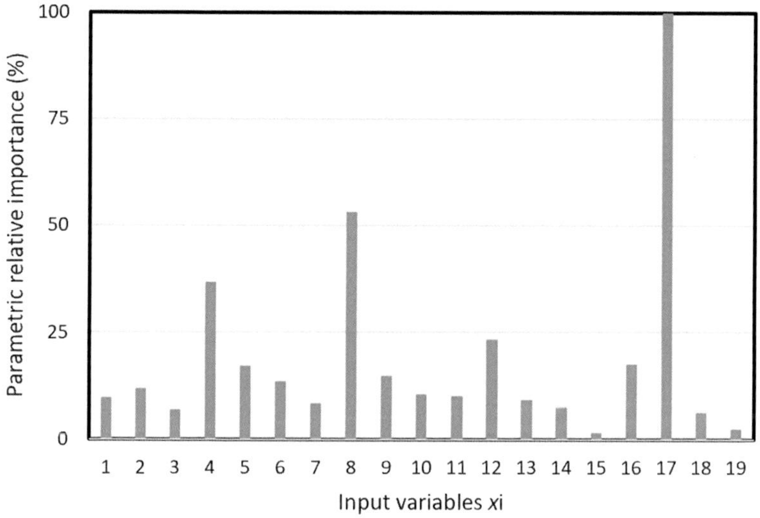

FIGURE 19.3   Relative importance of the input variables for the optimal BPNN model. *BPNN*, back-propagation neural network.

bias, and the transfer functions. The mathematical expression for pile settlement obtained by the optimal 17 + TT + TP analysis is shown in Appendix A. In addition, Appendix B provides the weights and bias values used for partitioning of BPNN weights for pile settlement. The specific procedures can be referred to Zhang (2013). For simplicity, this part has been omitted.

8. Summary and conclusion

A database containing 499 pile load test data sets with a total of 21 full variables is adopted to develop the BPNN model for prediction of load—settlement characteristics of piles. The predictive accuracy, model interpretability, and parametric sensitivity analysis of the developed BPNN pile settlement model are demonstrated. Performance measures indicate that BPNN model for the analyses of pile settlement provides reasonable predictions and can thus be used for predicting pile settlement.

Appendix A BPNN pile settlement model

The transfer functions used for BPNN output for pile settlement are *"logsig"* transfer function for hidden layer to output layer and *"tansig"* transfer function for output layer to target. The calculation process of BPNN output is elaborated in detail as follows:

From the connection weights for a trained neuron network, it is possible to develop a mathematical equation relating the input parameters and the single output parameter Y using

$$Y = f_{sig}\left\{b_0 + \sum_{k=1}^{h}\left[w_k f_{sig}\left(b_{hk} + \sum_{i=1}^{m} w_{ik}X_i\right)\right]\right\} \tag{A.1}$$

in which b_0 is the bias at the output layer, ω_k is the weight connection between neuron k of the hidden layer and the single output neuron, b_{hk} is the bias at neuron k of the hidden layer ($k=1,h$), ω_{ik} is the weight connection between input variable i ($i =1, m$) and neuron k of the hidden layer, X_i is the input parameter i, and f_{sig} is the sigmoid (logsig & *tansig*) transfer function.

Using the connection weights of the trained neural network, the following steps can be followed to mathematically express the BPNN model:

Step1: Normalize the input values for x_1, x_2,... and x_{19} linearly using

$$x_{norm} = 2(x_{actual} - x_{min})/(x_{max} - x_{min}) - 1$$

Let the actual $x_1 = X_{1a}$ and the normalized $x_1 = X_1$

$$X_1 = -1 + 2 \times (X_{1a} - 796.74)/(33106.34 - 796.74) \tag{A.2}$$

Let the actual $x_2 = X_{2a}$ and the normalized $x_2 = X_2$

$$X_2 = -1 + 2 \times (X_{2a} - 100)/(7854 - 100) \tag{A.3}$$

Let the actual $x_3 = X_{3a}$ and the normalized $x_3 = X_3$

$$X_3 = -1 + 2 \times (X_{3a} - 58.5)/(957.56 - 58.5) \tag{A.4}$$

Let the actual $x_4 = X_{4a}$ and the normalized $x_4 = X_4$

$$X_4 = -1 + 2 \times (X_{4a} - 5.5)/(56.39 - 5.5) \tag{A.5}$$

Let the actual $x_5 = X_{5a}$ and the normalized $x_5 = X_5$

$$X_5 = -1 + 2 \times (X_{5a} - 5.5)/(45 - 5.5) \tag{A.6}$$

Let the actual $x_6 = X_{6a}$ and the normalized $x_6 = X_6$

$$X6 = -1 + 2 \times (X6a - 0)/(10.38 - 0) \tag{A.7}$$

Let the actual $x_7 = X_7a$ and the normalized $x7 = X_7$

$$X_7 = -1 + 2 \times (X_7a - 0)/(274 - 0) \tag{A.8}$$

Let the actual $x_8 = X_{8a}$ and the normalized $x_8 = X_8$

$$X_8 = -1 + 2 \times (X_{8a} - 0.05)/(17.16 - 0.05) \tag{A.9}$$

Let the actual $x_9 = X_{9a}$ and the normalized $x_9 = X_9$

$$X_9 = -1 + 2 \times (X_{9a} - 1.83)/(275.5 - 1.83) \tag{A.10}$$

Let the actual $x_{10} = X_{10a}$ and the normalized $x_{10} = X_{10}$

$$X_{10} = -1 + 2 \times (X_{10a} - 0.3)/(31.54 - 0.3) \tag{A.11}$$

Let the actual $x_{11} = X_{11a}$ and the normalized $x_{11} = X_{11}$

$$X_{11} = -1 + 2 \times (X_{11a} - 1.615)/(618.7 - 1.615) \tag{A.12}$$

Let the actual $x_{12} = X_{12a}$ and the normalized $x_{12} = X_{12}$

$$X_{12} = -1 + 2 \times (X_{12a} - 0.25)/(33.37 - 0.25) \tag{A.13}$$

Let the actual $x_{13} = X_{13a}$ and the normalized $x_{13} = X_{13}$

$$X_{13} = -1 + 2 \times (X_{13a} - 4.421)/(1293 - 4.421) \tag{A.14}$$

Let the actual $x_{14} = X_{14a}$ and the normalized $x_{14} = X_{14}$

$$X_{14} = -1 + 2 \times (X_{14a} - 0.25)/(53.82 - 0.25) \qquad (A.15)$$

Let the actual $x_{15} = X_{15a}$ and the normalized $x_{15} = X_{15}$

$$X_{15} = -1 + 2 \times (X_{15a} - 7.99)/(559 - 7.99) \qquad (A.16)$$

Let the actual $x_{16} = X_{16a}$ and the normalized $x_{16} = X_{16}$

$$X_{16} = -1 + 2 \times (X_{16a} - 0.25)/(70.29 - 0.25) \qquad (A.17)$$

Let the actual $x_{17} = X_{17a}$ and the normalized $x_{17} = X_{17}$

$$X_{17} = -1 + 2 \times (X_{17a} - 0)/(30000 - 0) \qquad (A.18)$$

Let the actual $x_{18} = X_{18a}$ and the normalized $x_{18} = X_{18}$

$$X_{18} = -1 + 2 \times (X_{18a} - 0)/(1 - 0) \qquad (A.19)$$

Let the actual $x_{19} = X_{19a}$ and the normalized $x_{19} = X_{19}$

$$X_{19} = -1 + 2 \times (X_{19a} - 0)/(2 - 0) \qquad (A.20)$$

Step2: Calculate the normalized value (Y_1) using the following expressions:

$A_1 = 0.0796 + 5.0087 logsig(X_1) - 3.3926 logsig(X_2) + 6.8371 logsig(X_3)$
$75.6342 logsig(X_4) + 45.8013 logsig(X_5) + 13.0191 logsig(X_6) +$
$24.0145 logsig(X_6) - 96.1639 logsig(X_8) - 41.1331 logsig(X_9)$
$+14.57 logsig(X_{10}) + 24.0111 logsig(X_{11}) + 58.357 logsig(X_{12}) -$
$23.5117 logsig(X_{13}) - 21.0635 logsig(X_{14}) - 2.6677 logsig(X_{15}) +$
$36.8799 logsig(X_{16}) + 18.098 logsig(X_{17}) - 15.3542 logsig(X_{18})$
$+1.7168 logsig(X_{19})$ \qquad (A.21)

$A_2 = -61.9379 - 7.165 logsig(X_1) + 9.7258 logsig(X_2) + 4.0935 logsig(X_3) +$
$9.7937 logsig(X_4) + 1.3488 logsig(X_5) + 8.2361 logsig(X_6) + 0.1617 logsig(X_7)$
$-18.4019 logsig(X_8) + 0.705 logsig(X_9) + 4.9512 logsig(X_{10}) +$
$1.7347 logsig(X_{11}) + 3.1179 logsig(X_{12}) - 1.1133 logsig(X_{13}) -$
$0.4005 logsig(X_{14}) + 0.5711 logsig(X_{15}) + 4.4941 logsig(X_{16})$
$-84.7805 logsig(X_{17}) + 0.9767 logsig(X_{18}) + 1.6406 logsig(X_{19})$ \qquad (A.22)

$$B_1 = 2.6304 \times \tanh(A_1) \qquad (A.23)$$

$$B_2 = -3.0709 \times \tanh(A_2) \qquad (A.24)$$

$$C_1 = -1.3496 + B_1 + B_2 \qquad (A.25)$$

$$Y_1 = C_1 \qquad (A.26)$$

Step3: Denormalize the output to obtain pile settlement

$$\delta_m = 0 + (137.88 - 0) \times (Y_1 + 1)/2 \qquad (A.27)$$

Note: $logsig(x) = 1/(1 + exp(-x))$ while $\tanh(x) = 2/(1 + exp(-2x)) - 1$

Appendix B weights and bias values for BPNN pile settlement model

See Tables B.1–B.3.

TABLE B.1 Weights for inputs layer to the hidden layer.

Hidden neuron 1	Input 1	Input 2	Input 3	Input 4	Input 5	Input 6	Input 7	Input 8	Input 9	Input 10
	−5.009	−3.393	6.837	−75.63	45.801	13.019	24.015	−96.16	−41.13	14.57
	Input 11	Input 12	Input 13	Input 14	Input 15	Input 16	Input 17	Input 18	Input 19	
	24.011	58.357	−23.51	−21.06	−2.668	36.88	18.098	−15.35	1.717	
Hidden neuron 2	Input 1	Input 2	Input 3	Input 4	Input 5	Input 6	Input 7	Input 8	Input 9	Input 10
	−7.165	9.7258	4.0935	9.7937	1.3488	8.2361	0.1617	−18.40	0.705	4.9512
	Input 11	Input 12	Input 13	Input 14	Input 15	Input 16	Input 17	Input 18	Input 19	
	1.7347	3.1179	−1.113	−0.401	0.5711	4.4941	−84.78	0.9767	1.6406	

TABLE B.2 Bias for inputs layer to the hidden layer.

Theta	Hidden neuron 1	Hidden neuron 2
	0.0796	−61.9379

TABLE B.3 Weights for the hidden layer to output layer.

Weight	Hidden neuron 1	Hidden neuron 2
	2.6304	−3.0709

The bias value for the hidden layer to output layer is −1.3496.

Acknowledgments

The authors would like to express their appreciation to Nejad and Jaksa (2017) for making their pile load−settlement database available for this work.

References

Abu-Kiefa, M.A., 1998. General regression neural networks for driven piles in cohesionless soils. Journal of Geotechnical and Geoenvironmental Engineering 124 (12), 1177−1185.

Ahmad, I., El Naggar, H., Kahn, A.N., 2007. Artificial neural network application to estimate kinematic soil pile interaction response parameters. Soil Dynamics and Earthquake Engineering 27 (9), 892−905.

Alkroosh, I., Nikraz, H., 2011a. Correlation of pile axial capacity and CPT data using gene expression programming. Geotechnical and Geological Engineering 29, 725−748.

Alkroosh, I., Nikraz, H., 2011b. Simulating pile load-settlement behavior from CPT data using intelligent computing. Central European Journal of Engineering 1 (3), 295−305.

Ardalan, H., Eslami, A., Nariman-Zahed, N., 2009. Piles shaft capacity from CPT and CPTu data by polynomial neural networks and genetic algorithms. Computers and Geotechnics 36, 616−625.

Avasarala, S.K.V., Davidson, J.L., McVay, A.M., 1994. An evaluation of predicted ultimate capacity of single piles from spile and unpile programs. In: Proc Int Conf on Design and Construction of Deep Foundations, 2, pp. 12−723. Orlando: FHWA.

Ballouz, M., Nasr, G., Briaud, J.-L., 1991. Dynamic and Static Testing of Nine Drilled Shafts at Texas A&M University Geotechnical Sites. Res Rep. Civil Engineering, Texas A&M University, College Station, Texas, p. 127.

Berardi, R., Bovolenta, R., 2005. Pile settlement evaluation using field stiffness nonlinearity. Geotechnical Engineering 158, 35−44.

Briaud, J.L., Tucker, L.M., 1988. Measured and predicted axial capacity of 98 piles. J Geotech Eng 114 (9), 984−1001.

Brown, M.J., Hyde, A.F.L., Anderson, W.F., 2006. Analysis of a rapid load test on an instrumented bored pile in clay. Géotechnique 56 (9), 627−638.

Campnella, R.G., Gillespie, D., Robertson, P.K., 1981. Pore pressure during cone penetration testing. In: Proc of 2nd European Symp on Penetration Testing, Amsterdam, vol. 2, pp. 507–512.

Chan, W.T., Chow, Y.K., Liu, L.F., 1995. Neural network: an alternative to pile driving formulas. Computers and Geotechnics 17, 135–156.

Das, S.K., Basudhar, P.K., 2006. Undrained lateral load capacity of piles in clay using artificial neural network. Computers and Geotechnics 33 (8), 454–459.

Demuth, H., Beale, M., 2003. Neural Network Toolbox for MATLAB-User Guide Version 4.1. The Math Works Inc.

Fellenius, B.H., Harris, D.E., Anderson, D.G., 2004. Static loading test on a 45m long pile in Sanpoint, Idaho. Canadian Geotechnical Journal 41, 613–628.

Florida Department of Transportation (FDOT), 2003. Large Diameter Cylinder Pile Database. Research Management Center.

Gambini, F., 1985. Experience in Italy with centricast concrete piles. In: Proc Int Symp on Penetrability and Drivability of Piles, San Francisco, vol. 1, pp. 97–100.

Garson, G.D., 1991. Interpreting neural-network connection weights. AI Expert 6 (7), 47–51.

Goh, A.T.C., 1996. Pile driving records reanalyzed using neural networks. Journal of Geotechnical Engineering 122 (6), 492–495.

Goh, A.T., Kulhawy, F.H., Chua, C.G., 2005. Bayesian neural network analysis of undrained side resistance of drilled shafts. Journal of Geotechnical and Geoenvironmental Engineering 131 (1), 84–93.

Harris, E., Mayne, P., 1994. Axial compression behavior of two drilled shafts in piedmont residual soils. In: Proc Int Conf on Design and Construction of Deep Foundation, vol. 2. FWHA, Washington, DC, pp. 352–367.

Haustoefer, I.J., Plesiotis, S., 1988. Instrumented dynamic and static pile load testing at two bridges. In: Proc 5th Australia New Zealand Conf on Geomechanics – Prediction vs Performance, Sydney, pp. 514–520.

CH2M Hill, 1987. Geotechnical Report on Indicator Pile Testing and Static Pile Testing. Berths 225-229 at Port of Los Angeles. CH2M Hill, Los Angeles.

Horvitz, G.E., Stettler, D.R., Crowser, J.C., 1986. Comparison of predicted and observed pile capacity. In: Proc Symp on Cone Penetration Testing. ASCE, St. Louis, pp. 413–433.

Lee, I.M., Lee, J.H., 1996. Prediction of pile bearing capacity using artificial neural networks. Computers and Geotechnics 18 (3), 189–200.

Matsumoto, T., Michi, Y., Hirono, T., 1995. Performance of axially loaded steel pipe piles driven in soft rock. Journal of Geotechnical and Geoenvironmental Engineering 121 (4), 305–315.

McCabe, B.A., Lehane, B.M., 2006. Behavior of axially loaded pile groups driven in clayey silt. Journal of Geotechnical and Geoenvironmental Engineering 132 (3), 401–410.

Murthy, V.N.S., 2002. Principles and Practices of Soil Mechanics and Foundation Engineering. Marcel Dekker Inc.

Nejad, F.P., Jaksa, M.B., 2017. Load-settlement behavior modeling of single piles using artificial neural networks and CPT data. Computers and Geotechnics 89, 9–21.

Nottingham, L.C., 1975. Use of Quasi-Static Friction Cone Penetrometer Data to Predictload Capacity of Displacement Piles. PhD Thesis. Dept Civil Eng, University of Florida.

Omer, J.R., Delpak, R., Robinson, R.B., 2006. A new computer program for pile capacity prediction using CPT data. Geotechnical and Geological Engineering 24, 399–426.

O'Neil, M.W., 1988. Pile Group Prediction Symposium – Summary of Prediction Results. FHWA draft report.

Paik, K.H., Salgado, R., 2003. Determination of bearing capacity of open-ended piles in sand. Journal of Geotechnical and Geoenvironmental Engineering 129 (1), 46−57.

Poulos, H.G., Davis, E.H., 1980. Pile Foundation Analysis and Design. Wiley.

Poulos, H.G., Davis, A.J., 2005. Foundation design for the emirates twin towers, Dubai. Canadian Geotechnical Journal 42, 716−730.

Reese, J.D., O'Neill, M.W., Wang, S.T., 1988. Drilled Shaft Tests, Interchange of West Belt Roll Road and US290 Highway. Texas. Lymon C. Reese and Associates, Austin, Texas.

Shahin, M.A., 2010. Intelligent computing for modelling axial capacity of pile foundations. Canadian Geotechnical Journal 47 (2), 230−243.

Shahin, M.A., 2014. Load-settlement modeling of axially loaded steel driven piles using CPT-based recurrent neural networks. Soils and Foundations 54 (3), 515−522.

Shahin, M.A., Jaksa, M.B., 2006. Pullout capacity of small ground anchors by direct cone penetration test methods and neural networks. Canadian Geotechnical Journal 43 (6), 626−637.

Tarawneh, B., Imam, R., 2014. Regression versus artificial neural networks: predicting pile setup from empirical data. KSCE Journal of Civil Engineering 18 (4), 1018−1027.

Teh, C.I., Wong, K.S., Goh, A.T.C., Jaritngam, S., 1997. Prediction of pile capacity using neural networks. Journal of Computing in Civil Engineering 11 (2), 129−138.

Tucker, L.M., Briaud, J.L., 1988. Analysis of Pile Load Test Program at Lock and Dam 26 Replacement Project Final Report. US Army Corps of Engineering.

Tumay, M.Y., Fakhroo, M., 1981. Pile capacity in soft clays using electric QCPT data. In: Proc Conf on Cone Penetration Testing and Experience. ASCE, St Louis, pp. 434−455.

U.S. Department of Transportation, 2006. A Laboratory and Field Study of Composite Piles for Bridge Substructures. FHWA-HRT-04-043.

Vesic, A.S., 1977. Design of Pile Foundations. National Cooperative Highway Research Program. Transportation Research Board, Washington, DC. Synthesis of Practice No. 42.

Viergever, M.A., 1982. Relation between cone penetration and static loading of piles on locally strongly varying sand layers. In: Proc of 2nd European Symp on Penetration Testing, Amsterdam, vol. 2, pp. 927−932.

Zhang, W.G., 2013. Probabilistic Risk Assessment of Underground Rock Caverns. PhD thesis. Nanyang Technological University.

Zhang, W.G., Goh, A.T.C., 2016. Multivariate adaptive regression splines and neural network models for prediction of pile drivability. Geoscience Frontiers 7, 45−52.

Chapter 20

A Monte Carlo approach applied to sensitivity analysis of criteria impacts on solar PV site selection

Hassan Z. Al Garni[1], Anjali Awasthi[2]

[1]*Department of Electrical and Electronic Engineering Technology, Jubail Industrial College, Jubail Industrial City, Saudi Arabia;* [2]*Concordia University, CIISE, Montreal, QC, Canada*

1. Introduction

One of the fastest growing renewable energy source (RES) technologies worldwide is the solar photovoltaic (PV) technology. Recently, modules prices have dropped by 80% and are anticipated to continue falling in the coming years (Al Garni and Awasthi, 2017c; Al et al., 2019; Ferroukhi et al., 2014). One of the barriers in solar power development is the inconsistency and variability of solar irradiation, which can be geographically dissimilar from one site to another (Al Garni, Awasthi and Ramli, 2018; Al Garni and Garni, 2018). To select a site for such an installation, certain aspects must be investigated, such as the suitability of the PV power plant location for power generation and its proximity to existing infrastructures to minimize the project cost while maximizing power output from the solar panels. Solar site investigation is a strategic step to ensure a cost-effective and well-designed solar project. Several criteria affect such project site selection. Therefore, multiple-criteria decision-making (MCDM) methods are needed for site selection for PV solar energy systems. MCDM methods have been successfully applied in many energy planning projects. Mateo (Mateo, 2012), Pohekar and Ramachandran (Pohekar and Ramachandran, 2004), and Wang et al. (Wang et al., 2009) offered an excellent literature review on application of MCDM approaches in the RES planning. Recently, the geographical information system (GIS) has turned out to be popular for various site selection studies, particularly for energy planning (Al Garni and Awasthi, 2017a; Al Garni and Garni, 2018; Barrows et al., 2016; D and Satish, 2015;

Handbook of Probabilistic Models. https://doi.org/10.1016/B978-0-12-816514-0.00020-5

489

Delivand et al., 2015; Liu et al., 2017; Noorollahi et al., 2007; Rogeau et al., 2017; Sánchez-Lozano, García-Cascales and Lamata, 2016b, 2016a; Sánchez-Lozano et al., 2014; Xu et al., 2015; Yousefi-Sahzabi et al., 2011). National Renewable Energy Laboratory considered screening possible sites for solar projects as a key strategic step (Asakereh et al., 2014; Lopez et al., 2012; Mentis et al., 2015; Tisza, 2014). Noorollahi et al. (2007) structured the decision-making process for site selection into general phases as follows: development of criteria and constraints; model-based prioritization of selected potential sites; and finally, the sensitivity analysis to draw insights into the relevance of decision criteria.

An evaluation of five renewable sources in Saudi Arabia was conducted in Al Garni et al. (2016), considering 14 decision criteria, and using real data. The solar irradiation and air temperature data were obtained using ArcGIS software. The study concluded that solar PV technology is the most promising energy alternative. Additional economic and technical criteria, including solar irradiation, slope, land orientation, urban areas, protected areas, transmission lines, and road accessibility, were considered in a case study for solar PV site selection in Saudi Arabia (Al Garni and Awasthi, 2017b).

In this study, to account for the uncertainty in determining weather parameters and the large study area in Saudi Arabia, Monte Carlo simulation (MCS) was implemented as a probabilistic approach to analyzing the properties of sites and evaluating the impact of each decision criteria on the decision model results, i.e., the suitability scores of the potential sites. This chapter is organized as follows: in Section 2, a literature review of MCDM applications in different energy planning sectors is presented, including the methods for site selection of PV power plants; Section 3 describes the proposed approach to sensitivity analysis, where a MCS analysis is applied to study the case of Saudi Arabia; and finally, Section 4 presents the conclusions.

2. Literature review

2.1 Criteria and constraints

The decision criteria were selected based on the literature, the objective, and access to the database (see Table 20.1). The amount of solar energy shining on the site directly affects the solar power produced. It follows naturally that most solar site suitability studies deliberated solar irradiation as one of the most vital criteria.

In the context of solar irradiation, the global horizontal irradiance is the sum of direct normal irradiance (DNI), diffuse horizontal irradiation (DHI), and ground-reflected irradiation, as represented in Fig. 20.1. The DNI is the result of direct sunlight, while DHI encompasses the irradiation components scattered by clouds or other objects in the atmosphere. However, the irradiation reflected from the ground is considerably lesser than the other components

TABLE 20.1 Solar photovoltaic site suitability criteria (Al Garni and Awasthi, 2018).

Criteria	Criteria
Environmental	Land use
	Agrological capacity
Location	Distance to urban areas
	Distance to substations
	Land cover
	Population density
	Distance to main roads
	Distance to power lines
	Distance to historical areas
	Distance to wildlife designations
Economic	Land cost
	Construction cost
Climatic	Solar irradiation
	Average temperature
Orography	Slope
	Orientation (aspect angles)
	Plot area

and could be neglected. It should be noted that PV technology works using both DNI and DHI solar irradiation, unlike concentrated solar thermal technology which works only using DNI (Hurst, 2015).

Close proximity to utilities improves accessibility and diminishes the high cost of infrastructure construction as well as the harmful consequences to the environment. Furthermore, minimizing the distance to electric transmission lines helps to avoid the high cost of establishing new power lines and it minimizes the power loss in the transmission. Certain studies (Janke, 2010; Sánchez-Lozano et al., 2014; Uyan, 2013) consider locations more distant from cities more suitable for RES development as they avoid negative environmental impact on urban development and the not in "my back yard" (NIMBY) opposition. On the other hand, other studies (Aydin et al., 2013; Effat, 2013) indicate that sites near cities have greater economic advantages. To obtain valid decision results, the study area should be screened to eliminate

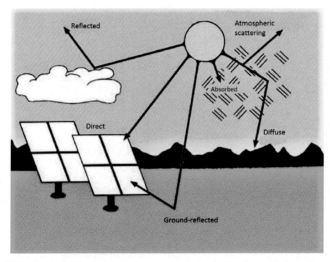

FIGURE 20.1 Solar irradiation Components intercepted by photovoltaic surface (National Renewable Energy Laboratory, 1992).

infeasible locations that pose hindrance to the installation of a utility-scale PV plants. Such unsuitable locations can be discarded using the GIS.

2.2 Methods in solar site selection

MCDM techniques assist decision makers (DMs) to select the best option among several alternatives considering various relevant criteria. These techniques have been frequently deployed in the planning of RES, especially for site selection considering environmental, technical, and economic factors. Furthermore, multiple DMs could have different opinions regarding the specific criteria or alternatives that should be involved in the decision framework. The selection of sites for RES based merely on one criterion is inadequate (Loken, 2007). Huang et al. (Huang et al., 1995) and Loken (Loken, 2007) propose site selection to be handled using MCDM, particularly having in mind the complexity of decisions in energy planning. Recently, the integration of the GIS with MCDM has become increasingly popular for various siting applications, such as landfill site selection (Akbari and Rajabi, 2008; Sener et al., 2010), urban planning (Chandio et al., 2013; Mohit, Mohammad Abdul and Ali, 2006), and the planning of RES sites (Delivand et al., 2015; Herva and Roca, 2013; Izadikhah and Saen, 2016; Pohekar and Ramachandran, 2004; Rumbayan and Nagasaka, 2012).

The GIS has proven its primary role in exploiting geographical data for developing a decision system for siting solar farms. The large amount of information from the GIS offers advantages in determining site suitability for utility-scale solar PV power plants (Arán Carrión et al., 2008). Integrating the

GIS and MCDM brings additional benefits and offers a method for sound decision-making for solar site selection. A number of MCDM approaches exist in the literature, whereas the research on GIS-MCDM has utilized fewer methods, including AHP (Charabi and Gastli, 2011; Effat, 2013; Sánchez-Lozano et al., 2015; Sánchez-Lozano et al., 2013; Uyan, 2013; Watson and Hudson, 2015), the weighted linear combination (Aydin et al., 2013), the technique for order of preference by similarity to ideal solution (Choudhary and Shankar, 2012), and elimination and choice translating reality (Sánchez-Lozano et al., 2014). Table 20.3 summarizes the applications of MCDM techniques in different countries for energy site selection (Table 20.2).

TABLE 20.2 Multiple-criteria decision-making techniques applied in different studies.

No.	Technique	Location	Reference
1	AHP	Indonesia	Rumbayan & Nagasaka (2012)
2	Elimination and Choice Translating Reality (ELECTRE)	Southeast of Spain	Juan M. Sánchez-Lozano et al. (2014)
3	AHP	Ismailia, Egypt	Effat (2013)
4	AHP	Konya, Turkey	Uyan (2013)
5	AHP—technique for order of preference by similarity to ideal solution (TOPSIS)	Southeast Spain	Juan M. Sánchez-Lozano et al. (2013)
6	AHP—Fuzzy TOPSIS and ELECTRE	Murcia, Spain	J.M. Sánchez-Lozano et al. (2015)
7	AHP—Fuzzy OWA	Oman	Charabi & Gastli (2011)
8	Fuzzy OWA	Western Turkey	Aydin et al. (2013)
9	Weighted linear combination	Colorado, USA	Janke (2010)
10	AHP	Central England	Watson & Hudson (2015)
11	Gray cumulative prospect theory	Northwest China	Liu et al. (2017)
12	TOPSIS-ELECTRE	Southeast of Spain	J.M. Sánchez-Lozano et al. (2016a)

TABLE 20.3 Distribution fitting evaluation for each criterion.

No.	Criteria	Distribution fitting	Parameters	NLogL
1	Solar irradiation (X1)	Loglogistic	'mu',3.8975,'sigma',0.0879	2.3043e + 03
2	Air Temperature (X2)	Normal	'mu',53.4706,'sigma',16.1327	2.7840e + 03
3	Slope (X3)	GeneralizedPareto	'k',-1.4591,'sigma',131.3199,'theta',10	−1.2685e + 03
4	Aspect (X4)	GeneralizedPareto	'k',-0.812,'sigma',73.1706,'theta',10	2.9707e + 03
5	Proximity to cities (X5)	GeneralizedPareto	'k',-0.8597,'sigma',77.955,'theta',10	2.9811e + 03
6	Proximity to routes (X6)	GeneralizedPareto	'k',-1.6106,'sigma',144.9533,'theta',10	1.1200e + 03
7	Proximity to grid (X7)	GeneralizedPareto	'k',-1.1105,'sigma',99.9450,'theta',10	2.6282e + 03

3. Sensitivity analysis using Monte Carlo simulation

The following sensitivity analysis was applied to an MCMD method considering seven selection criteria: solar irradiation, average temperature, slope, land aspects, proximity to cities, proximity to highways, and proximity to power lines, as described in Al Garni and Awasthi (2017a,b,c). In this analysis, all the selection criteria are assigned equal weights.

3.1 Parametric approach

In this approach, we make explicit assumptions about the distribution of the population of values of the criteria $X1, X2, X3, X4, X5, X6, X7$ by fitting each criterion with the available distributions. The distribution of the sample mean Y and sample variance is not assumed. Nevertheless, using a large number of sample simulations $(M = 1000)$, we expect the sample distributions of the mean and the variance to be close to a normal distribution. We proceed as follows (see Fig. 20.2):

1. The data of all alternatives (663) with their associated criteria scores imported to the Matlab $(X1, ..., X7)$
2. Use the function "allfitdist" in Matlab to evaluate all the criteria scores under 20 distribution functions to get the potential distributions of criteria scores $(s1, ..., s7)$.
3. Select the best-fit distribution with their parameters for each criterion and generate the function $(pd1, ..., pd7)$. Results of this step are summarized in Table 20.3.
4. Simulate 1000 times random sample for each criterion (using distribution found in Step 3).
5. Choose the weights for Y, then compute $Y = \frac{1}{7(sample_{x1}, ..., sample_{x7})}$, where $sample\_xi$ is the created random samples of the sampled distribution i.
6. Compute the index Y for each of the 1000 samples found in Step 5 $(sampleY[1000X663])$.
7. Calculate the mean of each sample from Step 6 to get a single sample of means for Y $[1X663]$.
8. Calculate the mean of each sample from Step 4 to get a single sample of means for each criterion.
9. Compute the correlation among Y and each criterion to determine the most correlated variables.

3.2 Monte Carlo simulation approach

MCS, which is considered as a methodical approach of doing *what-if* analysis, was used to measure the reliability of the MCDM analysis results and to draw insightful conclusions regarding the relationship between the variation in

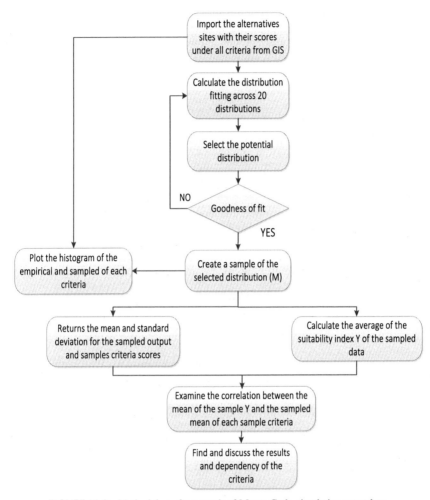

FIGURE 20.2 Methodology framework of Monte Carlo simulation procedure.

decision criteria values and the decision results. MCS, which is a very useful statistical tool for analyzing uncertain scenarios and providing deep analysis of multiple different scenarios, was first used by Jon von Neumann and Ulam in the 1940s. These days MCS represents any simulation that involves repeated random generation of samples and analyzing the behavior of statistical values over population samples (Martinez and Martinez, 2002). Information obtained from random samples is used to estimate the distributions and obtain statistical properties for different situations. In this research, the MCS approach is applied as depicted in Fig. 20.2, starting from the Step 4. "Create a sample of the selected distribution (m)." An explicit assumption was made about the distribution of the population of the criteria values $(X1, ..., X7)$ by fitting

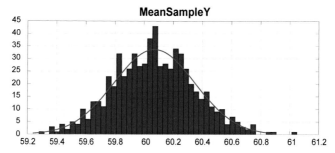

FIGURE 20.3   The mean of the suitability index after 1000 simulation.

each criterion with the available distributions using *Matlab* software, whereas no assumptions were made about the distribution of the sample mean (Y) and sample variance. Nevertheless, using a large number of sample simulations ($M = 1000$), the sample distributions of the mean and the variance are close to a normal distribution as depicted in Fig. 20.3.

Figs. 20.4 and 20.5 illustrates the histogram of samples means for criteria Xs and Y for all simulations, whereas the correlations between the simulated criteria and the suitability index are represented in Table 20.4. The computation of the correlation between Y and each criterion X quantifies the degree to which two quantitative values X and sampled mean, Y, are linked to each other. The correlation coefficient which ranges between -1 and 1 indicates that the strongest linear relationship between the suitability index and the decision criteria occurred for proximity to cities ($X5$) followed by the aspect of the site, whereas the weakest linear relationship is indicated by proximity to roads ($X6$) as shown in Table 20.4.

Furthermore, because the potential sites have different amounts of solar irradiation and different slopes, which both have significant impact on the final suitability index of the site, the correlation coefficients of these criteria have moderate positive correlations.

4. Conclusion

The MCS probabilistic analysis was applied to the MCDM site selection procedure, considering seven selection criteria of equal weights, and 663 solar PV site alternatives in Saudi Arabia. A method was developed to find the final suitability index for a given site. The MCS was used to evaluate how the variation in decision criteria scores of an alternative affects the decision results and thus to account for the uncertainty and variability of the spatial data. The strongest linear relationship between the site suitability and the criteria scores of the alternatives was found for proximity to cities, followed by the aspect criteria. As a future step in the research, the same type of sensitivity analysis could be applied to more complex MCDM models.

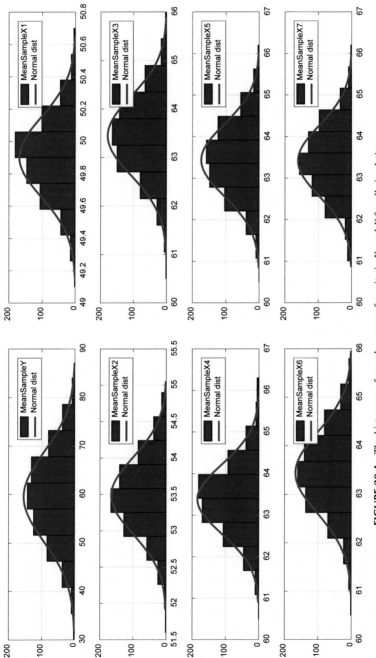

FIGURE 20.4 The histogram of samples means for criteria Xs and Y for all simulations.

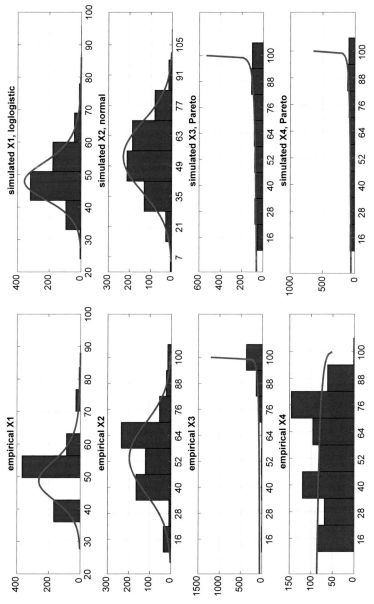

FIGURE 20.5 Comparison between the empirical data of criteria and their corresponding simulated values based on the selected distributions.

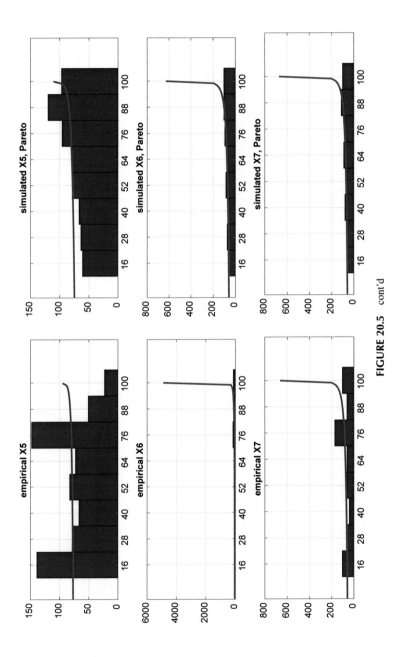

FIGURE 20.5 cont'd

TABLE 20.4 Correlation coefficient of the associated decision criteria.

Decision criteria	Correlation coefficient
Solar (X1)	0.0523
Temp. (X2)	0.0387
Slope (X3)	0.0553
Aspect (X4)	0.0642
Urban (X5)	0.0662
Roads (X6)	−0.009
Power (X7)	0.0164

References

Akbari, V., Rajabi, M., 2008. Landfill site selection by combining GIS and fuzzy multi-criteria decision analysis, case study: Bandar Abbas, Iran. World Applied Sciences Journal 3 (Suppl. 1), 39−47. Retrieved from: http://users.ugent.be/~schavosh/WorldAppliedScience,.pdf.

Al Garni, H.Z., Awasthi, A., 2017a. A fuzzy AHP and GIS-based approach to prioritize utility-scale solar PV sites in Saudi Arabia a fuzzy AHP and GIS-based approach to prioritize utility-scale solar PV sites in Saudi Arabia. In: 2017 IEEE International Conference on Systems, Man, and Cybernetics (SMC). Banff, Canada,: IEEE SMC.

Al Garni, H.Z., Awasthi, A., 2017b. Solar PV power plant site selection using a GIS-AHP based approach with application in Saudi Arabia. Applied Energy 206C, 1225−1240. http://doi.org/10.1016/j.apenergy.2017.10.024.

Al Garni, H.Z., Awasthi, A., 2017c. Techno-economic feasibility analysis of a solar PV grid-connected system with different tracking using HOMER software. In: 2017 the 5th IEEE International Conference on Smart Energy Grid Engineering, pp. 217−222 (Oshawa, ON, Canada). http://doi.org/10.1109/SEGE.2017.8052801.

Al Garni, H.Z., Awasthi, A., 2018. Solar PV power plants site selection. In: Advances in Renewable Energies and Power Technologies. Elsevier, pp. 57−75. http://doi.org/10.1016/B978-0-12-812959-3.00002-2.

Al Garni, H.Z., Awasthi, A., Ramli, M.A.M., 2018. Optimal design and analysis of grid-connected photovoltaic under different tracking systems using HOMER. Energy Conversion and Management 155C, 42−57.

Al Garni, H.Z., Garni, A., 2018. Optimal Design and Analysis of Grid-Connected Solar Photovoltaic Systems. Concordia University.

Al Garni, H.Z., Kassem, A., Awasthi, A., Komljenovic, D., Al-Haddad, K., 2016. A multicriteria decision making approach for evaluating renewable power generation sources in Saudi Arabia. Sustainable Energy Technologies and Assessments 16, 137−150. http://doi.org/10.1016/j.seta.2016.05.006.

Al, H.Z., Awasthi, A., Wright, D., 2019. Optimal orientation angles for maximizing energy yield for solar PV in Saudi Arabia. Renewable Energy 133, 538−550. http://doi.org/10.1016/j. renene.2018.10.048.

Arán Carrión, J., Espín Estrella, A., Aznar Dols, F., Zamorano Toro, M., Rodríguez, M., Ramos Ridao, A., 2008. Environmental decision-support systems for evaluating the carrying capacity of land areas: optimal site selection for grid-connected photovoltaic power plants. Renewable and Sustainable Energy Reviews 12 (9), 2358−2380. http://doi.org/10.1016/j.rser.2007.06. 011.

Asakereh, A., Omid, M., Alimardani, R., Sarmadian, F., 2014. Developing a GIS-based fuzzy AHP model for selecting solar energy sites in Shodirwan region in Iran. International Journal of Advanced Science and Technology 68, 37−48.

Aydin, N.Y., Kentel, E., Sebnem Duzgun, H., 2013. GIS-based site selection methodology for hybrid renewable energy systems: a case study from Western Turkey. Energy Conversion and Management 70, 90−106. http://doi.org/10.1016/j.enconman.2013.02.004.

Barrows, C., Mai, T., Haase, S., Melius, J., Mooney, M., Barrows, C., et al., 2016. Renewable Energy Deployment in Colorado and the West : A Modeling Sensitivity and GIS Analysis.

Chandio, I.A., Matori, A.N.B., WanYusof, K.B., Talpur, M.A.H., Balogun, A.-L., Lawal, D.U., 2013. GIS-based analytic hierarchy process as a multicriteria decision analysis instrument: a review. Arabian Journal of Geosciences 6 (8), 3059−3066. http://doi.org/10.1007/s12517-012-0568-8.

Charabi, Y., Gastli, A., 2011. PV site suitability analysis using GIS-based spatial fuzzy multi-criteria evaluation. Renewable Energy 36 (9), 2554−2561. http://doi.org/10.1016/j.renene. 2010.10.037.

Choudhary, D., Shankar, R., 2012. An STEEP-fuzzy AHP-TOPSIS framework for evaluation and selection of thermal power plant location: a case study from India. Energy 42 (1), 510−521. http://doi.org/10.1016/j.energy.2012.03.010.

Delivand, M.K., Cammerino, A.R.B., Garofalo, P., Monteleone, M., 2015. Optimal locations of bioenergy facilities, biomass spatial availability, logistics costs and GHG (greenhouse gas) emissions: a case study on electricity productions in South Italy. Journal of Cleaner Production 99, 129−139. http://doi.org/10.1016/j.jclepro.2015.03.018.

Effat, H.A., 2013. Selection of potential sites for solar energy farms in Ismailia Governorate, Egypt using SRTM and multicriteria analysis. Cloud Publications International Journal of Advanced Remote Sensing and GIS 2 (1), 205−220. Retrieved from: http://technical.cloud-journals.com/ index.php/IJARSG/article/view/Tech-125.

Ferroukhi, R., Gielen, D., Kieffer, G., Taylor, M., Nagpal, D., Khalid, A., 2014. REthinking energy: towards a new power system. The International Renewable Energy Agency (IRENA). Retrieved from: http://www.irena.org/rethinking/Rethinking_FullReport_web_print.pdf.

Herva, M., Roca, E., 2013. Review of combined approaches and multi-criteria analysis for corporate environmental evaluation. Journal of Cleaner Production 39, 355−371. http://doi. org/10.1016/j.jclepro.2012.07.058.

Huang, J.P., Poh, K.L., Ang, B.W., 1995. Decision analysis in energy and environmental modeling. Energy 20 (9), 843−855. http://doi.org/10.1016/0360-5442(95)00036-G.

Hurst, T., 2015. Renewable Power Costs Plummet. Retrieved from: http://www.irena.org/News/ Description.aspx?NType=A&mnu=cat&PriMenuID=16&CatID=84&News_ID=386.

Izadikhah, M., Saen, R.F., 2016. A new preference voting method for sustainable location planning using geographic information system and data envelopment analysis. Journal of Cleaner Production 137, 1347−1367. http://doi.org/10.1016/j.jclepro.2016.08.021.

Janke, J.R., 2010. Multicriteria GIS modeling of wind and solar farms in Colorado. Renewable Energy 35 (10), 2228−2234. http://doi.org/10.1016/j.renene.2010.03.014.

Lenin, D., Dr. Satish Kumar, J., 2015. GIS based multicriterion site suitability analysis for solar power generation plants in India. The International Journal of Science and Technoledge 3 (3), 197−202.

Liu, J., Xu, F., Lin, S., 2017. Site selection of photovoltaic power plants in a value chain based on grey cumulative prospect theory for sustainability: a case study in Northwest China. Journal of Cleaner Production 148, 386−397. http://doi.org/10.1016/j.jclepro.2017.02.012.

Loken, E., 2007. Use of multicriteria decision analysis methods for energy planning problems. Renewable and Sustainable Energy Reviews 11 (7), 1584−1595. http://doi.org/10.1016/j.rser.2005.11.005.

Lopez, A., Roberts, B., Heimiller, D., Blair, N., Porro, G., 2012. U.S. Renewable energy technical Potentials: a GIS-based analysis U.S. Renewable energy technical Potentials: a GIS- based analysis. National Renewable Energy Laboratory Document. Washington, D.C. Retrieved from: http://www.ifc.org/wps/wcm/connect/f05d3e00498e0841bb6fbbe54d141794/IFC+Solar+Report_Web+_08+05.pdf?MOD=AJPERES.

Martinez, W.L., Martinez, A.R., 2002. Computational Statistics Handbook with Matlab. New York, vol. 65. Crc, Hall, New York, NY, USA. http://doi.org/10.1111/j.1541-0420.2009.01208_14.x.

Mateo, J.R.S.C., 2012. Multip-Criteria Analysis in the Renewable Energy Industry. Springer Science & Business Media, Santander. http://doi.org/10.2174/97816080528511060101.

Mentis, D., Welsch, M., Fuso Nerini, F., Broad, O., Howells, M., Bazilian, M., Rogner, H., 2015. A GIS-based approach for electrification planning—a case study on Nigeria. Energy for Sustainable Development 29, 142−150. http://doi.org/10.1016/j.esd.2015.09.007.

Mohit, M.A., Ali, M.M., 2006. Integrating GIS and AHP for land suitability analysis for urban development in a secondary city of Bangladesh. Jurnal Alam Bina 8 (1), 1−20. Retrieved from: http://irep.iium.edu.my/id/eprint/33724.

National Renewable Energy Laboratory, 1992. Shining on: Aprimer on Solar Radiation Data. Denver. Retrieved from: http://www.nrel.gov/docs/legosti/old/4856.pdf.

Noorollahi, Y., Itoi, R., Fujii, H., Tanaka, T., 2007. GIS model for geothermal resource exploration in Akita and Iwate prefectures, northern Japan. Computers and Geosciences 33 (8), 1008−1021. http://doi.org/10.1016/j.cageo.2006.11.006.

Pohekar, S.D., Ramachandran, M., 2004. Application of multi-criteria decision making to sustainable energy planning—a review. Renewable and Sustainable Energy Reviews 8 (4), 365−381. http://doi.org/10.1016/j.rser.2003.12.007.

Rogeau, A., Girard, R., Kariniotakis, G., 2017. A generic GIS-based method for small Pumped Hydro Energy Storage (PHES) potential evaluation at large scale. Applied Energy 197, 241−253. http://doi.org/10.1016/j.apenergy.2017.03.103.

Rumbayan, M., Nagasaka, K., 2012. Prioritization decision for renewable energy development using analytic hierarchy process and geographic information system. In: Advanced Mechatronic Systems (ICAMechS), 2012 International Conference on. IEEE, 2012, pp. 36−41 (Tokyo, Japan).

Sánchez-Lozano, J.M., García-Cascales, M.S., Lamata, M.T., 2015. Evaluation of suitable locations for the installation of solar thermoelectric power plants. Computers and Industrial Engineering 87, 343−355. http://doi.org/10.1016/j.cie.2015.05.028.

Sánchez-Lozano, J.M., García-Cascales, M.S., Lamata, M.T., 2016a. Comparative TOPSIS-ELECTRE TRI methods for optimal sites for photovoltaic solar farms. Case study in Spain. Journal of Cleaner Production 127, 387−398. http://doi.org/10.1016/j.jclepro.2016.04.005.

Sánchez-Lozano, J.M., García-Cascales, M.S., Lamata, M.T., 2016b. GIS-based onshore wind farm site selection using Fuzzy Multi-Criteria Decision Making methods. Evaluating the case of Southeastern Spain. Applied Energy 171, 86–102. http://doi.org/10.1016/j.apenergy.2016. 03.030.

Sánchez-Lozano, J.M., Henggeler Antunes, C., García-Cascales, M.S., Dias, L.C., 2014. GIS-based photovoltaic solar farms site selection using ELECTRE-TRI: evaluating the case for Torre Pacheco, Murcia, Southeast of Spain. Renewable Energy 66, 478–494. http://doi.org/10. 1016/j.renene.2013.12.038.

Sánchez-Lozano, J.M., Teruel-Solano, J., Soto-Elvira, P.L., Socorro García-Cascales, M., 2013. Geographical information Systems and multi-Criteria decision making methods for the evaluation of solar farms locations: case study in south-eastern Spain. Renewable and Sustainable Energy Reviews 24, 544–556. http://doi.org/10.1016/j.rser.2013.03.019.

Sener, S., Sener, E., Nas, B., Karagüzel, R., 2010. Combining AHP with GIS for landfill site selection: a case study in the Lake Beyşehir catchment area (Konya, Turkey). Waste Management 30 (11), 2037–2046. http://doi.org/10.1016/j.wasman.2010.05.024.

Tisza, K., 2014. GIS-based suitability modeling and multi-criteria decision analysis for utility scale solar plants. In: Four States in the Southeast US. Clemson University. Retrieved from: http:// tigerprints.clemson.edu/all_theses.

Uyan, M., 2013. GIS-based solar farms site selection using analytic hierarchy process (AHP) in Karapinar region, Konya/Turkey. Renewable and Sustainable Energy Reviews 28, 11–17. http://doi.org/10.1016/j.rser.2013.07.042.

Wang, J.-J., Jing, Y.-Y., Zhang, C.-F., Zhao, J.-H., 2009. Review on multi-criteria decision analysis aid in sustainable energy decision-making. Renewable and Sustainable Energy Reviews 13 (9), 2263–2278. http://doi.org/10.1016/j.rser.2009.06.021.

Watson, J.J.W., Hudson, M.D., 2015. Regional Scale wind farm and solar farm suitability assessment using GIS-assisted multi-criteria evaluation. Landscape and Urban Planning 138, 20–31. http://doi.org/10.1016/j.landurbplan.2015.02.001.

Xu, J., Song, X., Wu, Y., Zeng, Z., 2015. GIS-modelling based coal-fired power plant site identification and selection. Applied Energy 159, 520–539. http://doi.org/10.1016/j.apenergy. 2015.09.008.

Yousefi-Sahzabi, A., Sasaki, K., Yousefi, H., Pirasteh, S., Sugai, Y., 2011. GIS aided prediction of CO_2 emission dispersion from geothermal electricity production. Journal of Cleaner Production 19 (17–18), 1982–1993. http://doi.org/10.1016/j.jclepro.2011.06.009.

Chapter 21

Stochastic analysis basics and application of statistical linearization technique on a controlled system with nonlinear viscous dampers

Aboubaker Gherbi[1], Mourad Belgasmia[2]

[1]*Department of civil engineering, Constantine University, Constantine, Algeria;* [2]*Department of civil engineering, Setif University, Setif, Algeria*

1. Introduction

A phenomenon is said to be deterministic, if it is perfectly known, without any uncertainty; this designation cannot be linked to reality. It is obvious that any simple event or experiment such as materials strength or environmental loading cannot be reproduced with the same precision. Most used design methodologies are based on a more simplified concept, where the load is represented with a maximum and the strength is represented with a minimum value, based on empirical and theoretical information.

To consider all parameters involved in terms of their statistical properties, the *Stochastic Analysis* had to be developed. As more information is regarded, loading and therefore the response are characterized in probabilistic terms.

This chapter aims primarily toward initiation to probability analysis, which admits a random behavior of certain elements. However, some limitation can be assumed because of the ability to control the construction process. Therefore, it will be concentrated on the prediction of the response of a deterministic linear system subjected to stochastic excitation, i.e., wind load. The relation between time and frequency domains is also explained, as well as the simplicity allowed in the latter for determining the response of linear systems, i.e., spectral analysis.

As a second major objective of this chapter, the control of structures with additional damping devices, which has been widely investigated in the last

Handbook of Probabilistic Models. https://doi.org/10.1016/B978-0-12-816514-0.00021-7

decade, however, the use of nonlinear viscous dampers is still ambiguous and their design can be onerous. Considering a nonlinear supplemental devices for a linear structure can be challenging for engineers, especially when seeking for simplicity, i.e., frequency domain methods. Thus, statistical linearization technique (SLT) is of utmost importance, which is a powerful tool for probabilistic analysis of nonlinear structural dynamics problem.

Wind is always turbulent, and it fluctuates randomly; its properties need to be visualized in a statistical manner. It also means that the flow is chaotic with random periods ranging from fractions of seconds to few minutes. Dyrbye and Hansen (1997) described the turbulent components by their standard deviation, time scales, integral length scale, and power spectral density (PSD) functions that define the frequency distribution and normalized cospectra that specify the spatial correlation. In the need to analyze structures under wind loads, engineers are most of the time forced to simulate wind time series. Digitally generated samples become more essential for time domain analysis of structures, i.e., Monte Carlo method; as it is easier to get artificial samples of loading, these samples are introduced into the governing equation to be solved for each specific problem. Kareem (2008) stated that simulations are mostly based on models describing the process to be simulated, and it can be an exact theoretical model, phenomenological or empirical model, or observed data. The distribution of turbulent energy in wind as a function of frequency can be described by the power spectrum of the fluctuating component of wind speed (Das, 1988).

Civil structures are usually designed to resist moderate to mild intensities; these structures may sustain significant damage and deformation during their functionality. To limit structural displacement and peak accelerations to acceptable levels, two main options are available for engineers to reduce wind effects, enhance buildings lateral stiffness by either increasing the structural size and material quantity or increasing the level of damping by means of supplemental devices (Duflot and Taylor, 2008). Fluid viscous dampers (FVDs) are energy dissipation systems, and their use for civil applications has emerged after their proven efficiency in various military branches. Many researchers (Constantinou et al., 1993; Soong and Costantinou, 2014) proceeded in their testing and proved their efficiency for improving behavior of full-scale civil structures.

In the following sections, both wind randomness and FVDs will be discussed, and the authors aim toward keeping the principles general and simple for understanding, also providing further reading materials for an advanced comprehension.

2. Power spectral density

Before getting in the definition of the PSD function, the authors taught that recalling some of the basics of probability is of utmost importance, with more readings mentioned for further details.

2.1 Probability density function

As a random process is a succession of values that are very close to each other, it is natural to characterize it in the same manner of a random variable (see detail of random variables (Denol, 2002; Newland, 2012)), first-order probability density $p(x,t)$ with random variable x that evolves in time. $P(x,t)$ dx presents the probability for the function to take a value between x and $x + dx$ at time t. To characterize the process more efficiently, the second-order probability density function can be considered $p(x_1,t_1;x_2,t_2)$; therefore, if more information about the process is needed, the probability density function order should be higher (Denol, 2002).

Another representation of a random process is done by defining the moment function, where they characterize the process and probability density of any order. The two first moments have a great importance: The average is given by

$$\mu(t) = E[x(t)] = \int_{-\infty}^{\infty} x p(x,t) dx \qquad (21.1)$$

2.2 The autocorrelation function

$$R_{xx} = E[x(t_1), x(t_2)] = \int_{-\infty}^{\infty} \int_{-\infty}^{\infty} x_1 x_2 p_x(x_1, t_1; x_2, t_2) dx_1 dx_2 \qquad (21.2)$$

These two moments can be sufficient for the characterization of a random Gaussian process. As shown in the previous equations, it is needed to determine the ensemble average $E[\cdot]$. Based on Ergodicity theorem, instead of defining the value for a set of samples at a chosen time, it is more convenient to consider only one sample and define the average varying time. The autocorrelation function of a stationary process on one sample is

$$R_{xx}(\Delta t) = \lim_{T \to \infty} \int_{-T/2}^{T/2} x(t) x(t + \Delta t) dt \qquad (21.3)$$

This function has some important properties; for instance, the value of the autocorrelation function at the origin is equal to the variance of the sample:

$$R_{xx}(0) = \lim_{T \to \infty} \int_{-T/2}^{T/2} \left[x(t)^2 \right] dt = \sigma_x^2 \qquad (21.4)$$

The autocorrelation function is an even function:

$$R_{xx}(0) = \lim_{T \to \infty} \int_{-T/2}^{T/2} x(t)x(t + \Delta t)dt = R_{xx}(\Delta t) \tag{21.5}$$

The autocorrelation function is nonnegative for any complex function $h(t)$ defined in an interval $[a,b]$:

$$\int_{a}^{b} \int_{a}^{b} R_{xx}(t_1 t_2)\overline{h(t_1)}h(t_2)dt_1 dt_2 \geq 0 \tag{21.6}$$

An important function that has a major role in the stochastic analysis of structure is the PSD; Eq. (21.6) shows that the autocorrelation function is nonnegative, which implies that the Fourier transform of this function is also positive, and it is known as the PSD.

$$S_{xx}(\omega) = \frac{1}{2\pi} \int_{-\infty}^{+\infty} R_{xx}(\tau)e^{j\omega\tau}d\tau \tag{21.7}$$

This function presents the frequency distribution of the average energy contained in the random process. It is a real positive function because the autocorrelation function is an even function Eq. (21.5).

These two functions constitute Fourier transform pairs. The PSD of a random process must check the relation:

$$S_{xx}(\omega) = \lim_{T \to \infty} \frac{2\pi}{T} E\left[|X_i(\omega, T)|^2\right] \tag{21.8}$$

where $X_i(\omega,T)$ is the Fourier transform of the samples.

An important feature of this function is that its integration yields the variance of the considered random process:

$$\int_{-\infty}^{+\infty} S_{xx}(\omega)d\omega = \sigma_x^2 \tag{21.9}$$

In wind analysis, different power spectral densities are used. To define this difference, the variance is calculated by the integration of the PSD over the frequency range; also, only positive frequencies are considered, as they have a physical significance and the integration boundaries become 0 to $+\infty$.

Fig. 21.1 represents the relationship between time and frequency domain, related by the well-known Fourier transform.

2.3 Spectral moments

As this chapter is concerned with the investigation of response of systems under random excitations, i.e., wind, these moments should be introduced. The spectral moments can be defined from this general relation:

$$m_i = \int_{-\infty}^{+\infty} |\omega|^i S(\omega) d\omega \tag{21.10}$$

where

- Spectral moment of order $i = 0$ is the variance of the process, which defines the displacement.
- Spectral moment of order $i = 2$ is the derivative of the variance of the process, which defines the velocity.
- Spectral moment of order $i = 4$ is the second derivative of the variance of the process, which defines the acceleration.

2.4 Ergodicity

A random or stochastic process is said to be ergodic if its statistical properties, i.e., PSD, can be estimated from a single sample:

$$S_{xx}(\omega) = \lim_{T \to \infty} \frac{2\pi}{T} |X_i(\omega, T)|^2 \tag{21.11}$$

which allows the determination of the autocorrelation function and PSD from a single known sample.

In practical applications, ergodicity is assumed when the process is stationary to estimate the mean and the correlation function from a single sample function (Roberts and Spanos, 2003).

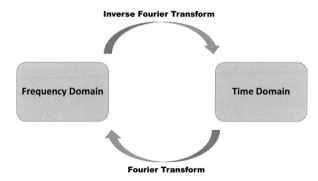

FIGURE 21.1 Time−frequency relationship.

3. Input–output relationship

Before introducing the principle of spectral analysis, one must recall some basics of structural dynamics. The response of an Single Degree of Freedom (sdof) system under a random loading $X(t)$ can be deduced by considering the loading as a succession of short impulses $X(\tau)$ (see Clough and Penzien, 1995, page 87), where its equation of motion can be written as follows:

$$M\ddot{y} + c\dot{y} + Ky = X(t) \tag{21.12}$$

Each impulse produces its own differential response; thus, the entire response is obtained by the summation of all the differential responses and can be written in a convolution integral form:

$$Y(t) = \int_0^t h(\tau)X(t - \tau)d\tau \tag{21.13}$$

where the function $h(t-\tau)$ is known as the unit impulse response function:

$$h(t - \tau) = \frac{1}{m\omega_d} sin\omega_d(t - \tau)e^{-\xi\omega(t-\tau)} \tag{21.14}$$

The integral of Eq. (21.12) exists in the mean square if the autocorrelation function R_{yy} exists. Considering two instants of time t and $t+\tau$ and taking the mathematical expectation $E[X(t)Y(t+\tau)]$ and $E[Y(t)Y(t+\tau)]$, one ends up with

$$R_{yy} = \int_0^{infty} h(\xi)R_{yx}(\tau + \xi)d\xi \tag{21.15}$$

This states that the autocorrelation of the response is given by the correlation integral of $R_{yx}(\tau)$ with the impulse response of the system. Because a convolution in a domain (time) corresponds to a product in the other domain (frequency), one can understand the focus of the authors on the PSD, as the response can be obtained by a simple multiplication in the frequency domain.

The Fourier transform gives the relationship between the PSD functions of the response S_{yy} and excitation S_{xx} as

$$S_{yx} = H(\omega)S_{xx}(\omega) \tag{21.16}$$

$$S_{yy} = H(\omega)^* S_{yx} = |H(\omega)|^2 S_{xx} \tag{21.17}$$

It can be seen that the output PSD at one frequency depends only on the PSD of the input and the frequency response function for that same frequency (see Preumont, 2013, page 79).

3.1 Frequency response function

The equation of motion of the system can be written in the following form (Preumont, 2013), when introducing the natural frequency ω_n and the damping ratio ξ:

$$\ddot{y} + 2\xi\omega_n\dot{y} + \omega_n^2 y = \frac{f(t)}{m} \tag{21.18}$$

The transfer function between the excitation and the response is

$$H(\omega) = \frac{1}{m} \frac{1}{(\omega_n^2 - \omega^2) + 2i\xi\omega_n\omega} \tag{21.19}$$

Knowing that $\omega_n = \sqrt{\frac{k}{m}}$, it becomes

$$H(\omega) = \frac{1}{k} \frac{1}{\left(1 - \dfrac{\omega^2}{\omega_n^2}\right) + 2i\xi\dfrac{\omega}{\omega_n}} \tag{21.20}$$

3.2 Example

As an illustration for the spectral analysis, a linear sdof system is considered, which will be subjected to a stochastic wind loading, characterized by a PSD, i.e., Davenport:

$$S_{xx}(\omega) = \frac{\omega\dfrac{2}{3}\left(\dfrac{L}{U}\right)^2 \sigma^2}{\left(1 + \left(\dfrac{\omega L}{U}\right)^2\right)^{4/3}} \tag{21.21}$$

where L is the turbulence length scale, U is the mean wind velocity, and σ^2 is the variance of the turbulent component of the wind velocity. A typical power density spectrum, referred to hereafter by PSD, has a peaked form as indicated in Fig. 21.2. In this example, Matlab was used to facilitate the calculations and to show graphically the response of the system.

As shown earlier, the response is obtained in the frequency domain by a simple multiplication of the PSD of the loading by the frequency response function; this latter defined in Eq. (21.20) takes the form as shown in Fig. 21.3, where the peak is identified in the natural frequency of the system.

Here is an example of the code used for this purpose:

```
for j = 1:Nmax
    n = Z(j);
```

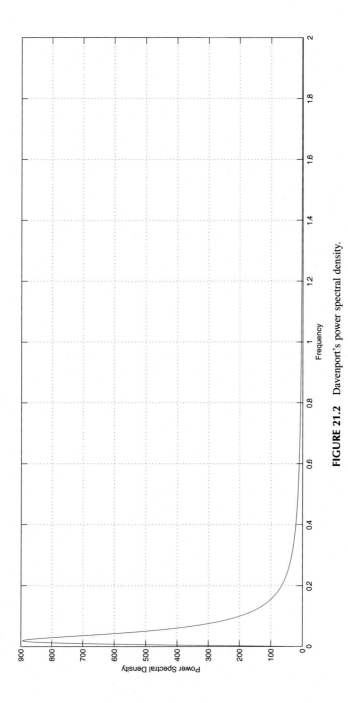

FIGURE 21.2 Davenport's power spectral density.

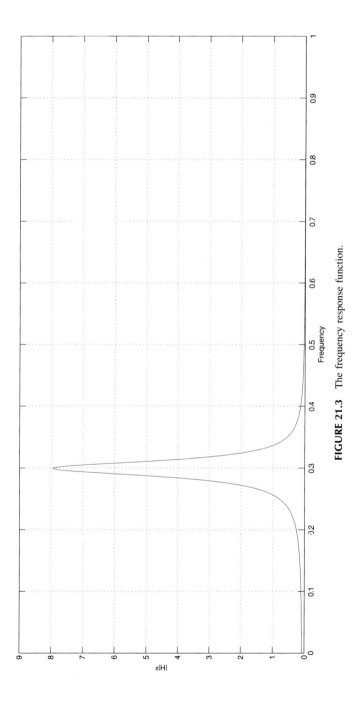

FIGURE 21.3   The frequency response function.

```
Sy(j) = n*2/3*(L/U)^2*sig.^2/((1+(n*L/U)^2)^(4/3))(1/
(K)^2)*1/((1-(n/f)^2)$^2+(2*ksi*n/f)^2);
end
plot (Z,Sy)
```

In this code, one can see the PSD of the loading and the frequency response function, both defined earlier, multiplied over a range of suitable frequencies (Denol, 2002). The response can be easily obtained (Fig. 21.4).

The area under the graph represents the variance of the displacement of the system. Two main contributions can be identified, a background and a resonant response (illustrated in Fig. 21.4. with two different graphs for a better understanding) (illustrated here with two different graphs for a better understanding); this decomposition can be used to obtain the result more efficiently, and this decomposition was initially proposed by Davenport (1961).

4. Monte Carlo simulation

Seeking accuracy in the estimation of the response of randomly excited system (linear or nonlinear), one must refer to Monte Carlo simulation. The theory behind this method assumes that the stochastic differential equation of motion can be interpreted as an infinite set of deterministic differential equation. For the case of this chapter (deterministic system and stochastic excitation), the method consists at generating a single sample of the random excitation, and the response is obtained using numerical integration method to solve the differential equation. The same work can be done all over again, and the response is used to update the statistics. Seemingly, for more accurate

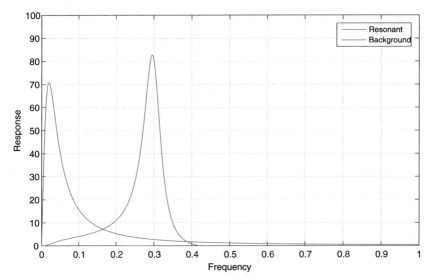

FIGURE 21.4 Response of an sdof system (background and resonant).

response, a larger number of simulated samples are needed, which can be challenging for multiple degree of freedom systems. However, thanks to the stationarity of wind excitation, and ergodicity principle allows the calculation of the statistical moments using single sample function of the response (Roberts and Spanos, 2003)

4.1 Wind sample generation

Engineers and researchers seek to verify the results obtained by the spectral analysis (probabilistic), which can be achieved by the generation of time series compatible to the PSD, for linear and nonlinear systems (Monte Carlo method).

The generation of time series equivalent with the PSD can be done by means of autoregressive filters (AR and ARMA) (Li and Kareem, 1990) and Fourier series, which will be used in this chapter. Readers are invited to consult further readings as this chapter will present a brief introduction about the method (Denol, 2002; Shinozuka, 1971).

Based on the ergodicity hypothesis seen previously, Eq. (21.11), the PSD of a stationary process can be obtained from a single sample. Thus, the generation could be done simply by choosing

$$X(\omega, T) = \sqrt{\frac{T}{2\pi}} \sqrt{S_{xx}(\omega_i)} e^{i\varphi j} \qquad (21.22)$$

where φ is a random phase angle taken between 0 and 2π.

The samples are then obtained by applying the inverse Fourier transform. An example of the generation code can be as follows:

```
Phi = 2*pi*rand(1,N/2);
for n = 0:DN:Nmax
    Sx = n*2/3*(L/U)^2*sig^2/((1+(n*L/U)^2)^(4/3));
    X = real (ifft(sqrt (T/(2*pi))* sqrt(Sx)*exp(i*Phi)));
end
```

Here, *rand* is a Matlab function that generates random numbers, which is used to create the random phase angles; *ifft* is also a Matlab function that applies the inverse Fourier transform. An example of the generated wind time history sample is presented in Fig. 21.5.

As discussed previously, using Fourier Transform, one can switch from a domain to another; based on this definition, both the loading and the response of the time domain analysis will be presented in the frequency domain along with the results already obtained (verification). First, the generated time history will be transformed for a frequency representation.

The response of the sdof (same characteristics as the previous example) subjected to wind time history (equivalent to the PSD) requires the use of step-by-step methods. From Fig. 21.6, the generated time history is well-matched

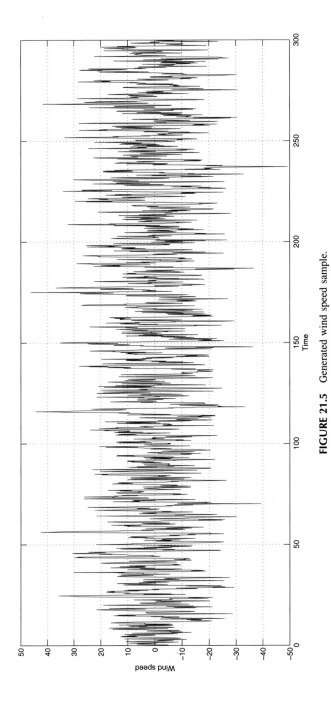

FIGURE 21.5 Generated wind speed sample.

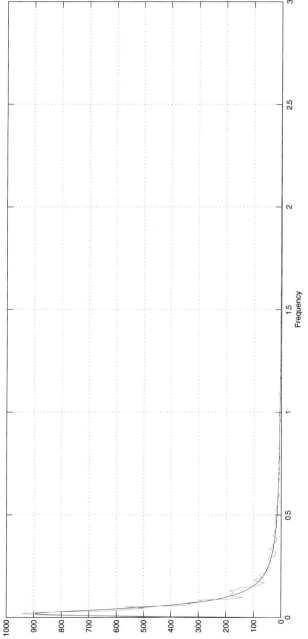

FIGURE 21.6 Frequency representation of the power spectral density and the generated sample.

with the equivalent Davenport PSD, and thus, a good match is expected when comparing the results; a linear Newmark method is used for this example (Chopra, 2001).

From Fig. 21.7 and as expected, a good match is achieved, proving the efficiency of the used method; this academic view of the problem (even for an sdof) can be a powerful tool for further research in this field. In the next section, uid viscous dampers (FVD) are presented and introduced in this sdof system to show their impact.

5. Fluid viscous dampers

When a structure is subjected to a dynamic excitation, a quantity of energy is diffused into the structure. This latter absorbs and dissipates this energy (through heat) by transforming it to kinetic and potential energy; this inherent damping which consists in a combination of strength, deformability, and flexibility allows the input energy to be extinguished. Additional energy dissipation devices may achieve a better structural behavior and improvement. In the last decade, structural control devices caught a lot of attention from the civil engineering community, thanks to their proven ability to reduce the damage induced by wind and seismic excitations. Great efforts were put to make the concept of additional damping a workable technology; nowadays, numerous civil structures are equipped with such devices.

FVDs operate on the principle of fluid flow through orifices. A stainless steel piston travels through chambers that are filled with silicone oil (inert, noninflammable, nontoxic, and stable); when the damper is excited, the fluid is forced to flow either around or through the piston. The input energy is transformed into heat when the silicone oil flows through orifices between the two chambers (Duflot and Taylor, 2008). In Constantinou et al. (1993); Soong and Costantinou (2014), it was demonstrated that energy dissipation systems are capable of producing significant reduction of interstory drift to structures and column bending moments to which they are installed (Fig. 21.8).

Many manufacturers supply a wide variety of devices (as illustrated in Fig. 21.8), operating at high or low fluid pressure and generally having nonlinear force−velocity relationship as follows (Paola et al., 2007):

$$F = C_D|\dot{y}|^{\alpha} sgn(\dot{y}) \tag{21.23}$$

in which C_D is the damping coefficient of the damper, α is a damping exponent in the range of 0.32, and $sgn(\cdot)$ is a signum function. The value of α depends on the shape of the piston head; when $\alpha = 2$ it means that the orifices are cylindrical and presents an unacceptable performance. For $\alpha = 0.3$, it represents an effective value for the case of high velocities. A linear behavior is achieved by a value of $\alpha = 1$, which is frequently used in case of seismic and wind energy. Fig. 21.9 shows the force−velocity relationship of FVD.

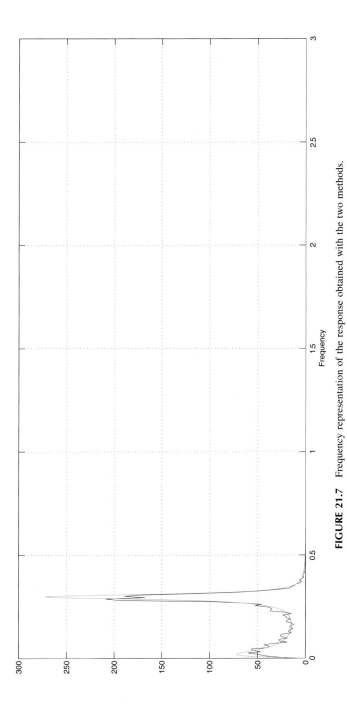

FIGURE 21.7 Frequency representation of the response obtained with the two methods.

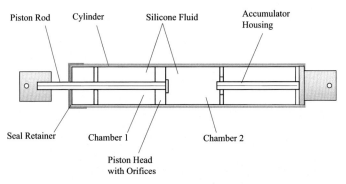

FIGURE 21.8 Typical fluid viscous damper.

Introducing this damper to the sdof system mentioned previously, the equation of motion becomes

$$M\ddot{y}(t) + c\dot{y}(t) + C_D|\dot{y}(t)|^\alpha sgn(\dot{y}(t)) + Ky(t) = X(t) \qquad (21.24)$$

In the case of a linear additional damper, one can easily resolve this equation and obtain the proper results. However, the choice of a nonlinear behavior for the damper requires a lot of attention as it becomes more complicated, and a linearization of the equation of motion must be achieved. Eq. (21.23) can be written in the following form:

$$\ddot{y}(t) + \eta|\dot{y}(t)|^\alpha sgn(\dot{y}(t)) + 2\xi^{(s)}\omega_n\dot{y}(t) + \omega_n^2 y(t) = X(t) \qquad (21.25)$$

where $\eta = C_D|cos|^{\alpha+1}/M, \xi^{(s)}$ is the structural damping ratio and ω_n is the natural radian frequency (Clough and Penzien, 1995; Paola et al., 2007).

5.1 Statistical linearization technique

This method was initially proposed by Caughey (1963) to solve nonlinear stochastic problems in structural dynamics, also known as *equivalent linearization* or *stochastic linearization*. The main idea of the SLT is to replace the governing nonlinear differential equation by a linear equation, in which the difference between the two equations must be minimized.

The objective is to replace Eq. (21.25) by a linear equivalent one:

$$\ddot{y}(t) + 2\xi^{(e)}\omega_n\dot{y}(t) + \omega_n^2 y(t) = X(t) \qquad (21.26)$$

$\xi^{(e)}$ is the equivalent damping ratio (structural and additional damper); this ratio is chosen in such way that the error between the two equation must be

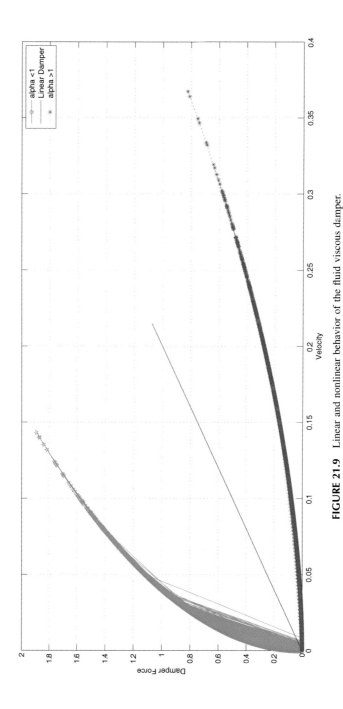

FIGURE 21.9 Linear and nonlinear behavior of the fluid viscous damper.

minimized. The usual method is to minimize the mean squared error with respect to $\xi^{(e)}$:

$$\xi^{(e)} = \xi^{(e)} + \eta \frac{E\left[|\dot{y}|^{\alpha+1}\right]}{E[\dot{y}^2]} \tag{21.27}$$

$E[\cdot]$ is the mean ensemble average, it follows

$$E\left[|\dot{y}|^{\alpha+1}\right] = \frac{2^{(1+\gamma)/2}\Gamma\left(\dfrac{1+\gamma}{2}\right)}{\sqrt{2\pi}}\sigma_{\dot{y}}^{\gamma} \tag{21.28}$$

$\Gamma(\cdot)$ is gamma function, $\gamma = \alpha+1$ and $\sigma_{\dot{y}}$ is the standard deviation of the velocity $\dot{y}$, the relation becomes

$$\xi^{(e)} = \xi^{(s)} + \eta \frac{2^{1+\alpha/2}\Gamma\left(1+\dfrac{\alpha}{2}\right)}{\sqrt{2\pi}}\sigma_{\dot{y}^{\alpha-1}} \tag{21.29}$$

It is obvious that the equivalent damping ratio $\xi^{(e)}$ for the linearized equation depends on $\sigma_{\dot{y}}$, which is still unknown because it essentially depends on $\xi^{(e)}$, recall

$$\sigma_{\dot{y}}^2 = \int\limits_0^{+\infty} \omega^2 |H(\omega)|^2 S(\omega)d\omega \tag{21.30}$$

Therefore, the definition of the value of $\xi^{(e)}$ consists of an iterative process.

After applying the SLT on the previous system, the results in the frequency representation are completely matched as shown:

Various parametric studies can be executed from this point; in Figs. 21.10 and 21.11, the impact of the added FVD on the systems displacement is shown:

The influence of the added FVD is obvious (Fig. 21.11), as the displacement decreases (the resonance peak) with the reduction of the damping exponent α from 1 to 0.4.

6. Conclusion

This chapter aimed toward initiation for the stochastic analysis, with some limitation and orientation to a target problem. The characterization of wind by means of the PSD was explained; also Monte Carlo simulation was briefly mentioned, as it has a great importance for further research and material of this work; accordingly, this chapter tries to represent the analysis procedure in both domains (time and frequency).

Secondly, energy dissipation system was discussed (FVD), and a concise review of their capacity was presented.

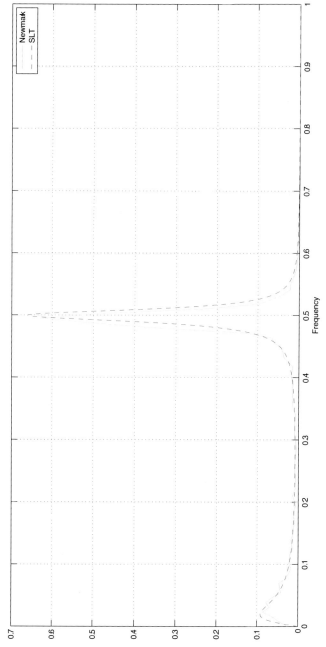

FIGURE 21.10 Frequency representation of statistical linearization technique and exact solution (NL-Newmark).

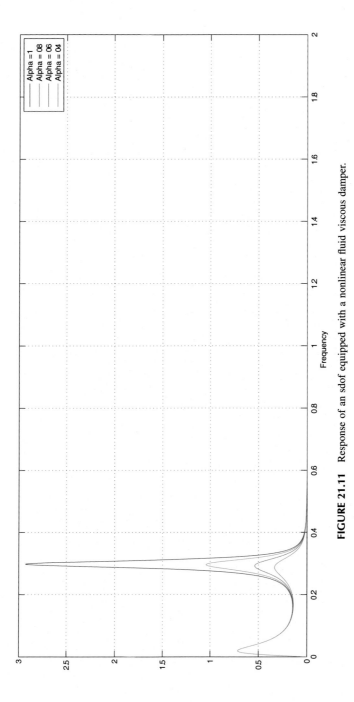

FIGURE 21.11 Response of an sdof equipped with a nonlinear fluid viscous damper.

This chapter also concentrated on the nonlinear behavior of this type of devices, which was included in a linear deterministic system, subjected to random wind excitation. The system was analyzed in the domains, using Monte Carlo simulation and generating the adequate time series for the time domain, and the response in the frequency domain was executed by means of SLT. The results show an accurate match, proving both the efficiency of the spectral analysis using SLT and thus analysis time are reduced considerably.

References

Caughey, T.K., 1963. Equivalent linearization techniques. Journal of the Acoustical Society of America 35 (11), 1706−1711.

Chopra, A.K., 2001. Dynamics of Structures: Theory and Applications.

Clough, R.W., Penzien, J.I., 1995. Dynamics of Structures. Computers I& Structures, Berkeley.

Constantinou, M.C., Symans, M.D., Tsopelas, P., I& Taylor, D.P., 1993. Fluid viscous dampers in applications of seismic energy dissipation and seismic isolation. Proceedings ATC 17 (1), 581−592.

Das, N.K., 1988. Safety Analysis of Steel Building Frames under Dynamic Wind Loading. Doctoral dissertation. Texas Tech University.

Davenport, A.G., 1961. The application of statistical concepts to the wind loading of structures. Proceedings − Institution of Civil Engineers 19 (4), 449−472.

Denol, V., 2002. Analyse de structures soumises au vent turbulent: de l'approche stochastique frquentielle au transitoire non linaire. Doctoral dissertation. University of Lige.

Duflot, P., Taylor, D.I., 2008. Experience and Practical Considerations in the Design of Viscous Dampers.

Dyrbye, C., Hansen, S.O., 1997. Wind Loads on Structures.

Kareem, A., 2008. Numerical simulation of wind effects: a probabilistic perspective. Journal of Wind Engineering and Industrial Aerodynamics 96 (10−11), 1472−1497.

Li, Y., Kareem, A.I., 1990. ARMA representation of wind field. Journal of Wind Engineering and Industrial Aerodynamics 36, 415−427.

Newland, D.E., 2012. An Introduction to Random Vibrations, Spectral and Wavelet Analysis. Courier Corporation.

Paola, M.D., Mendola, L.L., I& Navarra, G., 2007. Stochastic seismic analysis of structures with nonlinear viscous dampers. Journal of Structural Engineering 133 (10), 1475−1478.

Preumont, A., 2013. Random Vibration and Spectral Analysis, vol. 33. Springer Science I& Business Media.

Roberts, J.D., Spanos, P.D.I., 2003. Random Vibration and Statistical Linearization. Courier Corporation.

Shinozuka, M., 1971. Simulation of multivariate and multidimensional random processes. Journal of the Acoustical Society of America 49 (1B), 357−368.

Soong, T.T., Costantinou, M.C.I. (Eds.), 2014. Passive and Active Structural Vibration Control in Civil Engineering, vol. 345. Springer.

Chapter 22

A comparative assessment of ANN and PNN model for low-frequency stochastic free vibration analysis of composite plates

S. Naskar[1], Tanmoy Mukhopadhya[2], S. Sriramula[1]

[1]*School of Engineering, University of Aberdeen, Aberdeen, United Kingdom;* [2]*Department of Aerospace Engineering, Indian Institute of Technology Kanpur, Kanpur, India*

1. Introduction

Modern civil, mechanical, and aerospace industries exhaustively use composite structures due to various engineering advantages compared with conventional structural materials (Karsh et al., 2018a; Dey et al., 2018a; Kalita et al., 2019; Mukhopadhyay et al., 2019). Considerable amount of researches have been carried out for deterministic free vibration analysis of composite plates using classical laminated plate theory (CLPT) (Leissa, 1969), first-order shear deformation theory (Huang et al., 1994), higher-order shear deformation theory (Reddy, 1984), and three-dimensional theory elastic theory (Liew et al., 2001). A significant volume of literature is available on different analysis of laminated composites considering both deterministic (Tornabene et al., 2014; Dey et al., 2015a; Mukhopadhyay et al., 2015a, b; Huang et al., 2015) and stochastic (Shaker et al., 2008; Dey et al., 2015b, 2016a, 2017; Ganesan, 2005; Lal and Singh, 2009; Sephvand et al., 2010; Dey et al., 2018b, 2019) approaches. The coupled effect of matrix cracking damage and uncertainty is studied in a probabilistic framework (Naskar et al., 2017a, b; Naskar and Sriramula, 2017a, b, c). Based on higher-order theory, Naveenthraj et al. 1998 studied the linear static response of graphite-epoxy composite laminates with randomness in material properties by using combination of finite element analysis and Monte Carlo simulation (MCS). Furthermore, Salim et al. (1993) examined the effect of randomness in material properties such as elastic

Handbook of Probabilistic Models. https://doi.org/10.1016/B978-0-12-816514-0.00022-9

527

modulus Poisson's ratios, etc., on the response statistics of a composite plate subjected to static loading using CLPT in conjunction with first-order perturbation techniques (FOPTs). Onkar and Yadav (2003) investigated the nonlinear response statistics of composite-laminated flat panels with random material properties subjected to transverse random loading based on CLPT in conjunction with FOPT. Giunta et al. (2011) studied the free vibration of composite plates using refined theories accounting for uncertainties. Stochastic dynamic analysis of composite structures has received immense attention from the scientific community in last couple of years (Dey et al., 2016b, c, d; Mukhopadhyay et al., 2016a).

A nonintrusive method based on MCS for stochastic analysis of composite structures can obtain comprehensive probabilistic descriptions for the response quantities of interest (Dey et al., 2018a). However, the MCS—based approach of uncertainty quantification is computationally very expensive as it requires thousands of function evaluations. In general, for complex composite—laminated structures, the performance functions are not available as explicit functions of the random input variables (Mukhopadhyay and Adhikari, 2017a; Mukhopadhyay, 2017b; Mukhopadhyay and Adhikari, 2017b; Mukhopadhyay et al., 2017a; Mahata and Mukhopadhyay, 2018; Mukhopadhyay et al., 2018b; Mukhopadhyay et al., 2019; Mukhopadhyay et al., 2017b; Mukhopadhyay and Adhikari, 2016). Thus, the performance functions or responses (such as natural frequencies) of the structure can only be computed numerically at the end of an intensive structural analysis procedure (such as the FE method), which is often exorbitantly time-consuming and computationally expensive. A surrogate-based approach (Metya et al., 2017; Mukhopadhyay, 2017a; Mahata et al., 2016; Mukhopadhyay et al., 2016b; Karsh et al., 2018b; Mukhopadhyay et al., 2016c) can be adopted in such situation to achieve computational efficiency.

In this chapter, stochastic natural frequencies of laminated composite plates are analyzed in the presence of small random variation in the systems input variables. A neural network (NN) model is developed to reduce the computational time with adequate level of accuracy. The finite element formulation in conjunction with NN model is thereby utilized to map the variation of responses of interest for randomness in layer-wise input parameters with different level of noise. An exhaustive comparative assessment is presented for the accuracy and computational efficiency of two different NN models (artificial neural network [ANN] and polynomial neural network [PNN]).

2. Governing equations for composite plates

In this section, a brief description of the mathematical model for dynamic analysis of composite plates is provided (Dey et al., 2016e). Consider a rectangular composite-laminated cantilever plate of uniform length L, width b, and thickness t, which consists of three plies located in a three-dimensional Cartesian coordinate system (x, y, z), where the $x-y$ plane passes through the middle of the plate thickness with its origin placed at the corner of the

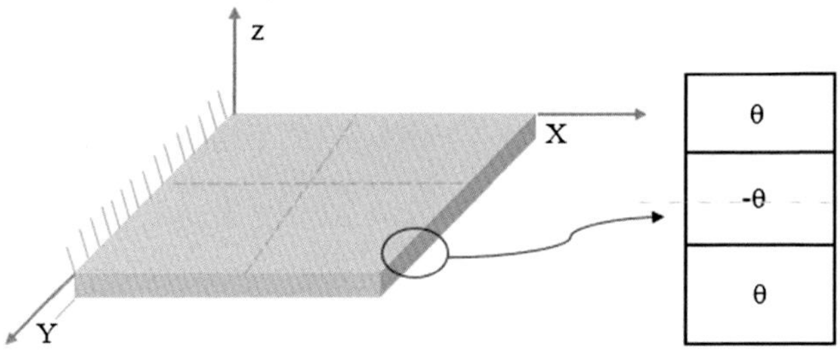

FIGURE 22.1 Laminated composite cantilever plate.

cantilever plate as shown in Fig. 22.1. The composite plate is considered with uniform thickness with the principal material axes of each layer being arbitrarily oriented with respect to midplane. If the midplane forms the $x-y$ plane of the reference plane, then the displacements can be computed as

$$u(x, y, z) = u^0(x, y) - z\theta_x(x, y)O_2 \tag{22.1}$$

$$v(x, y, z) = v^0(x, y) - z\theta_y(x, y)$$

$$w(x, y, z) = w^0(x, y) = w(x, y),$$

Assuming u, v, and w are the displacement components in x-, y-, and z-directions, respectively, and u^0, v^0, and w^0 are the midplane displacements, and θ_x and θ_y are rotations of cross sections along the x- and y-axes. The strain−displacement relationships for small deformations can be expressed as

$$\varepsilon_x = \frac{\partial u^o}{\partial x} - z\frac{\partial^2 w^o}{\partial x^2} \qquad \varepsilon_y = \frac{\partial v^o}{\partial y} - z\frac{\partial^2 w^o}{\partial y^2} \tag{22.2}$$

and $\gamma_{xy} = \frac{\partial u^o}{\partial y} + \frac{\partial v^o}{\partial x} - z\frac{2\partial^2 w^o}{\partial x\partial y}$ which in matrix form can be expressed as

$$\left\{ \begin{array}{c} \varepsilon_x \\ \varepsilon_y \\ \gamma_{xy} \end{array} \right\} = \left\{ \begin{array}{c} \varepsilon_x^o \\ \varepsilon_y^o \\ \gamma_{xy}^o \end{array} \right\} + z \left\{ \begin{array}{c} k_x \\ k_y \\ k_{xy} \end{array} \right\} \tag{22.3}$$

where $\varepsilon_x^0, \varepsilon_y^0, \gamma_{xy}^0$ are the strains in the reference plane and k_x, k_y, k_{xy} are the curvatures of reference plane of the plate. The random in-plane forces and moments acting on small element and the transverse shear forces (per unit length) are

$$N_x(\overline{\omega}) = \int_{-t_b(\overline{\omega})}^{t_t(\overline{\omega})} \sigma_x dz \quad N_y(\overline{\omega}) = \int_{-t_b(\overline{\omega})}^{t_t(\overline{\omega})} \sigma_y dz \quad N_{xy}(\overline{\omega}) = \int_{-t_b(\overline{\omega})}^{t_t(\overline{\omega})} \tau_{xy} dz \tag{22.4}$$

$$M_x(\overline{\omega}) = \int\limits_{-t_b(\overline{\omega})}^{t_t(\overline{\omega})} z\sigma_x dz \quad M_y(\overline{\omega}) = \int\limits_{-t_b(\overline{\omega})}^{t_t(\overline{\omega})} z\sigma_y dz \quad M_{xy}(\overline{\omega}) = \int\limits_{-t_b(\overline{\omega})}^{t_t(\overline{\omega})} z\tau_{xy} dz$$

$$V_x(\overline{\omega}) = \int\limits_{-t_b}^{t_t} \tau_{xz} dz \text{ and } V_y(\overline{\omega}) = \int\limits_{-t_b}^{t_t} \tau_{yz} dz$$

The stress–strain relationship for each ply can be expressed in matrix form

$$\left\{ \begin{array}{c} \sigma_x \\ \sigma_y \\ \tau_{xy} \end{array} \right\} = \left[\overline{Q}_{ij}(\overline{\omega}) \right] \left\{ \begin{array}{c} \varepsilon_x \\ \varepsilon_y \\ \gamma_{xy} \end{array} \right\} \tag{22.5}$$

where $\left[\overline{Q}_{ij}(\overline{\omega}) \right]$ is the stiffness matrix of the ply in x–y coordinate system and expressed as

$$\left[\overline{Q}_{ij}(\overline{\omega}) \right] = \begin{bmatrix} m^4 & n^4 & 2m^2n^2 & 4m^2n^2 \\ n^4 & m^4 & 2m^2n^2 & 4m^2n^2 \\ m^2n^2 & m^2n^2 & (m^4+n^4) & -4m^2n^2 \\ m^2n^2 & m^2n^2 & -2m^2n^2 & (m^2-n^2) \\ m^3n & mn^3 & (mn^3-m^3n) & 2(mn^3-m^3n) \\ mn^3 & m^3n & (m^3n-mn^3) & 2(m^3n-mn^3) \end{bmatrix} [Q_{ij}] \tag{22.6}$$

Here $m = \sin \theta(\overline{\omega})$ and $n = \cos \theta(\overline{\omega})$, wherein $\theta(\overline{\omega})$ is the random fiber orientation angle. However, laminate consists of a number of laminae wherein $[Q_{ij}]$ and $\left[\overline{Q}_{ij}(\overline{\omega}) \right]$ denote the on-axis elastic constant matrix and the off-axis elastic constant matrix, respectively. In matrix form, the in-plane stress resultant $\{N\}$, the moment resultant $\{M\}$, and the transverse shear resultants $\{Q\}$ can be expressed as

$$\{N\} = [A]\{\varepsilon^0\} + [B]\{k\} \text{ and } \{M\} = [B]\{\varepsilon^0\} + [D]\{k\} \tag{22.7}$$

$$\{Q\} = [A*]\{\gamma\}$$

$$\left[A_{ij}^* \right] = \int\limits_{-t_b(\overline{\omega})}^{t_t(\overline{\omega})} \overline{Q}_{ij} dz \quad \text{where } i,j = 4,5$$

The elasticity matrix of the laminated composite plate is given by

$$[Dt(\overline{\omega})] = \begin{bmatrix} A_{ij}(\overline{\omega}) & B_{ij}(\overline{\omega}) & 0 \\ B_{ij}(\overline{\omega}) & D_{ij}(\overline{\omega}) & 0 \\ 0 & 0 & S_{ij}(\overline{\omega}) \end{bmatrix} \qquad (22.8)$$

where

$$[A_{ij}(\overline{\omega}), B_{ij}(\overline{\omega}), D_{ij}(\overline{\omega})] = \sum_{k=1}^{n} \int_{z_{k-1}}^{z_k} [\overline{Q}_{ij}(\overline{\omega})]_k [1, z, z^2] dz \quad i,j = 1,2,6 \quad (22.9)$$

$$[S_{ij}(\overline{\omega})] = \sum_{k=1}^{n} \int_{z_{k-1}}^{z_k} \alpha_s [\overline{Q}_{ij}(\overline{\omega})]_k dz \quad i,j = 4,5 \qquad (22.10)$$

where α_s is the shear correction factor and is assumed as 5/6. Now, the mass per unit area is denoted by P and is given by

$$P(\overline{\omega}) = \sum_{k=1}^{n} \int_{z_{k-1}}^{z_k} \rho(\overline{\omega}) dz \qquad (22.11)$$

The element mass matrix is expressed as

$$[M_e(\overline{\omega})] = \int_{Vol} [N][P(\overline{\omega})][N]d(vol) \qquad (22.12)$$

The element stiffness matrix is given by

$$[K_e(\overline{\omega})] = \int_{-1}^{1} \int_{-1}^{1} [B(\overline{\omega})]^T [D(\overline{\omega})][B(\overline{\omega})]d\xi d\eta \qquad (22.13)$$

The Hamilton's principle is employed to study the dynamic nature of the composite structure. The principle used for the Lagrangian which is defined as $L_f = T - U - W$ where T, U, and W are total kinetic energy, total strain energy, and total potential of the applied load, respectively. The Hamilton's principle applicable to nonconservative system can be expressed as

$$\delta H = \int_{t_i}^{t_f} [\delta T - \delta U - \delta W] dt = 0 \qquad (22.14)$$

The energy function for Hamilton's principle is the Lagrangian (L_f) which includes kinetic energy (T) in addition to potential strain energy (U) of an elastic body. The expression for kinetic energy of an element is given by

$$T = \frac{1}{2}\{\dot{\delta}_e\}^T [M_e(\overline{\omega})] + [C_e(\overline{\omega})]\{\delta_e\} \qquad (22.15)$$

The potential strain energy for an element of a plate can be expressed as

$$U = U_1 + U_2 = \frac{1}{2}\{\delta_e\}^T [K_e(\overline{\omega})]\{\delta_e\} \qquad (22.16)$$

The Lagrange's equation of motion is given by

$$\frac{d}{dt}\left[\frac{\partial L_f}{\partial \dot{\delta}_e}\right] - \left[\frac{\partial L_f}{\partial \delta_e}\right] = \{F_e\} \qquad (22.17)$$

where $\{F_e\}$ is the applied external element force vector of an element and L_f is the Lagrangian function. Substituting $L_f = T - U$ and the corresponding expressions for T and U in Lagrange's equation, one obtains the dynamic equilibrium equation for free vibration of each element in the following form:

$$[M_e(\overline{\omega})]\{\ddot{\delta}_e\} + [K_e(\overline{\omega})]\{\delta_e\} = 0 \qquad (22.18)$$

After assembling all the element matrices and the force vectors with respect to the common global coordinates, the resulting equilibrium equation is obtained. For the purpose of this study, the finite element model is developed for different element types and finite element discretization and nodal positions of the driving point and measurement point. Considering randomness of input parameters like ply-orientation angle, thickness, elastic modulus, mass density, etc., the equation of motion of a linear free vibration system with n degrees of freedom can be expressed as

$$[M(\overline{\omega})]\ddot{\delta}(t) + [K(\overline{\omega})]\delta(t) = 0 \qquad (22.19)$$

where $K(\overline{\omega}) \in R^{n \times n}$ is the elastic stiffness matrix, $M(\overline{\omega}) \in R^{n \times n}$ is the mass matrix, and $\delta(t) \in R^n$ is the vector of generalized coordinates. The governing equations are derived based on Mindlin's theory incorporating rotary inertia, transverse shear deformation using an eight noded isoparametric plate bending element (Bathe, 1990). The composite cantilever plate is assumed to be under free vibration and the natural frequencies of the system are obtained as follows:

$$\omega_j^2(\overline{\omega}) = \frac{1}{\lambda_j(\overline{\omega})} \quad \text{where } j = 1, \ldots, n_r \qquad (22.20)$$

Here $\lambda_j(\overline{\omega})$ is the j-th eigenvalue of matrix $A = K^{-1}(\overline{\omega})\,M(\overline{\omega})$ and n_r indicates the number of modes retained in this analysis.

3. Artificial neural network

An artificial neuron or simply a neuron is the basic element of the ANN. A biological neuron collects information from other sources as input, combines them, performs a nonlinear operation, and gives output as result (Dey et al., 2016c; Karsh et al., 2018b; Manohar and Divakar, 2005). The ANN establishes relationships between the cause and effect by the training of multiple sets of input and output of the system, and hence it is also efficient for complex systems (Pareek et al., 2002). The main benefits of ANN over the response surface methods include that in case of ANN, any preexisting specification of fitting function is not required and it can approximate almost all types of nonlinear functions including quadratic functions, whereas response surface methodology is only used for quadratic functions (Desai et al., 2008). In case of computational modeling, a multilayer perceptron—based feed-forward ANN is used, which is based on a back propagation learning algorithm. The NN contains an input layer, one hidden layer, and an output layer. Each neuron summed together all incoming values, which acts as adding junction. After that, all values are filtered through an activation transfer function, and the output is forwarded to the next layer of neurons in the network. For the input and hidden layer nodes, the hyperbolic tangent was used as the transfer function. The transfer function is used as logistic function or hyperbolic tangent (*tanh*) because it generates the values nearer to zero if the argument of the function is substantially negative. Thus, the output of the hidden neuron can be obtained near to zero, thus for all subsequent weights, learning rate will be lowered, and finally learning will be stopped. The *tanh* function can also generate a value near to -1.0, and thus will maintain learning. Quick propagation (QP) algorithm is used to *train* ANN, which belongs to the gradient descent back propagation. It has been reported in the literature that for the training of all the ANN models, QP learning algorithm can be adopted (Ghaffari et al., 2008).

The working principle of ANN is just like a human neuron system, in which large numbers of parallel but distributed processors are connected for storing experimental knowledge and making the decision for next use (Jodaei et al., 2012). Weights, bias, and an activation function are the main units of the ANN. Fig. 22.2 shows the basic elements of ANN, where an input is given to each neuron, and weight function is multiplied with each input.

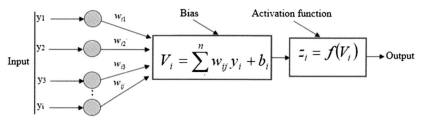

FIGURE 22.2 Basic elements of artificial neural network.

The bias (b_i) is a nonzero value which is added with the summation of inputs and corresponding weight given as

$$V_i = \sum_{i=1}^{n} w_{ij} y_i + b_i \qquad (22.21)$$

The summation V_i is transformed using activation function also called transfer function, where z_i gives a value called the unit's "activation"

$$z_i = f(V_i) \qquad (22.22)$$

The performance of the ANNs was statistically measured by the root mean squared error, the coefficient of determination (R^2), and the absolute average deviation obtained as follows:

$$\text{RMSE} = \left[\frac{1}{n} \sum_{i=1}^{n} (Y_i - Y_{id})^2 \right]^{0.5} \qquad (22.23)$$

$$R^2 = 1 - \frac{\sum\limits_{i=1}^{n} (Y_i - Y_{id})^2}{\sum\limits_{i=1}^{n} (Y_{id} - Y_m)^2} \qquad (22.24)$$

$$\text{AAD} = \left[\frac{1}{n} \sum_{i=1}^{n} \left| \frac{(Y_i - Y_{id})}{Y_{id}} \right| \right] \times 100 \qquad (22.25)$$

where n is the number of points, Y_i is the predicted value, Y_{id} is the actual value, and Y_m is the average of the actual values.

4. Polynomial neural network

In general, the PNN algorithm (Dey et al., 2016d; Karsh et al., 2018b; Mellit et al., 2010; Oh et al., 2003) is the advanced succession of group method of data handling method wherein different linear, modified quadratic, cubic polynomials are used. By choosing the most significant input variables and polynomial order among various types of forms available, the best partial description (PD) can be obtained based on selection of nodes of each layer and generation of additional layers until the best performance is reached. Such methodology leads to an optimal PNN structure wherein the input–output data set can be expressed as

$$(X_i, Y_i) = (x_{1i}, x_{2i}, x_{3i}, \ldots x_{ni}, y_i) \quad \text{where } i = 1, 2, 3 \ldots n \qquad (22.26)$$

By computing the polynomial regression equations for each pair of input variable x_i and x_j and output Y of the object system which desires to modeling

$$Y = A + Bx_i + Cx_j + Dx_i^2 + Ex_j^2 + Fx_ix_j \quad \text{where } i, j = 1, 2, 3 \ldots n \qquad (22.27)$$

where A, B, C, D, E, F are the coefficients of the polynomial equation. This provides $n(n-1)/2$ high-order variables for predicting the output Y in place of the original n variables $(x_1, x_2, ..., x_n)$. After finding these regression equations from a set of input-output observations, we then find out which ones to save. This gives the best predicted collection of quadratic regression models. We now use each of the quadratic equations that we have just computed and generate new independent observations that will replace the original observations of the variables $(x_1, x_2, ..., x_n)$. From these new independent variables, we will combine them exactly as we did before. That is, we compute all of the quadratic regression equations of Y versus these new variables. This will provide a new collection of $n(n-1)/2$ regression equation for predicting Y from the new variables, which in turn are estimates of Y from above equations. Now the best of new estimates is selected to generate new independent variables from selected equations to replace the old and combine all pair of these new variables. This process is continued until the regression equations begin to have a poorer predictability power than did the previous ones. In other words, it is the time when the model starts to become overfitted. The estimated output $\widehat{Y}_i$ can be further expressed as

$$
\begin{aligned}
\widehat{Y} &= \widehat{f}(x_1, x_2, x_3, ...x_n) \\
&= A_0 + \sum_{i=1}^{n} B_i x_i + \sum_{i=1}^{n}\sum_{j=1}^{n} C_{ij} x_i x_j + \sum_{i=1}^{n}\sum_{j=1}^{n}\sum_{k=1}^{n} D_{ijk} x_i x_j x_k + ... \quad (22.28)
\end{aligned}
$$
$$\text{Where } i, j, k = 1, 2, 3...n$$

where $X(x_1, x_2, ..., x_n)$ is the input variables vector and $P(A_0, B_i, C_{ij}, D_{ijk}, ...)$ is vector of coefficients or weight of the Ivakhnenko polynomials. Components of the input vector X can be independent variables, functional forms, or finite difference terms. This algorithm allows to find simultaneously the structure of model and model system output on the values of most significant inputs of the system. The following steps are to be performed for the framework of the design procedure of PNN:

Step1: *Determination of input variables:* Define the input variables as $x_i = 1, 2, 3, ..., n$ related to output variable Y. If required, the normalization of input data is also completed.

Step 2: *Create training and testing data*: Create the input−output data set (n) and divide into two parts, namely training data (n_{train}) and testing data (n_{test}) where $n = n_{train} + n_{test}$. The training data set is employed to construct the PNN model including an estimation of the coefficients of the PD of nodes situated in each layer of the PNN. Next, the testing data set is used to evaluate the estimated PNN model.

Step 3: *Selection of structure*: The structure of PNN is selected based on the number of input variables and the order of *PD* in each layer. Two kinds of PNN structures, namely a basic PNN and a modified PNN structure, are distinguished. The basic taxonomy for the architectures of PNN structure is given in Fig. 22.3.

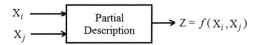

FIGURE 22.3   Taxonomy for architectures of polynomial neural network.

Step 4: *Determination of number of input variables and order of the polynomial*: Determine the regression polynomial structure of a *PD* related to PNN structure. The input variables of a node from n input variables x_1, x_2, x_3, ... x_n are selected. The total number of *PD*s located at the current layer depends on to the number of the selected input variables from the nodes of the preceding layer. This results in $k = n \ ! \ / \ (n \ ! \ - \ r \ !) \ r \ !$ nodes, where r is the number of the chosen input variables. The choice of the input variables and the order of a *PD* itself help to select the best model with respect to the characteristics of the data, model design strategy, nonlinearity, and predictive capability.

Step 5: *Estimation of coefficients of PD*: The vector of coefficients A_i is derived by minimizing the mean squared error between Y_i and $\widehat{Y}_i$.

$$PI = \frac{1}{n_{train}} \sum_{i=1}^{n_{train}} \left(Y_i - \widehat{Y}_i \right)^2 \tag{22.29}$$

where *PI* represents a criterion which uses the mean squared differences between the output data of original system and the output data of the model. Using the training data subset, this gives rise to the set of linear equations

$$Y = \sum_{i=1}^{n} X_i A_i \tag{22.30}$$

The coefficients of the *PD* of the processing nodes in each layer are derived in the form

$$A_i = \left[X_i^T X_i \right]^{-1} X_i^T Y \tag{22.31}$$

where $Y = \left[y_1, y_1, y_1, y_1, ... y_{n_{train}} \right]^T$

$X_i = \left[x_{1i}, x_{2i}, x_{3i} ... X_{ki} ... X_{traini} \right]^T$

$X_{ki}^T = \left[x_{ki1} x_{ki2} ... x_{kin} ... x_{ki1}^m x_{ki2}^m ... x_{kin}^m \right]^T$

$A_i = \left[A_{0i} A_{1i} A_{2i} ... A_{n'i} \right]^T$

with the following notations i as the node number, k as the data number, n_{train} as the number of the training data subset, n as the number of the selected input variables, m as the maximum order, and n' as the number of estimated coefficients. This procedure is implemented repeatedly for all nodes of the layer and also for all layers of PNN starting from the input layer and moving to the output layer.

Step 6: *Selection of PDs with the best predictive capability*: Each *PD* is estimated and evaluated using both the training and testing data sets. Then we compare these values and choose several PDs, which give the best predictive performance for the output variable. Usually a predetermined number *W* of PDs is utilized.

Step 7: *Check the stopping criterion*: The stopping condition indicates that a sufficiently good PNN model is accomplished at the previous layer, and the modeling can be terminated. This condition reads as $PI_j > PI^*$ where PI_j is a minimal identification error of the current layer, whereas PI^* denotes a minimal identification error that occurred at the previous layer.

Step 8: *Determination of new input variables for the next layer*: If PI_j (the minimum value in the current layer) has not been satisfied (so the stopping criterion is not satisfied), the model has to be expanded. The outputs of the preserved *PD* s serve as new inputs to the next layer.

5. Stochastic approach using neural network model

Layer-wise stochasticity in material and geometric properties are considered as input parameters for stochastic natural frequency analysis of composite plates. The individual and combined cases of layer-wise random variations considered in the present analysis are as follows:

(a) Variation of ply-orientation angle only (low dimensional input parameter space):

$$\theta(\overline{\omega}) = \{\theta_1\theta_2\theta_3...\theta_i...\theta_l\}$$

(b) Combined variation of ply-orientation angle, thickness, elastic modulus (longitudinal), and mass density (high dimensional input parameter space):

$$g\{\theta(\overline{\omega}), t(\overline{\omega}), E_1(\overline{\omega}), \rho(\overline{\omega})\} = \{\Phi_1(\theta_1\cdots\theta_l), \Phi_2(t\cdots t_l),$$
$$\Phi_3\left(E_{1(1)}\cdots E_{1(l)}\right), \Phi_4(\rho_1\cdots\rho_l)\}$$

where θ_i, t_i, $E_{1(i,)}$, and ρ_i are the ply-orientation angle, thickness, elastic modulus along longitudinal direction, and mass density, respectively, and "*l*" denotes the number of layer in the laminate. In the present study, it is assumed that the distribution for randomness of input parameters exists within a certain band of tolerance with their deterministic values. $\pm5°$ for ply-orientation angle and $\pm10\%$ tolerance for material properties and thickness from deterministic values are considered.

The flowchart of the proposed stochastic natural frequency analysis of composites using NN models is shown in Fig. 22.4. This flowchart is valid for

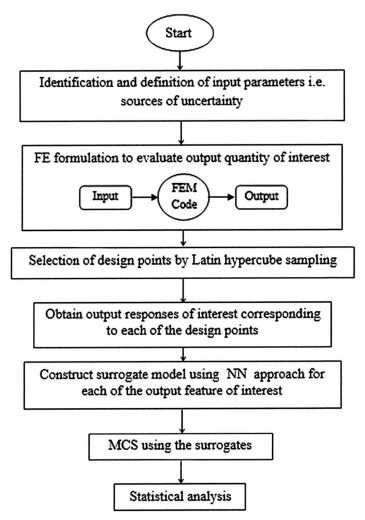

FIGURE 22.4 Flowchart of stochastic natural frequency analysis using neural network (NN) model.

both ANN and PNN. Latin hypercube sampling (Schetinin, 2001) is employed for generating sample points to ensure the representation of all portions of the vector space. In Latin hypercube sampling, the interval of each dimension is divided into m nonoverlapping intervals having equal probability considering a uniform distribution, so the intervals should have equal size. Moreover, the sample is chosen randomly from a uniform distribution with a point in each interval, in each dimension and the random pair is selected considering equal likely combinations for the point from each dimension.

6. Results and discussion

In this study, a three-layered graphite—epoxy symmetric angle-ply laminated composite cantilever plate is considered to obtain numerical results. The length, width, and thickness of the composite laminate considered in the present analysis are 1 m, 1 m, and 5 mm, respectively. Material properties of graphite—epoxy composite (Qatu and Leissa, 1991) considered with deterministic mean value as $E_1 = 138.0$ GPa, $E_2 = 8.96$ GPa, $G_{12} = 7.1$ GPa, $G_{13} = 7.1$ GPa, $G_{23} = 2.84$ GPa, and $\mu = 0.3$, $\rho = 1600$ kg/m^3. A discretization of (6 × 6) mesh on plan area with 36 elements and 133 nodes with natural coordinates of an isoparametric quadratic plate bending element are considered for the present FEM approach. The finite element mesh size is finalized using a convergence study as shown in Table 22.1.

The NN method is employed to develop a predictive and representative surrogate model relating each natural frequency to a number of input variables. Thus, it represents the result of structural analysis encompassing every possible combination of all stochastic input variables. From this mathematical model, thousands of combinations of all design variables can be created and performed using a pseudo analysis for each variable set, by adopting the corresponding predictive values. In the next paragraph, a comparative assessment of ANN and PNN is presented for the first three natural frequencies of composite plates.

For an angle-ply (45°/-45°/45°) composite plate, Figs. 22.5—22.8 present the percentage error of mean and standard deviation (SD) of first three natural frequencies between the two NN-based methods and MCS considering different sample sizes for a case of individual variation (ply-orientation angle $(\theta(\overline{\omega}))$ and combined variation of all the stochastic input parameters $(g(\overline{\omega}))$. The abbreviations used in the figures are FNF—first natural frequency, SNF—second natural frequency, and TNF—third natural frequency. As the

TABLE 22.1 Convergence study for nondimensional fundamental natural frequencies $[\omega = \omega_n \, L^2 \, \sqrt{(\rho/E_1 t^2)}]$ of three-layered $(\theta°/-\theta°/\theta°)$ graphite —epoxy composite plates, a/h = 1, b/t = 100, considering $E_1 = 138$ GPa, $E_2 = 8.96$ GPa, $G_{12} = 7.1$ GPa, and $\mu = 0.3$.

Ply angle (θ)	Results from literature (Qatu and Leissa, 1991)	Present FEM (4 × 4)	Present FEM (6 × 6)	Present FEM (8 × 8)	Present FEM (10 × 10)
0°	1.0175	1.0112	1.0133	1.0107	1.004
45°	0.4613	0.4591	0.4601	0.4603	0.4604
90°	0.2590	0.2553	0.2567	0.2547	0.2542

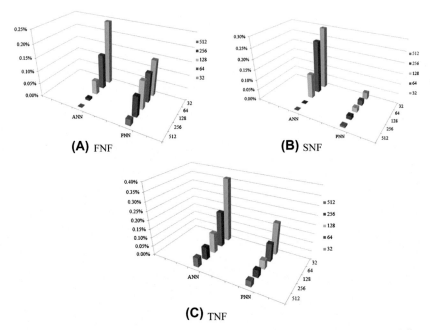

FIGURE 22.5 Error (%) of mean of first three natural frequencies between surrogate modeling methods and Monte Carlo simulation results with respect to different sample sizes for individual variation of ply-orientation angle $[\theta(\overline{\omega})]$ for angle-ply (45°/-45°/45°) composite plates. FNF - First natural frequency; SNF - Second natural frequency; TNF - Third natural frequency.

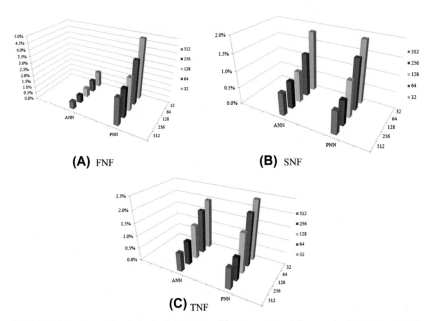

FIGURE 22.6 Error (%) of standard deviation of first three natural frequencies between surrogate modeling methods and Monte Carlo simulation results with respect to different sample sizes for individual variation of ply-orientation angle $[\theta(\overline{\omega})]$ for angle-ply (45°/-45°/45°) composite plates. FNF - First natural frequency; SNF - Second natural frequency; TNF - Third natural frequency.

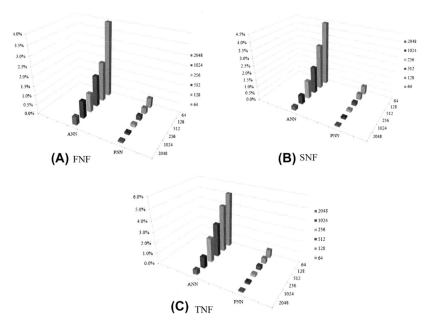

(A) FNF

(B) SNF

(C) TNF

FIGURE 22.7 Error (%) of mean of first three natural frequencies between surrogate modeling methods and Monte Carlo simulation results with respect to different sample sizes for the combined variation case $[g(\overline{\omega})]$ for angle-ply (45°/-45°/45°) composite plates. FNF - First natural frequency; SNF - Second natural frequency; TNF - Third natural frequency.

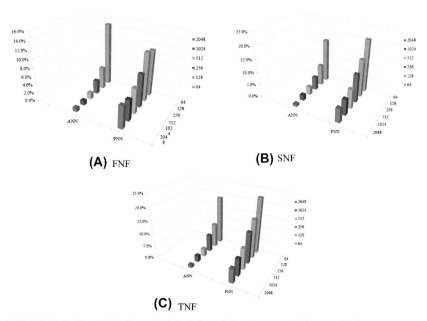

(A) FNF

(B) SNF

(C) TNF

FIGURE 22.8 Error (%) of standard deviation of first three natural frequencies between surrogate modeling methods and Monte Carlo simulation results with respect to different sample sizes for the combined variation case $[g(\overline{\omega})]$ for angle-ply (45°/-45°/45°) composite plates. FNF - First natural frequency; SNF - Second natural frequency; TNF - Third natural frequency.

sample size increases, the percentage of error of mean and SD of first three natural frequencies between surrogate modeling methods and MCS results is found to reduce irrespective of modeling methods. Based on the error analysis presented in the figures, sample sizes of 256 and 512 are chosen for individual and comparative effect of stochasticity.

Fig. 22.9 presents comparative results for the probabilistic descriptions of first three natural frequencies considering the selected sample sizes. The results can provide a clear understanding about the rate of convergence and deviation of

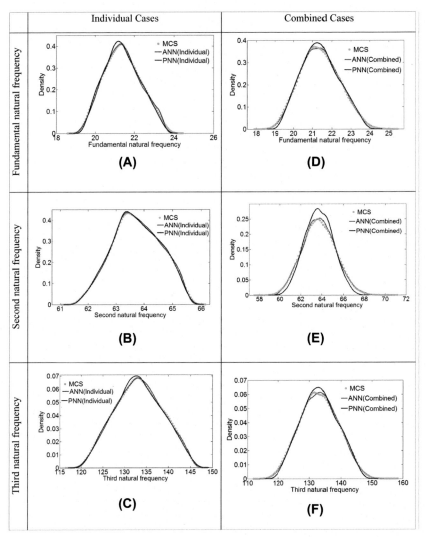

FIGURE 22.9 Probability density function for first three natural frequencies corresponding to combined and individual variation of input parameters considering artificial neural network (ANN) and polynomial neural network (PNN), respectively.

the first two statistical moments (mean and SD) from the direct MCS for low and comparatively high dimensional input parameter spaces. We have confined our analysis on two NN algorithms in this work. Future investigations can be carried out to investigate the performance of various other surrogate-based learning approaches (Kumar et al., 2019; Naskar, 2018; Naskar et al., 2019; Karsh et al., 2019; Mukhopadhyay et al., 2018; Naskar et al., 2018) for their prediction capability in the domain of structural dynamics.

7. Summary

Stochastic natural frequencies are analyzed considering layer-wise variation of individual and combined cases for random input parameters. The comparative performance of ANN and PNN is assessed with respect to direct MCS based on accuracy and computational efficiency. The computational time and cost is reduced by using the present approach compared with the direct MCS method. As the sample size increases, the percentage of error in mean and SD of first three natural frequencies between surrogate modeling methods and MCS results are found to reduce irrespective of modeling methods. The comparative study presented in this chapter showing the rate of convergence and deviation of the first two statistical moments from the direct MCS for low and comparatively high dimensional input parameter spaces can be useful for selecting an appropriate NN model for various other computationally intensive problems in future according to number of input parameters and relative degree of accuracy needed in the prediction of mean and SD.

Acknowledgments

SN and SS are grateful for the support provided through the Lloyd's Register Foundation Center. The Foundation helps to protect life and property by supporting engineering-related education, public engagement, and the application of research.

References

Bathe, K.J., 1990. Finite Element Procedures in Engineering Analysis. PHI, New Delhi, India.
Desai, K.M., Survase, S.A., Saudagar, P.S., Lele, S.S., Singhal, R.S., 2008. Comparison of artificial neural network (ANN) and response surface methodology (RSM) in fermentation media optimization: case study of fer-mentative production of scleroglucan, 41 (3), 266−273.
Dey, S., Mukhopadhyay, T., Naskar, S., Dey, T.K., Chalak, H.D., Adhikari, S., 2019. Probabilistic characterization for dynamics and stability of laminated soft core sandwich plates. Journal of Sandwich Structures and Materials 21 (1), 366−397. https://doi.org/10.1177/1099636217694229.
Dey, T.K., Mukhopadhyay, T., Chakrabarti, A., Sharma, U.K., 2015. Efficient lightweight design of FRP bridge deck. Proceedings of the Institution of Civil Engineers − Structures and Buildings 168 (10), 697−707.

Dey, S., Mukhopadhyay, T., Khodaparast, H.H., Kerfriden, P., Adhikari, S., 2015. Rotational and ply-level uncertainty in response of composite shallow conical shells. Composite Structures 131, 594–605.

Dey, S., Mukhopadhyay, T., Khodaparast, H.H., Adhikari, S., 2016. A response surface modelling approach for resonance driven reliability based optimization of composite shells. Periodica Polytechnica: Civil Engineering 60 (1), 103–111.

Dey, S., Mukhopadhyay, T., Sahu, S.K., Adhikari, S., 2016. Effect of cutout on stochastic natural frequency of composite curved panels. Composites Part B: Engineering 105, 188–202.

Dey, S., Mukhopadhyay, T., Spickenheuer, A., Gohs, U., Adhikari, S., 2016. Uncertainty quantification in natural frequency of composite plates – an Artificial neural network based approach. Advanced Composites Letters 25 (2), 43–48.

Dey, S., Naskar, S., Mukhopadhyay, T., Gohs, U., Spickenheuer, A., Bittrich, L., Adhikari, S., Heinrich, G., 2016. "Uncertain natural frequency analysis of composite plates including effect of noise – a polynomial neural network approach. Composite Structures 143, 130–142.

Dey, S., Mukhopadhyay, T., Spickenheuer, A., Adhikari, S., Heinrich, G., 2016. Bottom up surrogate based approach for stochastic frequency response analysis of laminated composite plates. Composite Structures (140), 712–727.

Dey, S., Mukhopadhyay, T., Adhikari, S., 2017. Metamodel based high-fidelity stochastic analysis of composite laminates: a concise review with critical comparative assessment. Composite Structures 171, 227–250.

Dey, S., Mukhopadhyay, T., Adhikari, S., 2018. Uncertainty Quantification in Laminated Composites: A Meta-Model Based Approach. CRC Press, Taylor & Francis Group. ISBN 9781498784450.

Dey, S., Mukhopadhyay, T., Sahu, S.K., Adhikari, S., 2018. Stochastic dynamic stability analysis of composite curved panels subjected to non-uniform partial edge loading. European Journal of Mechanics/A Solids 67, 108–122.

Ganesan, R., 2005. Free-vibration of composite beam-columns with stochastic material and geometric properties subjected to random axial loads. Journal of Reinforced Plastics and Composites 24 (1), 69–91.

Ghaffari, A., Abdollahi, H., Khoshayand, M.R., Soltani Bozchalooi, I., Dadgar, A., Rafiee-Tehrani, M., 2008. Performance comparison of neural networks. Environmental Science and Technology 42 (21), 7970–7975.

Giunta, G., Carrera, E., Belouettar, S., 2011. Free vibration analysis of composite plates via refined theories accounting for uncertainties. Shock and Vibration 18, 537–554.

Huang, C.S., McGee, O.G., Leissa, A.W., 1994. Exact analytical solutions for free vibrations of thick sectorial plates with simply supported radial edges. International Journal of Solids and Structures 31 (11), 1609–1631.

Huang, L., Sheikh, A.H., Ng, C.T., Griffith, M.C., 2015. An efficient finite element model for buckling of grid stiffened laminated composite plates. Composite Structures 122, 41–50.

Jodaei, A., Jalal, M., Yas, M.H., 2012. Free vibration analysis of functionally graded annular plates by state-space based differential quadrature method and comparative modeling by ANN. Composites Part B: Engineering 43 (B), 340–353.

Kalita, K., Mukhopadhyay, T., Dey, P., Haldar, S., 2019. Genetic programming assisted multi-scale optimization for multi-objective dynamic performance of laminated composites: The advantage of more elementary-level analyses. Neural Computing and Applications. https://doi.org/10.1007/s00521-019-04280-z.

Karsh, P.K., Mukhopadhyay, T., Dey, S., 2018a. Spatial vulnerability analysis for the first ply failure strength of composite laminates including effect of delamination. Composite Structures 184, 554–567.

Karsh, P.K., Mukhopadhyay, T., Dey, S., 2018b. Stochastic dynamic analysis of twisted functionally graded plates. Composites Part B: Engineering 147, 259–278.

Karsh, P.K., Mukhopadhyay, T., Dey, S., 2019. Stochastic low-velocity impact on functionally graded plates: probabilistic and non-probabilistic uncertainty quantification. Composites Part B: Engineering 159, 461–480.

Kumar, R.R., Mukhopadhyay, T., Pandey, K.M., Dey, S., 2019. Stochastic buckling analysis of sandwich plates: the importance of higher order modes. International Journal of Mechanical Sciences 152, 630–643. https://doi.org/10.1016/j.ijmecsci.2018.12.016.

Lal, A., Singh, B.N., 2009. Stochastic nonlinear free vibration of laminated composite plates resting on elastic foundation in thermal environments. Computational Mechanics 44 (1), 15–29.

Leissa, A.W., 1969. Vibration of Plates. NASA SP-160.

Liew, K.M., Ng, T.Y., Wang, B.P., 2001. Vibration of annular sector plates from three dimensional analysis. Journal of the Acoustical Society of America 110 (1), 233–242.

Mahata, A., Mukhopadhyay, T., 2018. Probing the chirality-dependent elastic properties and crack propagation behavior of single and bilayer stanene. Physical Chemistry Chemical Physics 20, 22768–22782.

Mahata, A., Mukhopadhyay, T., Adhikari, S., 2016. A polynomial chaos expansion based molecular dynamics study for probabilistic strength analysis of nano-twinned copper. Materials Research Express 3, 036501.

Manohar, B., Divakar, S., 2005. An artificial neural net- work analysis of porcine pancreas lipase catalysed Es- terification of anthranilic acid with methanol. Process Biochemistry 40 (10), 3372–3376.

Mellit, A., Drif, M., Malek, A., 2010. EPNN-based prediction of meteorological data for renewable energy systems. Revue Des Energies Renouvelables 13 (1), 25–47.

Metya, S., Mukhopadhyay, T., Adhikari, S., Bhattacharya, G., 2017. System reliability analysis of soil slopes with general slip surfaces using multivariate adaptive regression splines. Computers and Geotechnics 87, 212–228.

Mukhopadhyay, T., 2018a. A multivariate adaptive regression splines based damage identification methodology for web core composite bridges including the effect of noise. Journal of Sandwich Structures and Materials 20 (7), 885–903. https://doi.org/10.1177/1099636216682533.

Mukhopadhyay, T., 2017b. Mechanics of Quasi-Periodic Lattices. PhD thesis. Swansea University.

Mukhopadhyay, T., Adhikari, S., 2016. Free vibration analysis of sandwich panels with randomly irregular honeycomb core. Journal of Engineering Mechanics 142 (11), 06016008.

Mukhopadhyay, T., Adhikari, S., 2017. Stochastic mechanics of metamaterials. Composite Structures 162, 85–97.

Mukhopadhyay, T., Adhikari, S., 2017. Effective in-plane elastic moduli of quasi-random spatially irregular hexagonal lattices. International Journal of Engineering Science 119, 142–179.

Mukhopadhyay, T., Adhikari, S., Batou, A., 2019. Frequency domain homogenization for the viscoelastic properties of spatially correlated quasi-periodic lattices. International Journal of Mechanical Sciences 150, 784–806. https://doi.org/10.1016/j.ijmecsci.2017.09.004.

Mukhopadhyay, T., Dey, T.K., Chowdhury, R., Chakrabarti, A., Adhikari, S., 2015. Optimum design of FRP bridge deck: an efficient RS-HDMR based approach. Structural and Multidisciplinary Optimization 52 (3), 459–477.

Mukhopadhyay, T., Dey, T.K., Dey, S., Chakrabarti, A., 2015. Optimization of fiber reinforced polymer web core bridge deck — a hybrid approach. Structural Engineering International 25 (2), 173–183.

Mukhopadhyay, T., Naskar, S., Dey, S., Adhikari, S., 2016. On quantifying the effect of noise in surrogate based stochastic free vibration analysis of laminated composite shallow shells. Composite Structures 140, 798–805.

Mukhopadhyay, T., Chowdhury, R., Chakrabarti, A., 2016. Structural damage identification: a random sampling-high dimensional model representation approach. Advances in Structural Engineering 19 (6), 908–927.

Mukhopadhyay, T., Mahata, A., Dey, S., Adhikari, S., 2016. Probabilistic analysis and design of HCP nanowires: an efficient surrogate based molecular dynamics simulation approach. Journal of Materials Science and Technology 32 (12), 1345–1351.

Mukhopadhyay, T., Mahata, A., Adhikari, S., Asle Zaeem, M., 2017. Effective mechanical properties of multilayer nano-heterostructures. Scientific Reports 7, 15818.

Mukhopadhyay, T., Mahata, A., Adhikari, S., Asle Zaeem, M., 2017. Effective elastic properties of two dimensional multiplanar hexagonal nano-structures. 2D Materials 4, 025006.

Mukhopadhyay, T., Naskar, S., Karsh, P.K., Dey, S., You, Z., 2018a. Effect of delamination on the stochastic natural frequencies of composite laminates. Composites Part B: Engineering 154, 242–256.

Mukhopadhyay, T., Mahata, A., Adhikari, S., Asle Zaeem, M., 2018b. Probing the shear modulus of two-dimensional multiplanar nanostructures and heterostructures. Nanoscale 10, 5280–5294.

Mukhopadhyay, T., Naskar, S., Dey, S., Chakrabarti, A., 2019. Condition assessment and strengthening of aged structures: perspectives based on a critical case study. Practice Periodical on Structural Design and Construction 24 (3), 05019003.

Naskar, S., 2018. Spatial variability characterisation of laminated composites. PhD Thesis. University of Aberdeen.

Naskar, S., Sriramula, S., 2017a. Effective elastic property of randomly damaged composite laminates. In: Engineering Postgraduate Research Symposium, April 2017, Aberdeen, UK.

Naskar, S., Sriramula, S., 2017b. Vibration analysis of hollow circular laminated composite beams — a stochastic approach. In: 12th International Conference on Structural Safety & Reliability, August 2017, TU Wien, Austria.

Naskar, S., Sriramula, S., 2017c. Random field based approach for quantifying the spatial variability in composite laminates. In: 20th International Conference on Composite Structures (ICCS20), Sept 2017, Paris.

Naskar, S., Mukhopadhyay, T., Sriramula, S., Adhikari, 2017a. Stochastic natural frequency analysis of damaged thin-walled laminated composite beams with uncertainty in micromechanical properties. Composite Structures 160, 312–334.

Naskar, S., Mukhopadhyay, T., Sriramula, S., 2017b. Non-probabilistic analysis of laminated composites based on fuzzy uncertainty quantification. In: 20th International Conference on Composite Structures (ICCS20), Sept 2017, Paris.

Naskar, S., Mukhopadhyay, T., Sriramula, S., 2018. Probabilistic micromechanical spatial variability quantification in laminated composites. Composites Part B: Engineering 151, 291–325.

Naskar, S., Mukhopadhyay, T., Sriramula, S., 2019. Spatially varying fuzzy multi-scale uncertainty propagation in unidirectional fibre reinforced composites. Composite Structures 209, 940–967.

Naveenthraj, B., Iyengar, N.G.R., Yadav, D., 1998. Response of composite plates with random material properties using FEM and MCS. Advanced Composite Materials 7, 219–237.

Oh, S.K., Pedrycz, W., Park, B.J., 2003. Polynomial neural networks architecture: analysis and design. Computers and Electrical Engineering 29 (6), 703–725.

Onkar, A.K., Yadav, D., 2003. Non-linear response statistics of composite laminates with random material properties under random loading. Composite Structures 60 (4), 375–383.

Pareek, V.K., Brungs, M.P., Adesina, A.A., Sharma, R., 2002. Artificial neural network modeling of a multiphase photodegradation system. Journal of Photochemistry and Photobiology A 149, 139–146.

Qatu, M., Leissa, A., 1991. Vibration studies for laminated composite twisted cantilever plates. International Journal of Mechanical Sciences 33 (11), 927–940.

Reddy, J.N., 1984. A refined nonlinear theory of plates with transverse shear deformation. International Journal of Solids and Structures 20 (9–10), 881–896.

Salim, S., Yadav, D., Iyengar, N.G.R., 1993. Analysis of composite plates with random material characteristics. Mechanics Research Communications 20 (5), 405–414.

Schetinin, V., 2001. Polynomial neural networks learnt to classify EEG signals. In: NIMIA-SC2001 – 2001 NATO Advanced Study Institute on Neural Networks for Instrumentation, Measurement, and Related Industrial Applications: Study Cases Crema, Italy; 9–20 October.

Sephvand, K., Marburg, S., Hardtke, H.J., 2010. Uncertainty quantification in stochastic systems using polynomial chaos expansion. International Journal of Applied Mechanics 2 (2), 305.

Shaker, A., Abdelrahman, W.G., Tawfik, M., Sadek, E., 2008. Stochastic finite element analysis of the free vibration of laminated composite plates. Computational Mechanics 41 (4), 493–501.

Tornabene, F., Fantuzzi, N., Bacciocchi, M., 2014. The local GDQ method applied to general higher-order theories of doubly-curved laminated composite shells and panels: the free vibration analysis. Composite Structures 116, 637–660.

Index

Printed in the United States
By Bookmasters